NATO ASI Series

Advanced Science Institutes Series

A series presenting the results of activities sponsored by the NATO Science Committee, which aims at the dissemination of advanced scientific and technological knowledge, with a view to strengthening links between scientific communities.

The Series is published by an international board of publishers in conjunction with the NATO Scientific Affairs Division

| A | Life Sciences | Plenum Publishing Corporation |
| B | Physics | London and New York |

C	Mathematical and Physical Sciences	Kluwer Academic Publishers
D	Behavioural and Social Sciences	Dordrecht, Boston and London
E	Applied Sciences	

F	Computer and Systems Sciences	Springer-Verlag
G	Ecological Sciences	Berlin Heidelberg New York
H	Cell Biology	London Paris Tokyo Hong Kong

Series H: Cell Biology Vol. 43

The ASI Series Books Published as a Result of
Activities of the Special Programme on
CELL TO CELL SIGNALS IN PLANTS AND ANIMALS

This book contains the proceedings of a NATO Advanced Research Workshop held within the
activities of the NATO Special Programme on Cell to Cell Signals in Plants and Animals, running
from 1984 to 1989 under the auspices of the NATO Science Committee.

The books published as a result of the activities of the Special Programme are:

Vol. 1: Biology and Molecular Biology of Plant-Pathogen Interactions. Edited by J.A. Bailey. 1986.
Vol. 2: Glial-Neuronal Communication in Development and Regeneration.
        Edited by H.H. Althaus and W. Seifert. 1987.
Vol. 3: Nicotinic Acetylcholine Receptor: Structure and Function. Edited by A. Maelicke. 1986.
Vol. 4: Recognition in Microbe-Plant Symbiotic and Pathogenic Interactions.
        Edited by B. Lugtenberg. 1986.
Vol. 5: Mesenchymal-Epithelial Interactions in Neural Development.
        Edited by J.R. Wolff, J. Sievers, and M. Berry. 1987.
Vol. 6: Molecular Mechanisms of Desensitization to Signal Molecules.
        Edited by T.M. Konjin, P.J.M. Van Haastert, H. Van der Starre, H. Van der Wel,
        and M.D. Houslay. 1987.
Vol. 7: Gangliosides and Modulation of Neuronal Functions. Edited by H. Rahmann. 1987.
Vol. 9: Modification of Cell to Cell Signals During Normal and Pathological Aging.
        Edited By S. Govoni and F. Battaini. 1987.
Vol. 10: Plant Hormone Receptors. Edited by D. Klämbt. 1987.
Vol. 11: Host-Parasite Cellular and Molecular Interactions in Protozoal Infections.
        Edited by K.-P. Chang and D. Snary. 1987.
Vol. 12: The Cell Surface in Signal Transduction. Edited by E. Wagner, H. Greppin, and B. Millet. 1987.
Vol. 19: Modulation of Synaptic Transmission and Plasticity in Nervous Systems.
        Edited by G. Hertting and H.-C. Spatz. 1988.
Vol. 20: Amino Acid Availability and Brain Function in Health and Disease. Edited by G. Huether. 1988.
Vol. 21: Cellular and Molecular Basis of Synaptic Transmission. Edited by H. Zimmermann. 1988.
Vol. 23: The Semiotics of Cellular Communication in the Immune System.
        Edited by E.E. Sercarz, F. Celada, N.A. Mitchison, and T. Tada. 1988.
Vol. 24: Bacteria, Complement and the Phagocytic Cell. Edited by F.C. Cabello and C. Pruzzo. 1988.
Vol. 25: Nicotinic Acetylcholine Receptors in the Nervous System.
        Edited by F. Celementi, C. Gotti, and E. Sher. 1988.
Vol. 26: Cell to Cell Signals in Mammalian Development.
        Edited by S.W. de Laat, J.G. Bluemink, and C.L. Mummery. 1989.
Vol. 27: Phytotoxins and Plant Pathogenesis. Edited by A. Graniti, R.D. Durbin, and A. Ballio. 1989.
Vol. 31: Neurobiology of the Inner Retina. Edited by R. Weiler and N.N. Osborne. 1989.
Vol. 32: Molecular Biology of Neuroreceptors and Ion Channels. Edited by A. Maelicke. 1989.
Vol. 33: Regulatory Mechanisms of Neuron to Vessel Communication in the Brain.
        Edited by F. Battaini, S. Govoni, M.S. Magnoni, and M. Trabucchi. 1989.
Vol. 35: Cell Separation in Plants: Physiology, Biochemistry and Molecular Biology.
        Edited by D.J. Osborne and M.B. Jackson. 1989.
Vol. 36: Signal Molecules in Plants and Plant-Microbe Interactions. Edited by B.J.J. Lugtenberg. 1989.
Vol. 39: Chemosensory Information Processing. Edited by D. Schild. 1990.
Vol. 41: Recognition and Response in Plant-Virus Interactions. Edited by R.S.S. Fraser. 1990.
Vol. 43: Cellular and Molecular Biology of Myelination.
        Edited by G. Jeserich, H. H. Althaus, and T. V. Waehneldt. 1990.

Cellular and Molecular Biology of Myelination

Edited by

G. Jeserich
Universität Osnabrück, FB Biologie/Chemie
Barbarastraße 11, 4500 Osnabrück, FRG

H. H. Althaus
T. V. Waehneldt
Max-Planck-Institut für Experimentelle Medizin
Hermann-Rein-Straße 3, 3400 Göttingen, FRG

Springer-Verlag Berlin Heidelberg New York
London Paris Tokyo Hong Kong
Published in cooperation with NATO Scientific Affairs Division

Proceedings of the NATO Advanced Research Workshop on Cellular and Molecular Biology of Myelination, held at Monastery Ohrbeck near Osnabrück, FRG, August 28–September 2, 1989

ISBN-13:978-3-642-83970-2   e-ISBN-13:978-3-642-83968-9
DOI: 10.1007/978-3-642-83968-9

Library of Congress Cataloging-in-Publication Data. Cellular and molecular biology of myelination/edited by G. Jeserich, H.H. Althaus, T.V. Waehneldt. p. cm.—(NATO ASI series. Series H, Cell biology; vol. 43) "Proceedings of the NATO Advanced Research Workshop on Cellular and Molecular Biology of Myelination held at Monastery Ohrbeck near Osnabrück, FRG, August 28–September 2, 1989"—T.p. verso.
ISBN-13:978-3-642-83970-2 (U.S.)
1. Neuroglia—Growth—Molecular aspects—Congresses. 2. Myelination—Congresses. I. Jeserich, G. (Gunnar), 1950– . II. Althaus, Hans H. III. Waehneldt, T.V. (Thomas V.), 1932– . IV. NATO Advanced Research Workshop on Cellular and Molecular Biology of Myelination (1989: Osnabrück, Germany) V. Series. QP363.2.C45 1989 599'.0188—dc20 90-9587

This work is subject to copyright. All rights are reserved, whether the whole or part of the material is concerned, specifically the rights of translation, reprinting, re-use of illustrations, recitation, broadcasting, reproduction on microfilms or in other ways, and storage in data banks. Duplication of this publication or parts thereof is only permitted under the provisions of the German Copyright Law of September 9, 1965, in its current version, and a copyright fee must always be paid. Violations fall under the prosecution act of the German Copyright Law.

© Springer-Verlag Berlin Heidelberg 1990
Softcover reprint of the hardcover 1st edition 1990

2131/3140-543210 – Printed on acid-free-paper

PREFACE

The process of myelination is one of the key events during nervous system development and represent the ultimate step of cellular differentiation by which proper neuronal function is attained. Knowledge of the biochemistry and cell biology of the myelin-forming glia has advanced rapidly in recent years. This progress was to a large extent conveyed by the concerted application of new experimental tools, including improved cell culture systems, availability of specific immunological probes and the powerful development of recombinant DNA technology and cell transfection.

The multidisciplinary aspects of glial cell biology and myelination were comprehensively dicussed by leading neuroscientist from Europe, Israel, USA and Canada at a NATO Advanced Research Workshop held at Monastery Ohrbeck, near Osnabrück, August 28 - September 2, 1989.

The meeting concentrated on the following major topics:
A detailed characterization of the different steps of oligodendroglial differentiation ranging from a bipotential precurser to the actively myelinating oligodendrocyte was given. Subsequently the regenerative potential of mature oligodendrocytes after experimental lesioning was discussed and most recent progress made in the field of glial cell transplantation was reported.

Furthermore the role of specific growth factors and their receptors for glial cell proliferation and differentiation were evaluated in detail. Special emphasis was given to the involvement of proteinkinases A and C in the underlying transmembrane signalling events. New avenues of glial cell biology were pursued by means of genetically manipulated cells. In a subsequent session recent developments in the field of myelin membrane biochemistry were discussed. Another topic of major interest was related to the molecular structure and regulation of genes coding for myelin proteins. Detailed information was given as regards the structure of myelin basic protein and proteolipid protein genes as well as the chain of events underlying myelin gene expression in the peripheral nervous system.

The rather remote location of the meeting place and its intimate setting provided an appropriate frame for fruitful discussions and a stimulating exchange of ideas in a friendly and harmonious atmosphere. We hope that these positive experiences are reflected in the contents of this book.

The papers presented at the conference are combined in the present volume. The editors are very grateful to the participants for their cooperative efforts.

We are particularly grateful to the NATO Scientific Affairs Division who provided the major financial support of this conference. In addition the meeting was sponsored by the following companies: Becton-Dickinson (Heidelberg), Boehringer (Mannheim), Millipore (Eschborn), Sartorius (Göttingen), Schleicher & Schuell (Dassel).

Finally we would like to thank the administration and staff of Monastery Ohrbeck for their kind hospitality as well as our coworkers who contributed to the success of the meeting.

Gunnar Jeserich   Hans H. Althaus   Thomas V. Waehneldt

VII

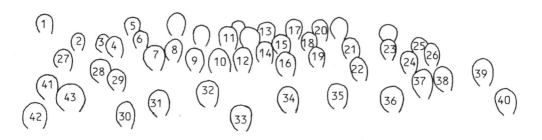

1 J.N. Larocca, 2 A. Stratmann, 3 D. Michaelis, 4 W.T. Norton, 5 J. Malotka, 6 M.G. Rumsby, 7 J.A. Black, 8 P. Morell, 9 D.A. Kirschner, 10 M. Dubois-Dalcq, 11 G. Almazan, 12 J.F. Poduslo, 13 M. Tosic, 14 P.E. Braun, 15 M. Gumpel, 16 S.E. Pfeiffer, 17 P. Honegger, 18 F. Kirchhoff, 19 J.-P. Zanetta, 20 S.K. Ludwin, 21 M. Schuch, 22 U. Walter, 23 G. Lemke, 24 B.D. Trapp, 25 H. Marahrens, 26 W.D. Richardson, 27 F.A. McMorris, 28 B. Zalc, 29 A. Roach, 30 A. Espinosa de los Monteros, 31 G. Jeserich, 32 C. Linington, 33 S. Szuchet, 34 H. Berlet, 35 W.F. Blakemore, 36 S.U. Kim, 37 F.X. Omlin, 38 D.R. Colman, 39 T.V. Waehneldt, 40 T. Rauen, 41 R.K. Yu, 42 M.V. Gardinier, 43 H.H. Althaus.

CELLULAR AND MOLECULAR BIOLOGY OF MYELINATION

August, 28 - September 2, 1989

Monastery Ohrbeck, FRG

Almazan, Guillermina
Department of Pharmacology
& Therapeutics
McGill University
3655 Drummond Street
Montreal, PQ H3G 1Y6
Canada

Althaus, Hans H.
Max-Planck-Institut
f. Experimentelle Medizin
Hermann-Reinstr. 3
3400 Göttingen
FRG

Black, Joel A.
Department of Neurology
Yale University of Medicine
333 Cedar Street
New Haven, Connecticut 06510
USA

Blakemore, William F.
Department of Clinical
Veterinary Medicine
University of Cambridge
Madingleyroad
Cambridge CB3 0ES
England

Braun, Peter E.
Department of Biochemistry
McGill University
3655 Drummond Street
McIntyre Medical Science Building
Montreal, PQ H3G 1Y6
Canada

Colman, David R.
Department of Anatomy
Columbia University
630 W 168th Street
New York, NY 10032
USA

Dubois-Dalcq, Monique
Laboratory of Molecular Genetics
National Institute of Neurology
and Communicative Disorders and
Stroke
National Institutes of Health
Bethesda, MD 20892
USA

Espinosa de los Monteros, Aracelie
Neuropsychiatric Institute
University of California
Rm 68-225
760 Westwood Plaza
Los Angeles, CA 90024-1759
USA

Gardinier, Minnetta V.
Laboratoire de Neurochimie
Service de Pédiatrie
Centre Hospitalier
Universitaire Vaudois
CH-1011 Lausanne
Switzerland

Gumpel, Madelaine
Laboratoire de Neurochimie
I.N.S.E.R.M. U.134
Hopital de la Salpétrière
47, Boulevard de l'Hopital
F-75651 Paris Cedex 13
France

Honegger, Paul
Institut de Physiologie
Université Lausanne
Faculté de Médicine
7, Rue du Bugnon
CH-1011 Lausanne
Switzerland

Jeserich, Gunnar
Universität Osnabrück
FB Biologie/Chemie
Abt. Zoophysiologie
Barbarastr. 11
4500 Osnabrück
FRG

Kim, Seung U.
Department of Medicine
Division of Neurology
University of British Columbia
Vancouver
Canada

Kirschner, Daniel A.
Neurology Research
Children's Hospital
Department of Neurology
Harvard Medical School
300 Longwood Avenue
Boston, MA 02115
USA

Kuo, J.F.
Department of Pharmacology
Amory University
Atlanta, Georgia 30322
USA

Larocca, Jorge N.
Albert Einstein College
Yeshiva University
Department of Neurology
1300 Morris Park Avenue
Bronx, NY 10461
USA

Lemke, Greg
Molecular Neurobiol. Laboratory
Salk Institute
10010 Torrey Pines Rd
La Jolla, CA 92037
USA

Linington, Christopher
University of Wales
Department of Medicine
Heath Park
Cardiff, CF4 4XN
England

Ludwin, Samuel K.
Department of Pathology
Queen's University
Kingston, Ontario K7L 3N6
Canada

McMorris, F. Arthur
The Wistar Institute and
Children's Hospital of
Philadelphia
36th Street at Spruce
Philadelphia, PA 19104
USA

Morell, Pierre
Biological Sciences Research
Center
Building 220H
University of North Carolina
School of Medicine
Chapel Hill, NC 27514
USA

Norton, William T.
Department of Neurology
Albert Einstein College
of Medicine
1300 Morris Park Ave
Building F Rm140
Bronx, NY 10461
USA

Omlin, Francois X.
Institut d'Histologie et
d'Embryologie
Université de Lausanne
Rue du Bugnon 9
CH-1011 Lausanne
Switzerland

Pfeiffer, Steven E.
Department of Microbiology
University of Connecticut
Health Center
Farmington, CT 06032
USA

Poduslo, Joseph F.
Membrane Biochemistry Laboratory
Departments of Neurology and
Biochemistry
Mayo Medical School and
Mayo Clinic
Rochester, MN 55905
USA

Richardson, William D.
Department of Zoology
University College London
Gower Street
London WC1E 6BT
England

Roach, Arthur
Mount Sinai Hospital
Research Institute
600 University Avenue
Toronto, Ontario M5G 1X5
Canada

Rumsby, Martin G.
Department of Biology
University of York
York, YO1 5DD
England

Schwartz, Michal
Department of Neurobiology
The Weizmann Institute
of Science
IL-Rehovot
Israel

Stoffel, Willy
Institut für
Physiologische Chemie
Universität Köln
Joseph Stelzmann Str. 52
5000 Köln-Lindental
FRG

Szuchet, Sarah
Department of Neurology
University of Chicago
BH Box 425
5841 South Maryland Avenue
Chicago, IL 60637
USA

Tosic, Mirianna
Laboratoire de Neurochimie
Service de Pédiatrie
Centre Hospitalier
Universitaire Vaudois
CH-1011 Lausanne
Switzerland

Trapp, Bruce D.
Johns Hopkins Medical School
Meyer 6-181
600 North Wolfe Street
Baltimore, MD 21205
USA

Waehneldt, Thomas V.
Max-Planck-Institut für
Experimentelle Medizin
Hermann-Reinstr. 3
3400 Göttingen
FRG

Walter, Ulrich
Labor f. Klinische Biochemie
Medizinische Universitätsklinik
Joseph-Schneiderstr. 2
8700 Würzburg
FRG

Yu, Robert K.
Department of Biochemistry
and Molecular Biophysics
Box 614
Virginia Commenwealth University
Richmond, Virginia 23298-0614
USA

Zalc, Boris
Laboratoire de Neurochimie
I.N.S.E.R.M. U.134
Hopital de la Salpétrière
47, Boulevard de l'Hopital
F-75651 Paris Cedex 13
France

Zanetta, Jean Pierre
C.N.R.S.
Centre de Neurochimie
5 Rue Blaise Pascal
F-67085 Strasbourg Cedex
France

TABLE OF CONTENTS

STEPS IN GLIAL CELL DIFFERENTIATION AND MYELINOGENESIS

THE 0-2A PROGENITOR DURING DEVELOPMENT AND REMYELINATION

M. Dubois-Dalcq, R. Armstrong, B. Watkins and R. McKinnon 3

REGULATION OF OLIGODENDROCYTE PROGENITOR DEVELOPMENT:
ANTIBODY-PERTURBATION STUDIES

S.E. Pfeiffer, R. Bansal, A.L. Gard and A.E. Warrington 19

OLIGODENDROCYTE DIFFERENTIATION: DEVELOPMENTAL AND
FUNCTIONAL SUBPOPULATIONS

A. Espinosa de los Monteros and J. de Vellis 33

LONG-TERM TISSUE CULTURES OF NEWBORN RAT OPTIC NERVE
CELLS: FUNCTIONAL DIFFERENTIATION OF GLIA AND APPEARANCE
OF NEURON-LIKE CELLS

F.X. Omlin 47

DISTRIBUTION OF MYELIN PROTEIN GENE PRODUCTS IN
ACTIVELY-MYELINATING OLIGODENDROCYTES

B.D. Trapp 59

IMMUNO-LOCALIZATION OF SODIUM CHANNELS IN AXON MEMBRANE
AND ASTROCYTES AND SCHWANN CELLS IN VIVO AND IN VITRO

J.A. Black, B. Friedman, A. Cornell-Bell, K.J. Angelides,
J.M. Ritchie and S.G. Waxman 81

DEMYELINATION, REMYELINATION AND GLIAL CELL TRANSPLANTATION

SOME ASPECTS OF MECHANISMS OF INFLAMMATORY DEMYELINATION
W.T. Norton, C.F. Brosnan, W. Cammer and E. Goldmuntz   101

TELLURIUM-INDUCED DEMYELINATION
P. Morell, M. Wagner-Recio, A.D. Toews, J. Harry and T.W. Bouldin   115

OLIGODENDROCYTE REACTION TO AXONAL DAMAGE
S.K. Ludwin   129

GLIAL CELL DIFFERENTIATION IN REGENERATION AND
MYELINATION
M. Schwartz, V. Lavie, M. Murray, A. Solomon and M. Belkin   143

AGGREGATING BRAIN CELL CULTURES: A MODEL TO STUDY
MYELINATION AND DEMYELINATION
P. Honegger and J.-M. Matthieu   155

THE TRANSPLANTATION OF GLIAL CELLS INTO AREAS OF
PRIMARY DEMYELINATION
W.F. Blakemore, A.J. Crang and R.J.M. Franklin   171

REMYELINATION OF A CHEMICALLY INDUCED DEMYELINATED
LESION IN THE SPINAL CORD OF THE ADULT SHIVERER MOUSE BY
TRANSPLANTED OLIGODENDROCYTES
O. Gout, A. Gansmuller and M. Gumpel   185

SIGNAL TRANSDUCTION AND REGULATORY EVENTS IN MYELIN-FORMING CELLS

cAMP-DEPENDENT PROTEIN KINASE: SUBUNIT DIVERSITY AND
FUNCTIONAL ROLE IN GENE EXPRESSION
M. Meinecke, W. Büchler, L. Fischer, S.M. Lohmann and U. Walter 201

PROTEIN KINASE C IN NEURONAL CELL GROWTH AND
DIFFERENTIATION
J.F. Kuo 217

OLIGODENDROCYTE-SUBSTRATUM INTERACTION SIGNALS CELL
POLARIZATION AND SECRETION
S. Szuchet and S.H. Yim 231

PROTEINKINASE A AND C ARE INVOLVED IN OLIGODENDROGLIAL
PROCESS FORMATION
H.H. Althaus, P. Schwartz, S. Klöppner, J. Schröter and V. Neuhoff 247

GROWTH FACTORS IN HUMAN GLIAL CELLS IN CULTURE
S.U. Kim and V.W. Yong 255

REGULATION OF OLIGODENDROCYTE DEVELOPMENT BY
INSULIN-LIKE GROWTH FACTORS AND CYCLIC AMP
F.A. McMorris, R.W. Furlanetto, R.L. Mozell, M.J. Carson and
D.W. Raible 281

PLATELET-DERIVED GROWTH FACTOR AND ITS RECEPTORS IN
CENTRAL NERVOUS SYSTEM GLIOGENESIS
I.K. Hart, E.J. Collarini, S.R. Bolsover, M.C. Raff and W.D. Richardson 293

TRANSFECTED CELLS AS A TOOL IN MYELIN RESEARCH

GENE TRANSFER OF RAT MATURE OLIGODENDROCYTES AND O-2A
PROGENITOR CELLS WITH THE ß-GALACTOSIDASE GENE
C. Goujet-Zalc, C. Lubetzki, C. Evrard, P. Rouget and B. Zalc   311

IMMORTALIZATION OF OLIGODENDROCYTE PRECURSORS FROM THE
OPTIC NERVE OF THE RAT WITH A TEMPERATURE-SENSITIVE
FORM OF THE SV40T ANTIGEN USING A RETROVIRUS VECTOR
G. Almazan   317

EXPRESSION OF NERVOUS SYSTEM cDNAS IN GLIAL AND NON-GLIAL
CELL LINES
B. Allinquant, S.M. Staugaitis, D. D'Urso, G. Almazan, S. Chin, P.J. Brophy and
D.R. Colman   329

PHYLOGENETIC ASPECTS OF MYELINATION

MYELIN AND MYELIN-FORMING CELLS IN THE BRAIN OF FISH -
A CELL CULTURE APPROACH
G. Jeserich, T. Rauen and A. Stratmann   343

MYELIN PROTEOLIPID PROTEIN: CLADISTIC TOOL TO STUDY
VERTEBRATE PHYLOGENY
T.V. Waehneldt, J. Malotka, C.A. Gunn and C. Linington   361

PHYLOGENETIC ASPECTS OF MYELIN STRUCTURE
H. Inouye and D.A. Kirschner   373

COMPONENTS AND STRUCTURES OF MYELIN

COMPOSITION AND METABOLISM OF MYELIN CEREBROSIDES
AND GANGLIOSIDES
R.K. Yu 391

RECEPTOR ACTIVITY AND SIGNAL TRANSDUCTION IN MYELIN
J.N. Larocca, F. Golly, M.H. Makman, A. Cervone and R.W. Ledeen 405

MYELIN OLIGODENDROCYTE GLYCOPROTEIN - A MODEL TARGET
ANTIGEN FOR ANTIBODY MEDIATED DEMYELINATION
S. Piddlesden and C. Linington 417

ROLE OF AN ENDOGENOUS MANNOSYL-LECTIN IN MYELINATION
AND STABILIZATION OF MYELIN STRUCTURE
J.-P. Zanetta, S. Kuchler, P. Marschal, M. Zaepfel, A. Meyer, A. Badache, A. Reeber,
S. Lehmann and G. Vincendon 433

MYELIN GENE REGULATION IN THE PERIPHERAL NERVOUS SYSTEM
J.F. Poduslo 451

GTP-BINDING PROTEINS ASSOCIATED WITH CNS MYELIN
P.E. Braun and L. Bernier 463

MOLECULAR BIOLOGY OF GENES CODING FOR MYELIN PROTEINS

MOLECULAR APPROACHES TO THE STUDY OF THE REGULATION OF
MYELIN SYNTHESIS
A. Roach 475

FROM MYELIN BASIC PROTEIN TO MYELIN DEFICIENT MICE
M. Tosic and J.-M. Matthieu 489

PROTEOLIPID PROTEIN: FROM THE PRIMARY STRUCTURE TO
THE GENE
W. Stoffel 505

DEVELOPMENTAL EXPRESSION OF THE MYELIN PROTEOLIPID
PROTEIN GENE
M.V. Gardinier and W.B. Macklin 517

THE MYELINATION CASCADE
G. Lemke, G. Weinmaster and E.S. Monuki 533

INDEX
. . . . 543

STEPS IN GLIAL CELL DIFFERENTIATION AND MYELINOGENESIS

The O-2A Progenitor during Development and Remyelination

Monique Dubois-Dalcq, Regina Armstrong, Bryn Watkins and Randall McKinnon

National Institute of Neurological Disorders and Stroke
Laboratory of Viral and Molecular Pathogenesis
Building 36, Room 5D-04
Bethesda, Maryland 20892

We are investigating the factors controlling growth and differentiation of oligodendrocytes precursor cells during CNS myelination and remyelination. These studies have stemmed from the description of O-2A progenitor cells in the newborn rat optic nerve (reviewed in Raff, 1989). These progenitor cells are bipotential since they can differentiate either into oligodendrocytes, which form myelin in the CNS, or type 2 astrocytes, which extend processes to nodes of Ranvier. Therefore, the linéage derived from this progenitor is called the oligodendrocyte-type 2 astrocyte (O-2A) lineage. O-2A progenitor cells are identified by immunolabeling with an antibody to cell surface gangliosides (A_2B_5) or glycoproteins (NSP4). O-2A progenitor cells purified and cultured from newborn rat brain (Behar et al., 1988) are shown in Figure 1.

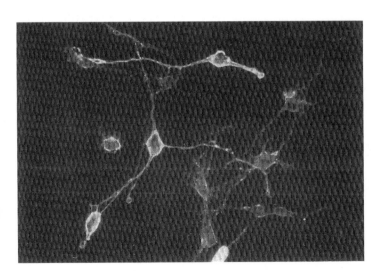

Figure 1. Purified brain O-2A progenitors in culture stained with NSP4 antibody. Cells are bipolar or multipolar with short processes.

In this short report of our ongoing studies, we describe how O-2A progenitor cells from the rodent CNS move along the oligodendrocyte differentiation pathway during development *in vitro,* and the role of polypeptide growth factors including platelet-derived growth factor (PDGF) and fibroblast growth factor (FGF) in the growth and rate of differentiation of this progenitor cell. We will also review the evidence for the persistence of O-2A progenitors in the rodent adult nervous system and how these cells react to a demyelinating episode and may play a role in remyelination.

1. DEVELOPMENT

Earlier studies on newborn rat optic nerve have shown that O-2A progenitor cells cultured in defined medium quickly develop into oligodendrocytes expressing galacto-cerebroside (GC), a major myelin glycolipid (Raff et al., 1983). In contrast, addition of 10% fetal bovine serum (FBS) drives the cell along the type 2 astrocyte pathway and induces glial fibrillary acidic protein (GFAP) expression. This situation is different *in vivo*: O-2A progenitors give rise to the first oligodendrocytes around birth while they keep proliferating for at least the next 2 weeks (Skoff et al., 1976). Conditioned medium (CM) derived from type 1 astrocytes, the first type of astrocyte emerging in the brain early before birth, restores this normal sequence of events *in vitro*: the progenitor will go through a number of divisions *in vitro* as they do *in vivo* before they differentiate (Raff et al., 1985). One question raised by these studies (reviewed in Raff, 1989) is whether the O-2A progenitor is committed to become an oligodendrocyte after a set number of divisions or whether it is a true stem cell, capable of self renewal in addition to generating oligodendrocytes. The best way to address this question is to perform cultures of single O-2A progenitors and analyse the progeny derived from clonal expansion of single cells.

A. Clonal Analysis of the O-2A progenitor

Initial studies by Temple and Raff (1986) on O-2A progenitors seeded singly on a layer of type 1 astrocytes indicated that the O-2A progenitor divides a set number of times during 8-10 days before its progeny become multipolar differentiated oligodendrocytes. In similar experiments, we seeded single progenitor cells from either optic nerve or cerebrum onto monolayers of type 1 astrocytes (Dubois-Dalcq, 1987; McMorris and Dubois-Dalcq, 1988). We selected the larger clones at 8 days and followed them for up to

3 weeks in culture. Clonal growth progressed steadily during the first week (~1 division per day) then slowed considerably. The majority of the clones were still expanding at 3 weeks and contained a variable ratio of multipolar cells reacting with A2B5 antibody but not expressing GC. In addition these GC negative cells stained with the O4 antibody, which recognized sulfatide and another glycolipid that are present on the surface of the O-2A progenitor at a later stage along the pathway. These O4+ cells had a more complex morphology and did not express vimentin. Gard and Pfeiffer (1989) have also described a population of O4+ multipolar cells which did not acquire GC 10 days after sorting and culture. Our clonal analysis demonstrated that the relative abundance of multipolar O4+GC - cells was increased when insulin was deleted from defined medium, while many more oligodendrocytes formed in the presence of insulin (5μg/ml) or IGF1 (100ng/ml) (McMorris and Dubois-Dalcq, 1988). In the presence of these factors, some clones contained only GC+ cells as described by Temple and Raff (1986). These studies indicated that progenitor cells can be expanded for a prolonged period of time *in vitro,* as *in vivo* studies had suggested. Furthermore IGF1 can promote proliferation of this O4+ progenitor and induces these cells to develop into oligodendrocytes (McMorris and Dubois-Dalcq,1988).

B. PDGF and the O-2A Lineage

The recent observations that type 1 astrocytes synthesize PDGF transcripts and protein and that anti-PDGF antibodies can neutralize the mitogenic effect of type 1 astrocyte CM have suggested a key role of PDGF in the growth of O-2A progenitors (Richardson et al., 1988). PDGF is active as a dimer and composed of 2 chains, designated A and B. The gene coding for PDGF-A chain in man is on chromosome 7, and the B chain gene is on chromosome 22 (reviewed in Ross et al., 1986). All three dimeric forms of PDGF have been identified and are biologically active, such as PDGF AB from human platelets, BB from porcine platelets and AA from osteosarcoma cells. Recent studies have identified both the PDGF ligands and PDGF receptors that are important for O-2A progenitor development in the rat CNS. These results are summarized in Table 1. As mentioned above, type 1 astrocytes are the major source of PDGF (A chain) in the rat, although it is not excluded that O-2A progenitor cells and or oligodendrocytes can also synthesize small amounts of PDGF. In the developing rat CNS, the A chain transcripts start to be expressed at E17 and

Table I. PDGF ligands and receptors in O-2A lineage cells[*]

	Type-1 Astrocyte	O-2A Progenitor	Oligodendrocyte
PDGF Ligands:			
A-chain mRNA	++		
A-chain protein	+		
B (sis) mRNA	(+)		
B (sis) protein			
PDGF Receptors:			
α-R mRNA	0	++	+
α–R protein	0	++	+
ß-R mRNA			
ß-R protein		0	0
Binding assays:	0	AA,AB,BB	AA,AB,BB
Mitogenicity:		AA,AB>BB	0

+ indicates the presence and relative abundance of PDGF ligands and receptors.
0 indicates that levels are non-detectable.
[*] see text for references.

increase in abundance thereafter, while low and constant levels of PDGF B transcripts are detected throughout rat brain development (Richardson et al., 1988). In addition, A chain transcripts are easily detected by *in situ* hybridization in the newborn rat optic nerve while the B chain transcripts are not detectable in the optic nerve parenchyma (Pringle et al.,1989).

The O-2A progenitor was examined recently for the expression of PDGF receptors, members of the tyrosine-kinase family of growth factors receptors. Two types of PDGF receptors (type-A or α; type-B or ß) have been characterized (Hart et al., 1988; Heldin et al., 1988) and cDNA copies cloned (Yarden et al., 1986; Matsui et al., 1989; Claesson-Welsh et al., 1989). The α receptor binds all 3 dimeric forms of PDGF (AA, AB, BB) whereas the ß receptor binds the BB dimers with high affinity and AB dimers with low affinity. Ligand binding studies have indicated that O-2A progenitor cells from optic nerve express the PDGF-α receptor (Hart et al.,1989), and PDGF AA dimers are a more potent mitogen than BB dimers in this system (Pringle et al., 1989) as discussed by Richardson et al. in this volume. The identity of the PDGF receptor on rat O-2A progenitor cells as the α–receptor has been confirmed by Northern blot analysis of the receptor mRNA transcripts in purified brain O-2A progenitor cells. Western blot analysis further determined that O-2A cells express abundant levels of the α-receptor but non-detectable levels of the PDGF-β receptor (McKinnon et al., 1989). The O-2A progenitor appears to be the first example of a mammalian cell expressing only the PDGF-α receptor.

PDGF drives the proliferation of O-2A progenitor cells from optic nerve and regulates the timing of their differentiation (Raff et al.,1988). We have examined the induction of myelin basic protein (MBP), a gene expressed in differentiated oligodendrocytes (Zeller et al., 1985), in cultures of purified brain progenitor cells treated with PDGF. PDGF delayed MBP transcript accumulation but did not prevent the expression of this differentiation gene (McKinnon et al., 1989).

C. A Possible role for FGF in the O-2A Lineage

Both acidic and basic FGFs are abundant in the adult CNS (reviewed by Barde, 1989) and have been shown to have mitogenic effects on oligodendrocyte precursors and possibly oligodendrocytes (Eccleston and Silberberg, 1985; Saneto and DeVellis, 1985; Besnard et al., 1989). The basic form of FGF (bFGF) has been purified from cell lysates of cultured astrocytes (Hatten et al., 1988; Ferrara et al., 1988). As shown in Fig 2, bFGF is a more potent mitogen than PDGF for purified brain O-2A progenitors as assayed by BUdR incorporation, using double immunofluorescence with A2B5 to identify progenitor cells with BUdR labeled nuclei.

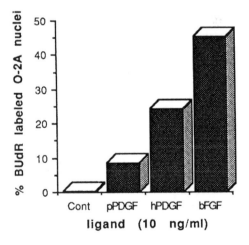

Figure 2. Mitogenic response of O-2A progenitor cells to porcine PDGF-BB (pPDGF), human PDGF-AB (hPDGF), and bFGF. Purified progenitors were cultured for 24 hrs with growth factors, pulsed 2 hrs with 10 μM BUdR, then stained with anti-A2B5 and anti-BUdR. Values represent the % A2B5 positive cells that incorporated BUdR.

We have also used Northern blot analysis of myelin-specific gene transcripts to examine the effects of bFGF on O-2A differentiation (McKinnon et al. 1989). Expression of MBP was completely inhibited by treatment of O-2A progenitors by 10 ng/ml bFGF, a situation different from that seen with PDGF (see above). When both FGF and PDGF were present for 2 days, the same inhibition of MBP expression was seen. The inhibition of myelin expression was reversible when FGF was removed from the cells. FGF also induced a rapid and dramatic increase in the levels of the PDGF-α receptor expressed by brain O-2A progenitors. We are presently examining whether this FGF induced increase in receptor levels results in an enhanced sensitivity of the progenitor cells to lower doses of the different dimeric forms of PDGF.

Our present results suggest that FGF can regulate the O-2A progenitor interactions with PDGF. Thus FGF and PDGF can influence the timing of oligodendrocyte differentiation *in vitro*. This indicates that a subtle interplay between these factors may set the timing of myelination in specific tracts *in vivo*. Since FGF appears to be increased locally in brain wounds (Finkelstein et al., 1988), this factor may also play a role in glial cell regeneration and remyelination (see below). bFGF may in fact be produced by neurons (Pettmann et al., 1986; Finkelstein et al., 1988; review by Barde, 1989) and FGFs have been shown to promote survival and neurite outgrowth of a variety of neurons (reviewed in Barde, 1989). Thus this polypeptide may successively have an effect on neuronal development and the timing of myelination of axons. Interestingly purified cultures of cerebellar interneurons secrete in their medium a factor which has mitogenic activity for O-2A progenitor cells and appears to arrest their differentiation (Levine, 1989). Whether cerebellar interneuron CM contains FGF-like factors is presently not known.

If FGF allows O-2A progenitor cells to divide for prolonged periods of time, one would postulate that other factors made in the CNS might inhibit PDGF induced mitosis. One possible candidate for this function would be transforming growth factor ß, TGF-ß (Van Obberghen-Schilling et al., 1987). TGFß is synthezised by type 1 astrocytes and can decrease the number of O-2A progenitors dividing in the presence of PDGF. The possibility of upregulation of the O-2A progenitor response to PDGF by one factor versus down regulation by another needs to be further explored. It would constitute an attractive mechanism by which the number of oligodendrocytes could be controlled with flexibility during development and regeneration. In addition the factors mentioned above may have a direct effect on the remyelination program (see below).

2. Remyelination

A. The O-2A progenitor in adult CNS

Evidence that rare O-2A progenitor cells exist in the normal adult rodent CNS has arisen from recent studies on optic nerve, rat brain and mouse spinal cord (ffrench-Constant and Raff, 1986; Armstrong et al., 1988; Hunter et al., 1988; Wolswijk and Noble, 1989). These investigators have successfully isolated progenitor cells by enzymatic and mechanical dissociation of the nervous tissue, sometimes followed by a Percoll gradient to separate the glial cells from the myelin. The *in vitro* properties of the progenitors isolated from adult CNS are compared to those of the newborn progenitors in table 2. The latter differ from the adult O-2A progenitor by their antigenic phenotype, cell shape and growth potential. Adult O-2A progenitors may extend branched processes which contrast with the simple bipolar morphology of the neonatal progenitor (Wolswijk and Noble, 1989). These adult O-2A progenitors simultaneously express the glycolipids recognized by the A2B5 and O4 antibodies whereas the neonatal O-2A progenitor stains only for A2B5 at birth and then moves on to express O4 one week later (Sommer and Noble, 1986). Vimentin is clearly present in the A2B5+ neonatal progenitor (Raff et al.,1984). However, vimentin is not detectable in adult progenitors of optic nerve. Similarly, O4+, A2B5+, vimentin- progenitors are present in 3 week old clones derived from single O-2A progenitors of newborn optic nerve and brain (Dubois-Dalcq, 1987; McMorris and Dubois-Dalcq, 1988).

Table 2. Comparison between neonatal and adult O-2A progenitor cells of rodent CNS[*].

	Neonatal Progenitor	Adult Progenitor
A2B5 (GQ Gangliosides)	+	+
O4 (sulfatide other lipid)	(-) → (+)	+
Vimentin	(+) → (-)	-
Galactocerebroside	—	—
Shape	Bipolar or small multipolar	Complex processes
Cell Cycle Time	18 h	65 h
Bipotentiality	+	+

[*] see text for references.

The adult O-2A progenitor of optic nerve has retained bipotentiality. It can evolve into either an oligodendrocyte or a type 2 astrocyte depending on the media conditions, but it differentiates more slowly (>5 days) than the neonatal progenitor (<3 days) (Wolswijk and Noble, 1989). O-2A progenitor cells in adult rodents can proliferate, although the cell cycle is lengthened relative to the neonatal O-2A progenitor (65 hours versus 18 hours) (Wolswijk, & Noble 1989). The properties of bipotentiality and mitogenic capacity identify the adult cell as a true progenitor of the O-2A lineage which continues to exist in fully myelinated tissue.

B. The O-2A Progenitor in Remyelination

The studies summarized above point to a potential role for the O-2A progenitor of the adult CNS in the regeneration of myelin after a demyelinating episode. The antigenic phenotype and biological properties of these precursors had not previously been described during a demyelinating disease. The fact that new oligodendrocytes are generated in reaction to a demyelinating episode of viral or toxic origin has been established in several electron microscopic studies combined with autoradiography (Herndon et al., 1977; Ludwin et al., 1981; Aranella and Herndon, 1984). Moreover, the emergence of myelin proteins and the expression of myelin genes during remyelination recapitulates the developmental program of myelination (Ludwin and Sternberger, 1984; Jordan et al., 1989a). This suggests that immature oligodendrocytes are the cells responsible for remyelinating denuded axons.

The demyelinating model we have studied is produced in C57Bl/6N mice at four weeks of age by intracerebral inoculation of the A-59 strain of mouse hepatitis virus. One week after intracranial inoculation (1 WPI), virus replication in glial cells is widespread throughout the spinal cord and results in a distinct decrease in all myelin-specific mRNA's prior to demyelination (Jordan et al., 1989b). Areas of demyelination develop by 3 WPI. Within the lesions the number of oligodendrocytes is reduced and total MBP transcripts are decreased by 75% compared to control white matter. Virus is cleared throughout most of the spinal cord by 4 WPI. At the earliest stage of remyelination, we have described an important increase in MBP transcripts containing exon 2 (encoding pre-large and pre-small forms of MBP), which are characteristic of myelination in early development (Jordan et al., 1989a). This increase in MBP transcripts containing exon 2 occurred in and around the lesions. This suggests that oligodendrocyte progenitor cells can repopulate the demyelinating lesions through of a process of mitosis, migration, and differentiation.

To examine this possibility further, we first attempted to detect directly *in vivo* O-2A lineage cells in the spinal cord of these infected mice at several stages of the demyelinating disease (Godfraind et al., 1989). We used triple-label immunofluorescence for O4, a progenitor marker, GFAP, the astrocyte marker, and 2'-3'-cyclic nucleotide 3'-phosphohydrolase (CNP), an early myelin protein which stains oligodendrocytes more reliably than GC in 1 micron frozen sections. While we found rare O4+ cells in control mice, these same cells were frequent in the spinal cord of infected mice at 1 and 2 WPI. Some of these cells were labeled with tritiated thymidine after a two hour *in vivo* pulse, indicating their ability to divide in response to a stimulus. Between 2 and 4 WPI, both O-2A progenitors (O4+) and type 2 astrocytes (O4+, GFAP+) were found in the lesions, and a proportion of these cells again incorporated thymidine (Godfraind et al., 1989). As remyelination proceeded, tritiated thymidine injected during the demyelinating phase appeared in CNP+ oligodendrocytes, indicating that precursor cells which divided earlier in the disease had generated new oligodendrocytes (Godfraind et al., 1989).

Encouraged by the identification of O-2A lineage cells *in vivo* in these animals, we then isolated these cells from the spinal cord of demyelinating and remyelinating mice (Armstrong et al., 1988). O-2A lineage cells identified in cultures from adult CNS included O-2A adult progenitor cells (figure 3), oligodendrocytes and type 2 astrocytes (figure 4).

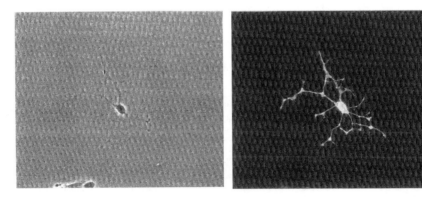

Figure 3. An O-2A progenitor cell (O4+ GC- GFAP-) isolated from spinal cords of 10 week old mice during remyelination (6 WPI). Cells were fixed at 3 days *in vitro*. Left: phase contrast. Right: O4 immunostain.

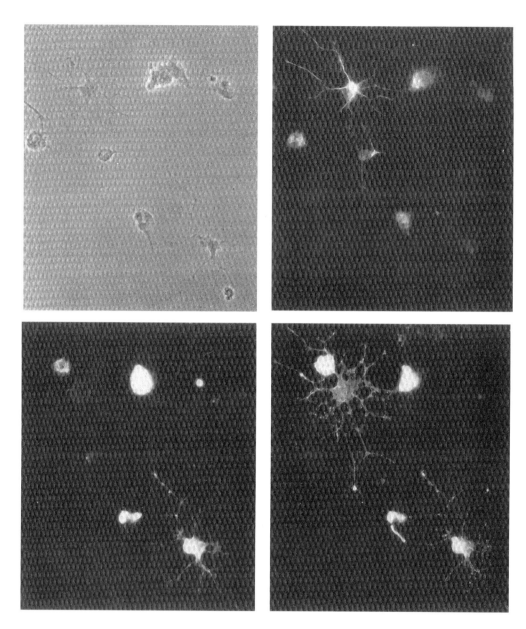

Figure 4. An oligodendrocyte (O4+ GC+ GFAP-) and a type 2 astrocyte (O4+ GC- GFAP+) isolated from spinal cords of 9 week old mice at the onset of remyelination (5 WPI). Cells were fixed at 3 days *in vitro* and processed for 3-color immunofluorescence. Top left: phase contrast. Top right: GFAP. Bottom left: GC. Bottom right: O4.

We found an important increase, relative to controls, in the number of O-2A lineage cells that could be isolated and cultured from demyelinating tissues at three to five WPI. Cultures derived from demyelinating tissues also contained numerous microglial cells which were rarely encountered in controls. Within the O-2A lineage population isolated from demyelinating tissues, there was a higher proportion of type 2 astrocytes and cells with a mixed oligodendrocyte-astrocyte phenotype, as compared to controls. The mixed phenotype cell, almost unique to remyelinating animals, was also identified *in vivo* in demyelinating animals, although rarely (Godfraind et al., 1989). This indicates a greater phenotypic plasticity in O-2A lineage cells from these diseased animals. Such

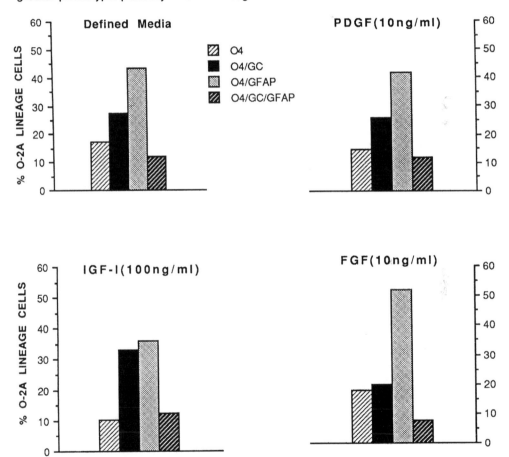

Figure 5. O-2A lineage cells were isolated from spinal cords of 9 week old mice (5 WPI). Growth factors were added to defined media after 1 day *in vitro*. Cells were fixed at 3 days *in vitro* and identified by 3-color immunofluorescence.

behavior could be related to the release of a number of polypeptides in the lesions; for instance ciliary neurotrophic factor is increased in CNS tissues during regeneration (Nieto-Sampedro et al., 1983) and may transiently induce GFAP in O-2A progenitors (Lillien et al., 1988). The increased number of O-2A lineage cells isolated during demyelination is at least partially due to proliferation. Cultures derived from spinal cords of infected mice contained an increased proportion of tritiated thymidine-labeled O-2A lineage cells after a short *in vivo* or *in vitro* pulse. Dividing cells were rarely seen in control cultures. Thymidine incorporation was not markedly enhanced by addition of human PDGF, basic FGF or IGF1 (Amstrong et al., 1989). In these cultures the ratio of oligodendrocytes relative to type 2 astrocytes was higher in the presence of IGF-1, and was lower in the presence of basic FGF (figure 5) as was seen with cultured O-2A progenitors isolated from early development (see above). Thus, in this *in vitro* system, we can now study the conditions for O-2A lineage cells to proliferate and express phenotypic plasticity as they do *in vivo* in the course of demyelination and remyelination.

3. Conclusions:

Studies on glial O-2A progenitor cells *in vitro* suggest that an interaction between several different polypeptide growth factors may affect the process of O-2A lineage differentiation *in vivo*. FGF and PDGF are mitogenic for O-2A progenitor cells, and PDGF appears to control the timing of O-2A differentiation while FGF may block differentiation into oligodendrocytes. In contrast, IGF-1 promotes oligodendrocyte development. An O4+ O-2A progenitor cell persists in the adult and may represent a self-renewing "stem" cell that proliferates in response to a demyelinating episode. The phenotype of adult O-2A lineage cells can also be modulated by some of these growth factors. Thus polypeptide growth factors that are important for normal glial cell development may also have a role in the process of remyelination. Understanding the mechanisms regulating expression of these factors during development and disease will further our ability to intervene in demyelinating diseases in man.

Acknowledgements:

We thank Ray Rustin for excellent technical assistance. R.McK was supported by a senior fellowship from the US National Multiple Sclerosis Society.

References:

Armstrong R, Friedrich VL Jr, Holmes KV, Dubois-Dalcq M (1988) In vitro analysis of neuroglial cells isolated during demyelination and remyelination. Soc Neurosci Abstr 14: 787

Armstrong R, Friedrich VL Jr, Holmes KV, Dubois-Dalcq M (1989) Proliferation and differentiation of neuroglial cells isolated during demyelination and remyelination. Ann NY Acad Sci (in press)

Aranella LS, Herndon RM (1984) Mature oligodendrocytes division following expeirmental demyelination in adult animals. Arch Neurology 41:1162-1165

Barde Y-A (1989) Trophic factors and neuronal survival. Neuron 2:1525-1534

Behar T, McMorris FA, Novotny EA, Barker JL, Dubios-Dalcq M (1988) Growth and differentiation properties of 0-2A progenitors purified from rat cerebral hemispheres. J Neurosci Res 21:168-180

Besnard F, Perraud F, Sensenbrenner M, Labourdette G (1989) Effects of acidic and basic fibroblast growth factors on proliferation and maturation of cultured rat oligodendrocytes. Int J Develop Neurosci in press

Claesson-Welsh L, Eriksson A, Westermark B, Heldin C-H (1989) cDNA cloning and expression of the human A-type platelet-derived growth factor (PDGF) receptor establishes structural similarity to the B-type PDGF receptor. Proc Natl Acad Sci USA 86:4917-4921

Dubois-Dalcq M (1987) Characterization of a slowly proliferative cell along the oligodendrocyte differentiation pathway EMBO 6:2587-2595

Eccleston PA, Silberberg DH (1985) Fibroblast growth factor is a mitogen for oligodendrocytes in vitro. Dev Brain Res 21:315-318

Ferrara N, Ousley F, Gospodarowicz D (1988) Bovine brain astrocytes express basic fibroblast growth factor, a neurotropic and angiogenic mitogen. Brain Research 462:223-232

Finkelstein SP, Apostolides PJ, Caday CG, Prosser J, Philips MF, Klagsbrun M (1988) Increased basic fibroblast growth factor (bFGF) immunoreactivity at the site of local brain wounds. Brain Res 460:253-259

ffrench-Constant C, Raff MC (1986) Proliferating bipotential glial progenitor cells in adult rat optic nerve. Nature 319:499-502

Gard AL, Pfeiffer SE (1989) Oligodendrocyte progenitors isolated directly from developing telencephalon at a specific phenotypic stage myelinogenic potential in a defined environment. Development 106:119-132

Godfraind C, Friedrich VL, Holmes KV, Dubois-Dalcq M (1989) In vivo analysis of glial cell phenotypes during a viral demyelinating disease in mice. J Cell Biology (in press)

Hart CE, Forstrom JW, Kelly JD, Seifert RA, Smith RA, Ross R, Murray MJ, Bowen-Pope DF (1988) Two classes of PDGF receptor recognize different isoforms of PDGF. Science 240:1529-1534

Hart IK, Richardson WD, Heldin C-H, Westermark B, Raff MC (1989) PDGF receptors on cells of the oligodendrocyte-type-2 astrocyte (O-2A) cell lineage. Development 105:595-603

Hatten ME, Lynch M, Rydel RE, Sanchez J, Joseph-Silverstein J, Moscatelli D, Rifkin DB (1988) In vitro neurite extension by granule neurons is dependent upon astroglial-derived fibroblast growth factor. Develop Biol 125:280-289

Heldin C-H, Backstrom G, Ostman A, Hammacher A, Ronnstrand L, Rubin K, Nister M, Westermark B (1988) Binding of different dimeric forms of PDGF to human fibroblasts: evidence for two separate receptor types. EMBO J 7:1387-1393

Herndon RM, Price DL, Weiner LP (1977) Regeneration of oligodendroglia during recovery from demyelinating disease. Science 195:693-694

Hunter SE, Seidel MF, Bottenstein JE (1988) Response of neonatal and adult glial progenitors to neuronal cell line-derived mitogens. Soc Neurosci Abstr 14:321

Jordan C, Friedrich VL, deFerra F, Weismiller D, Holmes K, Dubois-Dalcq M (1989a) Differential exon expression in myelin basic protein transcripts during CNS remyelination. J Cellular and Molecular Neurobiol (in press)

Jordan C, Friedrich VL, Godfraind C, Holmes KV, Dubois-Dalcq M (1989b) Expression of viral and myelin gene transcripts in a murine demyelinating disease caused by a corona virus. Glia (in press)

Lillien LE, Sendtner M, Rohrer H, Hughes SM, Raff MC (1988) Type-2 astrocyte development in rat brain cultures is initiated by a CNTF-like protein produced by type-1 astrocytes. Neuron 1:485-494

Levine JM (1989) Neuronal influences on glial progenitor cell development. Neuron 3:103-113

Ludwin SK, (1981) Pathology of demyelination and remyelination. In: Waxman SG, Ritchie JM (eds) Advances in neurology Vol 31. Demyelinating diseases. Raven Press, New York, 123-168

Ludwin SK, Sternberger NH (1984) An immunohistochemical study of myelin proteins during remyelination in the central nervous system. Acta Neuropathologica 63:240-248

Matsui T, Heidaran M, Miki T, Popescu N, La Rochelle W, Krous M, Peirce J, Aaronson S (1989) Isolation of a novel receptor cDNA establishes the existence of two PDGF receptor genes. Science 243:800-804

McKinnon RD, Matsui TA, Aaronson S, Dubois-Dalcq, M (1989) FGF inhibits myelin gene expression and induces the PDGF-a receptor in differentiating 0-2A glial progenitor cells. American Society for Cell Biology, 29th Annual Meeting

McMorris FA, Dubois-Dalcq M (1988) Insulin-like growth factor 1 promotes cell proliferation and oligodendroglial commitment in rat glial progenitor cells developing in vitro. J Neurosci Res 21:199-209

Nieto-Sampedro M, Manthorpe M, Barbin G, Varon S, Cotman CW (1983) Injury-induced neuronotrophic activity in adult rat brain: correlation with survival of delayed implants in the wound cavity. J Neuroscience 3:2219-2229

Pettmann B, Labourdette G, Weibel M, Sensenbrenner M (1986) The brain fibroblast growth factor is localized in neurons. Neuroscience Letters 68:175-180

Pringle N, Collarini EJ, Mosley MJ, Heldin C-H, B Westermark B, Richardson WD (1989) PDGF A chain homodimers drive proliferation of bipotential (O-2A) glial progenitor cells in the developing rat optic nerve. EMBO J 8:1049-1056

Raff MC (1989) Glial cell diversification in the rat optic nerve. Science 243:1450-1455

Raff MC, Abney ER, Fok-Seang J (1985) Reconstitution of a developmental clock in vitro: A critical role for astrocytes in the timing of oligodendrocytes. Cell 42:61-69

Raff MC, Lillien LE, Richardson WD, Burne JF, Noble MD (1988) Platelet-derived growth factor from astrocytes drives the clock that times oligodendrocyte development in culture. Nature 333:562-565

Raff MC, Miller RH, Noble M (1983) A glial progenitor cell that develops in vitro into an astrocyte or an oligodendrocyte depending on culture medium. Nature 303:390-396

Raff MC, Williams BP, Miller RH (1984) The in vitro differentiation of a bipotential glial progenitor cell. EMBO 3:1857-1864

Richardson WD, Pringle N, Mosley JD, Westermark B, Dubois-Dalcq M (1988) A role for platelet-derived growth factor in normal gliogenesis in the central nervous system. Cell 53:303-319

Ross R, Raines EW, Bowen-Pope DF (1986) The biology of platelet-derived growth factor. Cell 46:155-169

Saneto RP, deVellis J (1985) Characterization of cultured rat oligodendrocytes proliferating in a serum-free, chemically defined medium. Proc Natl Acad Sci USA 82:3509-3513

Skoff R, Price D, Stocks A (1976) Electron microscopic autoradiographic studies of gliogenesis in rat optic nerve I cell proliferation. J Comp Neurol 169:291-312

Sommer I, Noble M (1986) Plasticity and commitment in oligodendrocyte development. Soc Neurosci Abstr 12:1585

Temple S, Raff MC (1986) Clonal analysis of oligodendrocyte development in culture: Evidence for a developmental clock that counts cell divisions. Cell 44:773-779

Van-Obberghen-Schilling E, Behar T, Sporn MB, Dubois-Dalcq M (1987) Signalling between type 1 astrocytes and their glial 0-2A progenitors: Modulation by transforning growth factor-beta (TGFß). J Cell Biol 105 (4) (Part 2):318a

Wolswijk G, Noble M (1989) Identification of an adult-specific glial progenitor cell. Development 105:387-400

Yarden Y, Escobedo JA, Kuang W-J, Yang-Feng TL, Daniel TO, Tremble PM, Chen EY, Ando ME, Harkins RN, Francke U, Fried VA, Ullrich A, Williams LT (1986) Structure of the receptor for platelet-derived growth factor helps define a family of closely related growth factor receptors. Nature 323:226-232

Zeller NK, Behar TN, Dubois-Dalcq M, Lazzarini RA (1985) The timely expression of myelin basic protein gene in cultured rat brain oligodendrocytes is independent of continuous neuronal influences. J Neurosci 5:2955-2962

REGULATION OF OLIGODENDROCYTE PROGENITOR DEVELOPMENT: ANTIBODY-PERTURBATION STUDIES

S. E. Pfeiffer, R. Bansal, A. L. Gard, A. E. Warrington
Department of Microbiology and
Program in Neurological Sciences
University of Connecticut Health Center
Farmington, CT 06032
USA

INTRODUCTION

The differentiation of oligodendrocytes (OL) and their elaboration of myelin sheaths is a crucial event of brain development. The resultant saltatory nerve conduction allows for dramatic savings in energy consumption and space utilization. Disruption of myelinogenesis causes serious neurological deficits, many with chronic implications (Morell, 1984).

Programmed myelin gene expression in oligodendrocytes (OL) is regulated both by genetic factors intrinsic to these cells, and by environmental influences encountered as the cells proliferate and migrate from their sites of embryological origin to their target neurons (Pfeiffer, 1984). A combination of studies *in vivo* and in culture have delineated several steps in OL development (reviewed in Gard and Pfeiffer, 1989, and Gard et al., 1989) which are summarized in Figure 1.

The "O-2A" bipotential progenitor is a bipolar, proliferative, and migratory cell that differentiates into either a "Type II astrocyte" or an OL depending on environmental signals (Raff *et al.*, 1983; Levi *et al.*, 1987). The OL-specific lineage branch is characterized by the appearance of the O4 antigen (Sommer and Schachner, 1982; Dubois-Dalcq, 1987) to produce "proligodendrocytes" (Gard and Pfeiffer, 1989). Subsequently, proligodendrocytes differentiate into authentic OL, indicated first by the cell surface expression of galactocerebroside (GalCer), sulfatide synthesis and CNP activity, and then by the highly ordered temporal expression of additional myelin proteins, including proteolipid protein (PLP) and myelin basic protein (MBP). The developmental state of oligodendroglia within the lineage can be specified using panels of immunological and molecular probes (Pfeiffer and Gard, 1988).

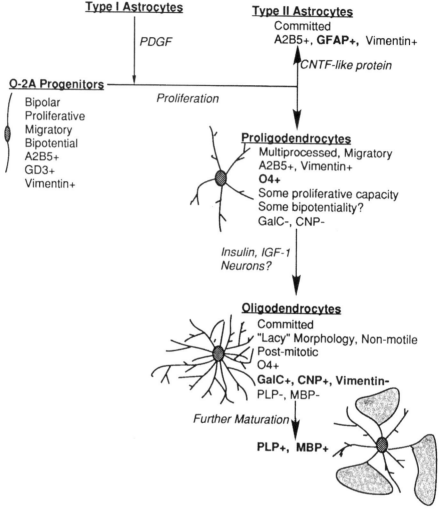

Figure 1. Working hypothesis for the sequence of stages in oligodendrocyte lineage progression, based on current literature (see text for references).

Here we shall review some of our recent work concerning (a) the characterization of three key monoclonal "anti-oligodendroglial" antibodies, (b) the immuno-isolation and developmental characterization of the proligodendrocyte, and (c) the application of these antibodies to study the oligodendrocyte lineage by perturbing normal developmental progression.

MATERIALS AND METHODS

The methods used in these studies have been descibed in detail in the referenced mansucripts, in particular those by Gard et al. (1988), Bansal et al. (1988), Gard and Pfeiffer (1989), Bansal and Pfeiffer (1989), and Bansal et al. (1989).

RESULTS

Characterization of Antibodies O1, O4 and R-mAb used in the Analysis of Oligodendrocyte Development (Bansal et al., 1989)

Antibodies against galactolipids, particularly galactocerebroside (GalCer) and sulfatide (SUL), have proven to be valuable tools in studies of oligodendrocytes and Schwann cells (Raff et al., 1978; Mirsky et al., 1980), providing a means to identify cell-type and lineage stage-specific milestones in their development (Pfeiffer and Gard, 1988), isolate subpopulations (Meier et al., 1982; McMorris and Dubois-Dalcq, 1988; Gard and Pfeiffer, 1989), and investigate antigen function by antibody perturbation (Ranscht et al., 1987; Dyer & Benjamins, 1988a,b; Bansal et al., 1988; Bansal and Pfeiffer, 1989). The conclusions derived from such applications depend, of course, on an understanding of antibody specificity.

Monoclonal antibodies O1 and O4 (generated by immunization with bovine corpus callosum; Sommer and Schachner, 1981), and monoclonal antibody R-mAb (prepared by immunizing with a synaptic plasma preparation; Ranscht et al., 1982) specifically label within the central nervous system the surfaces of oligodendrocytes and/or their immediate progenitors. $O4^+$ progenitors are detectable at birth, approximately 2-3 days before the appearance of O1 and R-mAb reactivities (Schachner et al., 1981; Singh and Pfeiffer, 1985; Gard and Pfeiffer, 1989). Previous studies showed that O1 and R-mAb reacts with GalCer, and O4 reacts with SUL (Schachner, 1982; Ranscht et al., 1982; Singh and Pfeiffer, 1985).

We have further characterized these three monoclonal antibodies with respect to their specificities for a number of purified lipids. The observed specificities were consistent regardless of how the antigens were presented to the antibodies.

O4 reacts with sulfatide, seminolipid and to some extend with cholesterol (Table 1). However, O4-positive "proligodendrocytes" (Figure 1, and below) which had not yet begun to express the O1 antigen failed to incorporate $^{35}SO_4$ or ^3H-galactose into sulfatide or seminolipid, the syntheses of which first appear in O1-positive cells. Therefore, O4 stains, in addition to sulfatide and seminolipid, an unidentified antigen (referred to here simply as "X") that appears on the surface of oligodendrocyte progenitors prior to the expression of sulfatide and galactocerebroside.

R-mAb reacted with galactocerebroside, monogalactosyldiglyceride, sulfatide, seminolipid, and psychosine. We found by several assay methods that R-mAb reacts nearly as well with sulfatide, indicating that it cannot be used as an unequivocal marker for GalCer, and that sulfatide reactivity may contribute to the staining of oligodendrocyte surfaces.

O1 reacted with galactocerebroside, monogalactosyldiglyceride and psychosine, and in addition labeled an unidentified species in rat brain extracts. Among the three antibodies tested, O1 would appear to be the choice for identifying the appearance of GalCer on the surfaces of developing oligodendrocytes.

Table 1. Summary of specificities of monoclonal antiodies O4, R-mAb, and O1.

O4	R-mAb	O1
Sulfatide	Galactocerebroside	Galactocerebroside
Seminolipid	Monogalactosyldiglyceride	Monogalactosyldiglyceride
(Cholesterol, +/-)	Sulfatide	Psychosine
"X"	Seminolipid	
	"Y?"	

In primary cultures of rat brain, developing O4$^+$ oligodendrocyte progenitors stained slightly earlier with R-mAb than with O1, and thus R-mAb transiently stained a larger population of oligodendrocytes than did O1 (Figure 2). Immunofluorescence studies of developing primary cultures of rat brain demonstrated the presence of both O4$^+$R-mAb$^-$ and R-mAb$^+$O1$^-$ cells. It appears, therefore, that at least in culture R-mAb reacts with an antigen (referred to here as "Y") that appears after the O4 antigen X, and slightly before the surface expression of GalCer (demonstrated by O1 reactivity) and sulfatide (Figure 3). The order of expression suggested by current data is summarized in Figure 4.

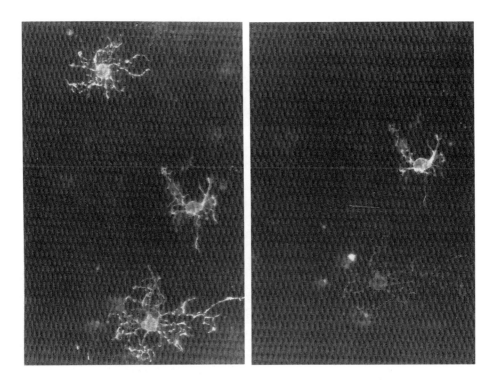

Figure 2. Immunofluorescent staining of 5-day old primary culture initiated from 19-21 day fetal rat telencephalon. Left panel: Cells stained with R-mAb. Right panel: Same field as in left panel, cells stained with O1.

In summary, O4, O1, and R-mAb have distinct specificities, each recognizing a unique subset of a group of related galactolipids. In addition, the data provide evidence for reactions with other, unidentified antigens. In particular, O4 appears to recognize a non-sulfolipid antigen that within the central nervous system specifically identifies oligodendrocytes and their immediate progenitors, and there is preliminary evidence for additional reactivities for R-mAb and O1. In addition to providing more complete, comparative characterizations of these three specific monoclonal antibodies, these data emphasize the need to extensively characterize anti-glycolipid antibodies using a panel of immuno-techniques. The identification of additional oligodendrocyte cell surface molecules can be expected to provide new insights into the determination of the oligodendrocyte cell lineage.

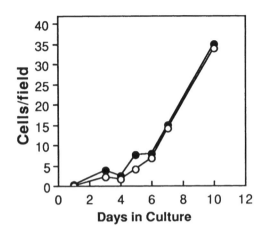

Figure 3. Time course of appearance of O1+ (open circles) and R-mAb+ (closed circles) cells in mixed primary cultures initiated from 19-21 day fetal rat telencephalon and grow in in DME plus 10% fetal calf serum. Cells were double-stained at various time intervals with monoclonal antibodies O1 and R-mAb. Field size, 1.45X10⁻³ cm².

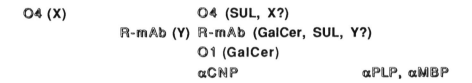

Figure 4. Proposed order of expression of antigens recognized by monoclonal antibodies O4, R-mAb and O1. Note that the predicted, unidentified antigens X and Y preceed the appearance of known lipid antigenic specificities. Whether X and Y are only transiently expressed, or continue to be expressed on mature oligodendrocytes, is not known. (Modified from Bansal et al., 1989).

Oligodendrocyte Progenitors Isolated directly from Developing Telencephalon at a Specific Phenotypic Stage (Gard and Pfeiffer, 1989)

Oligodendrocyte progenitors at a temporally narrow and well-defined phenotypic stage of development characterized by the expression of the O4 antigen, but the absence of GalCer, have been isolated directly from postnatal rat telencephalon. O4+GalCer⁻ progenitors appear at birth three days before O4+GalCer+ oligodendrocytes. A major subpopulation of O4+GalCer⁻ progenitors (80%), "proligodendrocytes," is fully committed to terminal OL differentiation.

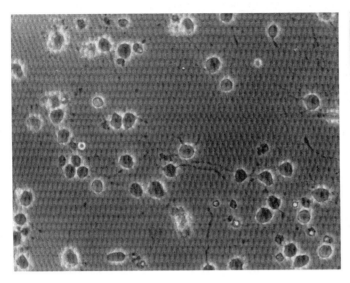
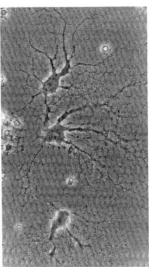

Figure 5. Left panel: O4+GalCer⁻ progenitors prepared by O4 immuno-isolation after one day in culture. Right panel: Morphological differentiation of a parallel culture after five days in culture.

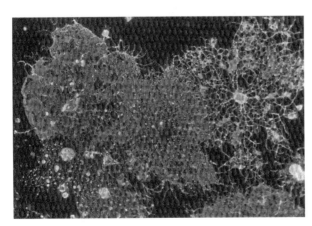

Figure 6. Isolated O4+GalCer⁻ progenitors grown in culture for 7 days. Note the extensive membrane formation in the cell on the left.

A small, maximal set of nutritional supplements is sufficient to support the myelinogenic cascade of differentiated gene expression in a temporally normal manner, in quantitatively significant amounts, in normal ratios of myelin protein isoforms, extending to the inclusion of myelin-specific products into membrane sheets (see Gard and Pfeiffer, 1989, for figures). Contact with other cell types, in particular neurons and astrocytes or unknown growth factors unique to these cell types, are not required. A developmentally quiescent subpopulation (20%) of O4+GalCer− cells is also present that may have significance for understanding "cryptic progenitors" in adult brain and remyelination. Two alternative models were proposed to explain the developmental occurence of both populations *in vivo*. In the first model, the quiescent cells represent a progenitor subpopulation distinguished by an alternative mode of myelinogenic activation. In the second model and simpler model, both O4+GalCer− subpopulations arise sequentially as the result of a critical regulatory event subdividing this phenotypic stage.

In summary, isolated O4+GalCer− cells demonstrate in culture the myelinogenic potential of an important developmental intermediate exisiting *in vivo,* and constitute a simplified system in which to further probe OL differentiation and growth control. In particular, while these cells growing in culture under highly defined conditions undergo a remarkable degree of development, they deviate from their counterparts *in vivo* in at least two significant ways. First, proligodendrocytes have significant proliferative capacity *in vivo*, but fail to multiply in the defined culture system (Gard et al., 1989a); the presence *in vivo* of a mitogen for these cells is indicated. Second, survival is apparently high *in vivo*, but poor in isolated proligodendrocyte cultures; thus the existence of a survival factor(s) is suggested (Gard et al., 1989b). The identification and analysis of these predicted growth factors is of considerable current interest.

Antibody-mediated Perturbation of Oligodendrocyte Development

Perturbation of myelinogenesis by monoclonal antibodies O4, R-mAb and O1 are being used to study the role of the corresponding antigens in oligodendrocyte differentiation.

O4 Stimulation of Myelinogenesis (Bansal *et al.,* 1988). In the first set of studies, dissociated primary cultures initated from 19-21 day fetal rat telencephala were grown in the presence of O4. In the presence of IgM O4, the oligodendrocytes formed aggregates connected by fasciculated processes. Immunofluorescence microscopy and biochemical analyses of treated cultures demonstrated 2-3 fold increases in the fraction of O4-positive cells expressing myelin basic protein, and in the levels of myelin basic protein RNA, myelin basic protein, 2',3'-cyclic nucleotide

3'-phosphohydrolase activity, and $^{35}SO_4$ incorporation into sulfatide. Greater than 90 percent of the cells positive for myelin basic protein in treated cultures were in aggregates. The specific activities of oligodendrocyte markers were unaffected in control cultures grown with non-specific myeloma IgM. Since there was no increase in the total number of O4-positive cells in treated cultures, the increases in the specific activities of the myelin protein markers appears to be due to an increase in the fraction of cells expressing these markers. Time course studies demonstrated that both the rate and extent of oligodendrocyte differentiation were enhanced in treated cultures. Neither N-CAM nor HNK-1 affected either the extent of oligodendrocyte aggregtion or developmental profiles. Possible mechanisms include direct effects of the antibody on cell physiology, or indirect responses to antibody-induced aggregation.

To further resolve these two mechanisms, isolated oligodendrocyte cultures were grown at a relatively low density to minimize the extent of antibody-induced aggregate formation. Interestingly, under these conditions, O4 treatment *interferred* with oligodendrocyte process and membrane formation, and myelin basic protein was largely restricted to the cell body. It appears, therefore, that the stimulation described above may be a secondary effect due to oligodendrocyte aggregration, leading perhaps to the accumulation of an autocrine factor within the aggregates (see also Holmes et al., 1989). We propose that a different mechanism, possibly involving interference with cell-substrate or cell receptor-ligand interactions, leads to the inhibitory effects.

<u>Reversible Inhibition of Oligodendrocyte Progenitor Differentiation by a Monoclonal Antibody R-mAb (Bansal and Pfeiffer, 1989)</u>. Mixed primary cultures were grown in medium containing R-mAb (Figure 7). The normal increase in the number of O4$^+$ cells in these cultures was not affected by R-mAb treatment, indicating that the differentiation and entry of precursors into the O4$^+$ OL-specifc pathway and O4$^+$ cell viability were not unchanged by R-mAb treatment. In contrast, the appearance of cells immuno-labeled with antibodies R-mAb and anti-MBP (Figure 7, squares), and with O1 and anti-CNP (not shown), was blocked. A significant reduction in the appearance of R-mAb$^+$ and MBP$^+$ cells occurs at antibody dilutions as low as 2 µg/ml total IgG$_3$, and inhibition is nearly complete at 10 µg/ml. Upon addition of R-mAb to R-mAb-positive cells, immuno-reactivity was lost over the succeeding two hours (Figure 8).

Several other antibodies that bind to surface antigens of OL and/or their progenitors, including anti-NCAM, HNK-1, O1, 1A9, and anti-cholesterol, were ineffective at blocking differentiation. Antibody-blocked cells had an immature morphology with sparsely branched processes.

R-mAb treatment also causes dramatic reductions in ^{3}H-galactose incorporation into GalCer and sulfatide, CNP activity, MBP protein content, and MBP RNA levels compared to untreated control cultures; therefore the block of MBP expression appears to be primarily at the level of MBP RNA synthesis or stability (see Bansal and Pfeiffer, 1989, for figures). In contrast, the levels of total protein and astrocytic glial fibrillary acidic protein in antibody-treated cultures remain nearly normal.

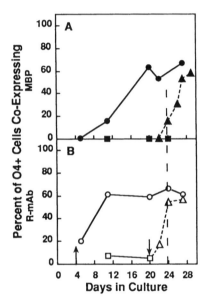

Figure 7. R-mAb induced inhibition and reinitiation of differentiation in mixed primary cultures from 19-21 day fetal rat telencephala. R-mAb was added to one set of cultures 4 days after plating (up arrow) and removed at 20 days (down arrow). Shown are the percent of O4+ cells also stained with anti-MBP (A) or R-mAb (B). Untreated control cultures (circles); cultures treated with R-mAb (squares), or treated with R-mAb and grown further in antibody-free medium (triangles). (From Bansal and Pfeiffer, 1989).

Figure 8. Time course of the loss of cell surface R-mAb immunoreactivity upon exposure to R-mAb. Mixed primary cultures treated with R-mAb at 37°C for various durations were subsequently immunostained at 4°C with additional R-mAb. Field size, 1.45X10^{-3} cm^2.

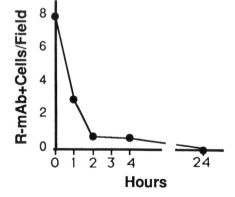

Upon replacing the R-mAb-containing medium with control medium, R-mAb+ cells began to reappear within one day (Figure 7B, triangles), while MBP+ cells (Figure 7A, triangles), MBP, and MBP-RNA (not shown) all began to reappear within 3 days. The normal lag between the temporal development of R-mAb and MBP expression was maintained. Therefore, the block may occur before the activation of a regulatory event leading to the ordered terminal differentiation sequence. ^{3}H-galactose incorporation into GalCer and sulfatide, and CNP activity, also reappeared efficiently after removal of the antibody.

Reverted mixed primary cultures differentiated more rapidly than control cultures, suggesting that the R-mAb block results in a synchronization of developmental progression along the oligodendrocyte lineage by accumulating progenitors at the inhibition point. Thus, the point in the oligodendrocyte lineage at which the inhibition occurs is restricted to the period after the appearance of the O4+GalCer⁻ progenitor through its differentiation into GalCer+ cells. During a prolonged treatment of dissociated primary cultures with R-mAb, cells developmentally younger than those at the block point continue to progress and become O4+, begin to express the R-mAb antigen, bind R-mAb, and become developmentally inhibited. The blocked cells continue to express vimentin and retain a relatively simple morphology, characteristics of O4+GalCer⁻ progenitors (Gard and Pfeiffer, 1989; Gard et al., 1989), but apparently accumulate at or just beyond the GalCer⁻/GalCer+ interface.

Similar results are obtained when cultures of isolated O4+GalCer⁻ proligodendrocytes were treated with R-mAb. Therefore, R-mAb must act directly on the OL progenitors themselves, not indirectly via interactions of progenitors with other cell types. Further, the reversion must be due to recovery of the blocked cells themselves, rather than repopulation by more immature progenitors still flowing into the developing population, because earlier stages of the OL lineage are removed in experiments using enriched O4+GalCer⁻ progenitors.

The identity of the antigen responsible for the perturbation is not firmly established. As described above, R-mAb reacts with both GalCer and sulfatide, as well as an unidentified antigen "Y". O1 mAb, which reacts with GalCer, did not inhibit OL differentiation. O4 mAb, which reacts with an unidentified antigen "X" and sulfatide, produced effects quite different from R-mAb (see above). These data suggest that the target could be antigen "Y".

In summary, the data indicate that the normal progression along the OL lineage from O4+GalCer⁻ progenitor to terminally differentiated OL is blocked when progenitors are grown in the presence of R-mAb. The continued presence of the R-mAb is required in order to maintain the develomental block. Removal of the perturbing antibody leads to rapid resumption of lineage progression. This reversible antibody-induced inhibition of myelinogenesis at a specific progenitor stage of development offers interesting experimental possibilities for studying the regulation of OL development, and has important implications for demyelinating diseases such as multiple sclerosis, in which a similar inhibition would impair remyelination by populations of developmentally quiescent progenitor cells (Wolswijk and Noble, 1989). (Supported by NIH grant NS10861, and a grant from the State of Connecticut Department of Higher Education for the Center for Neurological Sciences).

REFERENCES

Bansal R, Gard AL, Pfeiffer SE (1988) Stimulation of oligodendrocyte differentiation in culture by growth in the presence of a monoclonal antibody to sulfated glycolipid. J Neurosci Res 21:260-267.

Bansal R, Pfeiffer SE (1989) Reversible inhibition of oligodendrocyte progenitor differentiation by a monoclonal antibody against surface galactolipids. Proc Natl Acad Sci USA 86:6181-6185.

Bansal R, Warrington AE, Gard AL., Ranscht B, Pfeiffer SE (1989) Multiple and novel specificities of monoclonal antibodies O1, O4 and R-mAb used in the analysis of oligodendrocyte development. J Neurosci Res 24 (#4), in press.

Dubois-Dalcq M (1987) Characterization of a slowly proliferative cell along the oligodendrocyte differentiation pathway. EMBO J 6:2587-2595.

Dyer CA., Benjamins JA (1988a):Redistribution and internalization of antibodies to galactocerebroside by oligodendroglia. J Neurosci 8:883-891.

Dyer CA, Benjamins JA (1988b) Antibody to galactocerebroside alters organization of oligodendroglial membrane sheets in culture. J Neurosci 8:4307-4318.

Gard AL, Pfeiffer SE (1989) Oligodendrocyte progenitors isolated directly from developing telencephalon at a specific phenotypic stage: Myelinogenic potential in a defined environment. Development 106:119-132.

Gard AL, Bansal R, Pfeiffer SE (1989a) Developmental phenotype of late-stage oligodendrocyte progenitors in vivo. Trans Amer Soc Neurochem 20:248 (abstr)

Gard, AL, Gonye G, Pfeiffer SE (1989b) Oligodendroglia survival factor from primary heart culture. Soc Neurosci Abstr 15:709 (abstr #287.10).

Gard AL, Warrington AE, Pfeiffer SE (1988) Direct microculture enzyme-linked immunosorbent assay for studying neural cells: oligodendrocytes. J Neurosci Res 20:46-53.

Holmes, E, Hermanson, G, Cole, R, DeVellis J (1988) The developmental expression of glial specific mRNAs in primary cultures of rat brain visualized by in situ hybridization. J Neurosci Res 19:389-396.

Levi G, Aloisi, F, Wilkin GP (1987) Differentiation of cerebellar bipotential glial precursors into oligodendrocytes in primary culture: Developmental profile of surface antigens and mitotic activity. J Neurosci Res 18:407-417.

McMorris FA, Dubois-Dalcq M (1988) Insulin-like growth factor I promotes cell proliferation and oligodendroglial commitment in rat glial progenitor cells developing in vitro. J Neurosci Res 21:199-209.

Meier DH, Schachner, M (1982) Immunoselection of oligodendorcytes by magnetic beads. *In vitro* maintenance of immunoselected oligodendrocytes. J Neurosci Res 7:135-145.

Mirsky R, Winter J, Abney ER, Pruss RM, Gavrilovic J, Raff, MC (1980) Myelin-specific proteins and galactolipids in rat Schwann cells and oligodendrocytes in culture. J Cell Biol 84:483-494.

Morell P, editor (1984): Myelin. Plenum Press, New York.

Pfeiffer SE (1984) Oligodendrocyte development in culture systems. in: Oligodendroglia, Advances in Neurochemistry, vol. 5 1984, edited by WT Norton, pp. 233-298. Plenum Press, New York/London.

Pfeiffer SE and Gard AL (1988): Biochemical, immunological, and molecular cell-type specific markers of the central nervous system. in: Advances in Neuro-oncology 1988, edited by P. L. Kornblith and M. D. Walker, pp. 3-40. Futura, Mount Kisco, NY.

Raff MC, Miller RH and Noble MD (1983) A glial progenitor cell that develops in vitro into an astrocyte or an oligodendrocyte depending on the culture medium. Nature, 303:390-396.

Raff MC, Mirsky R, Fields KL, Lisak RP, Dorfman SH, Silberberg DH, Gregson NA, Liebowitz S, and Kennedy MC (1978): Galactocerebroside is a specific cell surface antigenic marker for oligodendrocytes in culture. Nature, 274:813-816.

Ranscht B, Wood PM, Bunge RP (1987) Inhibition of *in vitro* peripheral myelin formation by monoclonal anti-galactocerebroside. J Neurosci 7:2936-2947.

Ranscht B, Clapshaw PA, Noble M, Seifert W (1982) Development of oligodendrocytes and Schwann cells studied with a monoclonal antibody against galactocerebroside. Proc Natl Acad Sci USA 79:2709-2713.

Schachner M (1982) Cell type-specific surface antigens in the mammalian nervous system. J Neurochem 39:1-8.

Schachner, M, Kim SK, Zehnle R (1981) Developmental expression in central and peripheral nervous system of oligodendorycte cell surface antigens (O antigens) recognized by nonoclonal antibodies. Dev Biol 83:328-338.

Singh H, Pfeiffer SE (1985) Myelin-associated galactolipids in primary cultures from dissociated fetal rat brain: Biosynthesis, accumulation and cell surface expression. J Neurochem 45:1371-1381.

Sommer I, Schachner M (1981) Monoclonal antibodies (O1 to O4) to oligodendrocyte cell surfaces: An immunocytological study in the central nervous system. Dev Biol 83:311-327.

Sommer I, Schachner M (1982) Cells that are O4 antigen-positive and O1 antigen-negative differentiate into O1 antigen-positive oligodendrocytes. Neurosci Lett 29:183-188.

Wolswijk G, Noble M (1989) Identification of an adult-specific glial progenitor cell. Development 105:387-400.

Oligodendrocyte Differentiation: Developmental and Functional Subpopulations

A. Espinosa de los Monteros and J. de Vellis
Depts. of Anatomy, Cell Biology and Psychiatry, Brain Res. Inst., Mental
Retard. Res. Ctr., Lab. of Biomed. & Environmental Sci., UCLA Sch. of Med.
University of California, Los Angeles
760 Westwood Plaza, Rm. 68-225 NPI
Los Angeles, CA 90024-1759, USA

Oligodendrocytes represent one of the three major cell types in the central nervous system (CNS). Their main function has traditionally been recognized as the production of the myelin components which are added to the plasma membrane of olidogdendrocytes as they coil their processes around axons to form the elaborate multilayered myelin sheath.

History and Description of the Oligodendrocyte

The first description of this cell was reported by Robertson in 1899. He described it as a small cell with a few thin cell processes of variable length. Later in 1921, del Rio Hortega, on the basis of silver staining gave a complete description of the oligodendroglial cells and coined the term "oligodendrocyte" which means "a cell with few processes". He classified oligodendrocytes into three different types considering their location in the brain: (1) the satellite perineuronal oligodendrocytes, which are located near the neuronal cell body; (2) the perivascular oligodendrocytes and (3) the interfascicular oligodendrocytes, which are located among axons in the white matter tracts. This was the first evidence for the existence of heterogeneity in this cell population. Later, electron microscopy allowed Stensaas and Stensaas in 1968, to confirm the classification given by del Rio Hortega in 1921.

Genesis of the Oligodendroglial Cell

The first developmental study of oligodendrocytes was done in the rat brain by Caley and Maxwell (1968). These authors concluded that an undifferentiated cell type gives rise to an immature glial cell, called spongioblast. This cell is characterized by few cell processes and a prominent rough endoplasmic reticulum. These authors suggested that the spongioblast gives rise to either astroblasts or oligodendroblasts; two cell types with different morphological characteristics. A cell similar to the spongioblast was found in the optic nerve by Vaughn in 1969 and 1971.

Later on the basis of ultrastructural studies, Mori and Leblond (1970), classified oligodendrocytes as light, medium-dark and dark, suggesting that the electron density of the cell relates to the degree of maturation. The light oligodendrocytes were considered the most immature form of these cells. In the young rat, the proportion of these types was 6, 25 and 40 %, respectively, while in the adult, 90 % of oligodendrocytes were of the dark type (Ling et al., 1973; Ling and Leblond, 1973; Paterson et al., 1973). Mori and Leblond (1970) suggested that the clear oligodendrocyte would be the cell involved in active myelin synthesis, while the dark oligodendrocyte could represent a mature cell whose main role will be the maintenance of the myelin sheath.

The Oligodendrocyte and the Myelin Sheath

Since 1904, Hardesty suspected the connection between the cells later named oligodendrocytes and the myelin sheath, on the basis of light microscopic observations. Later the development of the electron microscope unequivocally demonstrated that the myelin sheath is an extension of the oligodendrocyte membrane (Bunge et al., 1962, and Peters in 1964). These authors also showed that one oligodendrocyte can myelinate several axons located at different distances from the cell body. Peters and Vaughn (1970), calculated that one oligodendrocyte may myelinate up to 40 axons. However, it seems that not all the interfascicular oligodendrocytes myelinate, and that only few of the satellite perineuronal oligodendrocytes do myelinate (Ludwin, 1979).

The Myelin Sheath

Myelin is a membrane-structure unique to the CNS and PNS, whose synthesis is performed by the oligodendrocyte and the Schwann cell, respectively. The myelin sheath is formed by a cell process that wraps around the axon. As maturation proceeds myelin is compacted. This process results in alternating intraperiodic and dense lines as observed in electron micrographs. The dense line results from the fusion of the internal faces of the cytoplasmic membrane, while the intraperiodic line results from the apposition of the external surface of the plasma membrane.

The myelin sheath is found along the axons in a segmented form called "in-

This category includes:

Galactocerebrosides (GC) and sulfatides: Both are localized to the oligodendrocyte (Sternberger et al., 1978) and on the surface of isolated myelin membranes (Dupoey et al., 1979). However, cerebrosides are also located in subependymal cells (Dupoey et al., 1979) and, the sulfatides in the arachnoid and Bergmann fibers (Zalc et al., 1981).

Cerebroside sulfotransferase is the enzyme which catalyzes cerebroside sulfation. It is present in the optic nerve and in kidney (Tennekoon et al., 1980).

Glycerol-3 phosphate dehydrogenase (GPDH), an enzyme that participates in glucose and phospholipid metabolism is localized to the cytoplasm of the oligodendrocyte and its processes (Leveille et al., 1980). This enzyme is also present in Bergmann fibers of mouse cerebellum (Fisher et al., 1981).

Carbonic anhydrase is present as the isoenzyme, CAII, in the CNS, liver, fat and muscle. It is localized in the oligodendrocyte cytoplasm (Roussel et al., 1979) and in the cytoplasmic loops of the myelin sheath (Kumpulainen & Nystrom, 1981). The enzyme is also found in small amounts in astrocytes and Müller cells (Kumpulainen and Korhonen, 1982).

Transferrin, the iron binding glycoprotein, is found in the adult rat brain. It is localized in oligodendrocytes, endothelial cells, and choroid plexus cells (Aldred et al., 1987; Bloch et al., 1985, 1987; Connor and Fine 1986, 1987). Transferrin has been recently established as an oligodendroglial marker in neural cell cultures (Espinosa and de Vellis, 1988).

During the last five years, evidence for transferrin gene expression in the CNS has been ascertained by the detection of Tf mRNA in brain extracts by Levin et al 1984, and by in situ hybridization in oligodendrocytes and choroid plexus (Bloch et al., Aldred et al., 1987). Immunocytochemical studies have shown in several species such as chicken and rat, that transferrin immunoreactivity appears transiently in different cell types in the CNS. Early in development, the cells that accumulate transferrin are neurons (Oh & Markelonis, 1986). Later after birth, this immunoreactivity declines with the concomitant rise of transferrin expression by oligodendrocytes (Bloch, et al., 1985). The presence of transferrin in several CNS cell types, suggests that this protein plays an important role during CNS development.

Transferrin is a glycoprotein that plays a major role in the transport of functional iron across the cell membrane, therefore, it is essential to much of the

ternodes" which are delimited by the node of Ranvier. In a myelinated fiber, the axonal membrane is in contact with the external environment only at the nodes of Ranier, but much of the axolema is surrounded by myelin that presents a very high electric resistance. When the axolema is depolarized at the node level, the local current may reach the next node where the axonal membrane gets depolarized. This phenomenon has been named "saltatory conduction", resulting in an increased conduction velocity. The amount of K^+ and Na^+ ions needed is lower than in a non-myelinated fiber.

In consequence, the energy required to re-establish the ionic gradient at the time of repolarization is much less. These two properties of the myelin sheath, explain the importance of the integrity of the myelin sheath.

Biochemical Composition and Metabolism of the Oligodendrocyte

The existing morphological evidence of the connection between the oligodendrocyte and the myelin sheath suggests that a main part of the metabolic activity of these cells is oriented towards the synthesis of myelin components. Most of these components are synthesized in the cell body although some substances such as myelin basic protein, are produced in the cytoplasm of the cell processes (Colman et al., 1982). The lack of space prevents us from describing the lipids and proteins present in myelin. (For review see Norton, 1984; Espinosa, 1987). However, this knowledge has allowed investigators to develop specific cell markers.

Oligodendrocyte Markers

Much of the data that describes the biochemical characteristics of oligodendrocytes has been obtained or confirmed by immunocytochemistry, both in vivo studying tissue sections or in vitro using cultured cells in mixed primary culture or in isolated cells (For review, see Norton, 1984).

Three categories of markers can be distinguished:

1) Antigens such as Wolfgram proteins and myelin basic proteins that are major markers for oligodendrocytes as well as myelin major components, all along the animal's life after birth.

2) Compounds or (antigens) present in oligodendrocytes, but also present in other cell types within the CNS.

ponents. Fig. 1 shows the double immunostaining pattern of MBP and Tf in primary glial cultures at 9, 15 and 20 days. We could note that in most of the cells these antigens do not colocalize. We observed that Tf+ cells are distributed around MBP+ aggregates and that some of the cells forming part of the clumps express both antigens.

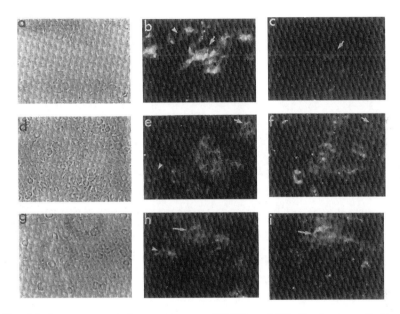

Fig. 1. Double immunostaining patterns of MBP and Tf of primary glial cultures using rabbit polyclonal transferrin and mouse monoclonal MBP antibodies. Panels a, e, and g show the typical morphology of mixed glial cultures under phase contrast microscopy at 9, 15 and 20 days in vitro, glial cultures using rabbit polyclonal transferrin and mouse monoclonal MBP antibodies. Panels a, d, and g show the typical morphology of mixed glial cultures under phase contrast microscopy at 9, 15 and 20 days in vitro, respectively. Panel b shows the Tf immunostaining pattern at 9 days. The staining pattern of the same area with anti-MBP is shown in panel c. Tf and MBP immunostaining at 15 days in vitro are shown in panels e and f, respectively. Panels g and h show the 20-day-old Tf and MBP immunostaining photomicrographs, respectively. The arrows point to an oligodendrocyte expressing both markers, and the arrowheads point to an oligodendrocyte stained only with anti-Tf. The negatively stained astrocytic bed serves as an internal control.

The percentage of cells which express these various antigens are represented in Fig. 2c. The total number of cells was obtained by counting cells from the

cell's metabolic activity, including its growth, proliferation and differentiation. The transferrin family of proteins includes several molecules, such as ovotransferrin, serotransferrin, lactoferrin and other Tf-like proteins. All of them are iron transport glycoproteins with a molecular weight of 76 to 80 kD. Each Tf molecule presents two iron binding sites, and then transferrin enters the cell by a specific receptor.

The use of the above immunocytochemical oligodendrocyte markers in combination with in situ hybridization of cell specific mRNAs, opens the possibility of studying developmental mechanisms and their regulation. In the following section, we summarize our recent findings concerning the oligodendrocyte considered until now as a myelinating cell, and its relation to transferrin, the main iron carrier protein in the body.

Results and Discussion

Recently, we have characterized a high affinity transferrin receptor in cultured oligodendrocytes. We found that the number of receptors remained constant during the different steps of oligodendrocyte maturation, such as proliferation, expression of myelin markers and synthesis of myelin-like membranes (Espinosa and Foucaud, 1987). The fact that the transferrin receptor presents these particular characteristics in oligodendrocytes, led us to investigate further transferrin and iron metabolism by these cells in culture. We found that in primary glial cultures established from newborn rat brain, transferrin appears earlier than galactocerebroside, the standard marker for oligodendrocytes. Furthermore, transferrin immunoreactivity does not colocalize with neuronal or astroglial markers. Therefore, we consider transferrin as the earliest marker for oligodendrocytes. All these findings, together with those previously described concerning Tf mRNA in oligodendrocytes, strongly support the notion of a new function for oligodendrocytes, i.e., production of transferrin.

To approach the question of the participation of oligodendrocytes in iron metabolism in the CNS, we have compared developmentally two myelin markers: galactocerebroside and myelin basic protein in relation to transferrin. These markers were visualized by the double immunofluorescence technique. Newborn rat brain primary cultures were used for a time course experiment. These cultures are a valid model to use for developmental studies because they recapitulate major developmental milestones, such as synthesis of myelin com-

phase contrast micrographs and then the number of immunostained cells for the specific markers were counted. At all the time points examined, three distinct phenotypes for oligodendrocytes could be distinguished:

T+ / MBP-, Tf+ / MBP+ and Tf- / MBP+ which we term oligodendrocytes Types T, TM and M, respectively. The existence of these three phenotypes suggests that some oligodendrocytes may not myelinate, hence perform another function.

Since galactocerebroside is a standard marker for oligodendrocytes, we also compared Tf immunoreactivity in relation to GC. We found that the staining pattern for GC and Tf is identical; These antigens always colocalized (Espinosa and de Vellis, 1988). Therefore, we consider that the colocalization of GC and Tf confirms that in our cultures, Tf positive cells are oligodendrocytes.

To verify the relevance of these findings in tissue culture, we decided to study the expression of Tf and MBP in vivo. For this purpose we used freshly dissociated rat brain cells from birth to one year of age (Espinosa et al., 1989). The results of these studies are represented in Fig. 2a.

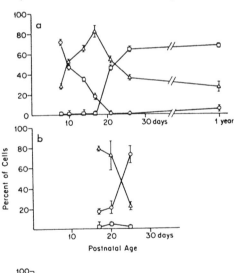

Fig. 2. Quantification of Tf+ and MBP+ cells from (a) freshly dissociated cells from normal rat brains (b) freshly dissociated cells from md rat brains and (c) primary glial cultures established from normal newborn rat brains. Data presented in (a) and (b) are from Espinosa et al., (1989). The culture data were replotted from original data published by Espinosa et al., (1988) to compare to the in vivo results.

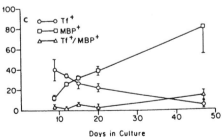

Table I

MARKER	IN VITRO (Cultures) 47 DIV	IN VIVO (Freshly dissociated) 1 year old
Tf %	4.5	5.0
Tf/MBP+ %	15.0 (95)*	26.8 (94.3)*
MBP+ %	80.0	67.5

*Cells potentially capable of myelination.

As in our cell culture findings, the freshly dissociated cell experiments showed that the cellular distribution pattern of MBP and Tf expression characterizes three different subpopulations of oligodendrocytes. Therefore, the observation that we made in the oligodendroglial cultures is confirmed and we conclude that during development as well as in adulthood, the three different oligodendroglial phenotypes T, TM, and M, are well represented. All three phenotypes followed the same developmental pattern found in cultured cells. Tf+ cells were first found in 3 days old brains, and their number increased with time, reaching a maximum at 9 days of age. Later the proportion of Tf+ cells decreased with time, reaching the plateau at 22 days of age and then remaining at that level (2-4 %) during adult life. The third phenotype Tf+ / MBP+ (TM) included a higher proportion of cells, when compared to cultured oligodendrocytes. This difference could be due to the presence of the blood-brain-barrier in vivo which provides an environment different from the in vitro model.

The observation of segregation of these markers in vivo as well as in vitro, suggests the existence of segregation of functions among the three phenotypes. We have presented a model for oligodendrocyte development, taking these data into account (Espinosa and de Vellis, 1988). This model includes both several developmental stages as well as functional subpopulations of adult oligodendrocytes. Although the relative proportion of the three cell populations is quantitatively different in vitro and in vivo, it is surprising to note that the cells potentially capable of myelination, i.e., MBP+ make up approximately 95 % of the total number of cells in both systems.

The presence of Tf mRNA, together with Tf in the cytoplasm of oligodendrocytes suggests active synthesis of Tf by these cells. However, in order to prove it, we have studied the synthesis and secretion of Tf by oligodendrocytes in vitro. Pure cultures of oligodendrocytes were prepared as described by McCarthy and de Vellis (1980). Cells were cultured in a chemically defined medium characterized by the absence of transferrin. Cells were incubated in ^{3}H-leucine medium for 20 hr to analyze the newly synthesized and secreted Tf. Conditioned medium was collected and concentrated. Transferrin was immunoprecipitated by a rabbit polyclonal rat Tf antibody. The precipitate was fractionated by electrophoresis in 10 % SDS polyacrylamide gel and the gel was fluorographed. A radiolabeled protein with a molecular weight of about 76-80 kD, identical to transferrin was identified (Espinosa et al., submitted). The characterization of this molecule is under further investigation.

In an attempt to elucidate the role of each oligodendrocyte phenotype and its relation to myelination, we compared the normal rat to the myelin deficient rat mutant (md) with respect to the expression of, Tf and MBP. The md rat fails to elaborate CNS myelin during development, although peripheral nerves are fully myelinated. This condition is carried by the Wistar rat lineage and is inherited as an X-linked recessive trait. Symptoms are first apparent at the end of the second week of life, and these symptoms progress to include generalized tonic-seizures occuring spontaneously at the end of their lives, around 24 to 26 days postnatally (Csiza and de Lahunta, 1979; Dentinger et al., 1982, 1985). Recently we have studied the developmental expression of neural cell type specific mRNA markers in the myelin deficient rat (Kumar et al., 1988; Gordon et al., 1989). We found that: glutamic acid decarboxylase (GAD), a neuronal marker, is normal in these mutants. Glial fibrillary acidic protein (GFAP) and glutamine synthetase (GS), both astroglial markers are also normal in the md rat. In contrast, mRNAs for several oligodendroglial markers, the cytoplasmic enzyme glycerol phosphate dehydrogenese, (GPDH) myelin basic protein (MBP) and proteolipid protein (PLP) were severely affected in these mutants. These markers fail to display the normal developmental increase in gene expression.

Since Tf is a developmental oligodendrocyte marker but is not a myelin protein, we considered the md rat as an excellent model to study Tf expression in relation to the myelin marker MBP, which we knew was severely affected in this mutant. We performed a time course experiment for Tf and MBP, using affected animals from 17 to 25 days old, just before the animal's death. The double immunofluorescence for Tf and MBP was performed in freshly dissociated cell suspension from md and unaffected littermates at 17, 20 and 25 days of age. In this case we observed that in (Fig. 2b), 17 days old md rats, oligodendrocytes seem to follow a regular timing for the expression of Tf and MBP. This phenomenon is similar to that previously observed for GPDH and MBP mRNAs. As cells continue maturing, an accumulation of Tf$^+$ cells is observed, about 20 % more than in normal animals at 20 days of age. At 25 days, just before the animal dies, the percentage of Tf$^+$ cells increased to 70 %. This value is equivalent to that obtained for MBP$^+$ cells at this age in normal animals. The percentage of Tf positive cells is increased considerably at the end of the life of the animal; at the same time very few MBP positive cells were found at all the time points studied. The proportion of cells expressing both markers Tf and MBP (Type TM) seemed to follow a regular pattern.

Summary and Conclusions

We have briefly reviewed some of the findings discovered during the last 90 years about oligodendrocytes and their relation to myelin. Recent findings from other authors as well as from our own work indicate that oligodendrocytes participate in an active way in iron metabolism in the CNS. However, the key role of the oligodendrocyte in this function during development, in adulthood and during the regeneration process, is still largely an enigma. We have observed that: 1) oligodendrocyte development is complex and progresses through four distinct developmental stages; 2) The presence of oligodendrocyte subpopulations in the brain, those that we have named as T, TM and M, suggest that these three phenotypes perform different functions. We have to consider from in vitro studies, that very early in development neurons becomes dependent during embryonic life on several external elements or factors, such as neuropeptides, growth factors, hormones and, in particular, transferrin to survive and grow (Aizenman et al., 1986). It seems that once these cells have been selected, they will be dependent on these factors throughout life. This phenomenon has been very well studied for neurons, however, there is some evidence that an equivalent mechanism exists for oligodendrocytes. Cell death has now been found to also be part of the developmental process for oligodendroglial cells (Skoff, 1989). Immunocytochemical studies from several authors in vivo (for review, see Espinosa et al., 1989) show that neurons first and oligodendrocytes later accumulate transferrin, and the plateau in both cell types coincides with early developmental activity.

Finally, we would like to speculate that under pathological conditions, a disruption of iron metabolism could contribute to cellular injury and hence to a very poor regeneration potential. It seems obvious that understanding the regulatory mechanisms and the factors involved in CNS development may allow us to create the necessary conditions to favor the regeneration process in adults or prevent neurodegenerative diseases. Furthermore, if we consider the oligodendrocyte T (which probably does not myelinate) as an immature form, which under certain conditions could proliferate and/or give rise to MBP+ cells just as it happened during development, the regenerative process will even be more successful.

In the future, by using both in vivo and in vitro models, it may be possible to provide the appropriate environmental conditions to favor the development of the cell type needed for regeneration. For instance, one approach to transplant into the CNS a given cell population selected under culture conditions that it will recapitulate as the developmental stages defined above.

Acknowledgements

We wish to thank Joyce Adler for her help in preparing the manuscript.

This work was supported by the Department of Energy Contract DE-FCO3-87-ER60615 operated for the United States Department of Energy by the University of California and the National Institute of Child Health and Human Development Grant HD 06576.

References

Aizenman Y, Weichsel ME Jr, de Vellis J (1986) Changes in insulin and transferrin requirements of pure neuronal cultures during embryonic development. Proc Natl Acad Sci 83:2263-2266.

Aldred AR, Dickson PW, Marley PD, Schreiber G (1987) Distribution of transferrin synthesis in brain and other tissues in the rat. J Biol Chem 262:5293-5297.

Bloch B, Popovici T, Lewin MJ, Tuil D, Kahn A (1985) Transferrin gene expression visualized in oligodendrocytes of the rat brain by using in situ hybridization and immunohistochemistry. Proc Natl Acad Sci 82:6706-6710.

Bloch B, Popovichi T, Chahman S, Levin M, Tuil D, Kahn A (1987) Transferrin gene expression in choroid plexus of the adult rat brain. Brain Res Bull 18:573-576.

Bunge MB, Bunge RP, Pappas GD (1962) Electron microscopic demonstrations of connections between glia and myelin sheaths in the developing mammalian central nervous system. J Cell Biol 12:448-453.

Caley D, Maxwell DS (1968) An electron microscopic study of the neuroglia during postnatal development of the rat cerebellum. J Comp Neurol 133:45-70.

Colman DR, Kreibich G, Frey AB, Sabatini DD (1982) Synthesis and incorporation of myelin polypeptides into CNS myelin. J Cell Biol 95:598-608.

Connor JR, Fine R (1987) Development of transferrin-positive oligodendrocytes in the rat central nervous system. J Neurosci Res 17:51-59.

Connor JR, Fine R (1986) The distribution of transferrin immunoreactivity in the rat nervous system. Brain Res 368:319-328.

Csiza CK, de Lahunta A (1979) Myelin deficiency (md). A neurologic mutant in the Wistar rat. Am J Pathol 95-215.

Del Rio Hortega P (1921) Posnatal gliogenesis in the mammalian brain. Int Rev Cytol 40:281-323.

Dentinger MP, Barron KD, Csiza CK (1982) Ultrastructure of the central nervous system in a myelin deficient rat. J Neurocytol 11:671.

Dentinger MP, Barron KD, Csiza CK (1985) Glial and axonal development in optic nerve of myelin deficient rat mutant. Brain Res 344:255.

Dupouey P, Zalc B, Lefroit-Jaby M, Gomes D (1979) Localization of galactosylceramide and sulfatide at the surface of the myelin sheath: an immunofluorescence study in liquid medium. Cell Mol Biol 25:269-272.

Espinosa de los Monteros A, Pena L, de Vellis J (1989) Does transferrin have a special role in the nervous system? In: J Neurosci Res (in press).

Espinosa de los Monteros A, Kumar S, Scully S, Cole R, de Vellis J (1989) Transferrin gene expression and synthesis by cultured neural cells (Submitted).

Espinosa de los Monteros A, de Vellis J (1988) Myelin basic protein and transferrin characterize different subpopulations of oligodendrocytes in rat primary glial cultures. J Neurosci Res 21:181-187.

Espinosa de los Monteros A, Foucaud B (1987) Effect of iron and transferrin on pure oligodendrocytes in culture: Characterization of high affinity transferrin receptor at different ages. Dev Brain Res 35:123-130.

Espinosa de los Monteros A (1987) Contribution a l'etude des oligodendrocytes in vitro. (Thesis) U.L.P. Strasbourg, France.

Fisher M, Gapp DA, Kozak LP (1981) Immunohistochemical localization of sn-glycerol-3-phosphate dehydrogenase in Bergmann glia and oligodendroglia in the mouse cerebellum. Dev Brain Res 1:497-518.

Gordon MN, Kumar S, Espinosa de los Monteros A, Scully S, Zhang M, Huber J, Cole R, de Vellis J (1989) Developmental regulation of myelin-associated genes in the normal and the myelin-deficient mutant rat. "In:" Lauder JM, Privat A, Giacobini E, Timiras PS, Vernadakis A (eds), Molecular Aspects of Development and Aging of the Nervous System (in press).

Hardesty I (1904) On the development and nature of the neuroglia. Am J Anat 3:229-268.

Kumar S, Gordon MN, Espinosa de los Monteros MA, de Vellis J (1988) Developmental expression of neural cell type-specific mRNA markers in the myelin-deficient mutant rat brain: Inhibition of oligodendrocyte differentiation.

Kumpulainen T, Nystrom SHM (1981) Immunohistochemical localization of carbonic anhydrase isoenzyme C in the central and peripheral nervous system of the mouse. J Histochem Cytochem 30:283-292.

Kumpulainen T, Korhonen LK (1982) Immunocytochemical localization of carbonic anhydrase isoenzyem C in the central and peripheral nervous system of the mouse. J Histo Cytochem 30:283-292.

Langley OK, Ghandour MS, Vincendon G, Gombos G (1980) Carbonic anhydrase: an ultrastructural study in rat cerebellum. Histochem J 12:473-483.

Leveille PJ, McGinnis JF, Maxwell DS, de Vellis J (1980) Immunocytochemical localization of glycerol-3-phosphate dehydrogenase in rat. Brain Res 196:287-305.

Levin MJ, Tvil D, Uzan G, Dreyfus J-C, Kahn A (1984) Expression of the transferrin gene during development of non-hepatic tissues: high level of transferrin mRNA in fetal muscle and adult brain. Biochem Biophys Res Comm 122:212-217.

Ling EA, Leblond CP (1973) Investigation of glial cells in semithin sections. II. Variation with age in the numbers of the various glial cell types in rat cortex and corpus callosum. J Comp Neurol 149:73-82.

Ling E, Paterson J, Privat A, Mori S, Leblond CP (1973) Investigation of glial cells in semithin sections. Identification of glial cells in the brain of young rats. J Comp Neurol 149:43-72.

Ludwin SK (1979) The perineuronal satellite oligodendrocyte. A possible role in myelination. Acta Neuropathol 47:49-53.

McCarthy KD, de Vellis J (1980) Preparation of separate astroglial and oligodendroglial cell cultures from rat cerebral tissue. J Cell Biol 85:890-902.

Mori S, Leblond CP (1970) Electron microscopic identification fo three classes of oligodendrocytes and a preliminary study of their proliferative activity in the corpus callosum of young rats. J Comp Neurol 139:1-30.

Norton WT (1984) Oligodendroglia, Plenum Press, New York.

Oh T, Markelonis GJ, Royal GM, Bregman BS (1986) Immunocytochemical distribution of transferrin and its receptor in the developing chicken nervous system. Dev Brain Res 30:207-220.

Paterson JA, Privat A, Ling EA, Leblond CP (1973) Investigation of glial cells semithin ections. III. Transformation of subependymal cells into glial cells, as shown by radioautography after ^{3}H-thymidine injection into the lateral ventricle of the brain of young rats. J Comp Neurol 149:83-102.

Peters A (1964) Observations on the connections between myelin sheath and glial cells in the optic nerves of young rats. J Anat 98:125-134.

Peters, A, Vaughn JE (1970) Morphology and development of the myelin sheath. "In:" Davison AN, Peters A (eds) Myelination. Charles C Thomas, Springfield Illinois, pp. 3-79.

Robertson JD (1899) On a new method of obtaining a block reaction in certain tissue elements of the central nervous system (Platinum method). Scott Med Surg J 4:23-30.

Roussel G, Delaunoy JP, Nussbaum JL, Mandel P (1979) Demonstration of a specific localization of carbonic anhydrase C in the glial cells of rat CNS by an immunohistochemical method. Brain Res 160:47-55.

Skoff R, Vermesch M, Studzinski D, Benjamin J (1989) Oligodendrocytes in vivo first express proteolipid protein three to four days after they have completed cell division. J Neurochemistry 52, supplement S45D.

Stensaas LJ, Stensaas SS (1968) Astrocytic neuroglial cells, oligodendrocytes and microgliacytes in the spinal cord of the toad. II. Electron Microscopy. Z Zellforsch Mikrosk Anat 86:184-213.

Sternberger NH, Itoyama Y, Kies M, Webster H de F (1978) Myelin basic protein demonstrated immunocytochemically in oligodendroglia prior to myelin sheath formation. Proc Natl Acad Sci USA 75:2521-2524.

Tennekoon GI, Kishimoto Y, Singh I, Noneka GI, Bourre JM (1980) The differentiation of oligodendrocytes in the rat optic nerve. Dev Biol 79:149-158.

Vaughn JE (1969) An electron microscopic analysis of gliogenesis in rat optic nerve. Z Zell-Forsch 94:292-324.

Vaughn J (1971) The morphology and development of neuroglial cells. "In:" Pease DC (ed) Cellular aspects of neural growth and differentiation. Univ of California Press, Berkeley, California, pp. 103-134.

Zalc B, Monge M, Dupouey P, Hauw JJ, Baumann N (1981) Immunohistochemical localization of galactosyl and sulfogalactosyl ceramide in the brain of the 30 day-old mouse. Brain Res 211:341-354.

LONG-TERM TISSUE CULTURES OF NEWBORN RAT OPTIC NERVE CELLS : FUNCTIONAL DIFFERENTIATION OF GLIA AND APPEARANCE OF NEURON-LIKE CELLS

F.X. Omlin
Institute of Histology and Embryology
University of Lausanne
Rue du Bugnon 9
CH-1005 Lausanne

Introduction

A large body of literature documents depelopmental, functional, pathological and experimental aspects of optic nerve glia and progenitor cells (Vaughn, 1969; Skoff et al., 1976; Tennekoon et al., 1977; Aguayo et al., 1978; Juurlink and Federoff, 1980; Privat et al., 1981; Raff et al., 1983).

In fact, for these interdisciplinary studies the optic nerve, as a primitive and anatomically well defined part of the central nervous system (CNS) offers several advantages: (i) This part of the CNS develops from the optic stalk. The latter is composed of an inner and outer wall of neuro-epithelial cells; the inner wall gives rise to both glia and progenitors. (ii) Within the nerve, the glial cells are in close contact to each other forming rows of cells, which represent a histotypic feature. (iii) The in situ optic nerve is free of differentiated neurons.

Previously we elaborated and described a long-term tissue culture system, using small pieces, called minisegments, of optic nerves from newborn rats (Omlin and Waldmeyer, 1986, 1989). The aim was to simultaneously investigate developmental aspects and functional properties of glia and progenitors in vitro, despite of the absence of ganglion cell axons but while respecting the three-dimensional organization of the cells. In the present *in vitro* study, we show that a stage of functional differentiation is attained and maintained (for more than one year) by optic nerve glia in culture. Furthermore, a neuron-like type of cell, which never occurs in the nerve of the animal *in vivo*, appears after 4 to 5 weeks *in vitro*. The importance of these findings are discussed briefly, possible applications and further experimental approaches elucidated.

Material and Methods

Preparations of optic nerve minisegments: Optic nerves of newborn Wistar rats were dissected and the meninges removed. In each culture experiment the nerves of 30-60 animals were dissected out and then cut into small pieces (5-7 per nerve), called minisegments. Totally 45 independent dissection experiments and series of cultures have been carried out. For the present study, minisegments were kept during 5 d up to 406 d as floating explants in a Dulbecco-Vogt modified Eagle's medium supplemented by 10 % FCS (Honegger and Richelson, 1976). At 2 - 60 d intervals, explants were removed from the flask for morphological, immunocytochemical and electro-physiological investigations.

Morphology and Immunocytochemistry: For light- and electron-microscopic investigations of minisegments, a fixative solution containing 1.5 % paraformaldehyde and 0.5 glutaraldehyde in 0.1 M phosphate buffer at pH 7.6 was used. Postfixation was carried out in 1 % osmium tetroxide followed by dehydration in ethanol and embedding in Epon. Monoclonal and polyclonal antibodies recognizing glial or neuron specific proteins served as immunocytochemical markers. Most of the immunostaining was performed on 1 μm thick plastic sections applying the PAP technique or on cryostat sections (8 - 12 μm thick) using the immunoflorescence procedure.

Electrophysiology and horse radish peroxidase (HRP) labeling: Minisegments cultured between 40 d and 120 d were used for intracellular recordings of astrocytes. From over 70 single astrocytes the resting membrane potential was determined by the use of glass microelectrodes filled with 2 M potassium acetate. HRP was then injected into some of these cells by passing depolarizing current pulses of 0.5 nA intensity (Kiraly et al., 1989). Those minisegments, which had been used for intracellular recordings, were chemically fixed or frozen for further morphological and/or immunocytochemical processing.

For more complete methodological details see Omlin and Waldmeyer (1986, 1989).

Results

Formation of a new nervous tissue: The most important finding in this long-term culture system concerns the formation of a <u>new, permanent and to a certain degree functional nervous tissue.</u> It is justified to call minisegments a new nervous tissue for two main reasons: first, the cells within the explants show an organization of a tissue, which differs enormously from that of the newborn rat optic nerve (Fig. 1), and second, within these explants a neuron-like type of cell appears, which never arises <u>in situ</u> (Figs. 5 and 6).

Cellular composition of minisegments in general (Fig. 2): Within the first 2-3 d in culture, degenerating axons and necrotic cells (the latter are close to the cutting edge of the explant) are eliminated from the minisegment, either by direct liberation into the medium or by phagocytosing activity of cells, which appear simultaneously with the axonal degeneration. Glia and precursor cells show, from 5 d up to 406 d <u>in vitro</u>, a three dimensional organization with specia-

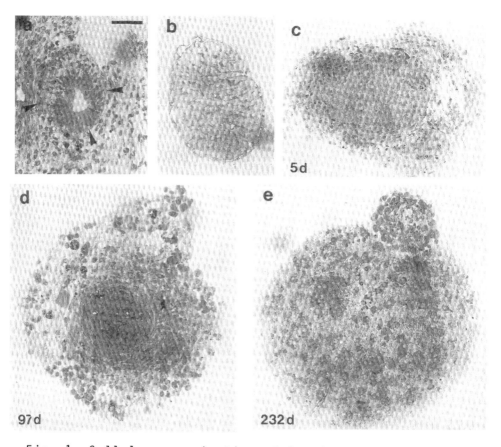

Fig. 1. Cellular organization of in situ optic stalk at E15 (a, arrowheads) and newborn optic nerve (b). In vitro minisegments were cultured for 5-(c), 97-(d) and 232-(e) days and illustrated on bright-field light micrographs of 1 μm thick plastic sections. Bar: 50 μm in a - e.

lized cell contacts and many processes, indicating a high degree of cellular interactions. Mitotic figures were observed during the whole period of cultivation. Furthermore, after 3 d in vitro and up to 406 d, minisegments did not show any sign of degeneration. It seems that after about 2-3 weeks in culture and up to 1 year, the cellular organization within the minisegment remains stable. At the periphery of the explant were most myelinating oligodendrocytes, some cell bodies of astrocytes (Fig. 4), neuron-like cells (Figs. 1, 5 and 6) and a highly complex network of loosely arranged cell

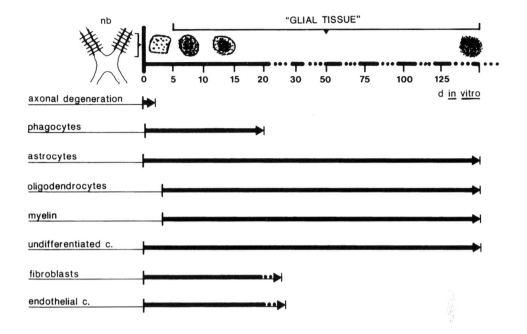

Fig. 2. Schematic drawing of cell-dynamics within minisegments during time of cultivation. Neuron-like cells appear after 4 to 5 weeks in vitro and seem to persist during the whole time in culture.

processes. In contrast, the center of the explant is compact and mainly formed by astrocytes, astrocytic processes and processes of other cell types.

Morphological, immunocytochemical and functional properties of glial cells: Oligodendrocytes proliferate and differentiate in vitro; they appear within the minisegments at day 3 and are present during the whole time of cultivation showing the typical fine structure, which corresponds to in situ oligodendrocytes. Immunocytochemical experiments revealed that they express myelin basic protein, myelin-associated glycoprotein, galactocerebroside and O_4, which are all specific proteins for this type of myelinating glia of mammals. Furthermore, lose and compact myelin, which is the most important functional expression of this type of macroglia, surrounds cell bodies, cell processes or empty spaces (Fig.3).

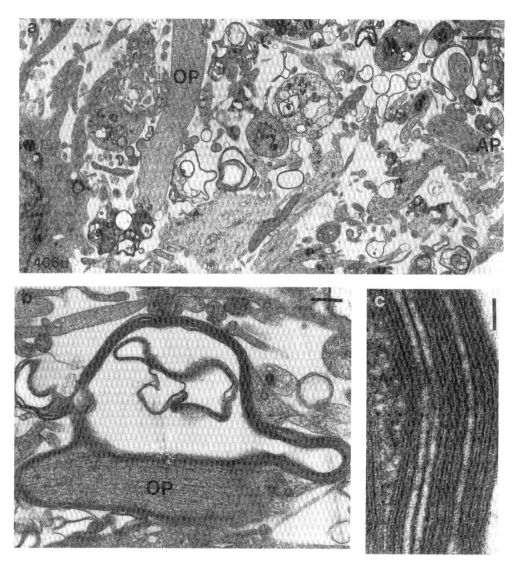

Fig. 3. Myelin figures which are formed by in vitro differentiated oligodendrocytes of minisegments cultured for 406 days. The sheaths surround empty spaces or cell processes. Electron micrographs show a myelin-field (a) and compact myelin (b and c); AP = astrocytic process, OP = oligodendrocytic process. Bars: 0.8 μm in a, 0.4 μm in b and 0.04 μm in c.

Measurements of the periodicity of compact myelin formed in the minisegments gave a mean of 11 nm which corresponds to that of in situ formed myelin.

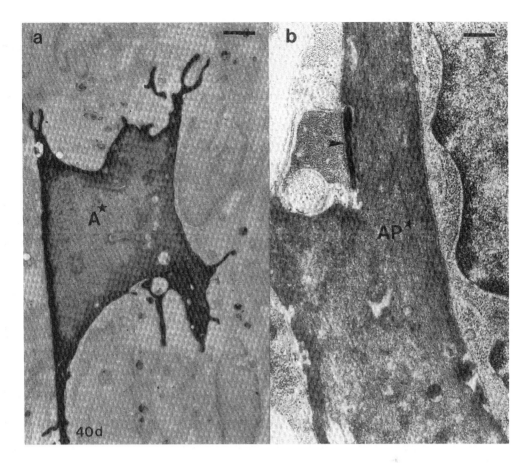

Fig. 4. Fine structure of a horse radish peroxidase labeled astrocyte (A*) after intracellular recordings. The electron micrographs show such a glia cell (a; thin section without counterstaining) with its intermediate filaments containing processes (b; AP* = labeled process; arrow-head point at junctional complex; thin section with counterstaining) from a minisegment cultured for 40 days. Bars: 3 μm in a and 0.4 μm in b.

Astrocytes show the typical structure of in situ glia. Their cytoskeleton is composed of gliofilaments and microtubules. Immunocytochemical characterization using several monoclonal and polyclonal antibodies (A2B5, glial fibrillary acidic protein, vimentin) has confirmed the identity of these cells. Furthermore, the mean resting membrane potential of these cells was -75 mV; a value which corresponds to that of

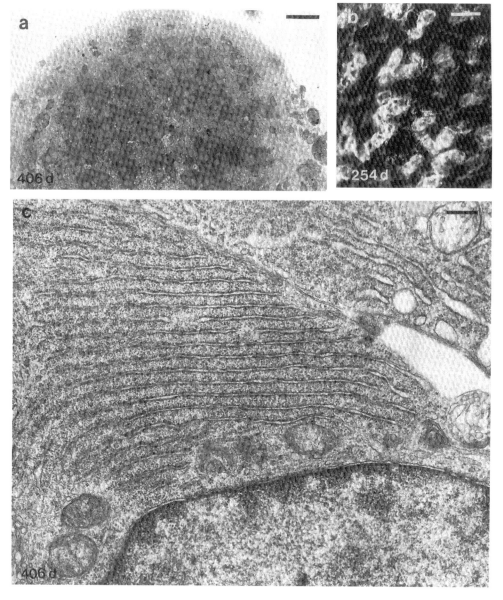

Fig. 5. Morphological (a), fine structural (c) and immunocytochemical characteristics of neuron-like cells, which were observed in minisegments cultured for 406 days (a and c) and 254 days (b) respectively: Bright-field light micrograph (a), cryostat section immunostained with monoclonal antibodies (SM32) directed against neurofilament proteins (b) and electron micrograph (c). Bars: 32 μm in a, 18 μm in b and 0.4 μm in c.

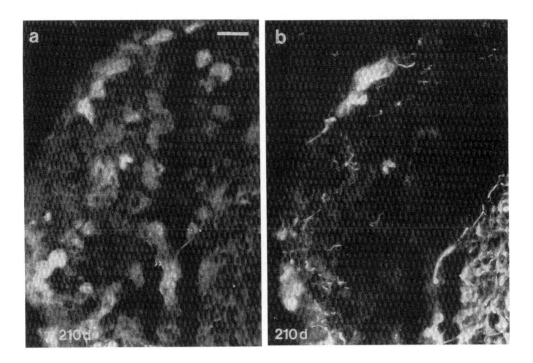

Fig. 6. Double immunostained cryostat section (a, monoclonal neurofilament antibodies, 2F11; b, polyclonal glial fibrillary acidic protein antibodies, GFAP) of a minisegment cultured for 210 d. Bar: 18 µm in a and b.

in situ astrocytes. Following recording, some of the cells were labeled with HRP, which allowed the identification of the cell type by means of morphological criteria (Fig. 4).

The neuron-like type of cell appears after about 4 - 5 weeks in culture. The fine structure of these cells shows a nucleus and a cytoplasmic organization, which are typical of neurons (Fig. 5). Furthermore, these cells develop processes which present differentiations resembling the fine structure of axons and growth cones. Immunocytochemical staining using neuron specific monoclonal and polyclonal antibodies (neurofilaments and neuron specific enolase) revealed immunoreaction products within these cells. Double immunostaining showed that while these cells were positive for neuronal markers, they were negative for glial specific markers (Fig. 6).

Concluding Remarks

It can be suggested that several factors together play a crucial role in the formation of a new, permanent and to a certain degree functional nervous tissue, which is exhibited by the optic nerve minisegments: (i) Tissue selection and preparation: The optic nerve of the newborn rat is an anatomically well defined structure, which allows other parts (e.g. brain, mesenchyma etc.) to be eliminated during dissection, in order to avoid contamination with cells of an origin other than the nerve. About 50 % of its cells are undifferentiated progenitors; differentiated myelin-forming oligodendrocytes appear in situ at day 6-7. (ii) Three dimensional organization of the cultured cells and lack of dissociation: From the very early beginning of the tissue culture up to the time-point of experimentation and final tissue fixation the three dimensional organization of the cells was respected. The cells maintained it during the whole period of cultivation forming a network of processes and establishing different types of intercellular junction. (iii) Size of minisegments: Not only the intercellular junctions but also the size of minisegments may allow a well equilibrated exchange of the medium and this permits the cells to survive for over one year. (iv) Fast elimination of cellular debris: Within the first 2-3 d in culture, almost all axonal and cell debris are liberated into the medium or eliminated by the phagocytosing activity of cells.

Both the functional differentiation of glia, the appearance of neuron-like cells and the very long survival time of the minisegments reflect an ideal tissue preparation and good culture conditions. With these long-term explant cultures, it will be possible to study cellular mechanisms involved in processes of degeneration/regeneration. Furthermore, since little is known about when neuronal and glial lineages diverge in the CNS, minisegments could provide a complementary tool to approach this problem.

There are two possible hypothesis dealing with the origin of the morphologically and immunocytochemically well characterized neuron-like cells. First, the optic nerve of the newborn rat still contains a small subpopulation of undifferentiated cells (neuroepithelial cells), which never produce glial descendants during normal _in situ_ development; these progenitors simply degenerate or migrate out of the nerve. During the first 2-3 days of culture, the minisegments show a complete degeneration of the axons and the formation of a complex glial tissue. It can be suggested, that these circumstances may be responsible for fundamental changes in both the microenvironment and the cell-to-cell interactions. Thus, the small subpopulation of undifferentiated cells survive _in vitro_ and even differentiate into neuron-like cells. Second, in the optic nerve of the newborn rat there exists a small cell population of progenitors with the potential to give rise to either glia or neurons. Under normal _in vivo_ conditions, these precursors differentiate into glia but not into neurons. This could be explained by a neuronal influence, which is provided by axons of the ganglion cells of the retina.

In relation to the finding that neuron-like cells appear within _in vitro_ developing minisegments obtained from newborn rat optic nerves, the following three main questions are attracting our interest: 1) Do neuron-like cells show functional properties of neurons? 2) What is the origin of these cells? 3) Are they derived from pre- or postmitotic cells?

Acknowledgements

This work was supported by grants of the NSF of Switzerland (Project 3100-009 237) and the Swiss MS Society.

References

Aguayo AJ, Dickson R, Trecarten J, Attiwell M, Bray GM, Richardson P (1978) Ensheathment and myelination of regenerating PNS fibres by transplanted optic nerve glia. Neurosci Lett 9:97-104

Honegger P, Richelson E (1976) Biochemical differentiation of mechanically dissociated mammalian brain in aggregating cell culture. Brain Res 109:335-354

Juurlink BHJ, Fedoroff S (1980) Differentiation capabilities of mouse optic stalk in isolation of its immediate in vivo environment. Dev Biol 78:215-221

Kiraly M, Maillard M, Omlin FX (1989) Functional characterization of astrocytes in rat optic nerve minisegments. Experimentia 45:A 19

Omlin FX, Waldmeyer J (1986) Minisegments of newborn rat optic nerves in vitro: Gliogenesis and myelination. Exp Brain Res 85:189-199

Omlin FX, Waldmeyer J (1989) Differentiation of neuron-like cells in cultured rat optic nerves: a neuron or common neuron-glia progenitor? Dev Biol 133:247-253

Privat A, Valat J, Fulcrand J (1981) Proliferation of neuroglial cell lines in the degenerating optic nerve of young rats. J Neuropath Exp Neurol 40:46-60

Raff MC, Miller RH, Noble M (1983) A glial progenitor cell that develops in vitro into an astrocyte or an oligodendrocyte depending on culture medium. Nature 303:390-396

Skoff RP, Price DL, Stocks A (1976) Electron microscopic autoradiographic studies of gliogenesis in rat optic nerve. 1. Cell proliferation. J Comp Neurol 169:291-312

Tennekoon GI, Cohen SR, Price DL, McKhann GM (1977) Myelinogenesis in optic nerve. A Morphological, autoradiographic and biochemical analysis. J Cell Biol 72:604-616

Vaughn JE (1969) An electron microscopic analysis of gliogenesis in rat optic nerves. Z Zellforsch 94:293-324

DISTRIBUTION OF MYELIN PROTEIN GENE PRODUCTS IN ACTIVELY-MYELINATING OLIGODENDROCYTES

Bruce D. Trapp
Department of Neurology
Johns Hopkins University School of Medicine
Meyer 6-181
600 N. Wolfe Street
Baltimore, Maryland 21205
United States

INTRODUCTION

The molecular and cellular biology of gene expression within myelin-forming oligodendrocytes currently offers one of the most fascinating and technically feasible areas of research in neurobiology. Myelination occurs throughout the neuroaxis and myelin-specific gene products are abundant and well-characterized. Translational products of myelin genes are assembled into ultrastructurally distinct membranes in an orderly and predictable manner. The spatial relationships between oligodendrocytes and their myelin internodes provide an excellent opportunity to study mechanisms of myelin protein transport and membrane assembly. In addition, functional diversity within various regions of the myelin internode, i.e., compact myelin versus paranodal regions versus peri-axonal regions, will be reflected by differences in their molecular composition. Myelin-forming cells are a good model for investigating cell-cell interactions. The genetic program of myelin-forming cells is regulated by axonal influences that are at present poorly understood. This regulation is likely to require physical contact between the axolemma and peri-axonal membrane and to involve specific adhesion molecules.

Oligodendrocytes have the potential to form several myelin internodes. Figure 1A illustrates the relationship between an oligodendrocyte perikaryon and one of its myelin internodes that has been "unrolled." The myelin internode is connected to the oligodendrocyte perikaryon by a cellular

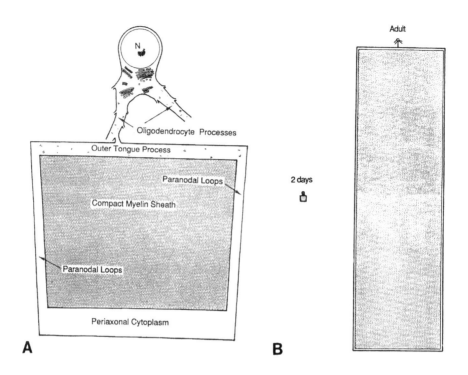

Figure 1. Schematic representation of the relationship between an oligodendrocyte perikaryon and an "unrolled" myelin internode (A) from a 2-day-old rat spinal cord. The relative size of this myelin internode is schematically compared with a myelin internode from the adult spinal cord (B).

process, which extends around the perimeter of the compact myelin and provides channels for the transport of molecules to various regions of the internode. The plasma membranes demarcating the various regions of these cytoplasmic channels are not mere extensions of compact myelin membranes. Rather, the periaxonal membrane, inner tongue process, paranodal loops, and outer tongue process can be considered as morphologically distinct regions of the myelin internode that have specialized functions. Many of these functions are likely to be mediated by specific molecules that are restricted to or enriched in these membranes.

Figure 1B schematically illustrates the massive expansion of myelin membranes that is required of a myelinating oligodendrocyte. This expansion is quite remarkable when considering that the oligodendrocyte forms several such myelin internodes. Since the biochemical composition (Lees and Brostoff, 1984) and ultrastructural features (Peters et al., 1976) of myelin membranes are well characterized, elucidation of myelin protein synthesis, and mechanics of protein sorting and transport are feasible. A general caudal-to-rostral gradient in the initiation and progression of myelination occurs during CNS development, and neighboring axons can be myelinated at different times. Because of this asynchrony, we have chosen to investigate molecular mechanisms of myelination by morphological approaches that use immunocytochemical and _in situ_ hybridization techniques.

This review considers how the subcellular localization of myelin proteins and their respective mRNAs can help elucidate the function of myelin proteins and how they are transported to the myelin sheath. Four myelin proteins--myelin basic protein (MBP), proteolipid protein (PLP), 2',3'-cyclic nucleotide 3'-phosphodiesterase (CNP), and myelin-associated glycoprotein (MAG)--that collectively comprise about 85% of the total proteins associated with isolated CNS myelin are discussed.

These molecules are part of or are associated with plasma membranes of the myelin internode, which implies that all of them are involved in formation and maintenance of myelin membrane. It is possible, however, that these molecules have additional functions within the cytoplasm of actively myelinating oligodendrocytes. All four molecules are present in multiple forms that are generated by alternate splicing of a single primary transcript. To date, it is not known if different forms of these molecules have different functions or locations within the myelin sheath. Unless otherwise specified in this review, members of these families will be referred to as single molecules.

Localization and Function of Myelin Proteins

The function of a molecule is determined by two factors: its structure and its location. Correlating the molecular and

biochemical properties of myelin proteins with their immunocytochemical localization in ultrastructurally-distinct myelin
membranes can provide clues to their possible role in formation and maintenance of the myelin sheath. Similarly, in
pathological conditions, correlation between the absence of a
myelin protein and ultrastructural alterations in myelin
membranes has strengthened the hypotheses that MAG participates in the formation and maintenance of the periaxonal space
and periaxonal cytoplasm of myelinating Schwann cells (Trapp
et al, 1984; Trapp, 1988), that proteolipid protein maintains
the intraperiod line of CNS myelin (Duncan et al., 1988;
1989), and that the myelin basic protein maintains the major
dense-line of CNS myelin (Privat et al., 1979). The location
of myelin proteins within the internode also defines the end-
points of myelin protein transport systems. The distribution
of MBP, PLP, CNP, and MAG with plasma membranes of the CNS
myelin internode and of the oligodendrocyte is illustrated
schematically in Fig. 2D. Some of their biochemical properties and potential functions are discussed below.

Myelin Basic Proteins. The myelin basic proteins consist
of a family of low-molecular-weight molecules that are produced from a single gene by alternate splicing (de Ferra et
al., 1985; Campagnoni, 1988). Four major forms of MBP (21.5
kD, 18.5 kD, 17 kD, 14 kD) exist in rodents (Barbarese et al.,
1978). They are extrinsic membrane proteins that are translated on free polysomes and located on the cytoplasmic side of
myelin membranes (Omlin et al., 1982). MBP represents about
30% of the total protein in purified CNS myelin and about 5%
in purified PNS myelin. Antibodies specific for MBP stain
compact CNS myelin (Sternberger et al., 1978; Hartman et al.,
1982) (Fig. 2A). To date no attempts have been made to
determine whether the various forms of MBP have discrete locations within the myelin internode. The absence of MBP in the
shiverer mouse (Nave et al., 1987) results in an altered
periodicity of compact myelin due to the lack of fusion of the
cytoplasmic leaflets (Privat et al., 1979). It is clear that
MBP plays a structural role in maintaining the close apposi-

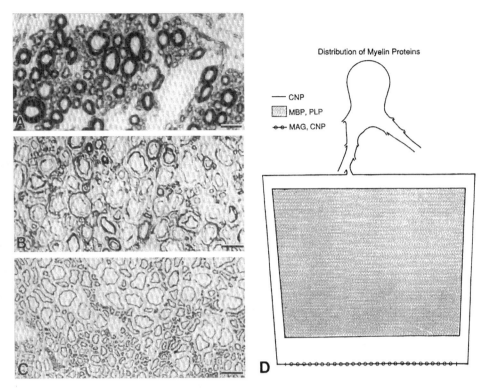

Figure 2. Immunocytochemical distribution of MBP (A), CNP (B), and MAG (C) in 1-μm-thick Epon section of spinal cord. Panel D summarizes the location of these molecules within CNS myelin. Scale bars = 10 μm. Fig. C taken from Trapp et al., 1989, with permission.

tion of the cytoplasmic leaflets or intraperiod line of compact CNS myelin.

Proteolipid Protein. PLP, an intrinsic membrane protein that represents about 50% of the total protein in isolated CNS myelin, is highly hydrophobic and 75% of its amino acids appear to exist in an α-helical conformation (Laursen et al., 1984; Stoffel et al., 1984; Hudson et al., 1989). PLP spans the lipid bilayer of myelin membranes; however, the precise orientation of the polypeptide within the membrane bilayer is unclear at present. A closely related molecule, DM-20, is produced by alternate splicing of the PLP gene (Nave et al., 1987). The PLP polypeptide is cotranslationally inserted into membranes of the rough endoplasmic reticulum (RER). It then passes through the Golgi prior to its transport and insertion into compact myelin. There is a substantial delay (≈30 min) between synthesis of the PLP polypeptide and its insertion into compact myelin (Benjamins et al., 1978; Colman et al., 1982). Antisera directed against proteolipid protein intensely stain compact CNS myelin and produce intense particulate staining of perinuclear regions of actively myelinating oligodendrocytes (Hartman et al., 1982). PLP is absent or dramatically reduced in the myelin-deficient rat (Yanagisawa et al., 1986) and jimpy mouse (Morello et al., 1986). These X-linked mutants have a severe hypomyelination of the CNS. Most of the compact myelin lamellae present in these mutants have fused intraperiod lines (Duncan et al., 1988;1989). PLP is likely, therefore, to play a role in maintaining the periodicity of the double leaflets of the intraperiod line of CNS myelin.

<u>2',3'-Cyclic Nucleotide 3'-Phosphodiesterase</u>. CNP consists of two polypeptides (46 kD and 48 kD) that represent about 5% of the total protein in purified CNS myelin (Kurihara and Tsukada, 1968; Sprinkle et al., 1978). Translated on free polysomes (Bernier et al., 1987), CNP becomes tightly associated with membrane shortly after synthesis (Karin and Waehneildt, 1985).

Recent immunocytochemical studies have established that CNP is not enriched in compact myelin membranes but is concentrated within specific regions of the oligodendrocyte and myelin internode (Figs. 2B and D). These include the plasma membranes of oligodendrocytes and their processes, the paranodal loops, the periaxonal membrane and inner mesaxon, the outer tongue process, and incisure-like membranes found in many larger CNS myelin sheaths (Trapp et al., 1988; Braun et al., 1988). This wide distribution suggests that CNP has a general function related to the maintenance of membranes that demarcate cytoplasmic channels within the myelin internode and plasma membranes of the oligodendrocyte. <u>In vitro</u> CNP possesses an enzymatic activity that catalyzes the hydrolysis of 2',3'-nucleotides into the corresponding 2'-nucleotide (Drummond et al., 1962). The <u>in vivo</u> function(s) of CNP remains elusive, however, because natural substrates for these enzymatic activities have not been found in brain. Since CNP appears to be the first myelin-related protein detected during development, it may have a crucial role in early stages of myelination.

Myelin-Associated Glycoprotein. MAG is a minor constituent of both CNS ($\approx$1%) and PNS ($\approx$0.1%) myelin (Quarles et al., 1973a; Figlewicz et al., 1981). It has an apparent molecular weight of 100 kD, of which 30% consists of carbohydrate. As an intrinsic membrane protein, MAG's polypeptide is cotranslationally inserted into membrane of the rough endoplasmic reticulum and it travels through the Golgi prior to its transport to the myelin internode. MAG's amino acid sequence, deduced from cDNA clones (Arquint et al., 1987; Lai et al., 1987; Salzer et al., 1987), predicts a single transmembrane domain, a large extracellular domain that contains five Ig-like regions and eight potential N-linked glycosylation sites, and one of two possible cytoplasmic domains that contain putative phosphorylation sites.

Two developmentally regulated MAG polypeptides (72 kD and 67 kD) are present in the CNS (Arquint et al., 1987; Lai et al., 1987; Salzer et al., 1987). These polypeptides are generated by alternate splicing of a single gene and have identical extracellular and transmembrane domains, but differ in their C-terminal cytoplasmic domain; the 67-kD polypeptide has 10 new amino acids and lacks 54 amino acids that are present in the 72-kD polypeptide. The 72-kD polypeptide is the principal form during early and active stages of CNS myelination, while the 67-kD polypeptide predominates in the mature CNS (Quarles et al, 1973b; Frail and Braun, 1984). In the peripheral nervous system the 67-kD polypeptide represents at least 95% of the total MAG at all stages of development (Frail et al., 1985; Tropak et al., 1988).

Biochemical experiments involving subfractionation of myelin from normal brain and dysmyelinating murine mutants indicated that MAG was enriched in oligodendrocyte membranes that were part of myelin sheaths but distinct from compact myelin (reviewed in Quarles, 1979). This was confirmed by immunocytochemical studies that localized MAG to periaxonal regions of CNS myelin internodes but not to compact myelin (Sternberger et al., 1979) (Fig. 2C). This periaxonal location provided the first evidence that MAG was involved in oligodendrocyte-axon interactions. Recent immunocytochemical studies have demonstrated MAG's presence in oligodendrocyte processes during initial ensheathment of CNS axons (Trapp et al., 1989b). The significant amino acid sequence homology between the extracellular domain of MAG and other cell-adhesion and ligand-binding molecules also supports the notion that MAG is involved in contact between myelin-forming cell and axon. Confirmation of this hypothesis awaits characterization of an axolemmal molecule that binds to MAG. The possibility that MAG serves as a membrane spacer that modulates cell adhesion would also be consistent with current data.

MAG is also enriched in periaxonal membranes of PNS myelin internodes (Trapp and Quarles, 1982; Martini and Schachner, 1986; Trapp et al., 1989a,b). A strict correlation between the presence of MAG in periaxonal membranes and the formation and maintenance of the periaxonal space during Schwann cell remyelination in Quaking mice indicates that MAG plays a role in maintaining contact between myelin-forming Schwann cells and axons (Trapp et al., 1984). Adhesion to the axolemma, however, cannot be the only function of MAG within the PNS, because MAG is enriched in membranes that are not involved in axonal attachment, i.e. Schmidt-Lanterman incisures, paranodal loops, and outer- and inner-mesaxons (Trapp and Quarles, 1982). The potential role of MAG in the formation and maintenance of PNS myelin internodes is discussed in detail elsewhere (Trapp, 1988) and has recently been reviewed (Quarles et al., in press).

Subcellular Sites of Myelin Protein Synthesis

When considering the mechanisms of myelin protein transport, it is essential to delineate the primary intracellular sites of synthesis, as they represent the starting points of the transport systems. This issue can be approached intuitively by considering the subcellular localizations of rough endoplasmic reticulum, free ribosomes, and Golgi membranes-- the organelles on which proteins are synthesized. The predominant site of intrinsic membrane protein synthesis will be in oligodendrocyte perinuclear regions where the majority of RER and Golgi membranes are found. Free polysomes on the other hand are found in the cytoplasm of 1) perinuclear regions, 2) processes extending to myelin internodes, and 3) the outer tongue process of myelin internodes. Potential translational sites of extrinsic membrane protein within the oligodendrocyte myelin unit, therefore, are much more extensive than are sites for intrinsic membrane protein synthesis. In situ hybridization has clearly shown the primary sites of PLP, MAG, MBP and CNP synthesis (Fig. 3).

The major site of synthesis for the intrinsic membrane proteins, PLP and MAG, is within perinuclear regions of oligodendrocytes (Trapp et al, 1987; Jordan et al., 1989; Higgins et al., 1989). Silver grains representing PLP (Fig. 3A) and MAG (Fig. 3B) mRNAs occur in clusters around nuclei of cells that have the appearance and distribution of oligodendrocytes. Since PLP and MAG polypeptides undergo post-translational processing in Golgi, these membranes are the starting point of PLP and MAG transport systems.

In contrast, mRNA encoding MBP is distributed diffusely over myelinated fibers within the central nervous system (Fig. 3C) (Kristensson et al., 1986; Trapp et al., 1987; Jordan et al., 1989), indicating that MBP mRNA is transported to and translated at sites near compact myelin during active stages of myelination. Evidence for this transport system was first presented by Colman and collaborators (Colman et al., 1982), whose studies showed a 20-fold enrichment of MBP mRNA in RNA extracts of purified myelin fractions when compared to RNA extracts of whole brain homogenates. Based on this finding and biochemical studies that showed MBP to enter myelin within a few minutes after synthesis, they proposed that the majority

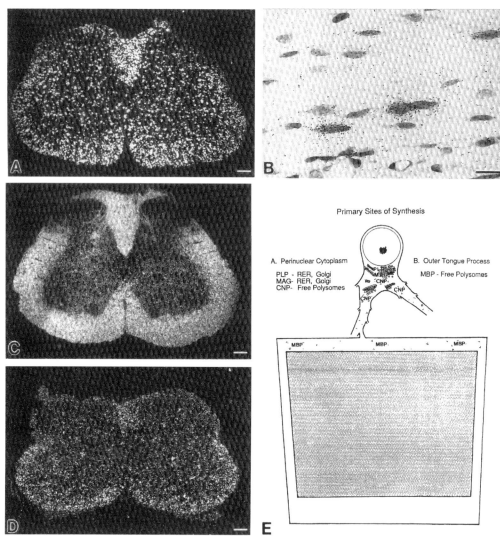

Figure 3. Comparison of the distributions of PLP (A), MAG (B), MBP (C), and CNP (D) mRNAs in tissue sections hybridized with ^{35}S cDNAs. PLP (A), MAG (B) and CNP (C) mRNAs are concentrated in perinuclear regions of oligodendrocytes (B, arrowheads). MBP mRNA is distributed diffusely over myelinated fiber tracts (C). Primary translation sites of these myelin proteins are summarized in E. Figs. A,C, and D are dark-field images of paraffin sections from rat spinal cord. Fig. D is a bright-field image of a paraffin section from rat corpus callosum. Scale bar, A,C,D = 200 μm; B = 20 μm. Figs. A and C reproduced from Trapp et al., 1987, with permission; Fig. D reproduced from Trapp et al., 1988, with permission.

of MBP synthesis occurs along the myelin internode on poly-somes that are present in the outer tongue process. This site of MBP synthesis also explains the dramatic reduction of MBP immunoreactivity in perinuclear regions of oligodendrocytes during peak periods of myelination (Sternberger et al., 1978). It is apparent that the major means of transporting MBP to the myelin internode is by translocation of its mRNA. Since MBP contains a large number of positively-charged amino acids, they bind nonspecifically to acidic lipids (Smith, 1978). Synthesizing the highly charged MBP molecule in the outer tongue process assures its immediate association with myelin as it forms and avoids the nonspecific binding of MBP to organelles within oligodendrocyte perinuclear regions.

In contrast to MBP mRNA, mRNA encoding CNP is concentra-ted around oligodendrocyte perinuclear regions during active stages of myelination (Trapp et al., 1988; Jordan et al., 1989). Since CNP is synthesized on free polysomes, the transport of MBP mRNA out along the myelin internode, or retention of CNP mRNA within perinuclear regions of the oligodendrocytes, is an active process. Site-specific trans-port or segregation of certain mRNAs occurs in a variety of cell types. The highly polarized myelin-forming cell is advantageous for investigating segregation and transport of mRNA because spatial separation of the various pools can be resolved easily by in situ hybridization or by comparing RNA extracts from homogenates of myelin and whole brain.

Cytoplasmic Localization and Mechanisms of Transport

The distribution of myelin proteins within the cytoplas-mic channels that connect their sites of synthesis and loca-tion in the myelin internode can provide information regarding potential mechanisms of transport. This issue has been ap-proached at both the light- and electron-microscopic levels by a variety of immunocytochemical procedures. Whereas these studies cannot address mechanisms of protein transport di-rectly, they can provide a baseline for more direct analysis of potential transport systems. From the data collected thus far, it appears that MBP, PLP, CNP, and MAG are transported to the myelin internode by independent mechanisms.

Myelin Basic Protein. Very little MBP can be detected immunocytochemically within perinuclear regions of oligodendrocytes during active stages of myelination (Sternberger et al., 1978). This is consistent with MBP translation occurring close to the myelin sheath. Nevertheless, significant amounts of MBP and MBP mRNA can be detected in oligodendrocyte perinuclear regions during early stages of myelination (Sternberger et al., 1978; Trapp et al., 1987), which raises the possibility that MBP has functions unrelated to myelin compaction. Isoforms of MBP that are more prominent during early stages of myelination and less abundant during active stages of myelination may be translated in perinuclear cytoplasm. The translocation of MBP mRNA within oligodendrocytes provides a unique model for investigating mechanisms of mRNA transport. Several questions that deserve further elucidation include: does MBP mRNA transport depend on non-coding regions of mRNA transcripts or on exons that are present on differentially spliced transcripts? Are MBP mRNAs transported attached to, or independent of, polyribosomes, and are ribonuclear proteins involved in MBP mRNA translocation?

CNP. Little can be said about the mechanisms of CNP transport at the present time. A substantial pool of CNP is distributed diffusely throughout oligodendrocyte cytoplasm (Trapp et al., 1988; Braun et al., 1988). This could represent CNP in transit, CNP functioning within oligodendrocyte cytoplasm, or both. Since extraction of CNP from CNS tissue requires detergent, it appears that CNP becomes membrane-associated shortly after its synthesis. If this is the case, vesicular transport would be the most likely means of translocating CNP. The possibility, however, that a substantial soluble pool of CNP is present in oligodendrocyte cytoplasm cannot be ruled out at this time.

PLP and MAG. As intrinsic membrane proteins, MAG and PLP transport is likely to occur via vesicles. The possibility that oligodendrocytes use mechanisms similar to those proposed for vesicular transport in the axon--i.e. energy-dependent, kinesin-mediated, vesicular transport along microtubules--is likely, but untested. The distribution of microtubules within the cytoplasmic domains of oligodendrocytes and myelin internodes (Peters et al., 1976) is appropriate for such a transport system. Subcellular subfractionation studies (Pereyra and Braun, 1983; Pereyra et al., 1983) and the ability of monensin to disrupt PLP delivery to myelin (Townsend and Benjamins, 1983) support a vesicular transport system for PLP. PLP and MAG have been detected in the same Golgi complex (BD Trapp, unpublished results). Because PLP and MAG go to different locations within the myelin internode, it is likely that they are segregated into separate transport vesicles at the Golgi.

Further evidence for separate transport systems comes from studies of oligodendrocytes cultured in the absence of neurons (Dubois-Dalcq et al., 1986). PLP and MAG do not co-localize extensively in these oligodendrocytes, and MAG is transported out of perinuclear regions several days before PLP. In addition to Golgi labeling, PLP and MAG antibodies produce diffuse labeling of oligodendrocyte cytoplasm in ultrathin cryosections. By analogy, antibodies directed against P_0 protein, the major intrinsic protein of PNS myelin, produce diffuse labeling of actively myelinating Schwann cell cytoplasm (BD Trapp, unpublished observation). It is likely therefore that this diffuse labeling represents the antero-grade transport of these intrinsic membrane proteins to the myelin internode. It remains to be determined what governs the site-specific transport and insertion of PLP to compact myelin, and MAG to the periaxonal membrane. Do PLP and MAG direct their own transport, or are there specific transport molecules that direct different vesicles to specific locations?

Recent immunocytochemical studies have indicated that MAG is associated with an endocytic pathway that originates in the periaxonal membrane during active stages of CNS myelination (Trapp et al., 1989b). MAG antibodies diffusely stain

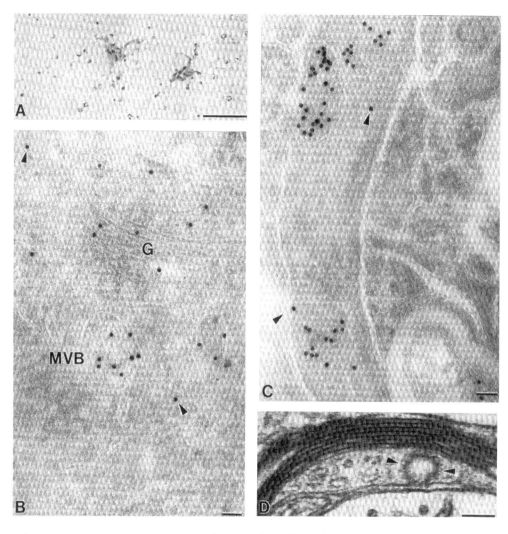

Figure 4. Immunocytochemical distribution of MAG in 7-day-old rat spinal cord (A-C). In light micrographs, oligodendrocyte cytoplasm contains intense particulate staining and less intense diffuse staining (A). In electron micrographs (B,C), MAG is enriched in Golgi membranes (G) and multivesicular bodies (MVB). Gold particles are also distributed randomly within oligodendrocyte cytoplasm (B,C, arrowheads). Clathrin-coated pits (D, arrowheads) are associated with oligodendrocyte periaxonal membrane during active stages of myelination. Scale bars: A = 20 μm; B-D = 0.1 μm. Figs. A-C reproduced from Trapp et al., 1989, with permission.

oligodendrocyte perinuclear cytoplasm and the thin processes
extending toward developing myelin internodes (Fig. 4A). In
addition to this diffuse staining, MAG antibodies produce a
more intense particulate staining of oligodendrocyte perinuclear regions. Ultrastructural correlates for this particulate staining are shown in Figs. 4B-D and consist of Golgi
membranes and relatively large membrane-bound organelles that
often contain internal membranes. These organelles have
ultrastructural characteristics of multivesicular bodies (MVB)

TABLE I

PRO-TEIN	SITE OF SYNTHESIS	LOCATION	FUNCTION	MODE OF TRANSPORT
MBP	Outer tongue process Free poly-somes	Compact myelin	Compaction of cyto-plasmic leaflets	mRNA trans-location
PLP	Perinuclear cytoplasm RER-Golgi	Compact myelin	Compaction of extra-cellular leaflets	Vesicular, microtubules
CNP	Perinuclear cytoplasm Free poly-somes	All surface membranes except com-pact myelin	Unknown	Vesicular? Soluble?
MAG	Perinuclear cytoplasm RER-Golgi	Periaxonal	Axonal inter-actions	A. Anterograde - Vesicular microtubules B. Retrograde - Endosomes

or endosomes. MAG-enriched MVBs are present in oligodendrocyte perinuclear cytoplasm (Fig. 4B) and processes extending
to myelin internodes (Fig. 4C) and along the myelin internode
in outer tongue processes and paranodal loops (Trapp et al.,
1989b). In adult oligodendrocytes and in myelinating Schwann
cells, MAG is not enriched in MVBs, suggesting that the
association of MAG with these organelles is specific for the
72-kD polypeptide. It is generally accepted that MVBs are
associated with retrograde transport of plasma membrane

components that are internalized via receptor-mediated endocytosis (reviewed in Brown et al., 1983; Helenius et al., 1983). Clathrin-coated pits and vesicles associated with cellular plasma membranes serve as intermediaries in such endocytosis. Once internalized, these vesicles lose their clathrin coat and enter a ubiquitous pre-lysosomal compartment referred to as endosomes. These observations suggest, therefore, that MAG is associated with an endocytic pathway that originates in the periaxonal membrane of CNS myelin internodes and terminates in oligodendrocyte perinuclear cytoplasm. This possibility is supported by the abundant and consistent association of clathrin-coated pits with oligodendrocyte periaxonal membranes during active stages of myelination (Fig. 4D).

It is not known if MAG is actively involved as a receptor for components present in the axolemma/periaxonal space, or passively involved as a structural protein of the periaxonal membrane without bound ligand. It is noteworthy that MAG has significant amino acid homologies with, and an overall structure similar to, the polyimmunoglobulin receptor, another member of the immunoglobulin gene superfamily actively involved in an endocytic process (Mostov et al., 1984). While it is generally accepted that axons play the major role in modulating the process of myelination, how this modulation occurs is unknown, although it is likely to be mediated by interactions between the axolemma and the oligodendrocyte periaxonal membrane. Since MAG and coated pits are associated with oligodendrocyte periaxonal membranes from the onset of axonal ensheathment, both are candidates for transducing potential axonal influences.

Table I summarizes the data reviewed in this chapter. The advent of molecular neurobiology has resulted in rapid advances in the identification and/or characterization of CNS-specific proteins. Through gene cloning techniques, new molecules are being identified and sequenced at a rapid rate and antibodies and cDNAs specific for these molecules are being produced. Shared amino acid homologies among proteins, or individual exons of proteins, will provide testable hypotheses for elucidating potential functions of individual molecules. Morphological studies such as those reviewed in this chapter should continue to contribute to our understanding of the cellular and molecular biology of myelination.

ACKNOWLEDGMENTS

I thank Dr. Pamela Talalay for helpful suggestions in writing this chapter; Dr. Grahame Kidd for drawing Figures 1, 2D, and 3E; Peter Hauer for preparing the micrographs; and Rod Graham for typing this chapter. The work was supported in part by grants from the National Institute of Health (NS22849) and the National Multiple Sclerosis Society (JF-2030 A-1). Dr. Trapp is a Harry Weaver Neuroscience Scholar of the National Multiple Sclerosis Society.

REFERENCES

Arquint M, Roder J, Chia LS, Down J, Wilkinson D, Bayley H, Braun P, Dunn R (1987) Molecular cloning and primary structure of myelin-associated glycoproteins. Proc Natl Acad Sci USA 84:600-604

Barbarese E, Carson JH, Braun PE (1978) Accumulation of the four myelin basic proteins in mouse brain during development. J Neurochem 31:779-782

Benjamins JA, Iwata R, Hazlett J (1978) Kinetics of entry of protein into the myelin membrane. J Neurochem 31:1077-1085

Bernier L, Alvarez F, Norgard EM, Raible DW, Mentaberry A, Schembri JG, Sabatini DD, Colman DR (1987) Molecular cloning of a 2',3'-cyclic nucleotide 3'-phosphodiesterase: mRNAs with different 5' ends encode the same set of proteins in neurons and lymphoid tissues. J Neurosci 7:2703-2710

Braun PE, Sandillon F, Edwards A, Matthieu J-M, Privat A (1988) Immunocytochemical localization by electron microscopy of 2',3'-cyclic nucleotide 3'-phosphodiesterase in developing oligodendrocytes of normal and mutant brain. J Neurosci 8:3057-3066

Brown MS, Anderson RGW, Goldstein JL (1983) Recycling receptors: the round-trip itinerary of migrant membrane proteins. Cell 32:663-667

Campagnoni AT (1988) Molecular biology of myelin proteins from the central nervous system. J Neurochem 57:1-14

Colman DR, Kreibich G, Frey AB, Sabatini DD (1982) Synthesis and incorporation of myelin polypeptide into CNS myelin. J Cell Biol 95:598-608 de Ferra F, Engh H, Hudson L, Kamholz J, Puckett C, Molineaux S, Lazzarini R (1985) Alternative splicing accounts for the four forms of myelin basic protein. Cell 43:721-727

Drummond GI, Iyer NT, Keith J (1962) Hydrolysis of ribonucleoside 2',3' cyclic phosphates by a diesterase from brain. J Biol Chem 237:3535-3539

Dubois-Dalcq M, Behar T, Hudson L, Lazzarini RA (1986) Emergence of three myelin proteins in oligodendrocytes cultured without neurons. J Cell Biol 102:384-392

Duncan ID, Hammang JP, Trapp BD (1988) Abnormal compact myelin in the myelin-deficient rat: Absence of proteolipid protein correlates with a defect in the intraperiod line. Proc Natl Acad Sci USA 84:6287-6291

Duncan ID, Hammang JP, Goda S, Quarles RH (1989) Myelination in the jimpy mouse in the absence of proteolipid protein. Glia 2:155-160

Figlewicz DA, Quarles RH, Johnson D, Barbarash GR, Sternberger NH (1981) Biochemical demonstration of the myelin-associated glycoprotein in the peripheral nervous system. J Neurochem 37:749-758

Frail DE, Braun PE (1984) Two developmentally regulated messenger RNAs differing in their coding region may exist for the myelin-associated glycoprotein. J Biol Chem 259:14857-14862

Frail DE, Webster H deF, Braun PE (1985) Developmental expression of myelin-associated glycoprotein in the peripheral nervous system is different from that in the central nervous system. J Neurochem 45:1308-1310

Hartman BK, Agrawal HC, Agrawal D, Kalmback S (1982) Development and maturation of central nervous system myelin: comparison of immunocytochemical localization of proteolipid protein and basic protein in myelin and oligodendrocytes. Proc Natl Acad Sci USA 79:4217-4220

Helenius A, Mellman I, Wall D, Hubbard A (1983) Endosomes. Trend Biochem Science 7:245-250

Higgins GA, Schmale H, Bloom FE, Wilson MC, Milner RJ (1989) Cellular localization of 1B236/myelin-associated glycoprotein mRNA during rat brain development. Proc Natl Acad Sci USA 86:2074-2078

Hudson LD, Friedrich VL Jr, Behar T, Dubois-Dalcq M, Lazzarini RA (1989) The initial events in myelin synthesis: Orientation of proteolipid protein in the plasma membrane of cultured oligodendrocytes. J Cell Biol 109:717-727

Jordan C, Friedrich V Jr, Dubois-Dalcq M (1989) In situ hybridization analysis of myelin gene transcripts in developing mouse spinal cord. J Neurosci 9:248-257

Karin NJ, Waehneldt TV (1985) Biosynthesis and insertion of Wolfgram protein into optic nerve membranes. Neurochem Res 10:897-907

Kristensson K, Zeller NK, Dubois-Dalcq ME, Lazzarini RA (1986) Expression of myelin basic protein gene in the developing rat brain as revealed by in situ hybridization. J Histochem Cytochem 34:467-473

Kurihara T, Tsukada Y (1968) 2',3'-Cyclic nucleotide 3'-phosphohydrolase in developing chick brain and spinal cord. J Neurochem 15:827-832

Lai C, Brow MA, Nave K-A, Noronha AB, Quarles RH, Bloom FE, Milner RJ, Sutcliffe JG (1987) Two forms of 1B236/myelin-associated glycoprotein (MAG), a cell adhesion molecule for postnatal neural development, are produced by alternative splicing. Proc Natl Acad Sci USA 84:4337-4341

Laursen RA, Samiullah M, Lees MB (1984) The structure of bovine brain myelin proteolipid and its organization in myelin. Proc Natl Acad Sci USA 81:2912-2916

Lees MB, Brostoff SW (1984) Proteins of myelin. In: Morell P (ed) Myelin. Plenum Publishing Corp, New York, p 197-224

Martini R, Schachner M (1986) Immunoelectron microscopic localization of neural cell adhesion molecules (LI, N-CAM, and MAG) and their shared carbohydrate epitope and myelin basic protein in developing sciatic nerve. J Cell Biol 103:2439-2448

Morello D, Dautigny A, Pham-Dinh D, Jolles P (1986) Myelin proteolipid protein (PLP and DM-20) transcripts are deleted in jimpy mutant mice. EMBO J 5:3489-3493

Mostov KE, Friedlander M, Blobel G (1984) The receptor for trans-epithelial transport of IgA and IgM contains multiple immunoglobulin domains. Nature 308:37-43

Nave K-A, Lai C, Bloom F, Milner RJ (1987) Splice site selection in the proteolipid protein (PLP) gene transcript and primary structure of the DM20 protein of central nervous system myelin. Proc Natl Acad Sci USA, 84:5665-5669

Omlin FX, Webster HdeF, Palkovitz GG, Cohen SR (1982) Immunocytochemical localization of basic protein in major dense line regions of central and peripheral myelin. J Cell Biol 95:242-248

Pereyra PM, Braun PE (1983) Studies on subcellular fractions which are involved in myelin membrane assembly: isolation from developing mouse brain and characterization by enzyme markers, electron microscopy, and electrophoresis. J Neurochem 41:957-973

Pereyra PM, Braun PE, Greenfield S, Hogan EL (1983) Studies on subcellular fractions which are involved in myelin assembly: labeling of myelin proteins by a double radioisotope approach indicates developmental relationships. J Neurochem 41:974-988

Peters A, Palay SL, Webster HdeF (1976) The fine structure of the nervous system - The neurons and supporting cell. W.B. Saunders Company, Philadelphia

Privat A, Jacque C, Bourre JM, Dupouey P, Baumann N (1979) Absence of the major dense line in myelin of the mutant mouse "shiverer." Neurosci Lett 12:107-112

Quarles RH, Everly JL, Brady RO (1973a) Evidence for the close association of a glycoprotein with myelin. J Neurochem 21:1177-1191

Quarles RH, Everly JL, Brady RO (1973b) Myelin-associated glycoprotein: A developmental change. Brain Res 58:506-509

Quarles RH (1979) Glycoproteins in myelin and myelin-related membranes. In: Margolis RU, Margolis RK (eds) Complex Carbohydrates of the Nervous System. Plenum, New York, pp 209-233

Quarles RH, Hammer JA, Trapp BD (1989) The immunoglobulin gene superfamily and myelination. In: Hashim EG (ed) Dynamic Interactions of Myelin Proteins. Alan R. Liss, New York, in press

Salzer JL, Holmes WP, Colman DR (1987) The amino acid sequences of the myelin-associated glycoproteins: homology to the immunoglobulin gene superfamily. J Cell Biol 104:957-965

Smith R (1978) Crosslinking of lipid bilayers by central nervous system myelin basic protein: aggregation of free and vesicle-bound protein. Adv Exp Med Biol 100:221-234

Sprinkle TJ, Zaruba ME, McKhann GM (1978) Activity of 2',3'-cyclic nucleotide 3'-phosphodiesterase in regions of rat brain during development: quantitative relationship to myelin basic protein. J Neurochem 30:309-314

Sternberger NH, Itoyama Y, Koco MW, Webster HdeF (1978) Myelin basic protein demonstrated immunocytochemically in oligodendroglia prior to myelin sheath formation. Proc Natl Acad Sci USA 75:2521-2524

Sternberger NH, Quarles RH, Itoyama Y, Webster H deF (1979) Myelin-associated glycoprotein demonstrated immuno-cytochemically in myelin and myelin-forming cells of developing rats. Proc Natl Acad Sci USA 76:1510-1514

Stoffel W, Hillen H, Giersiefen H (1984) Structure and molecular arrangement of proteolipid protein of central nervous system myelin. Proc Natl Acad Sci USA 81:5012-5016

Townsend LE, Benjamins JA (1983) Effects of monensin on post-translational processing of myelin proteins. J Neurochem 40:1333-1339

Trapp, BD, Quarles RH (1982) Presence of the myelin-associated glycoprotein correlates with alterations in the periodicity of peripheral myelin. J Cell Biol 92:877-882

Trapp BD, Quarles RH, Suzuki K (1984) Immunocytochemical studies of quaking mice support a role for the myelin-associated glycoprotein in forming and maintaining the periaxonal space and periaxonal cytoplasmic collar of myelinating Schwann cells. J Cell Biol 99:594-606

Trapp BD, Moench T, Pulley M, Barbosa E, Tennekoon G, Griffin JW (1987) Spatial segregation of mRNA encoding myelin-specific proteins. Proc Natl Acad Sci USA 84:7773-7777

Trapp BD (1988) Distribution of the myelin-associated glyco-protein and P_0 protein during myelin compaction in quaking mouse peripheral nerve. J Cell Biol 107:675-685

Trapp BD, Bernier L, Andrews SB, Colman DR (1988) Cellular and subcellular distribution of 2',3'-cyclic nucleotide 3'-phosphodiesterase and its mRNA in the rat central nervous system. J Neurochem 51:859-868

Trapp BD, Andrews SB, Wong A, O'Connell M, Griffin JW (1989a) Co-localization of the myelin-associated glycoprotein and the microfilament components, F-actin and spectrin, in Schwann cells of myelinated nerve fibers. J Neurocytol 18:47-60

Trapp BD, Andrews SB, Cootauco C, Quarles RH (1989b) The
    myelin-associated glycoprotein is enriched in
    multivesicular bodies and periaxonal membranes of
    actively myelinating oligodendrocytes. J Cell Biol,
    in press
Yanagisawa K, Duncan ID, Hammang JP, Quarles RH (1986) Myelin
    deficient rat: Analysis of myelin proteins. J Neurochem
    47:1901-1907

Immuno-Localization of Sodium Channels in Axon Membrane and Astrocytes and Schwann Cells in vivo and in vitro

J. A. Black[1], B. Friedman[1], A. Cornell-Bell[1], K. J. Angelides[3], J. M. Ritchie[2] and S. G. Waxman[1]

Departments of Neurology[1] and Pharmacology[2]
Yale University School of Medicine
333 Cedar Street
New Haven, CT 06510 USA
and V.A. Medical Center, West Haven, CT 06516 USA

Department of Molecular Physiology and Biophysics[3]
Baylor College of Medicine
One Baylor Plaza
Houston, TX 77030 USA

Introduction

Recently, our laboratory has developed methods for the ultrastructural immuno-localization of voltage-sensitive sodium channels within mammalian CNS tissue (Black et al. 1989a,b). With polyclonal antibody 7493, which is generated against purified rat brain sodium channels (Elmer et al. 1985), the axon membrane at the node of Ranvier displays intense sodium channel immunoreactivity. However, quite unexpectedly, processes of astrocytes associated with nodes of Ranvier (termed 'perinodal astrocyte processes') also exhibit intense immunostaining with antibody 7493. The present chapter will examine the spatial distribution of sodium channels within axon membrane, and within astrocytes and Schwann cells in vivo and in vitro. The observations suggest a regional specialization of sodium channels within glial cells, and imply a major role for glial cells in the distribution of sodium channels along myelinated axon membrane.

Sodium channels in axon membrane

The ultrastructural localization of sodium channels within axon membrane utilized antibody 7493 as an immuno-probe. Antibody 7493 was generated against purified rat brain sodium channels (Elmer et al. 1985), and several lines of evidence suggest that the antibody recognizes epitopes on the sodium channel molecule. The antibody selectively recognizes a molecule of Mr 260 kD, which corresponds in mobility and migration pattern to the alpha subunit of the sodium channel, in mixtures of glycoproteins from rat brain (Elmer 1988) and rat optic nerve (Black et al. 1989a) on immunoblots with anti-sera dilutions exceeding 1:1000. The antibody immunoprecipitates ^{32}P- or ^{125}I-labelled sodium channels from mixtures of solubilized brain membranes (Elmer 1988). Moreover, the antibody immunoprecipitates purified sodium channels saturated with ^{3}H-STX in a concentration-dependent manner (K. J. Angelides, unpublished observation). The available immunological evidence demonstrates that the antibody does not recognize cytoskeletal components that are often associated with the channel (Srinivasan et al. 1988). Finally, ultrastructural immunocytochemical observations (to be described below) demonstrate a specific immunostaining pattern in regions of known high sodium channel density (Black et al. 1989a,b). Thus, the available evidence strongly suggests a specific recognition of epitopes located on sodium channels by antibody 7493.

In myelinated fibers, nodes of Ranvier are the sites of greatest anticipated sodium channel density (Waxman and Ritchie 1985). Indeed, rat optic nerves incubated with antibody 7493 exhibit intense immunostaining within the axon membrane at nodes of Ranvier (Figures 1 and 2). The paranodal axon membrane immediately adjacent to nodal axolemma beneath the terminal oligodendroglial loops does not display immunoreactivity, nor does the internodal axolemma beneath the myelin sheath. Notably, the terminal paranodal loops of

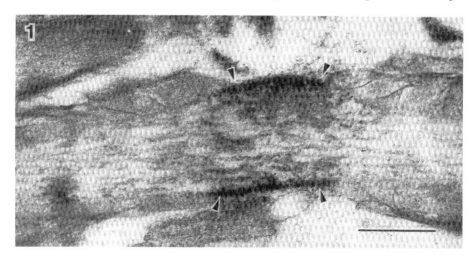

Figure 1
7493 immunoreactivity in adult rat optic nerve. Intense immunostaining is present within axon membrane at the node of Ranvier (between arrowheads). Axon membrane beneath the terminal paranodal loops does not exhibit immunoreactivity. Scale bar, 0.5 μm (from Black el al. 1989a).

oligodendrocytes, and other parts of the oligodendrocyte plasma membrane, are not immunostained. The axoplasm at the node of Ranvier displays a variable degree of immunoreactivity to antibody 7493, ranging from relatively light (Figure 1) to heavy (Figure 2). The immunostaining in the nodal axoplasm does not extend significantly into the axoplasm beneath the paranodal loops. This staining pattern may reflect the presence of an axoplasmic pool of sodium channels or channel precursors, possibly en route to the nodal axon membrane from a site of synthesis in the neuronal cell body, or of channel breakdown products.

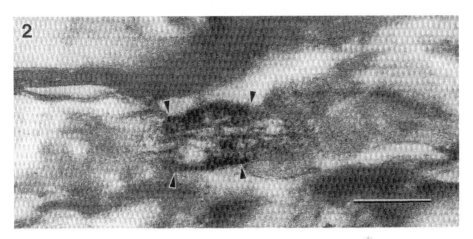

Figure 2
7493 immunostaining in adult rat optic nerve. The nodal membrane (between arrowheads) displays intense immunoreactivity. Note that immunostaining is present within nodal axoplasm. Scale bar, 0.5 μm (from Black et al. 1989a).

Control experiments demonstrate that the specificity of antibody 7493 for sodium channels, and the selective staining pattern exhibited by 7493, are not due to inaccessibility of the internodal axon to the immunoreagents. Optic nerve sections that were incubated with pre-immune sera, or with 7493 anti-sera that had been pre-adsorbed with purified sodium channel protein, did not exhibit immunostaining (Black et al. 1989a). Moreover, incubation of optic nerve sections with antibodies against neurofilaments resulted in immunostaining throughout the axoplasm, including staining of the axoplasm in the internodes beneath the myelin sheath (Black et al. 1989a).

The heterogeneous distribution of sodium channel immunoreactivity displayed by myelinated fibers is in contrast to that exhibited by premyelinated axons from neonatal rat optic nerves. Optic nerves from neonatal rats (0-2 days post-natal) incubated with antibody 7493 do not display focal regions of immunostaining along the axon membrane (Figure 3). In fact, there was a general absence of 7493 immunoreactivity within pre-myelinated axons. The lack of immunoreactivity to antibody 7493 was not due to a lack of accessibility by the immuno-reagents within in the nerve, as neonatal optic nerves incubated with antibodies directed against neurofilaments displayed intense immunostaining (data not shown).

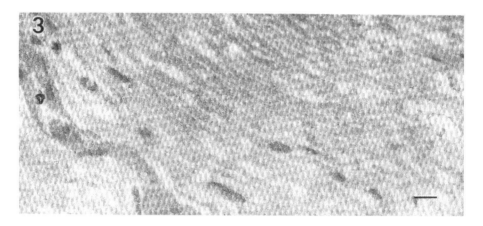

Figure 3
7493 immunostaining in neonatal rat optic nerve. The axon membrane of pre-myelinated fibers does not exhibit sodium channel immunoreactivity. Scale bar, 0.5 μm.

The results obtained with immuno-probes directed against sodium channels in pre-myelinated axons from neonatal rat optic nerves are consistent with previous descriptions of the macromolecular structure of pre-myelinated axon membrane. Freeze-fracture studies have demonstrated a homogeneous distribution of intra-membranous proteins (IMPs) along premyelinated axon membrane (Black et al. 1982). Moreover, the density of sodium channels in pre-myelinated axon membrane has recently been estimated from measurements of ^{3}H-STX binding and morphometric analysis of neonatal rat optic nerve (Waxman et al. 1989). The maximum saturable binding capacities correspond to a high affinity (K_d = 0.88 nm) STX-binding site density of $\sim 2/\mu m^2$ within pre-myelinated axon membrane. Thus, these observations are consistent with a very low density of sodium channels (too low to be detected by the present immunocytochemical methods) that are distributed homogeneously along pre-myelinated axon membrane. Interestingly, we have shown that this spatial organization of sodium channels within pre-myelinated axon membrane does support action potential conduction in neonatal rat optic nerves, as applications of 5 nM STX reversibly abolished the action potential (Waxman et al. 1989).

The heterogeneous distribution of sodium channels along myelinated axon membrane that is apparent with 7493 immunostaining is concordant with previous cytochemical (Quick and Waxman 1977; Waxman and Quick 1977), pharmacological (Ritchie and Rogart 1977) and electrophysiological (Chiu 1980; Chiu and Ritchie 1981; Brismar 1981) studies. These studies suggest a sodium channel density of $> 1000/\mu m^2$ within nodal axon membrane, while relatively few ($< 25/\mu m^2$) sodium channels are present within internodal membrane beneath the myelin sheath (see also Chiu and Schwarz 1987; Ritchie 1988; Shrager 1989). The widely different spatial organization of sodium channels observed between myelinated and pre-myelinated axon membrane demonstrates that myelination involves more than the acquisition of myelin sheaths by axons. As myelination proceeds, there is a segregation and maintenance of high densities of sodium channels within specific domains of axon membrane at the nodes of Ranvier. In this context, we have begun to examine mechanisms that modulate the distribution and stabililization of specific ion channels within developing and mature myelinated axon membrane. The remainder of this chapter will focus on the possible involvement of astrocytes and Schwann cells in the placement of sodium channels within axon membrane.

Sodium channels in astrocytes

In the CNS, a specific relationship has been demonstrated between astrocyte processes (termed 'perinodal astrocyte processes') and mature nodes of Ranvier (Hildebrand 1971; Berthold and Carlstedt 1977; Hildebrand and Waxman 1984). In optic nerve sections incubated with antibody 7493, perinodal astrocyte processes display distinct sodium channel immunoreactivity, in addition to the intense immunostaining present within nodal axon membrane (Figures 4 and 5). Sodium channel immunoreactivity is present on the plasmalemma of the astrocyte perinodal processes, as well as within the astrocyte cytoplasm. Glial filaments within the astrocyte cytoplasm are not stained with anti-sodium channel immunocytochemistry. The specificity of immunostaining within peri-

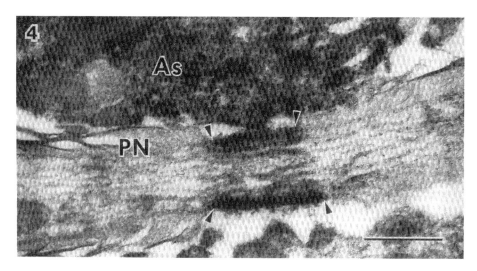

Figure 4
7493 immunoreactivity at node of Ranvier. The axon membrane and subjacent axoplasm at the node of Ranvier (between arrowheads) display intense immunostaining. Perinodal astrocyte processes (As) are extremely immunoreactive following incubation with 7493 antibody. Note that neither the axon membrane beneath the paranodal loops (PN) nor oligodendrocyte membrane are immunoreactive. Scale bar, 0.5 μm (from Black et al. 1989a).

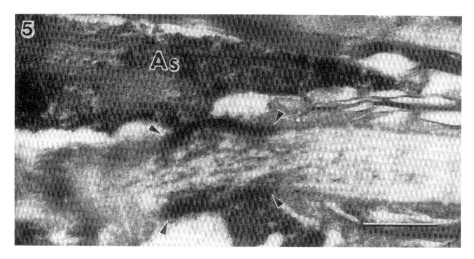

Figure 5
7493 immunostaining at node of Ranvier. The nodal axon membrane (between arrowheads) exhibits intense sodium channel immunoreactivity. The paranodal axon membrane does not display immunostaining. Adjacent perinodal astrocyte processes (As) are intensely immunoreactive following incubation with antibody 7493. Scale bar, 0.5 μm (from Black et al. 1989b).

nodal astrocyte processes was confirmed in experiments where tissue sections were incubated with pre-immune sera, or with anti-sera 7493 that had been pre-adsorbed with purified sodium channel protein; no immunoreactivity within perinodal astrocyte processes was detected under these conditions.

The intense sodium channel immunoreactivity displayed by perinodal astrocyte processes is in contrast to the antibody 7493 immunostaining pattern present within astrocyte processes forming the glia limitans and surrounding blood vessels. As shown in Figure 6, the generally limited staining of astrocyte processes within the glia limitans (Figure 6a) or around blood vessels (Figure 6b) can be compared to the intense sodium channel immunoreactivity observed for nearby perinodal astrocyte processes, and nodal axon membrane. The antibody 7493 immunoreactivity within the glia limitans and perivascular astrocyte processes is greatly reduced in comparison to perinodal astrocyte processes.

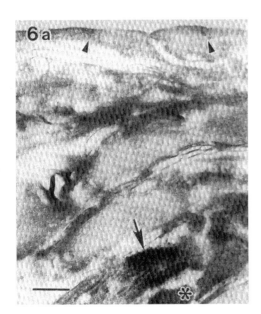

Figure 6
Antibody 7493 immunostaining of glia limitans (*a.*) and perivascular astrocyte processes (*b.*). *a.* The intensity of 7493 staining of the glia limitans (arrowheads) is much reduced compared to that displayed by a subjacent node of Ranvier (arrow), and perinodal astrocyte process (asterisk) associated with the node. *b.* Perivascular astrocyte processes (arrowheads) generally exhibit moderate 7493 immunoreactivity. Note the intense immunostaining displayed by an adjacent node of Ranvier (arrow) and perinodal astrocyte process (asterisk). Scale bar, 0.5 μm (from Black et al. 1989b).

The generally low sodium channel immunostaining present within in the glia limitans is contrasted with occasional regions of the glia limitans that exhibit more robust antibody 7493 immunoreactivity (Figure 7). The regions of 7493 staining are focal, and extend for several microns at most, with the regions being bounded by low 7493 immunoreactivity. Some of the sodium channel immunostaining at these focal sites appears to be associated with the astrocyte plasmalemma. Neither the basal lamina nor adjacent collagen fibrils exhibit antibody 7493 immunostaining at these sites of moderately dense immunoreactivity, or elsewhere along the glia limitans.

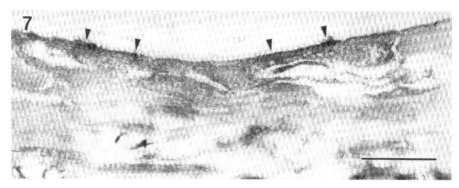

Figure 7
7493 immunoreactivity of glia limitans. Focal regions of the glia limitans display increased sodium channel staining (arrowheads). Scale bar, 1.0 μm (from Black et al. 1989b).

Sodium channel immunoreactivity is present in a patchy distribution within the cytoplasm of the astrocyte cell body (Black et al. 1989b). Focal accumulations of 7493 immunostaining are present in the cytoplasm, and these regions are adjacent to cytoplasm that lacks immunoreactivity. The 7493 immunoreactivity is not obviously associated with cellular organelles; 7493 immunostaining is not present within glial filaments.

The differential immunostaining pattern exhibited by astrocytes and their processes in situ with antibody 7493 may reflect a heterogeneous distribution of sodium channels within these cells. Astrocytes are known to establish and maintain spatially heterogeneous distributions of some molecules (Landis and Reese 1974; Black and Waxman 1984; ffrench-Constant et al. 1986). Conversely, the non-homogeneous staining of astrocytes may reflect the presence of subpopulations of astrocytes that express differing densities of sodium channels. In rat optic nerve, type 1 astrocytes (Raff et al. 1983) are thought to contribute perivascular and sub-pial processes, while type 2 astrocytes give rise to perinodal processes (Miller and Raff 1984; ffrench-Constant and Raff 1986). In addition, type 1 and type 2 astrocytes are reported to exhibit differing morphologies and locations within the rat optic nerve. Thus, subpopulations of astrocytes may express differing densities of sodium channels. It is also possible that different populations of astrocytes express different forms of sodium channels, which are differentially recognized by antibody 7493.

We have begun to examine factors that modulate the expression and spatial distribution of sodium channels within cultured astrocytes. Mammalian astrocytes in vitro have been shown to express voltage-sensitive sodium channels (Bevan et al. 1985; Nowak et al. 1987; Barres et al. 1988). We have examined 7493 immunostaining in astrocyte cultures derived from hippocampal regions CA1 and CA2 of neonatal (< P-2) rat pups, and which were maintained in culture for 14-21 days. The astrocytes were double-labelled for glial fibrillary acidic protein (GFAP) and sodium channels with a monoclonal anti-GFAP and polyclonal 7493 and appropriate rhodamine and fluorescein-labelled secondary antibodies. As shown in Figure 8a, the cultured astrocytes exhibit two morpho-

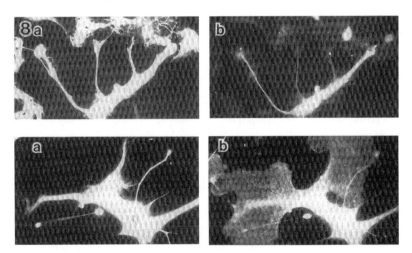

Figure 8
Cultured astrocytes double immunolabelled for GFAP (*a.*) and sodium channels (*b.*). *a.* Both flat, polygonal and stellate astrocytes exhibit intense GFAP staining (rhodamine-labelled secondary antibody). *b.* Sodium channel immunoreactivity (fluorescein-labelled secondary) is present in all GFAP-positive astrocytes. Stellate-shaped astrocytes appear to be more intensely 7493 immunoreactive than flat, polygonal astrocytes.

logies: some astrocytes are flat and polygonal, while others are stellate with branched processes. Both forms are intensely labelled for GFAP. The staining is expecially prominent in the astrocyte processes and around the nucleus. Double labelling experiments demonstrate that sodium channel immunostaining is present in GFAP-positive cells. Interestingly, the stellate astrocytes appear to exhibit more intense antibody 7493 immunoreactivity than the flat, polygonal astrocytes. For both astrocyte morphologies, the sodium channels appear to be relatively widely distributed, and not localized at specific sites within the processes or cytoplasm.

The observations of sodium channel staining within cultured astrocytes are consistent with an astrocytic synthesis of sodium channels in vitro, since neural sodium channels have been reported to have half-lives of 26-75 hours, and in vitro astrocytes exhibit 7493 immunoreactivity even after 21 days in culture without neuronal association. It is possible that at least some of the astrocytic sodium channels present in situ are glial, and not neuronal, in origin. It is especially interesting, then, that the differing astrocyte morphologies are associated with differential 7493 immunostaining. In this regard, two previous studies are particularly pertinent. First, Yarowsky and Krueger (1989) have demonstrated that the change in astrocyte morphology, from primitive flat polygonal cells to highly branched stellate forms, is coincident with the appearance of a population of saxitoxinsensitive (high affinity) sodium channels. Second, Barres et al. (1989) have recently reported that type 2 astrocytes, which have a branched, process-bearing morphology, express primarily a "neuronal" form of sodium channel ($\sim$ 85-90 % of total sodium channels), while type 1 astrocytes, which have a flat primitive appearance, express exclusively a "glial" form of sodium channel. These observations suggest that the expression of specific forms of sodium channels is associated with morphological differentiation of astrocytes. The available immuno-cytochemical evidence shows that antibody 7493 recognizes sodium channels in a variety of astrocyte morphologies, as well as channels in the nodal membrane.

Sodium channels in Schwann cells

In the PNS, perinodal Schwann cell processes come into close association with the axon membrane at nodes of Ranvier, and have been considered to constitute part of the 'paranodal apparatus' (Berthold 1968; Landon 1981; Rydmark and Berthold 1983). This relationship is reminiscent of the association of perinodal astrocyte processes with nodal axon membrane at central nodes. In sciatic nerve sections incubated with antibody 7493, sodium channel immunoreactivity is present within the nodal axon membrane (Figure 9). Sodium channel staining is not present within the axon membrane beneath the terminal para-

Figure 9
7493 immunostaining at node of Ranvier in rat sciatic nerve. The nodal axon membrane (between arrowheads) exhibits moderate 7493 immunoreactivity, and there is slight immunostaining of the subjacent axoplasm. The Schwann cell perinodal processes (arrows) exhibit 7493 immunoreactivity. Scale bar, 0.5 μm.

nodal loops, or elsewhere along the axolemma beneath the myelin sheath. The perinodal Schwann cell microvillar processes, which are derived from the outermost cytoplasm-filled lamellae of the myelin sheath, exhibit moderate 7493 immunostaining. The sodium channel staining of the perinodal processes appears to be associated both with the Schwann cell plasma membrane and the cytoplasm of the processes. Neither the adjacent basal lamina nor collagen fibrils exhibit 7493 immunostaining. The sodium channel immunoreactivity present within nodal axon membrane and associated glial processes at peripheral nodes is less than at central nodes. The reduced staining may reflect a more limited access of the immunoreagents, although control experiments with antibodies to neurofilaments, which immunostain the axoplasm, argue against this possibility. Alternatively, antibody 7493 may not recognize PNS sodium channels as well as CNS sodium channels. In this regard, antibodies directed against purified sodium channels from rat brain have been shown to discriminate antigenic differences between sodium channels from CNS, PNS and muscle tissues (Wollner and Catterall 1985).

Studies in progress in our laboratory demonstrate 7493 immunoreactivity in spinal root Schwann cells that are associated with both myelinated, and nonmyelinated axons. It will be of interest to determine whether sodium channel immunostaining is present in Schwann cells in situ prior to contact with axons, or following axonal transection and degeneration.

As with the expression of sodium channels by astrocytes, experiments with cultured Schwann cells demonstrate that association with neurons is not required for the expression of sodium channels by these cells. In physiological and pharmacological studies, cultured mammalian Schwann cells have been shown to express functional sodium channels (Chiu et al. 1984; Shrager et al. 1985; Gray and Ritchie 1985; Ritchie 1988).

In our studies, Schwann cells cultured from neonatal rabbit sciatic and vagus nerves exhibit sodium channel immunoreactivity with antibody 7493 (Figure 10). The cells were maintained for 7 days in culture prior to indirect immuno-fluorescent labelling. It appears that all fusiform-shaped cells are immunoreactive to antibody 7493. The intensity of staining does not appear to be

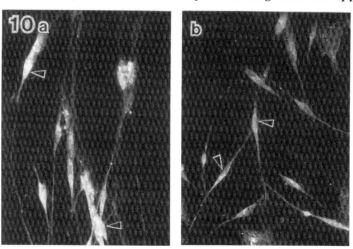

Figure 10
Antibody 7493 immunoreactivity in cultured Schwann cells from sciatic (*a.*) and vagus (b.). Schwann cells (arrowheads) are immunostained following incubation with antibody 7493. The immunoreactivity is more intense in the cytoplasm surrounding the nucleus than along the extended processes.

homogeneous within the Schwann cells, but appears to be more intense in the cytoplasm surrounding the nuclei than within the extended processes. Sodium channel immunoreactivity is present, and has a similar staining pattern, in Schwann cells from adult rabbit sciatic and vagus nerves, studied after 2 days in culture. The pattern of immunostaining may reflect a greater cytoplasmic pool of sodium channels in the perikaryal region than in the processes.

In a recent study utilizing whole-cell recordings from the body region of Schwann cells, Chiu (1987) observed tetrodotoxin-sensitive fast currents in Schwann cells which he believed to be associated with unmyelinated axons; he was unable to record currents from myelinating Schwann cells. The observations of sodium channel immunoreactivity within perinodal Schwann cell processes, and within cultured Schwann cells from sciatic nerves, are not necessarily in conflict with this report. From the immunoultrastructural results available to date, it is not clear whether sodium channels are actually incorporated within the plasma membrane of perinodal processes. Moreover, it is uncertain whether whole-cell recordings would be able to detect localized aggregations of sodium channels within the membrane of perinodal processes. Studies are in progress to examine the expression and spatial organization of sodium channels within myelinating, as well as nonmyelinating, Schwann cells.

Conclusions

We have demonstrated intense sodium channel immunoreactivity within the axon membrane at central and peripheral nodes of Ranvier, and within perinodal astrocyte and Schwann cell processes. Schwann cells and astrocytes in vitro also display sodium channel immunostaining.

While the role of sodium channels in electrogenesis within neurons is well-established (Hodgkin and Huxley 1952), it is not clear what functional role(s) is (are) played by sodium channels within astrocytes and Schwann cells. The density of sodium channels incultured rat astrocyte plasmalemma has been estimated to be approximately $10/\mu m^2$ (Ritchie 1986), while in cultured rabbit Schwann cell membrane approximately $30/\mu m^2$ has been suggested (Bevan et al. 1985). It has yet to be established whether these densities of sodium channels are uniformly distributed over the plasma membrane of the glial cells, or whether these densities of sodium channels would support electrogenesis or the propagation of action potentials. Indeed, if the glial cells do support electrogenesis, the utility to the cell is not yet apparent.

It has been suggested that Schwann cells and astrocytes may function as sites for the extra-neuronal synthesis of sodium channels which are subsequently transferred to the axon (Bevan et al. 1985; Gray and Ritchie 1985; Shrager et al. 1985). Augmentation of the neuronal machinery for channel synthesis would be expected to reduce the biosynthetic load on the neuron, since sodium channels may have a relatively rapid turnover rate. The estimated half-life of sodium channels incorporated into cultured CNS neurons is 50 hours (Schmidt and Catterall 1986), and is 2.2 days for cultured Schwann cells (Ritchie 1986). Moreover, synthesis in the neuronal cell body, of sodium channels destined for the node, would impose added workloads on the neuron, in terms of transport of channels and targeting to the nodes of Ranvier. Perinodal astrocyte/Schwann cell processes are ideally located for this latter function. The results presented here are consistent with a role for astrocytes and Schwann cells in establishing and replenishing aggregations of sodium channels at nodes of Ranvier.

Alternatively, sodium channels within perinodal astrocyte and Schwann cell processes may be involved in the ionic homeostasis of neural extracellular spaces (Boudier et al. 1988). In chick skeletal muscle, activation of sodium channels modulates Na^+/K^+-ATPase activity (Wolitzky and Fambrough 1986). As electrical activity at nodes would be expected to raise the extracellular levels of K^+ within the nodal-perinodal process microenvironment (Ransom et al. 1986), the activity may lead to a depolarization of the membrane of the glial processes. Sodium channels could be activated under these conditions, and might provide a signal for the regulation of Na^+/K^+-ATPase activity, with subsequent clearing of excess K^+ form the perinodal space. It has been demonstrated, in this regard, that an electrogenic pump is activated by impulse activity at physiological frequencies in rat optic nerve (Gordon et al. 1989). Consistent with this proposed role for glial sodium channels, Na^+/K^+-ATPase has been localized within the perinodal processes of astrocytes (Ariyasu et al. 1985) and Schwann cells (Ariyasu and Ellisman 1987).

Finally, the immunostaining of cytoplasm within perinodal astrocyte and Schwann cell processes raises questions regarding factors that regulate the incorporation of sodium channels into the plasmalemma, where presumably they will have physiological roles. Our observations suggest that there may be several pools of sodium channels or channel precursors (including cytoplasmic and membrane-inserted pools) in some cells. Mechanisms that maintain, and modulate, these channel/channel precursor pools (and possibly their dynamic equilibrium) may participate in the specification of specialized membrane properties in glial cells and neurons.

Acknowledgements

The authors thank Tina McKay and Maura Ford for excellent technical assistance. Work reported in this chapter has been supported by grants from the National Institutes of Health, the National Multiple Sclerosis Society and the Medical Research Service, Veterans Administration.

References

Ariyasu RG and Ellisman MH (1987) The distribution of (Na^+/K^+)ATPase is continuous along the axolemma of unensheathed axons from spinal roots of 'dystrophic' mice. J Neurocytol 16:239-248

Ariyasu RG, Nichol JA and Ellisman MH (1985) Localization of sodium/potassium adenosine triphosphatase in multiple cell types of the murine nervous system with antibodies raised against the enzyme from kidney. J Neurosci 5:2581-2596

Barres BA, Chun LLY and Corey DP (1988) Ion channel expression by white matter glia: I. type 2 astrocytes and oligodendrocytes. Glia 1:10-30

Barres BA, Chun LLY and Corey DP (1989) Glial and neuronal forms of the voltage-dependent sodium channel: characteristics and cell-type distribution. Neuron 2:1375-1388

Berthold C-H (1968) Ultrastructure of node-paranode region of mature feline ventral lumbar spinal-root fibres. Acta Soc Med Upsal 73(suppl 9):37-78

Berthold C-H and Carlstedt T (1977) Observations on the morphology at the transition between the peripheral and the central nervous system in the cat. Acta Physiol Scand 446:43-60

Bevan S, Chiu SY, Gray PTA and Ritchie JM (1985) The presence of voltage-gated sodium, potassium and chloride channels in rat cultured astrocytes. Proc R Soc Lond B 225:299-313

Black JA, Foster RE and Waxman, SG (1982) Rat optic nerve: freeze-fracture studies during development of myelinated axons. Brain Res 250:1-20

Black JA, Friedman B, Waxman SG, Elmer LW and Angelides, KJ (1989a) Immuno-ultrastructural localization of sodium channels at nodes of Ranvier and perinodal astrocytes in rat optic nerve. Proc R Soc Lond B (in press)

Black JA and Waxman, SG (1984) Specialization of astrocyte membrane at glia limitans in rat optic nerve: freeze-fracture observations. Neurosci Lett 55:371-378

Black JA, Waxman SG, Friedman B, Elmer LW and Angelides, KJ (1989b) Sodium channels in astrocytes of rat optic nerve in situ: immunoelectron microscopic studies. Glia (in press)

Boudier J-L, Jover E and Cau P (1988) Autoradiographic localization of voltage-dependent sodium channels of the mouse neuromuscular junction using ^{125}I-alpha scorpion toxin. I. Preferential labeling of glial cells on the presynaptic side. J Neurosci 8:1469-1478

Brismar T (1981) Specific permeability properties of demyelinated rat nerve fibres. Acta Physiol Scand 113:167-176

Chiu SY (1980) Asymmetry currents in the mammalian myelinated nerve. J Physiol 309:499-519

Chiu SY (1987) Sodium currents in axon-associated Schwann cells from adult rabbits. J Physiol 386:181-203

Chiu SY and Ritchie, JM (1981) Evidence for the presence of potassium channels in the internodal region of acutely demyelinated mammalian single nerve fibres. J Physiol 313:415-437

Chiu SY and Schwarz, W (1987) Sodium and potassium currents in acutely demyelinated internodes of rabbit sciatic nerves. J Physiol 391:631-649

Chiu SY, Shrager P and Ritchie JM (1984) Neuronal-type Na^+ and K^+ channels in rabbit cultured Schwann cells. Nature 311:156-157

Elmer LW (1988) Mammalian sodium channel physiochemical characterization and immunocytochemical localization and interaction with the neuronal cytoskeleton as a mechanism of restricted distribution. PhD Dissertation. University Microfilms

Elmer LW, O'Brien B, Nutter TJ and Angelides, KJ (1985) Physiochemical characterization of alpha-peptide of the sodium channel from rat brain. Biochemistry 24:8128-8137.

ffrench-Constant C, Miller RH, Kruse J, Schachner M and Raff MC (1986) Molecular specialization of astrocyte processes at nodes of Ranvier in rat optic nerve. J Cell Biol 102:844-852

ffrench-Constant C and Raff MC (1986) The oligodendrocyte-type 2 astrocyte cell lineage is specialized for myelination. Nature 323:335-338

Gordon TR, Kocsis JD and Waxman SG (1989) Electrogenic pump activity in rat optic nerve. Soc Neurosci Abstr (in press)

Gray PTA and Ritchie JM (1985) Ion channels in Schwann and glial cells. TINS 8:211-215

Hildebrand C (1971) Ultrastructural and light-microscopic studies of the nodal region in large myelinated fibres cf the adult feline spinal cord white matter. Acta Physiol Scand 364:43-71

Hildebrand C and Waxman SG (1984) Postnatal differentiation of rat optic nerve fibers: electron microscopic observations of the development of nodes of Ranvier and axoglial relations. J Comp Neurol 224:25-37

Hodgkin AL and Huxley AF (1952) A quantitative description of membrane current and its application to conduction and excitation in nerve. J Physiol 117:500-544

Landis DMD and Reese TS (1974) Arrays of particles in freeze-fractured astrocyte membrane. J Cell Biol 60:316-320

Landon DN (1981) Structure of normal peripheral myelinated nerve fibres. In: Advances in Neurology, Vol 31 (Demyelinating diseases - basic and clinical electrophysiology). SG Waxman and JM Ritchie (eds) Raven Press, New York, pp 25-39

Miller RH and Raff MC (1984) Fibrous and protoplasmic astrocytes are biochemically and developmentally distinct. J Neurosci 4:585-592

Nowak L, Ascher P and Berwald-Netter Y (1987) Ionic channels in mouse astrocytes in culture. J Neurosci 7:101-109

Quick DC and Waxman SG (1977) Specific staining of the axon membrane at nodes of Ranvier with ferric ion and ferrocyanide. J Neurol Sci 31:1-11

Raff MC, Abney ER, Cohen J, Lindsay R and Noble M (1983) Two types of astrocytes in cultures of developing rat white matter: differences in morphology, surface gangliosides, and growth characteristics. J Neurosci 3:1289-1300

Ransom BR, Carlini WG and Connors BW (1986) Brain extracellular space: developmental studies in rat optic nerve. Ann NY Acad Sci 481:87-105

Ritchie JM (1986) Distribution of saxitoxin-binding sites in mammalian neural tissue. Ann NY Acad Sci 479:385-401

Ritchie JM (1988) Sodium channel turnover in rabbit cultured Schwann cells. Proc R Soc Lond B 233:423-430

Ritchie JM and Rogart RB (1977) Density of sodium channels in mammalian myelinated nerve fibers and nature of the axonal membrane under the myelin sheath. Proc Natl Acad Sci USA 74:211-215

Rydmark M and Berthold C-H (1983) Electron microscopic serial section analysis of nodes of Ranvier in lumbar spinal roots of the cat: a morphometric study of nodal compartments in fibres of different sizes. J Neurocytol 12:537-565

Schmidt JW and Catterall WA (1986) Biosynthesis and processing of the alpha subunit of the voltage-sensitive sodium channel in rat brain neurons. Cell 46:437-445

Shrager P (1989) Sodium channels in single demyelinated mammalian axons. Brain Res 483:149-154

Shrager P, Chiu SY and Ritchie JM (1985) Voltage-dependent sodium and potassium channels in mammalian cultured Schwann cells. Proc Natl Acad Sci USA 82:948-952

Srinivasan Y, Elmer L, Davis J, Bennett V, and Angelides KJ (1988) Ankyrin and spectrin associated with voltage-dependent sodium channels in brain. Nature 333:177-180

Waxman SG, Black JA, Kocsis JD, and Ritchie JM (1989) Low density of sodium channels supports action potential conduction in axons of neonatal rat optic nerve. Proc Natl Acad Sci USA 86:1406-1410

Waxman SG and Quick DC (1977) Cytochemical differentiation of the axon membrane in A- and C-fibers. J Neurol Neurosurg Psychiatr 40:379-386

Waxman SG and Ritchie JM (1985) Organization of ion channels in the myelinated nerve fiber. Science 228:1502-1507

Wolitzky BA and Fambrough DM (1986) Regulation of the Na^+/K^+-ATPase in cultured chick skeletal muscle. Modulation of expression by the demand for ion transport. J Biol Chem 261:9990-9999

Wollner DA and Catterall WA (1985) Antigenic differences among the voltage-sensitive sodium channels in the peripheral and central nervous systems and skeletal muscle. Brain Res 331:145-149

Yarowsky PJ and Krueger BK (1989) Development of saxitoxin-sensitive and insensitive sodium channels in cultured neonatal rat astrocytes. J Neurosci 9:1055-1061

DEMYELINATION, REMYELINATION AND GLIAL CELL TRANSPLANTATION

SOME ASPECTS OF MECHANISMS OF INFLAMMATORY DEMYELINATION

William T. Norton,*# Celia F. Brosnan,+# Wendy Cammer*# and Ellen Goldmuntz#1
Departments of *Neurology, +Pathology and #Neuroscience
Albert Einstein College of Medicine
1300 Morris Park Avenue
Bronx, New York 10461

INTRODUCTION

The most extensively studied inflammatory demyelinating diseases are multiple sclerosis (MS) and the animal model, experimental autoimmune encephalomyelitis (EAE). Both are characterized by perivascular inflammation, edema, demyelination and reactive gliosis. In MS the formation of the lesion, which contains T-lymphocytes and macrophages, is generally acknowledged to be related to a cell-mediated immune (CMI) response, and it is known that EAE involves CMI mechanisms requiring T-cell sensitization to a central nervous system antigen. For the past dozen years we have used EAE to explore mechanisms by which such CMI reactions might initiate inflammation, edema and demyelination.

POSSIBLE ROLE OF ACTIVATED MACROPHAGES

The accepted sequence of cellular reactions in inflammation generated by a CMI response is that sensitized T-lymphocytes invade the tissue and react with the antigen on an antigen-presenting cell. As a consequence the T-cells release a number of factors, known as lymphokines. Monocytes are "called in" by these lymphokines, and are stimulated by them to undergo a marked metabolic change known as activation, and thus to become fully differentiated macrophages. Macrophages are a major component of the perivascular cuffs in both MS and EAE (Adams RD, 1959; Adams CWM, 1977; Raine, 1983). In the latter disease cellular infiltration precedes demyelination and demyelination is only observed in areas where macrophages are present. Macrophages can be seen penetrating between myelin lamellae (Lampert, 1965; Raine et al, 1974).

---

(1) Current Address: Children's National Medical Center, 111 Michigan Avenue N.W., Washington, D.C. 20010.

Additional observations suggest that macrophages can damage nearby myelin sheaths not actually in contact with them. In EAE extensive vesicular disruptions of myelin are seen in proximity to macrophages (Raine et al, 1974; DalCanto et al, 1975). Similar morphological observations in MS are the net-like disruptions of myelin sheaths in acute MS, the apparent melting away of myelin in contact with macrophages in active plaques, and the presence of myelin debris in the extracellular space as well as within macrophages (Prineas et al, 1984; Raine, 1983).

While these observations suggested a crucial role for macrophages in demyelination it was not clear how they effected it. Immune mechanisms have been proposed that required the direct involvement of a specific immuno-competent T-cell in addition to the non-specific macrophage, while others believed that myelin damage was caused by proteinases and other hydrolytic enzymes. This latter concept was supported by the many studies showing that both acid and neutral proteinases are elevated in MS plaques and EAE lesions (see Norton and Cammer, 1984, and Smith and Benjamins, 1984 for reviews).

The phenomenon called bystander demyelination supported a role for the non-immunocompetent macrophage (Wisniewski and Bloom, 1975a,b). This phenom-enon is induced by sensitizing an animal to mycobacterial antigens (complete Freund's adjuvant) and then introducing the tubercule bacillus antigen into the central nervous system. The resultant CMI reaction causes inflammation and primary demyelination similar to that in EAE, even though it is directed toward non-brain foreign antigens.

These morphological and biochemical observations took on more signif-icance with the discovery that activated macrophages secrete a number of products that mediate inflammation and are capable of tissue destruction. These include several neutral proteinases, among which is plasminogen activator (Unkeless et al, 1974).

The data summarized above led to our hypothesis that inflammatory de-myelination could be initiated by secretion products of activated macrophages (Cammer et al, 1978). We proposed that infiltrating lymphocytes "called in" and activated macrophages, which would secrete plasminogen activator. Since the blood-brain barrier is destroyed at the site of the inflammatory lesion, serum proteins, including plasminogen, are present in the CNS parenchyma.

The plasminogen activator catalyzes the formation from plasminogen of plasmin, a trypsin-like neutral proteinase, which could degrade myelin proteins. Since plasmin is also a plasminogen activator there is a potential for considerable amplification of proteolytic activity. Other secretion products, such as lipases, other proteinases and oxygen radicals might also be involved. We suggested that the specificity of this mechanism for myelin might lie in the extreme sensitivity of myelin basic protein to proteolysis, and that this mechanism would be common to cell-mediated demyelination independent of antigen.

Five approaches were designed to explore this hypothesis: (1) The effects of macrophage depletion on EAE, (2) The reaction of macrophage secretion products on myelin _in vitro_; (3) The effects of complement plus proteinases on myelin _in vivo_, (4) The induction of demyelination _in vivo_ with plasminogen activator, and (5) The suppression of EAE with proteinase inhibitors.

Effects of macrophage depletion: Brosnan et al (1981) have shown that EAE can be prevented if macrophages are depleted selectively by injecting silica dust intraperitoneally. All Lewis rats innoculated with guinea pig spinal cord in complete Freund's adjuvant develop severe clinical signs by days 14-15. Eighty percent of animals given I.P. injections of 200 mg of silica on days 8 and 11 failed to develop clinical signs for at least 4 weeks, and showed little inflammation and no demyelination. Lymphocytes taken from the silica-treated animals were, however, capable of passively transferring EAE to naive recipients.

These results strongly support the conclusion that macrophages function as the effector cells in the clinical and pathological expression of EAE. Thus, while the sensitized lymphocyte is the necessary immunocompetent cell for the initiation of the CMI response, the activated macrophage actually mediates the tissue damage.

Effects of macrophage secretion products on myelin: We showed that macrophage secretion products could degrade the basic protein in lyophilized myelin _in vitro_ and that this degradation could be considerably enhanced by the addition of plasminogen (Cammer et al, 1978). These results suggested that plasmin was effective in hydrolyzing basic protein. To verify it we showed that urokinase, a known plasminogen activator, together with purified plasminogen, could degrade basic protein in myelin. These data showed that myelin is susceptible to degradation by products of activated macrophages, and indicated that our hypothesis is feasible.

Effects of complement: Although macrophage secretion products alone, or together with plasminogen, could degrade myelin proteins in lyophilized myelin, they were ineffective when either freshly isolated myelin or fresh white matter was used as a substrate. We reasoned that some additional factor was needed to disrupt the myelin structure making it accessible to the proteinases. Two possible candidates for such factors were phospholipases and complement, both known to be secreted by macrophages under appropriate conditions (reviewed in Cammer et al, 1986). The addition of phospholipase or lysolecithin to plasmin, or to macrophage conditioned media plus plasminogen, potentiated the degradation of basic protein in fresh bovine myelin from none to 35-90% (Cammer et al, 1986).

Perhaps more significant was the finding that pretreatment of fresh myelin with complement was also effective in potentiating the action of plasmin, or of macrophage conditioned media plus plasminogen. Complement depleted (heated) sera, C_3-deficient sera or C_4-deficient sera were ineffective (Cammer et al, 1986). These data indicated that complement, activated by the classical pathway, could render the basic protein in fresh myelin vulnerable to proteolytic enzymes.

The properties and availability of complement strongly support its hypothetical role in demyelination. Once activation of complement has taken place, there is generation of the C5b-9 membrane-attack complex, which can form pores in membranes. Furthermore, myelin is known to be able to activate complement in the absence of antibody or immune complexes (Cyong et al, 1982; Vanguri et al, 1982), and this activation, by means of the classical pathway, results in the incorporation of the membrane-attack complex into the myelin (Liu et al, 1983; Silverman et al, 1984).

There is some evidence for a role of complement in inflammatory demyelination in vivo. In guinea pigs decomplemented with cobra venom factor EAE is delayed and the symptoms attenuated (Pabst et al, 1971; Morari and

Dalmasso, 1978). While this finding is commonly interpreted to show a role for antibody in EAE, an alternative explanation could involve a direct alteration of myelin by complement. The finding of a subnormal amount of complement component C_9 in the CSF of multiple sclerosis patients has suggested that the complement membrane attack complex might be involved in the damage to myelin that occurs in this disease (Morgan et al, 1984). However, the hypothetical role for complement as a potentiator of the action of proteinases _in vivo_ should be distinguished from its role in antibody-mediated complement-dependent demyelination _in vitro_ (Appel and Bornstein, 1964; Bornstein and Raine, 1977) and from the presumably similar role of complement in augmenting antibody-mediated demyelination _in vivo_ (Lassman et al, 1983; Saida et al, 1978). The present hypotheses does not require a role for antibody. Since complement and plasminogen could enter the CNS through lesions in the blood-brain barrier, we believe that the combined effects of complement, plasminogen and macrophage-secreted plasminogen activator provide an attractive hypothetical mechanism for the initiation of demyelination in inflammatory lesions.

Induction of demyelination with plasminogen activator: If our central hypothesis, that proteinases initiate demyelination, is correct then plasmin itself should be able to cause demyelination in the absence of a CMI reaction. We tested this possibility in the rabbit eye, which is unique in that it has a strip of myelinated fibers lying on the surface of the retina exposed to the vitreous humor. Urokinase, a plasminogen activator was injected into the posterior chamber of the eye (Brosnan et al, 1980a). Demyelination was consistently observed in the superficial layers of the retina. A mononuclear cell infiltrate was also present, and demyelination was observed in the vicinity of the invading cells. Urokinase inactivated by diisopropylfluorophosphate was not effective in causing demyelination or inflammation.

The observations support our postulate that neutral proteases play a role in inflammatory demyelination. We are not sure why an inflammatory response occurred in these animals. It was not induced by inactive urokinase, and it is possible that it was a response secondary to tissue destruction caused by the protease.

Suppression of EAE with proteinase inhibitors: If neutral proteinases, and particularly plasmin, have a key role in initiating the clinical signs and pathology of the inflammatory demyelinating diseases it might be expected that inhibitors of these enzymes would protect sensitized animals against EAE. We induced EAE in male Lewis rats with guinea pig spinal cord in complete Freund's adjuvant. Starting on day 6 proteinase inhibitors were injected intraperitoneally twice daily, and sensitized control animals were injected with saline. Control animals consistently became ill on days 12 to 14.

We found that aminomethylcyclohexane carboxylic acid (AMCA), ϵ-amino-caproic acid (EACA), and p-nitrophenylquanidinobenzoate (NPGB), all of which are inhibitors of plasminogen, plasmin and other neutral proteinases, gave significant protection against the clinical expression of EAE (Brosnan et al, 1980b). Pepstatin, an inhibitor of acid proteinases, was also effective in protecting some of the animals. Leupeptin and antipain were ineffective at the dose used. Trasylol (aprotinin) actually exacerbated the disease. AMCA and NPGB were most effective in decreasing the extent of weight loss. Histological examination of clinically well animals treated with NPGB and pepstatin showed that both perivascular infiltration and submeningeal inflammation were markedly reduced. In asymptomatic animals treated with AMCA and EACA perivascular infiltration was reduced only slightly; however in AMCA-treated animals, the degree of demyelination in the vicinity of the inflammatory cells was reduced considerably.

Smith and Amaducci (1982) carried out an extensive independent study of EAE protection using the same drugs we have tested. Their results are essentially in agreement with ours, including the finding that Trasylol exacerbates EAE.

It is not possible at present to explain all of these data. The protection afforded by AMCA, EACA and NPGB offer support for a significant role of serine proteinases in EAE. The finding that AMCA inhibits demyelination even though infiltrating mononuclear cells are present supports the hypothetical participation of the macrophage-secreted Plg-activator in the pathogenesis of EAE. On the other hand, the fact that some inhibitors also suppressed inflammation suggest that proteinases may also be important for development of the inflammatory response.

All of the experimental work described here strongly supports our original postulate. These data show that macrophages are necessary for the development of EAE, an inflammatory demyelinating disease known to involve CMI reactions; that macrophage secretion products, especially in the presence of plasminogen and complement, are capable of degrading myelin proteins; that demyelination can be induced by a plasminogen activator; and that EAE can be suppressed by inhibitors of plasminogen activator and plasmin.

BLOOD-BRAIN BARRIER, EDEMA AND INFLAMMATION

Our hypothesis outlined above depends upon increased vascular permeability, permitting leakage into the brain parenchyma of the serum proteins, plasminogen and complement. It is known that alteration in blood-brain barrier (BBB) permeability is an early and significant event in EAE. Several studies have shown that clinical signs of disease correlate more closely with the extent of edema in the spinal cord than with histological evidence of inflammation (Leibowitz and Kennedy, 1972; Kerlero de Rosbo et al, 1985) or with the degree of demyelination (Raine et al, 1981; Moore et al, 1984). In fact, there is little or no demyelination in animals in which EAE is induced with pure myelin basic protein (MBP), although the clinical signs are as pronounced as in EAE animals induced with whole white matter, where demyelination is obvious (Raine et al, 1981). The factors involved in this alteration of BBB function are not well understood, but in the mouse the histamine sensitizing factor of B. pertussis appears to play a major role (Linthicum and Frelinger, 1982). Furthermore, in the Lewis rat EAE can be reactivated by resensitizing the animal with MBP plus complete Freund's adjuvant plus B. pertussis (Waxman et al, 1982). In both the mouse model and the reactivated rat model (but not in the primary clinical episode in the rat) antagonists of histamine can suppress clinical signs of disease, thus implicating this vasoactive amine in the disease process. A possible requirement for vasoactive amines in the development of a delayed-type hypersensitivity (DTH) reaction has also been proposed (Gershon et al, 1975). In DTH reactions the augmenting inflammatory response is known to play a major role in the initiation of tissue damage. This response consists primarily of cells of the monocyte-macrophage lineage that do not normally leave the blood. Gershon et al (1975) have suggested that vasoactive amines cause constriction of the endothelial cells thus facilitating egress of bone marrow derived cells from the circulation.

If this hypothesis is correct then one would predict that only the specifically sensitized T-cell would elicit lymphokine production that results in vasoconstriction, edema, and a non-specific augmentation of the immune response. This hypothesis also involves the concept that all types of activated T-cells cross the endothelium into the parenchyma, and there is preliminary evidence from work with the T-cell lines that this is indeed the case. Thus, there would be two levels at which the response could be blocked: the receptor that mediates perivascular transit of activated T-cells, and the action of lymphokines on the vascular endothelium. Our work has focussed on the vascular response.

Preliminary studies indicated that antagonists of histamine and serotonin could not significantly suppress the development of clinical signs of EAE in the Lewis rat, and therefore we turned our attention to the catecholamines.

In most of the vasculature norepinephrine and epinephrine mediate vaso-constriction via the α-receptor and vasodilation via the β-receptor. We therefore tested the ability of antagonists of α and β adrenergic receptors to modulate the expression of EAE in the Lewis rat. Our results showed that significant suppression of the clinical and histological expression of EAE in the Lewis rat could be obtained by treatment with prazosin, a specific antagonist of α_1-adrenoceptors (Brosnan et al, 1985). Analysis of the effect of other adrenergic receptor antagonists supports the conclusion that the suppressive effect of prazosin is a consequence of blockade of the α_1-receptor since treatment with either the α_2-antagonist yohimbine or the β-antagonist propanolol exacerbated the disease, whereas treatment with the mixed α_1/α_2-antagonist phenoxybenzamine had some suppressive activity (Brosnan et al, 1985). Treatment with prazosin was also able to suppress clinical and histological signs of EAE in animals sensitized by adoptive transfer with activated spleen or lymph node cells. Since the presynaptic α_2-adrenoceptor exerts a negative feedback control on the release of norepinephrine, antagonism of this receptor could lead to unrestrained release of NE, which could account for the exacerbation of the disease observed in animals treated with yohimbine.

Further studies designed to examine the effect of prazosin on vascular permeability in EAE have shown that in both actively induced and passively transferred disease, treatment with prazosin significantly suppresses leakage of serum proteins into the spinal cord and delays the expression of the inflammatory response (Goldmuntz et al, 1986a).

Catecholamines modulate a diverse array of cellular functions through interaction with cell surface receptors, and we have considered two, not necessarily exclusive, sites of action to account for the suppressive effect of prazosin: the immune response and the vasculature. In order to explore the mechanism of action of prazosin further we have also tested the effect of prazosin on the early, inductive phase and on the late, effector phase of the disease (Brosnan et al, 1986). Additional experiments have also explored the effect of lymphocyte responses to mitogen and antigen in vitro.

The results of these studies have shown that treatment with prazosin has no effect on the early, inductive phase of EAE but can still significantly suppress disease when treatment is begun at the time of onset of early clinical signs (day 10). Leakage of serum proteins and perivascular inflammation were also suppressed in these animals, particularly in the early stages of the acute response. Lymphocytes, obtained from both treated animals and from sensitized animals incubated in the presence of prazosin in vitro, showed that prazosin had no effect on lymphocyte responses to antigen or mitogen.

We have examined the astrocyte reaction as another indicator of disease. Smith et al (1983) have shown that in Lewis rats, sensitized to develop EAE, enhanced immunostaining for glial fibrillary acidic protein (GFAP) is evident early in the disease (10-12 days post-inoculation). The staining intensity increases with time and the reactive astrocytes are found throughout the spinal cord, unrelated to sites of inflammation. We confirmed these results and showed that this enhanced immunostaining for GFAP was delayed in rats in which the clinical signs of EAE had been suppressed by prazosin treatment (Goldmuntz et al, 1986b).

To summarize the prazosin studies: we find that prazosin treatment suppresses clinical signs and edema, only partially suppresses and delays perivascular inflammation and delays the astrocyte response. Clinical signs, therefore, correlate well with increased vascular permeability and not with inflammation, whereas the onset of the astrocyte response correlates well with the onset of inflammation. It is also clear that enhanced GFAP staining does not correlate with clinical signs, since the glial reaction is still intense at 65 dpi, long after animals have recovered, and enhanced staining occurs in prazosin-treated animals in the absence of clinical signs.

Prazosin is a common antihypertensive drug. Its activity is the result of vasodilation, believed to be caused by blockade of vascular α_1-adrenergic receptors. Our results support the hypothesis that prazosin suppresses EAE through a direct vascular effect. This drug also allows us to distinguish the development and effects of inflammation from those of vascular permeability and provides a tool with which to explore the factors involved in edema.

CONCLUSION

The studies reviewed here concern several aspects of the induction and development of the cell-mediated immune lesion in EAE - from the opening of the BBB to the molecular mechanisms that may be responsible for demyelination. Although we recognize that EAE is an imperfect model of multiple sclerosis, the similarity in the pathology of the lesions in these two conditions justifies the use of this model disease to investigate what we believe may be universal mechanisms of inflammatory demyelination.

From our work and that of others, the following sequence of events can be proposed. Sensitized T-cells either invade the CNS and react with antigen (myelin basic protein in the case of EAE) presented by microglia or astrocytes or react with antigen presented by the endothelial cell. Lymphokines are released which amplify the response by "calling in" monocytes and other lymphocytes. Cellular products from one or more of these inflammatory cell types induce the release of vasoactive amines which cause vasospasm leading to an increase in vascular permeability. Our evidence indicates that in EAE vasospasm is maintained by agonists acting on α_1-adrenergic receptors. The resultant breakdown in the blood-brain barrier leads to vasogenic edema and increased perivascular transit of inflammatory cells. We have some evidence from our studies of passively-transferred disease that the increase of edema slightly precedes the increase in cellular infiltration.

The effector stage of disease, discussed above, leads subsequently to the augmented inflammatory response and a greatly increased infiltration of inflammatory cells. We believe that the activated macrophages are largely responsible for the tissue damage (demyelination) in the lesion, which is initiated by secretion products of these cells.

We have shown that this sequence of events can be interferred with at several stages, leading to suppression of various manifestations of the disease. For example, blockade of the α_1-receptor by prazosin during the inductive phase has no effect, suggesting that it has no effect on the immune response. However, blockade during the effector stage suppresses edema and clinical signs, but only delays inflammation and the astrocyte response. We are thus able to show that clinical disease correlates better with edema than with inflammation, whereas the astrocyte response is related to inflammation. We have also shown previously that clinical signs can be suppressed by depletion of macrophages and by treatment with proteinase inhibitors.

It is our hope that detailed studies of the pathogenesis of the model disease, EAE, will lead to logical ways of interfering with the progression of multiple sclerosis.

ACKNOWLEDGEMENTS

The work described here was supported by grant RG 1089 from the National Multiple Sclerosis Society and by United States Public Health Service grants NS 02476, NS 23247, NS 11920, NS 12890 and T32 GM 7288. We thank Ms. Renee J. Sasso for secretarial assistance.

REFERENCES

Adams CWM (1977) Pathology of multiple sclerosis: Progress of the lesion. Br Med Bull 33:15-20.
Adams RD (1959) A comparison of the morphology of the human demyelination disease and experimental "allergic" encephalomyelitis. In: Kies MW and Alvord EC (eds) Allergic Encephalomyelitis. Charles C Thomas, Springfield, Illinois, p 183.
Appel SH, Bornstein MB (1964) The application of tissue culture to the study of experimental allergic encephalomyelitis. II. Serum factors responsible for demyelination. J Exp Med 119:303-312.
Bornstein MB, Raine CS (1977) Multiple sclerosis and EAE: Specific demyelination of CNS in culture. Neuropathol Appl Neurobiol 3:359-367.
Brosnan CF, Cammer W, Bloom BR, Norton WT (1980a) Initiation of primary demyelination in vivo by a plasminogen activator (urokinase). J Neuropathol Exper Neurol 39:344.

Brosnan CF, Cammer W, Norton WT, Bloom BR (1980b) Proteinase inhibitors suppress the development of experimental allergic encephalomyelitis. Nature 285:235-237.

Brosnan CF, Bornstein MB, Bloom BR (1981) The effects of macrophage depletion on the clinical and pathologic expression of EAE. J Immunol 126:614-620.

Brosnan CF, Goldmuntz EA, Cammer W, Factor SM, Bloom BR, Norton WT (1985) Prazosin, an $alpha_1$-adrenergic receptor antagonist, suppresses experimental autoimmune encephalomyelitis in the Lewis rat. Proc Natl Acad Sci USA 82:5915-5919.

Brosnan CF, Sacks HJ, Goldschmidt RC, Goldmuntz EA, Norton WT (1986) Prazosin treatment during the effector stage of disease suppresses experimental autoimmune encephalomyelitis in the Lewis rat. J Immunol 137:3451-3456.

Cammer W, Bloom BR, Norton WT, Gordon S (1978) Degradation of basic protein in myelin by neutral proteases secreted by stimulated macrophages: A possible mechanism of inflammatory demyelination. Proc Natl Acad Sci USA 75:1554-1558.

Cammer W, Brosnan CF, Basile C, Bloom BR, Norton WT (1986) Complement potentiates the degradation of myelin proteins by plasmin: Implications for a mechanism of inflammatory demyelination. Brain Res 364:91-101.

Cyong C-J, Witkin SS, Rieger B, Barbarese E, Good RA, Day NK (1982) Antibody-independent complement activation by myelin via the classical complement pathway. J Exp Med 155:587-598.

Dal Canto MC, Wisniewski HM, Johnson AB, Brostoff SW, Raine CS (1975) Vesicular disruption of myelin in autoimmune demyelination. J Neurol Sci 24:313-319.

Gershon RK, Askenase PW, Gershon MD (1975) Requirement for vasoactive amines for production of delayed-type hypersensitivity skin reactions. J Exp Med 142:732-747.

Goldmuntz EA, Brosnan CF, Norton WT (1986a) Prazosin treatment suppresses increased vascular permeability in both acute and passively transferred experimental autoimmune encephalomyelitis in the Lewis rat. J Immunol 137:3444-3450.

Goldmuntz EA, Brosnan CF, Chiu F-C, Norton WT (1986b) Astrocyte reactivity and intermediate filament metabolism in experimental autoimmune encephalomyelitis: The effect of suppression with prazosin. Brain Res 397:16-26.

Kerlero de Rosbo N, Bernard CCA, Simons RD, Carnegie PR (1985) Concomitant detection of changes in myelin basic protein and permeability of blood-spinal cord barrier in acute experimental autoimmune encephalomyelitis by electroimmunoblotting. J Neuroimmunol 9:349-361.

Lampert PW (1965) Demyelination and remyelination in experimental allergic encephalomyelitis: Further electron microscopic observations. J Neuropathol Exp Neurol 24:371-385.

Lassman H, Stemberger H, Kitz K, Wisniewski HM (1983) In vivo demyelinating activity of sera from animals with chronic experimental allergic encephalomyelitis. J Neurol Sci 59:123-137.

Leibowitz S, Kennedy L (1972) Cerebral vascular permeability and cellular infiltration in experimental allergic encephalomyelitis. Neurology 22:859-869.

Linthicum DS, Frelinger JA (1982) Acute autoimmune encephalomyelitis in mice. II. Susceptibility is controlled by the combination of H-2 and histamine sensitization genes. J Exp Med 155:31-40.

Liu WT, Vanguri P, Shin ML (1983) Studies on demyelination in vistro: The requirement of membrane attack components of the complement system. J Immunol 313:778-782.

Moore GRW, Traugott U, Farooq M, Norton WT, Raine CS (1984) Experimental autoimmune encephalomyelitis: Augmentation of demyelination by different myelin lipids. Lab Invest 51:416-424.

Morari MA, Dalmasso AP (1978) Experimental allergic encephalomyelitis in cobra venom factor-treated and C4-deficient guinea pigs. Ann Neurol 4:427-430.

Morgan BP, Campbell AK, Compston DAS (1984) Terminal component of complement (C_9) in cerebrospinal fluid of patients with multiple scoerosis. Lancet 1:251-264.

Norton WT, Cammer W (1984) Chemical pathology of diseases involving myelin. In: Morell P (ed) Myelin, 2nd edn. Plenum Press, New York, p 369.

Pabst H, Day NK, Gewurz H, Good RA (1971) Prevention of experimental allergic encephalomyelitis with cobra venom factor. Proc Soc Exp Biol Med 136:555-560.

Prineas JW, Kwon EE, Cho E-S, Sharer LR (1984) Continual breakdown and regeneration of myelin in progressive multiple sclerosis plaques. Ann NY Acad Sci 436:11-32.

Raine CS (1983) Multiple sclerosis and chronic relapsing EAE: Comparative ultrastructural neuropathology. In: Hallpike JF, Adams CWM, Tourtellotte WW (eds) Multiple Sclerosis, Pathology, Diagnosis and Management. Chapman and Hall, London, p 413.

Raine CS, Snyder DH, Valsamis MP, Stone SH (1974) Chronic experimental allergic encephalomyelitis in inbred guinea pigs: An ultrastructural study. Lab Invest 31:369-380.

Raine CS, Traugott U, Farooq M, Bornstein MB, Norton WT (1981) Augmentation of immune-mediated demyelination by lipid haptens. Lab Invest 45:174-182.

Saida K, Saida T, Brown MJ, Silberberg DH, Asbury AK (1978) Antiserum mediated demyelination in vivo. A sequential study using intraneural injection of experimental allergic neuritis serum. Lab Invest 39:449-462.

Silverman BA, Carney DF, Johnston CA, Vanguri P, Shin ML (1984) Isolation of membrane attack complex of complement from myelin membranes treated with serum complement. J Neurochem 42:1024-1029.

Smith ME, Amaducci LA (1982) Observations on the effect of protease inhibitors on the suppression of experimental allergic encephalomyelitis. Neurochem Res 7:541-554.

Smith ME, Benjamins JA (1984) Model systems for study of perturbations of myelin metabolism. In: Morell P (ed) Myelin, 2nd edn. Plenum Press, New York, p 441.

Smith ME, Somera FP, Eng LF (1983) Immunocytochemical staining for glial fibrillary acidic protein and the metabolism of cytoskeletal proteins in experimental allergic encephalomyelitis. Brain Res 264:241-253.

Unkeless JC, Gordon S, Reich E (1974) Secretion of plasminogen activator by stimulated macrophages. J Exp Med 139:834-850.

Vanguri P, Koski CL, Silverman B, Shin ML (1982) Complement activation by isolated myelin: Activation of the classical pathway in the absence of myelin-specific antibodies. Proc Natl Acad Sci USA 79:3290-3294.

Waxman FJ, Bergman RK, Munoz JJ (1982) Abrogation of resistance to the reinduction of experimental allergic encephalomyelitis by pertussigen. Cell Immunol 72:375-383.

Wisniewski HM, Bloom BR (1975) Primary demyelination as a nonspecific consequence of a cell-mediated immune reaction. J Exp Med 141:346-359.

Wisniewski HM, Bloom BR (1975) Experimental allergic optic neuritis (EAON) in the rabbit. J Neurol Sci 24:257-263.

TELLURIUM—INDUCED DEMYELINATION

Pierre Morell[†§], Maria Wagner—Recio[†], Arrel Toews[†§], Jean Harry[§], & Thomas W. Bouldin[§‡]
Departments of Biochemistry[†] and Pathology[‡] and Biological Sciences Research Center[§]
University of North Carolina
Chapel Hill, NC, USA 27599–7250

Inclusion of elemental tellurium in the diet of young rats produces a peripheral neuropathy characterized by a highly synchronous demyelination. This is one of the few such non—immunological models available. Other models may result in an asynchronous patch—work of demyelinating and remyelinating segments (*e.g.*, lead intoxication), or are localized to very discrete regions (*e.g.*, demyelination induced by topical application of lysolecithin or diphtheria toxin). Morphological studies of tellurium intoxication suggest that the initial damage is to the myelinating Schwann cell (Lampert and Garrett, 1971; Hammang et al., 1986) and that vulnerability is proportional to the length of the internodal segment supported by the Schwann cell; the more myelin supported by the Schwann cell, the greater the vulner—ability (Bouldin et al., 1988). It is thus uniquely suited for biochemical studies of primary demyelination. Even in the continued presence of tellurium, morphologically observable demyelination diminishes after a few days and, a week after initiation of treatment, the nerve is well launched into a phase of rapid remyelination. We are currently investigating this model with the goal of gaining a better understanding of what metabolic processes are directly affected by tellurium and, assuming there are discrete metabolic alterations, why the resultant pathology is expressed only or preferentially in myelinating Schwann cells.

Characterization of the Model

Male Long—Evans hooded rats are weaned at seventeen days of age. On day twenty, the experimental animals are placed on a diet containing 1.25% elemental tellurium (tellurium powder added to Purina milled rodent chow with sufficient corn oil added to prevent separa—tion of the mixture) for seven consecutive days; control animals are placed on the same diet, but lacking tellurium. As we (Harry et al., 1989) and others (Lampert and Garrett, 1971; Duckett et al., 1979; Said et al., 1981; Takahashi, 1981; Bouldin et al., 1988) have noted, ultrastructural abnormalities develop rapidly (see Fig. 1). By one day, myelinating Schwann

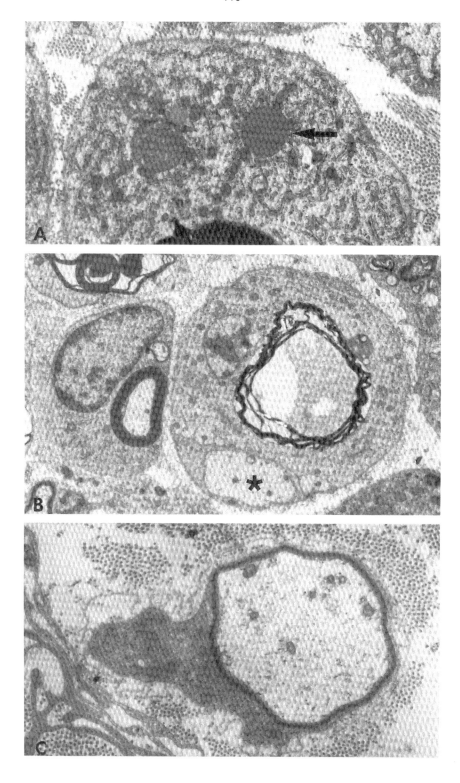

cells contain small lipid droplets, and after several days, many of these cells contain a number of such droplets. Demyelination is observable by day two, and in the next few days a large percentage of fibers are so involved. By a week, remyelination is observed even in the face of continued presence of tellurium in the diet.

Incorporation of Radioactive Precursors into Lipids of Sciatic Nerve

In an attempt to gain a better understanding of the metabolic alterations associated with tellurium neuropathy, the capacity of sciatic nerve to synthesize lipids was examined using *in vitro* incubation techniques (Wiggins and Morell, 1980; Toews et al., 1987). Tellurium treatment caused a rapid and dramatic decrease in acetate incorporation (Fig. 2). Acetate, *via* conversion to acetyl–CoA, is the precursor for both cholesterol and fatty acids, with the latter going on to form the long chain acyl residues of glycerolipids and sphingolipids. The decrease in acetate incorporation presumably reflects, to a large extent, the decrease in synthesis of lipids needed for formation of myelin. This assumption was verified by analysis of incorporation of radioactive acetate into individual lipids. As reported previously (Harry et al., 1989), incorporation of label into each individual class of lipids of sciatic nerve was decreased with a temporal pattern similar to that of inhibition of incorporation into bulk lipids. The extent of inhibition, however, differed for the various lipids. Incorporation of label into lipids enriched in myelin (cerebroside and ethanolamine plasmalogen) was preferentially decreased relative to incorporation of label into phosphatidylcholine, a lipid distributed preferentially into the non–myelin fraction of sciatic nerve. Interestingly, incorporation of labeled acetate into cholesterol was decreased more profoundly than incorporation into any other lipid (discussed in more detail below).

In view of the results obtained with acetate as precursor, we were surprised to note that incorporation of radioactive glycerol was greatly stimulated by tellurium treatment. Closer examination of the temporal sequence of events indicates that glycerol incorporation is increased during the time corresponding to remyelination. Glycerol would be expected to form the backbone of glycerolipids and its incorporation into lipids should be directly related to incorporation of fatty acyl groups. The explanation we have offered (Harry et al., 1989) for

Figure 1: **(A)** A myelinating Schwann cell from sciatic nerve shows cytoplasmic lipid droplets (arrow) 1 day after initiation of a 1.25% tellurium diet; x15,990. **(B)** A demyelinated nerve fiber is adjacent to an intact myelinated fiber in sciatic nerve of a rat fed tellurium for 5 consecutive days. The degenerating myelin sheath within the Schwann–cell cytoplasm no longer envelopes the demyelinated axon (*); x7,380. **(C)** A myelin sheath, inappropriately thin for the diameter of its axon, indicates that this is a remyelinating nerve fiber in sciatic nerve of a rat fed tellurium for 7 consecutive days; x13,940.

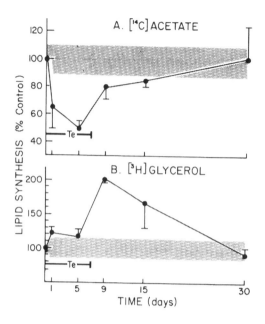

Figure 2: At the times noted, animals were sacrificed and two consecutive 5 mm segments of sciatic nerve distal to the sciatic notch were dissected and incubated in 0.5 ml of Krebs–Ringer buffer (pH 7.4) containing 10 mM glucose and 20 mM HEPES. Radioactive precursor (12.5 μCi of either [^{14}C]acetate or [^{3}H]glycerol) was added and the tubes gassed with 95% O_2/5% CO_2 and incubated for one hour at 37°C. Lipids were extracted by a modification (Benjamins et al., 1976) of the method of Folch et al. (1957) and radioactivity determined.

this anomaly is that, even though the amount of newly synthesized radioactive free fatty acid is low, the total pool of free fatty acids is greatly increased because of the breakdown of pre—existing myelin lipids. The fatty acid pool is, therefore, large and of low specific activity. Much of this free fatty acid pool is eventually reincorporated into newly formed glycerolipids. Because of the low specific radioactivity of the free fatty acid pool, relatively little acetate—derived radioactivity is incorporated into glycerolipids as a result of this synthesis. In contrast, the glycerol–backbone precursor pool (glycerol–3–phosphate and/or dihydroxy—acetone phosphate) is very small and turns over rapidly. This pool is, therefore, rapidly labeled with radioactive glycerol during the *in vitro* incubation and much glycerol–derived radioactivity enters into the newly resynthesized phospholipids. Thus, incorporation of radioactive glycerol is increased in nerves from tellurium–treated animals, relative to controls.

Metabolic Block Induced by Tellurium

In the course of the study of distribution of radioactivity into individual lipid classes, we observed that *in vitro* incubations of sciatic nerves from tellurium–treated animals resulted in marked increases in incorporation of radioactive acetate into a neutral lipid migrating near the solvent front on the thin–layer chromatography plate (Fig. 3) The fraction of total incorporated lipid radioactivity which was present in this neutral lipid was very

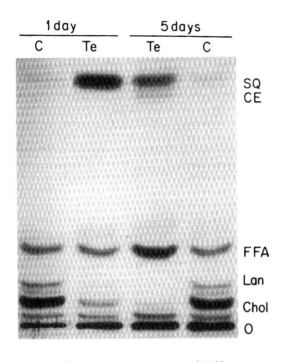

Figure 3: Sciatic nerve sections from rats exposed to a tellurium diet for one or five days, as well as from age-matched controls, were incubated with [^{14}C]acetate for one hour. Lipids were extracted and separated from each other by thin-layer chromatography in hexane/ethyl ether/acetic acid (90:10:1). The autoradiograph indicates the distribution of radioactivity among these lipids. Abbreviations (from bottom to top) are for origin, cholesterol, lanosterol, free fatty acid, cholesterol ester, and squalene, respectively.

high in tellurium-treated animals (23 % of the total radioactivity on the plate, as compared to approximately 1% in control animals). The radioactive substance was collected from the thin-layer chromatography plate and characterized as resistant to alkaline hydrolysis (therefore not containing fatty acids in ester linkage). It was identified as the cholesterol precursor, squalene, on the basis of its co-migration with an authentic squalene standard on two other thin-layer chromatography systems (Fig. 4), and by mass spectrographic analysis (data not shown). As is evident from the autoradiogram shown in Figure 3, the pile-up of radioactive squalene is at the expense of the final product in this metabolic pathway, cholesterol. A quantitative evaluation of this result is presented in Table I. The onset of squalene accumulation is rapid and then its rate of accumulation, as measured by acetate incorporation, declines rapidly (Fig. 5). This time course is compatible with the observation that after a few days remyelination is initiated, even in the face of continued challenge with dietary tellurium. We interpret the above data to suggest a primary metabolic block due to tellurium at the level of conversion of squalene to its product in the cholesterol synthetic pathway, squalene epoxide.

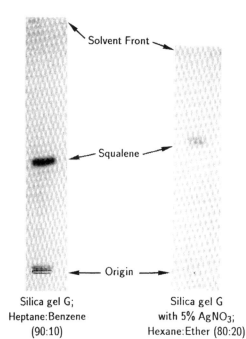

Figure 4: Radioactive lipids were extracted from the "squalene" region of chromatography plates such as those described in Figure 3, and the recovered lipid further characterized as squalene by chromatography in two systems, as indicated in the figure. Radioactivity, as visualized by autoradiography, co-migrated with an authentic squalene standard in each system.

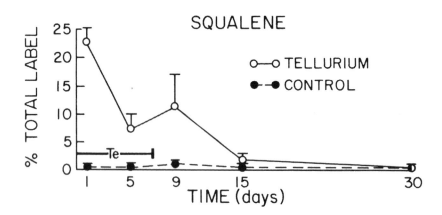

Figure 5: Percent of total lipid radioactivity from [^{14}C]acetate in squalene at various times after beginning a 7–day exposure to a 1.25% tellurium diet. Values were calculated from the distribution of radioactivity on chromatography plates, and are means ± S.E.M. for 4–8 determinations.

Table I: Radioactivity from [^{14}C]acetate in cholesterol and squalene

Organ	Treatment	Cholesterol	Squalene
		(% total lipid label)	
Nerve	Control	19.2 ± 0.2	1.2 ± 0.3
	Tellurium	2.8 ± 0.5	16.8 ± 0.3
Brain	Control	16.7 ± 4.0	3.0 ± 0.7
	Tellurium	4.8 ± 1.21	16.0 ± 2.2
Liver	Control	46, 62	2, 3
	Tellurium	1, 2	75, 69

Animals maintained for one day on tellurium–containing or control diets were sacrificed. The specified organs were removed, cut into 0.4 mm slices on a McIlwain tissue chopper (except for sciatic nerve which was incubated as intact 5 mm segments), and incubated with [^{14}C]acetate as indicated in the legend to Figure 2. Lipids were extracted and separated from each other by thin–layer chromatography. Radioactivity incorporated into cholesterol and into squalene was determined as a percentage of the radioactivity applied to the plate. Data obtained is presented as mean ± S.E.M. (n=3 or 4) except for liver for which two separate experiments were conducted.

Tellurium treatment blocks squalene metabolism in different tissues

The model of tellurium intoxication we utilize does not impart grossly observable pathology to tissues other than peripheral nerve. One possibility for the sparing of other tissues is that tellurium treatment is tissue–specific with respect to the blocking of choles–terol synthesis. The effect of *in vivo* exposure to tellurium on the metabolic block as assayed *in vitro* was determined, and some relevant data are presented in Figure 6. It is evident that conversion of squalene to sterols is blocked in other tissues as well. Our initial hypothesis had been that the metabolic block would be expressed more completely in sciatic nerve than in other tissue. We assumed that the more complete the metabolic block, the more squalene would be piled up relative to cholesterol. This is not the case; the metabolic block seems to be most profound in liver (Table I). Control data from the experiment described in Table I, and from several other experiments, was also analyzed with respect to the radioactivity incorporated into lipid per mg tissue protein. In all control incubations, the rate of incorpo–ration of label into sciatic nerve tissue was 1 to 2 orders of magnitude higher than in other tissues (this is as expected; similar data have been reported from other laboratories –Rawlins and Smith, 1971).

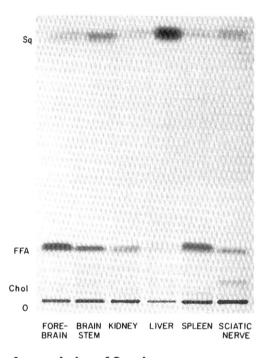

Figure 6: Animals maintained for one day on tellurium—containing or control diets were sacrificed. The specified organs were removed, cut into 0.4 mm slices on a McIlwain tissue chopper (except for sciatic nerve which was incubated as intact 5 mm segments), and incubated with [^{14}C]acetate as indicated in the legend to Figure 2. Lipids were extracted and separated from each other by thin—layer chromatography. The autoradiograph indicates the distribution of radioactivity among these lipids. Abbreviations (from bottom to top) are for origin, cholesterol, free fatty acids, and squalene, respectively.

Accumulation of Squalene

The primary metabolic pathway for utilization of squalene involves its further conversion to squalene epoxide and then, through a metabolic chain of several sterols, to cholesterol. Since radioactivity accumulated in squalene and there was a block in accumulation of radioactivity in sterols, we made the assumption that squalene was actually accumulating in the sciatic nerve of tellurium treated animals. This was tested by direct visualization following ashing of lipids separated on thin—layer chromatography plates. As is evident from Figure 7, there is indeed accumulation of bulk squalene in the various tissues. A crude evaluation of squalene accumulation was conducted by densitometric evaluation of the distribution of the charred lipid on thin layer chromatography plates. At one day, about 1 μg of squalene per μg of lipid phosphorus was present in sciatic nerve of tellurium—treated animals; this is at least an order of magnitude greater than levels of squalene in controls. The actual accumulation of squalene in liver may be some two—fold greater than is the case for sciatic nerve.

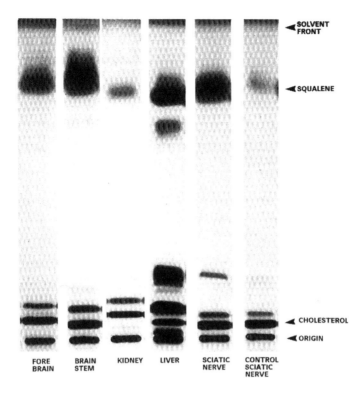

Figure 7: Twenty day–old rats were placed on a tellurium–containing diet for one day; animals were then sacrificed and lipids were extracted from various dissected tissues and separated by thin–layer chromatography. Lipids were visualized by spraying the plates with an acidic solution and then charring at 180°C. Each lane contains 50 μg of lipid phosphorus.

Subcellular distribution of squalene

We considered the possibility that tellurium exposure preferentially causes demyelination of the peripheral nervous system because squalene preferentially accumulates in peripheral myelin, thereby causing destabilization of this structure. A series of experiments (data not shown) indicated that for both sciatic nerve and brain samples, the squalene formed subsequent to tellurium feeding was distributed between myelin, microsomal and mitochondrial fractions roughly in proportion to the amount of material in each fraction. This was so whether distribution of the most recently synthesized (radioactive) squalene was determined or whether mass distribution was assayed by quantitation of charred thin–layer chromatography plates. The nonspecific distribution might reflect that achieved *in vivo*. We consider, however, that it is more likely that the distribution of squalene reflects an artifact of the preparation of tissue for subcellular fractionation; exogenous squalene added either prior to or following homogenization of control tissue was distributed in the same manner as squalene accumulating because of tellurium treatment. Neither interpretation offers support for the hypothesis that squalene is preferentially associated with myelin of the peripheral nervous system. The interpretation from the morphological data (Fig. 1), that squalene accumulates in droplets in cell cytoplasm, still seems reasonable.

The active species of tellurium

Elemental tellurium is almost completely insoluble in aqueous solvents, raising the question as to the active species which inhibits squalene metabolism. Tellurium has valence states of -2, 0, $+2$, $+4$, and $+6$. We assume that inhibition of metabolic activity involves an oxidized valence state ($+4$ or $+6$) soluble in aqueous solutions at reasonable concentrations. As indicated in Figure 8 and Table II, it is the $+4$ state, tellurite, that is active in inhibiting conversion of squalene to squalene epoxide. The concentration of tellurium needed to maintain inhibition of enzyme activity over the one hour incubation, 10 mM, is much higher than expected for a metabolic poison effective *in vivo*. This is because the *in vitro* assay requires the presence of NADPH to maintain the reducing potential required for assay of squalene epoxidase, a monooxygenase. In the crude assay mixture utilized (a post–mitochondrial supernatant prepared from rat liver), this reducing potential is also linked to reduction of Te^{+4} to some insoluble, more reduced state (Te^{+2} or Te^{0}). This can be observed

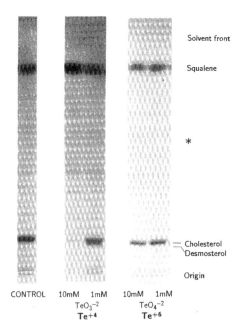

Figure 8: Inhibition of conversion of [^{14}C]squalene to sterols by various tellurium species in an *in vitro* assay system utilizing a post–mitochondrial supernatant obtained from rat liver. See legend to Table II for details. The asterisk (*) in the figure indicates the position of a squalene 2,3–epoxide standard. Lack of accumulation of label in this compound indicates that the metabolic block is at the squalene epoxidase step, and not at the subsequent squalene epoxide cyclase step.

Table II: Inhibition of sterol synthesis *in vitro* by various tellurium compounds

INHIBITOR	CONCENTRATION (mM)	STEROLS (% control)
K_2TeO_3 (Te^{+4})	10	11
	1	87
Na_2TeO_3 (Te^{+4})	10	10
	1	99
Na_2TeO_4 (Te^{+6})	10	101
	1	103
$H_2TeO_4\ 2H_2O$ (Te^{+6})	10	84
	1	101

Animals (21 days of age) were sacrificed and livers were removed, homogenized, and a 10,000 x g post—mitochondrial supernatant prepared. The incubation mixture was prepared by emulsifying 10,000 cpm of $[^{14}C]$squalene in Tween—80 so that, upon dilution to the final incubation volume of 0.5 ml in 20 mM phosphate buffer at pH 7.4, the detergent concentration was 0.005%. The incubation mixture also contained 1 mM EDTA, 1mM NADPH, 0.01 mM FAD, and 10 mg of protein. The incubation was carried out for 60 minutes under 95% O_2/5% CO_2. The reaction was terminated by alkaline methanolysis, and lipids were extracted with petroleum ether and separated from each of other by thin layer chromatography in benzene—acetone (90:1). The distribution of radioactivity on the chromatography plate was visualized by autoradiography (see Figure 8), and radioactivity in products subsequent to squalene in the cholesterol synthetic pathway (almost all was in cholesterol) was determined. Data presented are the mean of two experiments differing from each other by less than 10%.

---

visually during the *in vitro* assay; a post—mitochondrial supernatant will, in the presence of NADPH, slowly convert tellurite to a dark precipitate. Thus, the toxic species is being constantly removed from the assay mixture during the *in vitro* incubation. Experiments involving reconstitution of the assay system indicate that the factor linking NADPH to tellurite reduction is a cytoplasmic enzyme. We are studying ways to purify the assay system sufficiently to eliminate this interfering enzyme activity and allow for analysis of the action of tellurite on squalene metabolism.

This *in vitro* experiment, and a series of similar experiments involving a time course, also strongly suggested that the inhibition of squalene metabolism is a direct result of tellurite action on the enzymatic machinery involved, rather than secondary to repression of synthesis of the necessary proteins. We assume that the direct target of the interaction with tellurite is either the enzyme squalene epoxidase or the squalene carrier protein (because of the extreme insolubility of squalene, a specific carrier protein is required to present it to the enzyme).

Discussion

The data presented are compatible with the following proposed sequence of events. Upon ingestion of elemental tellurium, a small fraction is oxidized (by unknown mechanisms) in the gastrointestinal tract and enters the circulation as tellurite. This then distributes into various tissues, including those cells myelinating peripheral nerves. There is an immediate block in conversion of squalene to squalene epoxide. The resultant accumulation of squalene and/or lack of cholesterol, sets in motion a process which feedback inhibits *de novo* synthesis of all lipids. The lack of lipid substrates for synthesis not only blocks accumulation of myelin, but also results in extensive destruction of preformed myelin. After a few days, the capacity of ingested tellurium to bring about a block in the metabolism of squalene is decreased and the pathological series of events starts to reverse. Myelin synthesis is re—initiated. An interesting metabolic feature is that the fatty acids from degraded myelin lipids are rapidly re—esterified so that synthesis of glycerolipids is actually higher in recovering animals than in controls.

Tellurium—induced demyelination is a relatively reproducible model, studies of which have much potential for yielding insights into metabolism of myelinating Schwann cells. Among the obvious questions raised by our study concerning tellurium—induced demyelina—tion are: (i) since the metabolic block is induced in all tissues, why is the most prominent site of pathology the myelinating Schwann cell? (ii) why is there a defined developmental period during which rats are susceptible to tellurium—induced neuropathy? (iii) connected with the previous question is the matter of why there is remyelination even in the face of continued challenge with dietary tellurium and (iv) what is the relationship between the block in synthesis induced by tellurium and the destruction of pre—existing myelin?

Relevant to these questions may be the normally very high rate of lipid synthesis in sciatic nerve, relative to other tissues, in developing animals. We assume that this reflects the relative rates of membrane synthesis, *i.e.*, the Schwann cells are depositing vast amounts of myelin. It may be the high demand for cholesterol, or more specifically the extent to which this demand is not met because of the toxic effects of tellurium, which is responsible for the preferential expression of the pathological process in sciatic nerve. If cholesterol is not added to myelin at the same high rate as are the other lipids, the membrane may become destabilized and fall apart. The developmental time window of sensitivity to tellurium intoxi—cation would then correspond to the period of rapid lipid synthesis characteristic of sciatic nerve during development (Smith, 1971). The model of administration of tellurium in solid food precludes probing the lower end of the developmental window, since it is restricted by the age at weaning. A recent study (Jackson et al., 1989) involving administration of tellurium through the milk of lactating dams suggests, however, that the period of peripheral nerve vulnerability to tellurium intoxication also covers the pre—weaning period of rapid myelin deposition. Quantitative studies correlating the rate of metabolism with sensitivity to tellurium poisoning are currently in progress.

Acknowledgments

These studies were supported in part by USPHS grants ES—01104 and HD—03110. We thank Nelson D. Goines for assistance with the photomicrography.

References

Benjamins JA, Miller SL, and Morell P (1976) Metabolic relationships between myelin subfractions: Entry of galactolipids and phospholipids. *J Neurochem* **27**:565–570.
Bouldin TW, Samsa G, Earnhardt T, and Krigman MR (1988) Schwann—cell vulnerability to demyelination is associated with internodal length in tellurium neuropathy. *J Neuro—pathol Exp Neurol* **47**:41–47.
Duckett S, Said G, Streletz LG, White RG, and Galle P (1979) Tellurium—induced neuropathy: Correlative physiological, morphological and electron microprobe studies. *Neuropathol Appl Neurobiol* **5**:265–278.
Folch J, Lees M, and Sloane—Stanley GH (1957) A simple method for the isolation and purification of total lipids from animal tissues. *J Biol Chem* **226**:479–509.
Hammang JP, Duncan ID, and Gilmore SA, (1986) Degenerative changes in rat intraspinal Schwann cells following tellurium intoxication. *Neuropathol Appl Neurobiol* **12**:359–370.
Harry GJ, Goodrum JF, Bouldin TW, Wagner—Recio MW, Toews AD, and Morell P (1988) Tellurium—induced neuropathy: Metabolic alterations associated with demyelination and remyelination in rat sciatic nerve. *J Neurochem* **52**:938–945.
Jackson KF, Hammang JP, Worth SF, Duncan ID (1989) Hypomyelination in the neonatal rat central and peripheral nervous systems following tellurium intoxication. *Acta Neuropathol* (Berl) **78**:301–309.
Lampert PW and Garrett RS (1971) Mechanism of demyelination in tellurium neuropathy. Electron microscopic observations. *Lab Invest* **25**:380–388.
Rawlins FA and Smith ME (1971) Myelin synthesis *in vitro*: A comparative study of central and peripheral nervous tissue. *J Neurochem* **18**:1861–1870.
Said G, Duckett S, and Sauron B (1981) Proliferation of Schwann cells in tellurium—induced demyelination in young rats. A radioautographic and teased nerve fiber study. *Acta Neuropathol* (Berl) **53**:173–179.

Takahashi T (1981) Experimental study on segmental demyelination in tellurium neuropathy. *Hokkaido J Med Sci* **56**:105–131.

Toews AD, Fischer HR, Goodrum JF, and Morell P (1987) Metabolism of phosphate and sulfate groups modifying the P_0 protein of PNS myelin. *J Neurochem* **48**:883–887.

Wiggins RC and Morell P (1980) Phosphorylation and fucosylation of myelin proteins *in vitro* by sciatic nerve from developing rats. *J Neurochem* **34**:627–634.

OLIGODENDROCYTE REACTION TO AXONAL DAMAGE

S.K. Ludwin
Department of Pathology
Queen's University
Kingston, Ontario
K7L 3N6

INTRODUCTION

Most experimental and clinical studies of remyelination have in the past concentrated on the mechanism of formation of the myelin sheath and the behaviour of the oligodendrocyte (Ludwin 1987). It has become clear, that this process is composed of a variety of important steps, the understanding of each of which is essential in any future attempts to influence the possible extent of remyelination both experimentally and hopefully clinically. Attempts to enhance this remyelination potential are of importance not only in demyelinating diseases affecting mainly myelin and oligodendrocytes, but will increasingly assume importance in destructive diseases of the brain. With greater success in stimulating regeneration of axons and neurons in these situations, there will be a need for remyelination of these newly formed nerve processes. The importance of the oligodendrocyte has been well documented (Ludwin 1988, 1989), and it has been well emphasised that successful remyelination depends on an adequate number of oligodendrocytes, which may be provided by maturation of immature cells, possibly by proliferation of mature oligodendrocytes (Ludwin 1985, 1988), and more recently through the provision of exogenous oligodendrocytes to demyelinated areas (Blakemore & Crang, 1988). Those disease processes in which oligodendrocyte numbers remain high tend to have more successful remyelination, whereas diseases in which continual exposure to the demyelinating agent decreases the number of oligodendrocytes tend to have less successful remyelination (Ludwin 1987, Raine et al 1988). In recent years there has been, in addition, a great amount of interest in two other processes involving the oligodendrocyte, namely cell migration, and oligodendrocyte axon adhesion. Using the Shiverer mouse, it has been shown that the capacity for oligodendrocytes to migrate through tissue to reach unmyelinated and demyelinated tissue (Gout et al, 1988; Gumpel, 1989) is much greater than had been previously believed. However, even with the presence of adequate numbers of oligodendrocytes, and the proximity through migration of these cells to axons, it is still not clear what in the relationship between axons and oligodendrocytes triggers the process of remyelination. Axons have been shown to be mitogenic to oligodendrocytes (Wood & Bunge, 1986), and certainly in tissue culture oligodendrocytes preferentially adhere to axonal beds (Ludwin & Szuchet, in Preparation). Previous in vivo studies on the relationship between axons and oligodendrocytes, which have been mainly carried out in young animals, have tended to suggest that oligodendrocytes either do not differentiate, or else die when deprived of axonal stimuli (David et al, 1984; Valat et al 1988). This has profound implications for oligodendrocyte function and capacity for remyelination both in demyelinating diseases, and in diseases such as trauma or stroke which will depend on axonal regeneration. The purpose of this study was to examine, using cell markers, the potential for oligodendrocytes to remain in an axon-deprived environment, as well as to investigate the process of phagocytosis in the optic nerve in Wallerian degeneration, a subject that has been controversial for a long period of time.

MATERIALS AND METHODS

Mature male rats 3-4 months old were anesthetised and their left eyes enucleated. The animals were then sacrificed and their optic nerves removed for study. Survival periods post enucleation (p.e.) ranged from a few days to 22 months. In each animal the unoperated right eye served as a control. Animals were prepared in different ways depending on the type of investigation to be performed. For routine electron microscopy animals were perfused with Karnovsky's solution whereas for ultrastructural immunocytochemical studies, others were perfused with 4% paraformaldehyde. Optic nerves for light microscopic immunocytochemistry were immersion fixed in mercuric chloride fixative.

Oligodendrocytes were identified by immunochemistry methods at a light microscopic level using anti-sera to carbonic anhydrase (CA) (Ghandour & Skoff, 1988) and at an ultrastructural level using antisera to myelin-oligodendrocyte glycoprotein (MOG) as described by Linington et al (1988). Microglia and/or macrophages were identified at a light microscopic level by lectin histochemistry binding to Griffonia simplicifolia (Streit &

Kreutzberg, 1987), or by immunochemical staining with EDI monoclonal antibody (Sminia et al, 1987; Dijkstra et al, 1985).

For autoradiographic assessment of cell division animals were injected at various periods with tritiated thymidine, and after being perfused with Karnovsky's solution, the optic nerves were examined by ultrastructural autoradiography to detect labelled proliferating cells (Ludwin, 1985). For these experiments, time periods covering the first 3 weeks post enucleation were utilised.

RESULTS

Examination of the optic nerves following enucleation revealed a pattern consistent with that described many times in the literature. The earliest signs of degeneration were evident late within the first week, and were seen as subtle changes in the axons, with the presence of axonal densities and vacuoles. With further degeneration, loosening of the myelin lamellae, and intramyelin edema, accompanied progressive darkening of the axon, the presence of numerous dense bodies, and accumulation of filaments, mitochondria and phagocytic vacuoles in the axons. Phagocytosis was evident within 2 to 3 weeks post-enucleation. Most of the phagocytic activity appeared to be concentrated within the cell processes, which were well seen both surrounding and within myelin sheaths engulfing axons. This phagocytic activity within the processes was marked by tremendous dilatation of endoplasmic reticulin. It was not often clear at these early stages which cells were responsible for phagocytosis, because in the vast majority of instances, only processes and not the perikarya of these phagocytic cells were seen. From their general appearance they resembled most closely those of microglial processes. Phagocytic microglia were clearly identified, and in many cases cells were noted which had the appearance of oligodendrocytes. The process of removal of myelin and degeneration was extremely slow, and even by 2 to 3 months, large amounts of myelin was still seen, albeit in a degenerated and altered state. By this stage however most of the axons appeared to have completely degenerated. Concomitant with the decrease in myelin, there was a steady increase in astrocytic processes which insinuated themselves between the profiles of degenerating myelin and axons. There did not appear to be a major increase in the number of astrocyte perikarya as seen ultrastructurally. By 6 months post-enucleation, most of the degenerated axons and myelin had been removed, although large numbers of myelin profiles were still present mainly within phagocytic profiles of cell processes. The optic nerve consisted of a dense network of interlacing bundles of astrocytic processes lying within which were numerous cell perikarya (Fig 1).

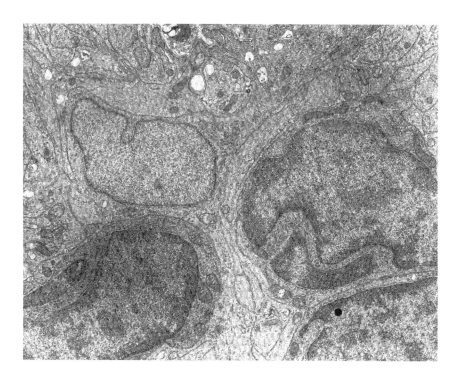

Fig. 1. Optic nerve 6 months p.e. Four cells are present in a mat of astrocytic fibers. The upper left cell is an astrocyte. The other cells, although atypical have some features of oligodendrocytes. (See text) X 10,200

The residual cells were either astrocytes, microglia, or cells resembling modified oligodendrocytes. These latter cells had many of the features of oligodendrocytes such as darker cytoplasm, prominent Golgi apparatus and a well defined microtubular content. They were however altered in that the external cytoplasmic membrane was smooth and rounded, cell processes appeared to be very few, and the nuclei appeared notched with clumped chromatin. Cytoplasmic volume also was markedly reduced. During the actual process of degeneration, it was often difficult to detect normal oligodendrocytes, because of the distortion of the nerve during the removal of myelin, and the apparent ingestion of myelin by many of these cells. Oligodendrocyte cell death could not be identified convincingly at any stage during the process of Wallerian degeneration. With further time periods, up to almost 2 years post-enucleation, the morphology of the nerve did not change significantly except that myelin profiles became more sparse, and the astrocytic network of processes became more established. The morphology of the residual cells remained similar.

With immunocytochemical methods, the identity of many of these cells could be definitively established. Staining with antiserum to GFA revealed, as expected, a dense background fibrillary network of astrocytic processes. Of interest was the fact that the number of cell perikarya stained with GFA did not appear to be significantly different from that seen in the early time periods or in normal astrocytic nerves. Using antisera to carbonic anhydrase (CA), numerous cells were identified as oligodendrocytes in the nerves remaining after enucleation (Fig. 2).

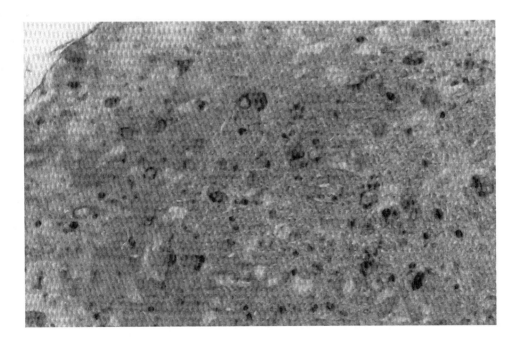

Fig. 2. Optic nerve 13 months p.e. stained with antibodies to carbonic anhydrase. Numerous cells are stained. No counter stain. X 200

Although this was carried out at light microscopic level, the morphology and distribution of these cells was consistent with those of putative oligodendrocytes seen on ultrastructural examination. Further evidence that these cells were indeed modified oligodendrocytes was seen using ultrastructural immunocytochemistry with antisera to myelin oligodendrocyte glycoprotein (MOG). This demonstrated staining of the external cytoplasmic membrane. Astrocytes and their processes were unstained (Fig. 3). Throughout the optic nerve in between the non-staining astrocytic processes numerous short profiles of membranes, tentatively identified as oligodendrocyte processes, were stained.

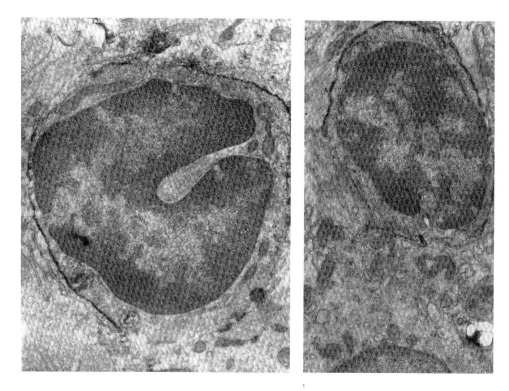

Fig. 3 a) Optic nerve 13 months p.e. stained with antibodies to MOG. A cell similar to those in Fig. 1 shows cell membrane staining, whereas the surrounding astrocytic processes are unstained. X 15,300

b) Similar preparation to that in a. The astrocyte on the bottom is unstained. X 11,550

EDI antibodies and GSA lectin both avidly stained a population of cells, which differed from that seen with CA and MOG. Many of these cells appeared to be large cells resembling foamy macrophages, whereas in other areas it was mainly debris, presumably in phagolysosomes that were staining. Although the peak staining for these 2 antisera was approximately 4 to 8 weeks, positively stained macrophages could be identified at all states of the experiment including the most longstanding cases at 20 to 22 months (Fig. 4).

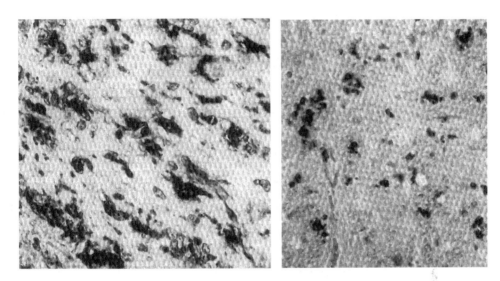

Fig. 4 a) Optic nerve 4 weeks p.e. stained with lectin Griffonia simplicifolia. Numerous microglia/macrophages are present. Hematoxylin counterstain. X 200
b) Optic nerve 8 weeks p.e. stained with EDI antibody, no counterstain. A large number of stained cellular profiles are still present. X 200

In addition throughout the period of observation, on ultrastructural examination, phagocytosed debris was seen in cells and cell processes resembling oligodendrocytes. This often appeared to represent retraction of axonal and myelin debris within the terminal processes of oligodendrocytes. Scattered oligodendrocyte perikarya in addition also showed the presence of cellular debris. Using anti-MOG antibody many of these processes were definitively established as oligodendrocytes, which stained the outer membranes clearly (Fig. 5).

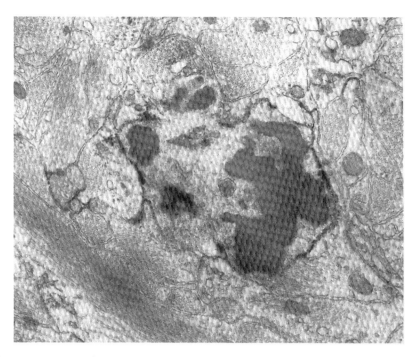

Fig. 5 Optic nerve 13 months p.e. A cell process filled with debris is stained with antibodies to MOG. X 20,200

In addition many more classical phagocytes with lipid laden debris and which resembled macrophages did not stain positively with MOG. On a quantitative level it was clear that most of the phagocytosis was carried out by macrophages/microglia, but that the uptake of debris, and/or the rectraction of debris into processes, was a ubiquitous finding in oligodendrocytes.

Examination of the optic nerves labelled for ultrastructural autoradiography revealed most of the early proliferating cells to be microglia/macrophages. These proliferating cells were seen from approximately 5 days on and continued for the next 2 to 3 weeks following which their rate of proliferation was not significantly increased above background. Somewhat surprisingly, only small numbers of the proliferating cells appeared to be astrocytes, and occasional cells clearly identified as oligodendrocytes were noted to be labelled. In many instances the identification of these cells was not entirely clear, but comparing these cells to similar ones stained for MOG helped confirm the oligodendroglial nature of some of these proliferating cells. Many of the labelled oligodendrocytes appeared to be located at the periphery of the nerve close of the pial surface.

DISCUSSION

These experiments demonstrated that following Wallerian degeneration in the adult animal, oligodendrocytes remained in significant numbers for long periods of time even up to 20 months post-enucleation. This was amply demonstrated using 2 markers for oligodendrocytes which have proved to be useful in tissue sections. Carbonic anhydrase 2 (Ghandour & Skoff 1988) is a valuable way of demonstrating oligodendrocytes at a light microscopic level, even though there have been some reports that under certain circumstances it may stain astrocytes. At an electron microscopic level, oligodendrocytes were well demonstrated with surface staining using antibodies against MOG as described by Linington. The ultrastructural staining of cells resembling, but not typical of oligodendrocytes, is interesting. Many of the cells that were stained had some features that microscopically were suggestive of oligodendrocytes, such as a dark cytoplasm, an absence of extensive Golgi apparatus, and a characteristic nuclear type. These cells however differed from normal oligodendrocytes in that they appeared to be very rounded, and had a smooth cytoplasmic membrane, with relatively scanty cytoplasm, and there did not appear to be many processes extending from the cells. In addition the nuclei were notched and convoluted. It is difficult to be sure about the significance of these changes, but it does suggest that the oligodendrocyte may be present in an altered relatively dormant state. There is an analogy with the Schwann cell following Wallerian degeneration (Politis, 1982) where it has been shown that this cell fails to express many of the myelination proteins until it is re-exposed to a growing regenerating axon. Under these circumstances the cell then undergoes mitosis and starts to become functional. It is tempting to think that the same situation may occur in the oligodendrocyte, but this will still have to be tested at a functional level, as the concept is important for C.N.S. regeneration. This finding is also of some importance, because it is different to the experience of others who have dealt with Wallerian degeneration in developing animals (Valat et al, 1988; Miller et al, 1989; David et al, 1984). These authors have generally shown that in the developing animal, deprivation of axonal stimuli by section of axons prevents full differentiation of oligodendrocytes, and further leads to gradual death of those young cells which have already differentiated into oligodendrocytes. Survival of oligodendrocytes was noted by Vaughn and Pease, (1970), who demonstrated cells similar to those in our experiments after long term degeneration in mature animals. Their observations were based on electron microscopic findings, and the present studies have the advantage of confirming the oligodendrocytic nature (and indeed a certain degree of residual production of a myelin associated protein) by immunochemical methods. Although it was difficult to be certain, there did appear to be slightly fewer oligodendrocytes in long term degenerated nerves compared with normal, but no evidence was found suggesting cell death during the process or migration of oligodendrocytes to blood vessels and into the blood stream.

The present experiments also have helped in an understanding of the role of oligodendrocytes and macrophages in phagocytosis during Wallerian degeneration. This has been matter of controversy for many years. Cook and Wisniewski, (1975) suggested a major role in phagocytosis for the oligodendrocytes, with subsequent loss of oligodendrocytes due to movement of the lipid laden cells towards the blood stream. No evidence was found for oligodendrocyte phagocytosis by Fulcrand and Privat (1977), and Skoff (1975) thought that endogenous microglia were active in Wallerian degeneration, with no contribution from hematogenous cells. The distinction between microglia and circulating macrophages is a matter of some controversy at the present time, and the readers are referred to other sources for a fuller discussion of this (Hickey & Kumura 1988; Perry and Gordon, 1988). In the present study, 2 different ways of detecting microglia/macrophages were used. The lectin Griffonia simplicifolia has been thought by some authors to stain brain microglia (Streit & Kreutzberg, 1987), and by other authors it has been felt to be a marker for circulating macrophages. In our experiments, lectin stained cells resembling microglia were seen in the normal adult nerve, suggesting that some endogenous cells were stained. These had the typical appearance of microglia. In addition, however, once degeneration began, the nerve was filled with lectin stained cells, suggesting that the microglia had either proliferated extensively, or that other cells also staining with the lectin (hematogenous macrophages) had passed into the nerve. Using EDI which stains macrophages (Sminia et al, 1978; Dijkstra et al 1985), no positively stained cells were found in the normal nerve, but during Wallerian degeneration extensive staining in the same distribution as GSA was noted. These experiments would tend to suggest that both endogenous and hematogenous cells are responsible for phagocytosis. The findings in this particular part of the study are in agreement with those of Stoll et al (1989), who also detected a similar staining pattern during Wallerian degeneration in the nerve. In addition, however, the present experiments demonstrated a small but definite involvement of the oligodendrocyte in removing myelin debris. Both cells and processes containing to degenerating myelin profiles were stained with the anti-MOG antibody, confirming their oligodendrocytic nature. This helps explain the longstanding controversy in the literature. Although our results are in agreement with those of Stoll et al (1989), we were further able to detect the oligodendrocytic component, because the present experiments utilised oligodendrocyte markers, whereas those of Stoll et al concentrated only on macrophage markers. The slow nature of the process, and the difficulty at times in distinguishing microglia from oligodendrocytes, especially when the latter contain debris, may help to explain the previous difficulties in the literature in trying to sort out this problem at an ultrastructural level.

Finally, the results of the experiments on cell proliferation are generally in accord with those in the literature. The early significant cell proliferation was seen in macrophages and microglia, and a small percentage of the labelled cells were identified as being either oligodendrocytes or astrocytes. Previous studies on proliferation during Wallerian degeneration have been done in young animals (Privat et al, 1981) although Skoff's (1975) study demonstrated proliferation in mature animals; the studies of Vaughn et al (1970) and Vaughn and Pease (1970) have shown that the macroglia populations remain relatively stable while the macrophages (either microglia or what they term multipotential glia) are the cells that account for most of the proliferation. Quite surprisingly although some astrocytes did take up tritiated thymidine, in spite of the tremendous reactive gliosis, there did not appear to be a great increase in the cell number. This is in accord with the findings of Skoff, and of Vaughn and Pease (1970), and Vaughn et al (1970). Occasional oligodendrocytes, early in the course of Wallerian degeneration did demonstrate labelled nuclei, suggesting a response to axotomy. Again there is an analogy with the Schwann cell, where severance of axons in vivo and in vitro (Salzer & Bunge 1980; Politis et al, 1982) causes mitosis of Schwann cells.

In summary then it is clear that oligodendrocytes from mature animals are able to survive the loss of their axons for long periods of time, and respond to Wallerian degeneration with a small degree of cell proliferation. Limited phagocytosis is possible although the major phagocytosis is carried out by the microglia/macrophage. Many of these findings are analogous to the behaviour of Schwann cells, although the degree and tempo of these reactions are probably of a lower order. This does however, raise the possibility that oligodendrocytes, under the right circumstances, may display many of the regenerative capabilities of Schwann cells following destruction of the axon.

ACKNOWLEDGMENTS

The authors would like to thank Ms. Mirta Chiong and Mr. Jim Gore for technical assitance and Mrs. Patricia Scilley for secretarial assistance.

This work was supported by a grant from the Medical Research Council of Canada, MA 5818.

REFERENCES

Blakemore WF, Crang AJ (1988) Extensive oligodendrocyte remyelination following injection of cultured central nervous system cells into demyelinating lesions in adult central nervous system. Dev Neurosci 10: 1-11

Cook RD, Wisniewski HM (1973) The role of oligodendroglia and astroglia in Wallerian degeneration of the optic nerve. Brain Res 61: 191-206

David S, Miller RH, Patel R, Raff MC (1984) Effects of neonatal transection on glial cell development in the rat optic nerve: evidence that the oligodendrocyte - type 2 astrocyte cell lineage depends on axons for its survival. J Neurocytol 13: 961-974

Dijkstra CD, Doepp EA, Joling P, Kraal G (1985) The heterogeneity of mononuclear phayocytes in lymphoid organs: distinct macrophage subpopulations in the rat recognized by monoclonal antibodies ED1, ED2, and ED3. Immunology 54: 589-599

Fulcrand J, Privat A (1977) Neurological reaction to Wallerian degeneration in the optic nerve of the postnatal rat: ultrastructural and quantitative study. J Comp Neurol 176: 189-224

Ghandour MS, Skoff RP (1988) Expression of galactocerebroside in developing normal and jimpy oligodendrocytes in situ. J Neurocytol 17: 485-498

Gout O, Gansmuller A, Baumann N, Gumpel M (1988) Remyelination by transplanted oligodendrocytes of a demyelinated lesion in the spinal cord of the adult Shiverer mouse. Neurosci Lett 87: 195-199

Gumpel M, Gout O, Lubetzki C, Gansmuller A, Baumann N (1989) Myelination and remyelination in the central nervous system by transplanted oligodendrocytes using the Shiverer model. Discussion on the remyelinating cell population in adult mammals. Dev Neurosci 11: 132-139

Hickey WF, Kumura H (1988) Perivascular microglial cells of the CNS are bone marrow-derived and present antigen in vivo. Science 239: 290-292

Linington C, Bradl M, Lassmann H, Brunner C, Vass K (1988) Augmentation of demyelination in rat acute allergic encephalomyelitis by circulating mouse oligodendrocyte glycoprotein. Am J Pathol 130: 443-454

Ludwin SK (1985) The reaction of oligodendrocytes and astrocytes to trauma and implantation: A combined autoradiographic and immunohistochemical study. Lab Invest 52: 20-30

Ludwin SK, Bakker DA (1988) Can oligodendrocytes attached to myelin proliferate? J Neurosci 8: 1239-1244

Ludwin SK (1988) Remyelination in the central nervous system and the peripheral nervous system; in Waxman, Functional recovery in neurological disease pp215-254 (Raven Press, New York)

Ludwin SK (1987) Remyelination in demyelinating diseases of the central nervous system; in CRC Critical Reviews in Neurobiology pp1-20 (CRC Press Inc)

Ludwin SK (1989) Evolving concepts and issues in remyelination. Dev Neurosci 11: 140-148

Miller RH, ffrench-Constant C, Raff MC (1989) The macroglia cells of the optic nerve. Ann Rev Neurosci 12: 517-534

Perry VH, Gordon S (1988) Macrophages and microglia in the nervous system. TINS 11: 273-277

Politis MJ, Sternberger N, Ederle K, Spencer PS (1982) Studies on the control of myelinogenesis. IV Neuronal induction of Schwann cell myelin-specific protein synthesis during nerve fiber regeneration. J Neurosci 2: 1252-1266

Privat A, Valat J, Fulcrand J (1981) Proliferation of neuroglial cell lines in the degenerating optic nerve of young rats. A radiographic study. J Neuropath Exp Neurol 40: 46-60

Raine C, Moore GR, Hintzen R, Traugott N (1988) Induction of oligodendrocyte proliferation and remyelination after chronic demyelination. Relevance to multiple sclerosis. Lab Invest 59: 467-476

Salzer JL, Bunge RP (1980) Studies of Schwann cell proliferation. I An analysis in tissue culture of proliferation during development, Wallerian degeneration, and direct injury. J Cell Biol 84: 739-752

Skoff RP (1975) The fine structure of pulse labeled (^{3}H-Thymidine) cells in degenerating rat optic nerve. J Comp Neur 161: 595-612

Sminia T, de Groot CJA, Dijkstra CD, Koetsier JC, Polman CH (1987) Macrophages in the central nervous system of the rat. Immunobiol 174: 43-50

Streit WJ, Kreutzberg GW (1987) Lectin binding by resting and reactive microglia. J Neurocytol 16: 249-260

Stoll G, Trapp BD, Griffin JW (1989) Macrophage function during Wallerian degeneration of rat optic nerve: clearance of degenerating myelin and IA expression. J Neurosci. 19: 2327-2335

Valat J, Privat A, Fulcrand J (1988) Experimental modifications of post-natal differentiation and fate of glial cells related to axo-glial relationships. Int J Devel Neurosci 6: 245-260

Vaughn JE, Pease D (1970) Electron microscopic studies of Wallerian degeneration in optic nerves II Astrocyte, oligodendrocytes and adventitial cells. J Comp Neur 140: 207-226

Vaughn JE, Hinds PL, Skoff RP (1970) Electron microscopic studies of Wallerian degeneration in rat optic nerves. I The multipotental glia. J Comp Neur. 140: 175-206

Wood PM, Bunge RP (1986) Evidence that axons are mitogenic for oligo-dendrocytes isolated from adult animals. Nature 320: 756-758

GLIAL CELL DIFFERENTIATION IN REGENERATION AND MYELINATION

M. Schwartz*, V. Lavie, A. Cohen, M. Murray, A. Solomon and M. Belkin

Department of Neurobiology
The Weizmann Institute of Science
Rehovot, Israel

Spontaneous growth of axons after injury is extremely limited in the mammalian central nervous system (CNS). It is now clear, however, that injured CNS axons can be induced to elongate when provided with a suitable environment.

The poor regenerative ability has been attributed, at least in part, to the presence of mature oligodendrocytes, which are non-permissive for axonal growth, and to the scar forming astrocytes, which are non-supportive for axonal growth.

A spontaneously regenerating system such as fish optic nerves, can be used as a source of soluble substances that cause alterations in the post-traumatic response to axonal injury, when applied to a non-regenerative injured nerves. We have induced apparent regenerative growth of injured optic nerves in adult rabbits by supplying them with soluble substances originating from regenerating fish optic nerves. This treatment was coupled with a daily irradiation with He-Ne laser, which appears to delay degenerative changes in the injured axons. The combination of treatments modalities resulted in appearance of unmyelinated and thinly myelinated axons, some of them presumably in an active process of remyelination extending to the lesion site and distal to it. Morphological and immunocytochemical evidence indicates that these thinly myelinated and unmyelinated axons are growing in close association with central nervous system glial cells. These newly growing axons traverse the site of injury and extend into the distal stump of the nerve, which contains degenerating axons. Axons of this type could be detected distal to the lesion only in nerves subjected to the combined treatment. Unmyelinated or thinly myelinated axons, at the site of injury and in association with glial cells, were not seen distal to it at 6 or 8 weeks post-operatively in nerves which were not treated or in nerves in which the two stumps were completely separated (disconnected). A temporal analysis of treated nerves indicates that axons have grown as far as 6 mm distal to the

* To Whom Correspondence Should be Addressed

Footnote: M. Murray, Department of Anatomy, The Medical College of Pennsylvania, Philadelphia, PA, USA; M. Belkin and A. Solomon, Tel Aviv University, Sackler School of Medicine, Tel-Hashomer, Israel

site of injury by 8 weeks post-injury.

Anterograde labeling with horse radish peroxidase, injected intraocularly, indicates that some of these newly growing axons arise from retinal ganglion cells.

We attributed the observed growth, at least in part, to the effect of the soluble substances derived from the regenerating fish optic nerves which we have shown to have a cytolytic effect on oligodendrocytes, thus eliminating their inhibitory effect on growth.

Introduction

Mammalian central nervous system (CNS) neurons have a negligible capacity to regenerate after lesions (see Review by Kiernan, 1979). In contrast, central neurons of lower vertebrates reliably regenerate after axotomy. In the goldfish visual system, the retinal ganglion cells regenerate severed axons and make functional connections with their appropriate targets (Ramón y Cajal,1959; Attardi and Sperry, 1963; Guth and Windle, 1970; Kiernan,1979). The regenerated axons become myelinated (Murray, 1976) and form their normal pattern of synaptic contacts with their targets (Murray and Edwards, 1982). Optic nerve injury in mammals leads, instead, to the death of most of the axotomized neurons and the failure of the surviving cells to regrow their axons (Grafstein and Ingoglia, 1982; Misantone *et al.*, 1984).

The prevailing hypothesis has been that all axotomized neurons, mammalian as well as non-mammalian, are capable of making a regenerative response after injury (see Ramón y Cajal, 1959). Therefore the success or failure of regeneration will depend upon the response of the non-neuronal cells (astrocytes, oligodendrocytes, microglia, macrophages) to injury. Astrocytes form glial scars and thus contribute to a barrier to growth (Reier *et al.*,1983). Mature oligodendrocytes inhibit axonal growth which is likely to impede the early regenerative responses; later, however, as Wallerian degeneration proceeds, the oligodendrocytes begin to die (David *et al.*, 1984) and their inhibitory properties may diminish (Schwab and Thoenen,1985; Schwab and Caroni, 1988; Caroni and Schwab, 1988). This suggests that lack of regeneration in mammalian CNS, at least in part, evolved from lack of temporal coordination between the neuronal cell body of the injured axon, the surrounding astrocytes and oligodendrocytes, which are not simultaneously in the conditions required for growth. In a successfully regenerating system, the responses of non-neuronal cells to the injury may differ from those of mammalian CNS either quantitatively, in time course or extent, or qualitatively, like in the nature of substances secreted. All or some of the conditions needed for regeneration might exist in regenerative systems and thus lead to a situation in which non-neuronal cells, such as in CNS of lower vertebrates or mammalian PNS, are capable of supporting the intrinsic regenerative capacity of their injured axons.

Axotomized mammalian central neurons can regenerate their axons over long distances under special conditions, for example those provided by replacement of the optic nerve by a segment of autologous nerve (Aguayo *et al.*, 1978; Vidal-Sanz *et al.*, 1987; Kierstead *et al.*, 1989). More recently we have shown that the CNS of lower vertebrates, specifically the regenerating fish optic nerve, is a source of factors which, when applied at the appropriate time and in appropriate amounts to injured mammalian adult optic nerves, can support regenerative axonal growth, perhaps as a result of modifying the mammalian neuronal environments, so as to make it more conducive to axonal growth (Schwartz *et al.*, 1985; Hadani *et al.*, 1984; Lavie *et al.*, 1987). Such factors in the form of conditioned medium (CM) elicited changes in both the neurons and their environment (Schwartz *et al.*, 1985; Lavie *et al.*, 1987; Zak *et al.*, 1987; Stein-Izsak *et al.*, 1985). Application of this CM alone is not adequate to promote marked axonal growth. Application of low energy laser irradiation to crushed rabbit and rat optic nerves has been shown to delay the development of post-traumatic degenerative processes (Schwartz *et al.*, 1987; Assia *et al.*, 1989). Combination of the two treatments, irradiation and CM, does elicit marked axonal growth in the injured rabbit optic nerves (Schwartz *et al.*, 1988; Lavie, *et al.*, 1989). Here, we will summarize the results from a series of experiments where significant axonal growth into the nerve's own environment was achieved. This environment, when treatment is omitted, is hostile for growth. We shall also provide information as to how the astrocytes and the oligodendrocytes might be affected by the treatment. We believe these studies can provide some indications of the mechanisms responsible for the effects of the combined treatment with CM and irradiation and thus elucidate some of the requirements for achieving regeneration of central neurons in adult mammals.

Description of the Surgical Approach

All of the *in vivo* studies were carried out in rabbits, whose optic nerves were injured 2-8 weeks prior to sacrifice. All animals were deeply anesthetized with xylazine (6 mg/kg) and ketamine (35 mg/kg) during surgery. The left optic nerve was exposed, as previously described (Solomon *et al.*, 1985) and visualized with the aid of a Zeiss operating microscope at a distance of 5-6 mm from the eyeball. The nerve was transected almost completely using a sharpened dissecting needle. The meningeal membrane was intentionally partially spared to ensure continuity of the nerve and to permit proper placement and retention of a nitrocellulose film. In some cases, this resulted in sparing of axons immediately subjacent to the meningeal covering; these spared axons could always be identified by their heavy myelin sheaths. The animals were then divided into two groups, an operated control group and an experimental group. In all operated nerves, a piece of nitrocellulose 2 to 5 mm long by 1 mm wide, soaked in medium for 1 h, was inserted at the site of the injury. In the operated control group, nitrocellulose was soaked in serum free medium (DMEM, Gibco) while in the experimental

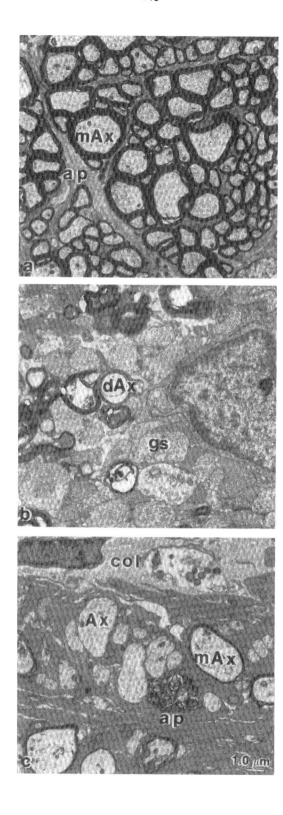

group the nitrocellulose was soaked in conditioned medium (CM; 100 μg protein/ml) from regenerating goldfish nerves (Lavie *et al.*, 1989). The experimental animals were irradiated daily for 10 min with He-Ne low energy laser (35 mW), for 10 days.

Results and Discussion

Combined application of CM and laser irradiation encourages survival and growth of injured optic nerve axons. This combined procedure supports axonal growth across the site of the lesion. The axons are identifiable as newly growing ones because they are unmyelinated or thinly myelinated. These axons, originating from the retinal ganglion cells, traverse the site of injury and survive for long periods. This growth does not occur in the absence of the combined treatment (Lavie *et al.*, 1989).

The normal rabbit optic nerve is composed almost entirely (98-99%) of heavily myelinated axons (Fig. 1a; see also Vaney and Hughes,1976). When the nerve is injured but not treated, the nerve distal to the injury is composed of degenerating axons and glial scar from 4 weeks post-injury (Fig. 1b,c).

In all animals subjected to the combined treatment, the optic nerves distal to the lesion site contains large numbers of unmyelinated axons most of which are closely associated with astrocytic cytoplasm (Fig. 2). At least some of these axons present distal to the lesion originate from the retina and are not peripheral axons which may have invaded at various points along the injured nerves. This was shown by intraocular injection of horseradish peroxidase which labeled axons arising from retinal ganglion cells (Fig. 2a,c). The astrocytic nature of the processes in which these axons are embedded was demonstrated using antibodies directed against glial fibrillary acidic protein (GFAP) (Fig. 2b,d).

The various relationships of these axons to their environment are shown in Figure 3. Occasionally, unmyelinated axons are seen within a collagenous environment, with minimal contact with non-neuronal cells (Fig. 3a). Other axons, either singly (Fig. 3b) or in small

Figure 1. Electron micrographs showing characteristic features of intact (a), injured untreated (b) and injured optic nerves subjected to the combined treatment (c). a. Normal rabbit optic nerve 5 mm distal to the globe. Myelinated axons (mAx) are organized in fascicles separated by astrocytic processes (Ap). X5040. b. Operated control nerve, 2 mm distal to the lesion site and insertion of nitrocellulose paper soaked in serum free medium, 6 weeks post-injury. Collagenous tissue and dense glial cell processes form a glial scar (gs) in which degenerating axons (dAx) are embedded. X9720. c. Optic nerve 2 mm distal to the lesion and insertion of nitrocellulose soaked with conditioned media and further treated with laser irradiation. The cross section is composed, for the most part, of collagenous tissue (Col), glial scar and degenerating fibers and thus resembles the nerve seen in Figure 1b. In one compartment of the nerve, many unmyelinated (Ax) and thinly myelinated axons (mAx) are seen in close association with astrocytic processes (AP). X5040.

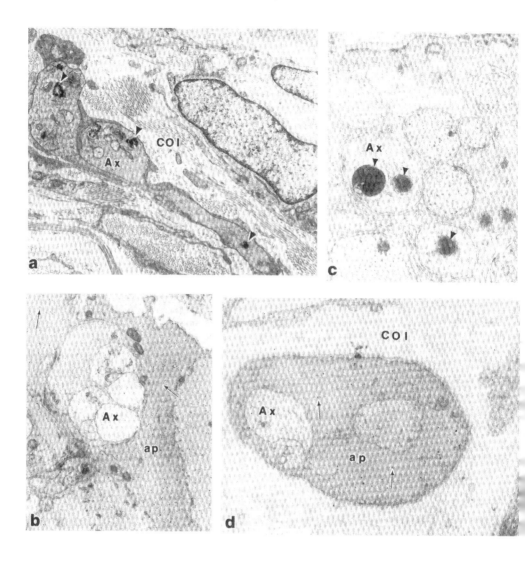

Figure 2. Axons in injured rabbit optic nerve subjected to combined treatment are associated with astrocytic cytoplasm. a,c. Wheat germ agglutinin horse radish peroxidase (WGA-HRP) was injected intraocularly 48 h before sacrifice. WGA-HRP containing vesicles (arrows) in unmyelinated axons distal to the lesion indicates the retinal origin of these axons. a. X9030; c. X31850. b,d. Post-embedding immunohistochemistry using antibodies against GFAP. Arrows indicate gold particles which label GFAP positive cells. The labeling demonstrates the astrocytic nature of the glial cells with which the axons are associated. b. X10320; d. X16320.

groups (Fig. 3c), are ensheathed by thin astrocytic processes, in close contact with collagenous connective tissue. Most of the unmyelinated and thinly myelinated axons are found more deeply embedded within a rather dense and compact tissue, which is identified by the presence of glial filaments as being astrocytic (Fig. 3d). Thus the axons appear to maintain a variable, but usually intimate, contact with astrocytes under these conditions. The large number of axons in close contact with astrocytes, seen as far as 6 to 7 mm from the site of injury, indicates that the astrocytes do not form a barrier to growth nor to survival of axons distal to the lesion.

Axonal growth in the treated nerves appeared to be largely co-existing with the nitrocellulose, extending no more than 1.5 mm distal to the last section in which the nitrocellulose could be identified (Lavie *et al.*, 1989). Based on the EM observations, it seems that the growing axons do not use the nitrocellulose as a substrate; nevertheless, the effects of the CM on the cells surrounding the injured axons may be dependent upon the accessibility to the CM provided by the nitrocellulose. Thus, the extent of growth may be determined and limited by the length of the piece of nitrocellulose or by the site at which the nitrocellulose is inserted. Alternatively it is possible that axonal growth under our experimental conditions may be slow, and therefore longer survival periods might reveal axonal growth farther distal to the nitrocellulose into the degenerating nerve. We therefore prepared two additional groups of animals subjected to the combined treatment. In one group, 3 mm pieces of nitrocellulose were inserted into the lesion. In the second group, 5 mm pieces of nitrocellulose were inserted. In these animals, the extent of axonal growth was related to the length of the nitrocellulose, never extending more than 1-2 mm beyond the last section containing nitrocellulose. Thus, these experiments indicate that while the axons do not use the CM-soaked nitrocellulose as a substrate, the extent of the axonal growth largely parallels the nitrocellulose and is unrelated to the survival time. The nitrocellulose may therefore act as a source of diffusible factors that affect adjacent non-neuronal cells. Non-neuronal cells which do not encounter factors diffusing from the nitrocellulose would not be modified by the treatment and would not support axonal growth.

The unmyelinated axons are found in intimate association with astrocytes at a distance of several mm from the site of injury. Type-1 astrocytes are present in the developing optic nerve of rats at the time of optic axon outgrowth (Noble *et al.*, 1984); the glial cells which are involved in scar formation in the adult are also type-1 astrocytes (Miller *et al.*, 1986). Type-1 astrocytes may thus become modified during normal maturation from growth supporting cells to those which form scars and do not support outgrowth. Our *in vivo* and *in vitro* results are in line with this hypothesis (Zak *et al.*, 1987; Cohen and Schwartz, 1989). This could indicate that the combined treatment may act to induce dedifferentiation of type-1 astrocytes. In the present study, we show not only axonal growth but also thinly myelinated axons, presumably as a result of an active process of remyelination. It is conceivable that the same type-1 cells are needed to support axonal growth and remyelination. Indeed, it has been recently proposed that type-1 astrocytes are also required for remyelination by oligodendrocytes (Blakemore and Crang, 1989).

Recent studies have implicated mature oligodendrocytes as cells which can inhibit axonal growth (Schwab and Thoenen, 1985; Caroni and Schwab, 1988). Our recent results show that among the factor(s) within the CM of the regenerating fish optic nerves, there are those which have a selective cytotoxic effect on oligodendrocytes (Cohen et al., 1989). The presence of these cytotoxic factor(s) is associated with injury since the toxic effect of CM from non-injured fish optic nerves was significantly lower (Cohen et al., 1989). The observed inhibitory effect is not mediated via any mitogenic activity, like that of platelet derived growth factor (PDGF) or ciliary neurotrophic factor (CNTF), which indirectly causes a reduction in the number of mature oligodendrocytes. This cytotoxicity might, however, be related to that observed in fetal calf sera (Bologa et al., 1988).

Regeneration of the fish optic nerves takes place spontaneously after injury. Recent works have also demonstrated that normal fish optic nerves can provide, in vitro, a permissive environment for mammalian CNS neurons to send out neurites (Carbonetto et al., 1987), indicating that either the normal fish optic nerves are free of the myelin-associated neuronal growth inhibitors, observed in mammals, or they have them but at a lower level. The question that fish oligodendrocytes might possess inhibitory proteins typical of mammalian oligodendrocytes, was recently raised (Beckmann et al., 1989). If the fish has a low level of axonal growth inhibitors, it would imply that the inhibitory/cytotoxic factor(s) observed in the regenerating fish optic nerves are also present in the intact fish optic nerves, constitutively, at levels which are close to or below the threshold levels detectable by our bioassay. It is suggestive that mammalian CNS either miss the immediate post-traumatic accessibility to such factors or that they do not have these factors at all. While during development of the rat optic nerve, there is a temporal separation between axonal growth (peak at embryonic day 20) and maturation of oligodendrocytes, manifested by myelin formation (post-natal day 6), the post-traumatic growth at the adult stage requires elimination of oligodendroglia cells soon after injury. Elimination of mature oligodendrocytes might allow the newly regenerating axons to penetrate an adult environment. The source of the cytotoxic agent in the regenerating fish optic nerves is presently unknown but its presence might depend, directly or indirectly, upon macrophages, activated resident microglial cells, or activated astrocytes. If the macrophages rapidly invade the injured optic nerve of fish, as they do in regenerating peripheral nerves of mammals (Perry et al., 1987), they could be responsible for the observed activity and for contributing to the elimination of the non-permissive mature oligodendrocytes. This would create an environment permissive for growth while neurons are still in an injury-induced mode

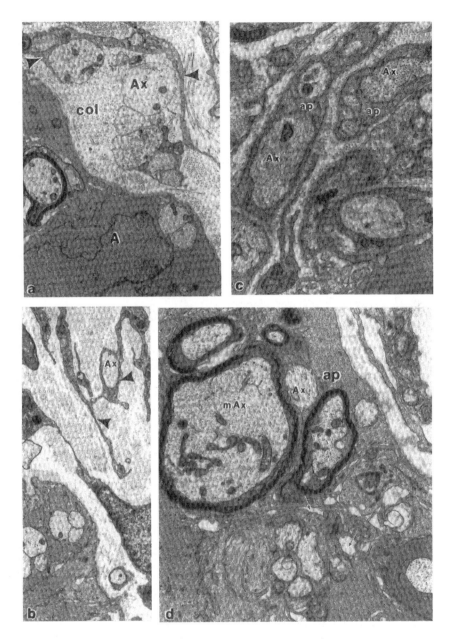

Figure 3. Relationship of unmyelinated axons to astrocytic processes. a. Group of unmyelinated axons (Ax) in collagenous connective tissue contacted by glial processes (arrow head). These axons are in close proximity to the compartment of viable axons containing unmyelinated and thinly myelinated axons embedded in astrocytes (A). X9030. b. A single unmyelinated axon (Ax) surrounded by glial process (arrow head), within collagenous tissue in close proximity to the compartment of viable axons. X9030. c. Groups of unmyelinated axons (Ax) surrounded by astrocytic processes (AP) close to the compartment of viable axons. X13230. d. The compartment of viable axons showing unmyelinated (Ax) and thinly myelinated (mAx) axons embedded in astrocytic processes (AP). X16320.

of growth. It is possible that the macrophages or the resident microglia cells are not the producers of the cytotoxic factor but are responsible for activating other resident cells, such as astrocytes, to produce the factor. That astrocytes can indeed produce a factor functionally similar to tumor necrosis factor (TNF) was shown in vitro in A23187 activated rat astrocytes (Robbins *et al.*, 1987). Other studies have reported the *in vitro* activity of TNF on myelinated cultures of mouse spinal cord tissue as well (Selmaj *et al.*, 1987).

References

Aguayo AJ, Samuel D, Richardson P, Bray G (1978) Axonal elongation in peripheral and central nervous system transplantations. Adv Cell Neurobiol 3:215-221

Assia E, Rosner M, Belkin M, Solomon A, Schwartz M (1989) Temporal parameters and low energy laser irradiation for optimal delay of post-traumatic degeneration of rat optic nerve. Brain Res 476:205-212

Attardi DG, Sperry RW (1963) Preferential selection of central pathways by regenerating optic nerve. Exp Neurol 7:46-64

Beckmann M, Bamstever M, Stuermer CAO (1989) Interaction of goldfish retinal axons with fish oligodendrocytes in vitro. Soc Neurosci Abstr 19:331

Blakemore WF, Crang AJ (1989) The relationship between type-1 astrocytes, Schwann cells and oligodendrocytes following transplantation of glial cell cultures into demyelinating lesions in the adult rat spinal cord. J Neurocytol 18:519-528

Bologa L, Cole R, Chiappelli F, Saneto RP, deVellis J (1988) Serum contains inducers and repressors of oligodendrocyte differentiation. J Neurosci Res 20:182-188

Carbonetto S, Evans D, Cochard P (1987) Nerve fiber growth in culture on tissue substrates for central and peripheral nervous system. J Neurosci 7:610-620

Caroni P, Schwab ME (1988) Antibody against myelin-associated inhibitor of neurite growth neutralizes non-permissive substrate properties of CNS white matter. Neuron 1:85-96

Cohen A, Sivron T, Duvdevani R, Schwartz M (1989) (submitted) Oligodendrocyte cytotoxic factor associated with fish optic nerve regeneration: Implication for mammalian CNS regeneration.

Cohen A, Schwartz M (1989) Conditioned media of regenerating fish optic nerves modulates laminin levels in glial cells. J Neurosci Res 22:269-273

David S, Miller RH, Patel R, Raff M (1984) Effect of neonatal transection on glial cell development in the rat optic nerve: Evidence that the oligodendrocyte - type 2 astrocyte lineage depends on axons for its survival. J Neurocytol. 13:961-974

Grafstein B, Ingoglia NA (1982) Intracranial transection of the optic nerve in adult mice: Preliminary observations. Exp Neurol 76:318-330

Guth L, Windle WF (1970) Regeneration in the vertebrate central nervous system. Exp Neurol 5:1-43

Hadani M, Harel A, Solomon A, Belkin M, Lavie V, Schwartz M (1984) Substances originating from the optic nerve of neonatal rabbit induce regeneration associated response in the injured optic nerve of adult rabbit. Proc Natl Acad Sci USA 81:7965-7969

Kiernan JA (1979) Hypothesis concerned with axonal regeneration in the mammalian nervous system. Biol Rev 54:155-197

Kierstead SA; Rasminsky M, Fukuda Y, Carter DA, Aguayo AJ, Vidal-Sanz M (1989) Electrophysiological responses in Hamster Superior Colliculus evoked by regenerating retinal axons. Science 246:255-257

Lavie V, Harel A, Doron A, Solomon A, Lobel D, Belkin M, Ben-Bassat S, Sharma S, Schwartz M (1987) Morphologic response of injured adult rabbit optic nerves to implant containing media conditioned by growing optic nerves. Brain Res 419:166-172

Lavie V, Murray M, Solomon A, Ben-Bassat S, Rumlet S, Belkin M and Schwartz M (1989) (submitted) Growth of injured rabbit optic axons within the degenerating optic nerve.

Miller RH, Abney ER, David S, ffrench-Constant C, Lindsay R, Petel R, Stone R, Raff MC (1986) Is reactive gliosis a property of a distinct subpopulation of astrocytes? J Neurosci 6:22-29

Misantone LJ, Gershenbaum M, Murray M (1984) Viability of retinal ganglion cells after optic nerve crush in adult rats. J Neurocytol 13:449-465

Murray M (1976) Regeneration of retinal axons into the goldfish optic tectum. J Comp Neurol 168:175-196

Murray M, Edwards ME (1982) A quantitative study of the reinnervation of the goldfish optic tectum following optic nerve crush. J Comp Neurol 209:363-373

Noble M, Fok-Seang J, Cohen J (1984) Glia are a unique substrate for the *in vitro* growth of central nervous system neurons. J Neurosci 4:1892-1903

Perry VH, Brown MC, Gordon S (1987) The macrophage response to central and peripheral nerve injury: A possible role for macrophages in regeneration. J Exp Med 165:1218-1223

Ramón y Cajal S (1959) Degeneration and regeneration of the central nervous system. Vol. 1. May RM, transl, New York, Hafner Publ

Reier PJ, Stensaas LJ, Guth L (1983) The astrocytic astrocytes scar as an impediment to regeneration in central nervous system. Kan CG, Bunge RP, Reier PJ (eds) *Spinal Cord Reconstruction*. New York, Raven Press, pp 163-195

Robbins DS, Shirazi Y, Drysdale B, Liberman A, Shin HS, Shin ML (1987) Production of cytotoxic factor for oligodendrocytes by stimulated astrocytes. J Immun 139:2593-2597

Schwab ME, Thoenen H (1985) Dissociated neurons regenerate into sciatic but not optic nerve explants in culture irrespective of neurotrophic factors. J Neurosci 5:2415-2423

Schwab ME, Caroni P (1988) Oligodendrocytes and CNS myelin are non-permissive substrates for neurite growth and fibroblast spreading *in vitro*. J. Neurosci 8:2381-2393

Schwartz M, Belkin M, Harel A, Solomon A, Lavie V, Hadani M, Rachailovich I, Stein-Izsak C (1985) Regenerating fish optic nerve and a regeneration-like response in injured optic nerve of adult rabbits. Science 228:600-603

Schwartz M, Belkin M, Solomon A, Lavie V, Ben-Bassat S, Murray M, Rosner M, Harel A, Rumlet S (1988) A novel treatment causes regeneration of axons among astrocytic processes within the adult rabbit optic nerve. Soc Neurosci Abstr 14:1198

Schwartz M, Doron A, Ehrlich M, Lavie V, Ben-Bassat S, Belkin M, Rochkind S (1987) Effects of low energy He-Ne laser irradiation on post-traumatic degeneration of adult rabbit optic nerve. Laser Surg Med 7:51-55

Selmaj KW, Ramie C S (1988) Tumor necrosis factor mediates myelin and oligodendrocyte damage *in vitro*. Ann Neurol 23:339-346

Solomon A, Belkin M, Hadani M, Harel A, Rachailovich I, Lavie V, Schwartz M (1985) A new transorbital surgical approach to the rabbit optic nerve. J Neurosci Meth 12:259-262

Stein-Izsak C, Harel A, Solomon A, Belkin M and Schwartz M (1985) Alterations in mRNA translation products are associated with regenerative response in the retina. J Neurochem 45:1754-1760

Vaney DI, Hughes A (1976) The rabbit optic nerve: Fiber diameter spectrum, fiber count, and comparison with a retina ganglion cell count. J Comp Neurol 170:241-252

Vidal-Sanz M, Bray MB, Villegas-Pérez MP, Thanos S, Aguayo AJ (1987) Axonal regeneration and synapse formation in superior colliculus by retinal ganglion cells in the adult rat. J Neurosci 7:2894-2909

Zak N, Harel A, Bawnik Y, Ben-Bassat S, Vogel Z, Schwartz M (1987) Laminin immunoreactive sites are induced by growth-associated triggering factors in injured rabbit optic nerve. Brain Res 408:263-266

Lavie V, Solomon A, Murray M, Ben-Bassat S, Rumlet S, Belkin M and Schwartz M (1989) (submitted) Growth of injured rabbit optic axons within the degenerating optic nerve.

Miller RH, Abney ER, David S, ffrench-Constant C, Lindsay R, Petel R, Stone R, Raff MC (1986) Is reactive gliosis a property of a distinct subpopulation of astrocytes? J Neurosci 6:22-29

Misantone LJ, Gershenbaum M, Murray M (1984) Viability of retinal ganglion cells after optic nerve crush in adult rats. J Neurocytol 13:449-465

Murray M (1976) Regeneration of retinal axons into the goldfish optic tectum. J Comp Neurol 168:175-196

Murray M, Edwards ME (1982) A quantitative study of the reinnervation of the goldfish optic tectum following optic nerve crush. J Comp Neurol 209:363-373

Noble M, Fok-Seang J, Cohen J (1984) Glia are a unique substrate for the *in vitro* growth of central nervous system neurons. J Neurosci 4:1892-1903

Perry VH, Brown MC, Gordon S (1987) The macrophage response to central and peripheral nerve injury: A possible role for macrophages in regeneration. J Exp Med 165:1218-1223

Ramón y Cajal S (1959) Degeneration and regeneration of the central nervous system. Vol. 1. May RM, transl, New York, Hafner Publ

Reier PJ, Stensaas LJ, Guth L (1983) The astrocytic astrocytes scar as an impediment to regeneration in central nervous system. Kan CG, Bunge RP, Reier PJ (eds) *Spinal Cord Reconstruction*. New York, Raven Press, pp 163-195

Robbins DS, Shirazi Y, Drysdale B, Liberman A, Shin HS, Shin ML (1987) Production of cytotoxic factor for oligodendrocytes by stimulated astrocytes. J Immun 139:2593-2597

Schwab ME, Thoenen H (1985) Dissociated neurons regenerate into sciatic but not optic nerve explants in culture irrespective of neurotrophic factors. J Neurosci 5:2415-2423

Schwab ME, Caroni P (1988) Oligodendrocytes and CNS myelin are non-permissive substrates for neurite growth and fibroblast spreading *in vitro*. J. Neurosci 8:2381-2393

Schwartz M, Belkin M, Harel A, Solomon A, Lavie V, Hadani M, Rachailovich I, Stein-Izsak C (1985) Regenerating fish optic nerve and a regeneration-like response in injured optic nerve of adult rabbits. Science 228:600-603

Schwartz M, Belkin M, Solomon A, Lavie V, Ben-Bassat S, Murray M, Rosner M, Harel A, Rumlet S (1988) A novel treatment causes regeneration of axons among astrocytic processes within the adult rabbit optic nerve. Soc Neurosci Abstr 14:1198

Schwartz M, Doron A, Ehrlich M, Lavie V, Ben-Bassat S, Belkin M, Rochkind S (1987) Effects of low energy He-Ne laser irradiation on post-traumatic degeneration of adult rabbit optic nerve. Laser Surg Med 7:51-55

Selmaj KW, Ramie C S (1988) Tumor necrosis factor mediates myelin and oligodendrocyte damage *in vitro*. Ann Neurol 23:339-346

Solomon A, Belkin M, Hadani M, Harel A, Rachailovich I, Lavie V, Schwartz M (1985) A new transorbital surgical approach to the rabbit optic nerve. J Neurosci Meth 12:259-262

Stein-Izsak C, Harel A, Solomon A, Belkin M and Schwartz M (1985) Alterations in mRNA translation products are associated with regenerative response in the retina. J Neurochem 45:1754-1760

Vaney DI, Hughes A (1976) The rabbit optic nerve: Fiber diameter spectrum, fiber count, and comparison with a retina ganglion cell count. J Comp Neurol 170:241-252

Vidal-Sanz M, Bray MB, Villegas-Pérez MP, Thanos S, Aguayo AJ (1987) Axonal regeneration and synapse formation in superior colliculus by retinal ganglion cells in the adult rat. J Neurosci 7:2894-2909

Zak N, Harel A, Bawnik Y, Ben-Bassat S, Vogel Z, Schwartz M (1987) Laminin immunoreactive sites are induced by growth-associated triggering factors in injured rabbit optic nerve. Brain Res 408:263-266

AGGREGATING BRAIN CELL CULTURES: A MODEL TO STUDY MYELINATION AND DEMYELINATION

P. Honegger[1] and J.-M. Matthieu[2]

[1] Institut de Physiologie,
Université de Lausanne
1005 Lausanne

[2] Laboratoire de Neurochimie,
Service de Pédiatrie
Centre Hospitalier Universitaire Vaudois
1011 Lausanne
Switzerland

INTRODUCTION

Cell cultures of the central nervous system (CNS) provide the investigator with a system to study selected factors influencing development without the complexity of _in vivo_ studies. Rotation-mediated aggregating cell cultures of dissociated fetal brain follow the pattern of normal brain development (Matthieu et al., 1978; Seeds and Haffke, 1978; Trapp et al., 1979). Within a one month culture period, the aggregates develop from agglomerates of undifferentiated cells to highly organized populations of fully differentiated neurons, astrocytes and oligodendroglia (Matthieu et al., 1978; Trapp et al., 1979). In these cultures, myelinated axons and synapses are present (Seeds and Vatter, 1971; Matthieu et al., 1978; Trapp et al., 1979).

The original techniques of Moscona and Seeds (Moscona, 1965; Seeds, 1973) have been improved by replacing the initial enzymatic dissociation of fetal rat CNS with a simple mechanical dissociation (Honegger and Richelson, 1976). The system was further improved with the use of a chemically defined, serum-free medium (Honegger et al., 1979). These modifications allow the production of a large number of identical cultures each containing a very homogeneous population of aggregates.

The ease with which large amounts of aggregates can be prepared provides an _in vitro_ system which is invaluable for morphological, biochemical and molecular genetic investigations. Each culture flask contains several thousand spherical aggregates, thus samples can be harvested sequentially during the course of an experiment. Another important aspect is the three-dimensional organization of the aggregates, allowing cellular interactions and migrations. This three-dimensional structure, lacking in most other cell culture systems, seems to be required for optimal differentiation (Seeds, 1973).

The aim of this article is to review findings from our laboratories showing the usefulness of this culture system to study the influence of growth and differentiation factors on myelin gene expression and its regulation during myelinogenesis. This system seems also promising for investigations on demyelination using antibodies to myelin and oligodendroglial constituents.

PREPARATION AND CHARACTERISTICS OF SERUM-FREE AGGREGATING BRAIN CELL CULTURES

Aggregating brain cell cultures are prepared routinely from 15-day fetal rat telencephalon (for technical details, Honegger, 1985). The dissected tissue is dissociated mechanically in ice-cold sterile solution D (modified Puck's D solution according to Wilson et al., 1972), with the aid of two separate nylon mesh bags with pores of 200 μm and 115 μm, respectively. The dissociated cells are washed in solution D by centrifugation, and they are finally resuspended in ice-cold serum-free culture medium (Table I). Aggregate formation is initiated by placing 4 ml aliquots of the single-cell suspension (approx. 10^7 cells/ml) into 25 ml De Long flasks (i.e., modified Schott Duran Erlenmeyer flasks) and incubating them at 37°C in an atmosphere of 10% CO_2 and 90% humidified air, under constant gyratory agitation (68 rpm). The aggregates are transferred to 50 ml De Long flasks two days after culture initiation, and the volume of culture medium per flask is increased to 8 ml. Within the first week, the rotation speed is progressively increased to 78 rpm. Medium is replenished by replacing 5 ml of medium with fresh culture medium every 3 days until

Table I. Composition of the chemically defined culture medium[a]

Ingredient	Concentration (M)		Ingredient	Concentration (M)	
D-Glucose	2	$\times 10^{-2}$	Choline chloride[b]	1	$\times 10^{-3}$
L-Arginine·HCl	4	$\times 10^{-4}$	L-Carnitine·HCl[c]	1	$\times 10^{-5}$
L-Cystine	2	$\times 10^{-4}$	Biotin[c]	4	$\times 10^{-6}$
L-Glutamine	4	$\times 10^{-3}$	D-calcium pantothenate[b]	1	$\times 10^{-5}$
Glycine	4	$\times 10^{-4}$	Folic acid[b]	1	$\times 10^{-5}$
L-Histidine·HCl·H_2O	2	$\times 10^{-4}$	i-Inositol[b]	5	$\times 10^{-5}$
L-Isoleucine	8	$\times 10^{-4}$	Nicotinamide[b]	4	$\times 10^{-5}$
L-Leucine	8	$\times 10^{-4}$	Pyridoxal·HCl[b]	2.5	$\times 10^{-5}$
L-Lysine·HCl	8	$\times 10^{-4}$	Riboflavine[b]	1.3	$\times 10^{-6}$
L-Methionine	2	$\times 10^{-4}$	Thiamin·HCl[b]	1.5	$\times 10^{-5}$
L-Phenylalanine	4	$\times 10^{-4}$	Vitamin B_{12}[c]	1	$\times 10^{-6}$
L-Serine	4	$\times 10^{-4}$	Linoleic acid (Na^+)[c]	1	$\times 10^{-5}$
L-Threonine	8	$\times 10^{-4}$	Thioctic acid[c]	1	$\times 10^{-6}$
L-Tryptophan	8	$\times 10^{-5}$	Triiodothyronine[c]	3	$\times 10^{-8}$
L-Tyrosine	4	$\times 10^{-4}$	Hydrocortisone-21-P[c]	2	$\times 10^{-8}$
L-Valine	8	$\times 10^{-4}$	Insulin (bovine)[c]	8	$\times 10^{-7}$
Transferrin (human)[c]	1.3	$\times 10^{-8}$	Retinol[c,d]	2	$\times 10^{-6}$
Phenol red	4	$\times 10^{-5}$	DL-α-Tocopherol[c,d]	2	$\times 10^{-6}$

a Basal medium: Dulbecco's modified Eagle's medium (DMEM; GIBCO, no. 041-019
b Concentration increased with respect to the basal medium
c Supplement to the basal medium
d Added to the medium by sonication and sterile filtration. Only the initial
  is given; the final concentration has not been determined

day 11 and every other day until day 19. Thereafter, cultures are either split (i.e., the aggregates of each flask are distributed equally to several flasks), or medium replenishment is performed more frequently in order to meet the increasing metabolic rate of the cultures.

The developmental characteristics of these cultures have been studied by using both morphological and biochemical criteria (for review, Honegger, 1985, 1987). It was found that mature cultures contain about equal numbers of neurons and glial cells. While most neurons enter terminal differentiation within the first few days in vitro, the majority of oligodendrocytes and astrocytes continue to proliferate during the first two weeks in culture. Thereafter, most of the glial cells are differentiating.

In these cultures, oligodendrocytes develop from a small pool of progenitor cells. During the first week in vitro, most of these cells are mitotically active. By the end of the second week, almost all oligodendrocytes show postmitotic terminal differentiation. Galactocerebroside (GC) staining is very

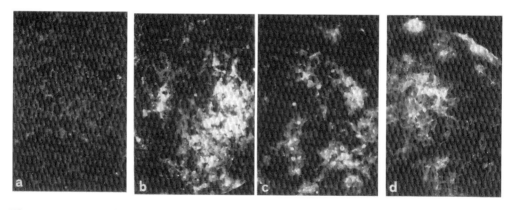

Figure 1. Immunofluorescence staining for MBP in serum-free aggregating brain cell cultures. Aggregates were examined on days 7 (a), 14 (b), 23 (c) and 37 (d). Very scarce staining is found on day 7, whereas from day 14 onwards clusters of intensely MBP-immunoreactive cells are found throughout the aggregate. Immunofluorescence studies were performed on acetone-fixed cryostat sections (16 μm) of cryoform-embedded aggregates, as described previously (Monnet-Tschudi and Honegger, 1989). Polyclonal rabbit antibodies directed against human MBP were used at a 1:100 dilution. Magnification = 185 x. Courtesy of Dr. F. Monnet-Tschudi.

scarce and punctuated during the first week, followed by a significant increase during the second week _in_ _vitro_ (Monnet-Tschudi and Honegger, 1989). Myelin basic protein (MBP) content measured by radioimmunoassay shows a rapid rise between days 19 and 30 _in_ _vitro_, in parallel with the period of intense myelin formation (Almazan et al., 1985a). These findings are also in good agreement with the developmental appearance of immuno-reactive MBP (Fig. 1).

Compact myelin around axons is found from the fourth week onwards (Honegger, 1987). Myelin isolated and purified from serum-free aggregate cultures shows a chemical composition sim-ilar to (but not identical with) myelin isolated from adult rat brain. By comparison, the amount of myelin produced in aggre-gate cultures is relatively low, and the proportion of poorly compacted myelin seems to be relatively high (Honegger and Matthieu, 1980). Nevertheless, the histotypic structure of the cultures including myelinated axons is maintained during sever-al months in serum-free culture conditions.

RESULTS

The effect of growth factors on myelination of brain cell
aggregates in culture

The possibility of studying CNS development in cell aggre-gates grown in a serum-free, chemically defined medium (Honegger et al., 1979) has opened new avenues for investigat-ing growth factor effects on the different steps of brain ma-turation. Among the different factors studied, only insulin and insulin-like growth factor (IGF-I) stimulated mitogenic activ-ity of both neuronal and glial precursor cells (Lenoir and Honegger, 1983). This mitogenic effect prevailed at low concen-trations (0.5-50 ng/ml), whereas high IGF-I concentrations (240 ng/ml) also stimulated protein synthesis, mimicking the anabolic action of insulin (Lenoir and Honegger, 1983). The effect of IGF-I on oligodendrocyte proliferation was confirmed by McMorris et al. (1986). In a recent study, McMorris and Dubois-Dalcq (1988) found that IGF-I promoted proliferation of precursor cells and induced precursors to mature into oligodendrocytes.

Although receptors for <u>epidermal growth factor</u> (EGF) have been shown to be associated mainly with astrocytes (Leutz and Schachner, 1982), some binding sites are present on purified oligodendrocytes (Simpson et al., 1982). The effect of EGF on oligodendrocyte differentiation was demonstrated by significant increases in MBP levels and 2',3'-cyclic nucleotide 3'-phosphodiesterase (CNP)-specific activities, which were observed in cell aggregates treated with EGF (Honegger and Guentert-Lauber, 1983; Almazan et al., 1985b). This effect was concentration-dependent (Fig. 2). EGF exerted a prolonged effect on both MBP and CNP, since a short treatment on days 2 and 5 resulted in significantly increased expression of both myelin markers on day 19. The pattern of stimulation strongly suggested that EGF accelerates the differentiation, rather than the proliferation of oligodendrocytes (Almazan et al., 1985b).

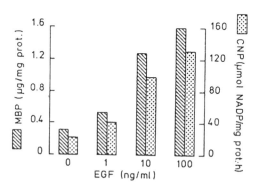

Figure 2. Concentration-dependent EGF stimulation of MBP content and CNP-specific activity in 12-day aggregating brain cell cultures. EGF as added on culture days 2 and 5.

The literature concerning <u>corticosteroid</u> effects on brain development is controversial. Dexamethasone treatment of brain cell aggregates has a complex effect on MBP accumulation. We observed an early increase (at 15 DIV), followed by a significant reduction (25 DIV) (Almazan et al., 1986). These results do not confirm in vivo studies using large doses of exogenous hormone, however, they are in agreement with in vitro studies using hydrocortisone (Warringa et al., 1987). In addition rats, which had been adrenalectomized at a similar developmental stage as in our study, exhibited reduced amounts of CNS myelin, whereas DNA content and CNP activity were not modified (Preston and McMorris, 1984). In contrast, increases in myelin deposition and CNP activity were observed 7 to 8 weeks after adrenalectomy (Meyer, 1983). These data suggest that glucocorticoids can be either stimulatory or inhibitory (Bau and Vernadakis, 1982). These opposing effects are apparently dependent on the stage of brain development. During the early myelination phase in brain cell aggregates, as well as in vivo, dexamethasone stimulates oligodendrocytes and the deposition of myelin. By contrast, at a later developmental stage, dexamethasone inhibits myelinogenesis.

Fetal brain cell aggregates maintained in the continued presence of bovine growth hormone (bGH) supplemented culture

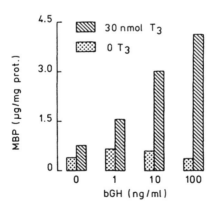

Figure 3. Concentration-dependent bGH stimulation of MBP content in 19-day aggregating brain cell cultures. bGH was added throughout the culture period. No stimulatory effect was found in the absence of T_3.

medium exhibit a large accumulation of MBP (Almazan et al., 1985b). This effect is concentration-dependent and requires the presence of T_3 (Fig. 3). A similar thyroid hormone dependency was observed with the activity of ornithine decarboxylase during brain development (Roger and Fellows, 1979). Our results demonstrate that bGH has a direct effect on myelination of the CNS. Thymidine incorporation and DNA content were not affected by addition of bGH, suggesting that bGH increases oligodendrocyte differentiation.

The crucial role of <u>thyroid hormones</u> on brain development has long been recognized (for review, Lelong et al., 1989). In order to study thyroid hormone action at the molecular level, investigators have abandoned <u>in vivo</u> models for cell culture systems. Thus, rotation-mediated aggregating fetal rat brain cell cultures have shown their usefulness. In this system, thyroid hormone effects on different cell types can be observed, and in particular, myelinogenesis can be investigated since compact myelin sheaths are formed around axons.

In aggregating fetal brain cell cultures, neurons respond to <u>triiodothyronine</u> (30 nM), as indicated by increased activities of choline acetyltransferase, acetylcholinesterase and glutamic acid decarboxylase (Honegger and Lenoir, 1980). Undifferentiated cultures are more responsive to T_3 than mature cultures. These observations are in agreement with observations <u>in vivo</u>. The response of neurons to T_3 was limited to a relatively short period in development, and no increased cellular proliferation was detected (Honegger and Lenoir, 1980).

Numerous <u>in vivo</u> and <u>in vitro</u> studies have shown that thyroid deficiency during early developmental stages leads to a reduction in myelin formation, whereas thyroid hormone administration accelerates myelination. The addition of T_3 to aggregating cultures of fetal rat brain cells increases the activities of two enzymes involved in myelinogenesis: galactosylceramide sulfotransferase (a myelin lipid synthesizing enzyme) and CNP (a myelin-oligodendroglial marker) (Honegger and Matthieu, 1980). In a more extensive study, MBP and CNP, myelinogenesis indicators, respond to T_3 at 19 days in a concentration-dependent manner (Almazan et al., 1985a). The continuous presence of

T_3 is necessary to maintain normal levels of MBP (Fig. 4) and CNP during development (Almazan et al., 1985a). In the absence of T_3, the steady-state levels of MBP and proteolipid protein (PLP) mRNAs are reduced drastically (Matthieu et al., 1987; Shanker et al., 1987; Pieringer et al., 1989). In contrast, two astrocytic markers, glutamine synthetase and glial fibrillary acidic protein, are not affected by T_3 deprivation between days 16 and 28 in culture (Honegger and Matthieu, unpublished results).

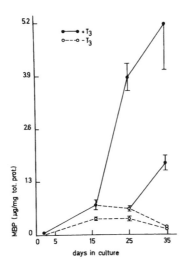

Figure 4. Influence of T_3 on the MBP content in serum-free aggregating brain cell cultures. During different developmental periods, the cultures were grown either in the presence (solid lines) or in the absence (dotted lines) of T_3. (From Almazan et al., 1985a, with permission of S. Karger AG, Basel).

The recent demonstration of nuclear T_3 receptors in oligodendrocytes indicates that T_3 affects the regulation of myelin protein gene expression specifically and directly (Sarliève et al., 1989; Yusta et al., 1988). Recently, we studied the regulation of myelin gene expression and found a good correlation between in vitro translation of MBP and steady-state MBP mRNA levels. This finding indicates that the reduced MBP concentration, in the absence of T_3, is not due to impaired translational efficiency or to post-translational events (Matthieu et al., in press). Thyroid hormone is known to increase the transcriptional activity of several genes by binding to a chromatin-associated receptor (Samuels et al., 1989). In addition, T_3 can also affect gene expression by mRNA stabilization (Shapiro et al., 1987). We found that T_3 does not stimulate MBP transcription but influences significantly the stability of MBP mRNAs (Matthieu et al., in press).

Demyelination in aggregating brain cell cultures

The presence of myelinated axons in aggregating brain cell cultures and the use of serum-free, chemically defined medium make this system ideally suited to investigate the potentially neurotoxic effects of antibodies directed against myelin and/or oligodendroglial constituents, thus avoiding complex systemic interactions. Antibodies raised against MBP show no demyelinating activity even in the presence of complement, as previously reported by Raine (1984) using explant cultures. In the presence of complement, antibodies directed against the specific myelin lipid, galactocerebroside, induce demyelination in aggregates (Honegger et al., 1989) confirming previous reports (Dubois-Dalcq et al., 1977; Dorfman et al., 1978; Raine et al., 1981).

A monoclonal antibody (8-18 C5) directed against a glycoprotein exposed on the external surface of myelin and oligodendrocytes (myelin-oligodendroglial glycoprotein; MOG) (Brunner et al., 1989) has a strong demyelinating effect in vivo (Lassmann

Table II. **Complement-mediated neurotoxicity of anti-MBP and anti-MOG antibodies on myelin and oligodendroglia markers**

	MBP	CNP
Complement	100	100
Anti-MBP	124	105
Anti-MOG	25	61

Twenty-five µl of guinea-pig serum was used as source of complement. Purified IgG (62.5 µg/ml) and complement were added to aggregating cultures for 48 hours on day 25. Cultures were analyzed on day 29. Values are expressed in percent of control cultures (e.g. cultures supplemented with complement alone) (Kerlero de Rosbo et al., submitted).

and Linington, 1987), as well as in aggregating brain cell cultures (Honegger et al., 1989; Kerlero de Rosbo et al., submitted). Anti-MOG antibodies show a more pronounced effect on MBP than on CNP (Table II). This could indicate a specific destruction of the myelin sheath with a relative preservation of oligodendrocytes. This hypothesis is currently under investigation in our laboratories. The results obtained so far suggest that MOG is a good target for complement-mediated demyelination, and that anti-MOG antibodies could play a role in demyelinating diseases.

DISCUSSION

Serum-free aggregating brain cell cultures are able to reproduce a series of critical developmental events known to occur in vivo. Thus, oligodendrocyte development can be studied in vitro from the early stage of progenitor cell multiplication through the formation of compact myelin sheaths around axons. The early events in oligodendrocyte maturation appear to follow a genetically determined program, whereas myelin formation is slightly delayed as compared to the situation in vivo. Furthermore, myelin sheaths around axons in culture tend to be less compact and fewer in number (Matthieu et al., 1979). It seems likely that culture conditions can be further improved to meet the specific requirements of myelinating oligodendrocytes.

A growing list of epigenetic factors have been shown to enhance various stages of oligodendrocyte development in vitro. In addition to the stimulatory factors mentioned in this review, several polypeptides have been shown to modulate oligodendrocyte proliferation and/or differentiation in various in vitro systems. These peptides include platelet-derived growth factor (Noble et al., 1988; Raff et al., 1988), fibroblast growth factor (Eccelston and Silberberg, 1985; Saneto and deVellis, 1985), lymphokines (Merrill et al., 1984; Saneto et al., 1986) and possibly undefined neuronal and astroglial factors. However, the physiological role of most of these agents remains to be elucidated. This task is complicated by the fact that in addition to soluble (endocrine or paracrine) growth and differentiation factors, extracellular matrix components (Cardwell and Rome, 1988) and cell membrane constituents may participate in the control of normal oligodendrocyte development and function. Furthermore, some of these factors may influence oligodendrocytes only during a limited period of development, whereas others may be required continuously. Aggregating brain cell cultures have the advantage that they form three-dimensional structures, thus providing the oligodendrocytes with natural cell-cell contacts and extracellular matrix components appropriate for all developmental stages. In addition, the great developmental potential of these cultures permits the study of regulatory factor actions and their interplay over a wide range of ontogenetic events.

Since compact myelin is formed around axons and maintained for several months in aggregating brain cell cultures, this culture system offers a unique model to study demyelinating processes. A strong complement-mediated demyelination was induced by treating cultures with anti-MOG antibodies, in good agreement with observations *in vivo*. Therefore, it will be possible in further studies to use this demyelinating reaction as a paradigm to examine ways to influence both the course of demyelination, as well as processes leading to remyelination.

Acknowledgements

We would like to thank Dr. Minnetta Gardinier for her critical appraisal of the manuscript. The work reported in this review article has received the continuous support of the Swiss National Science Foundation and the Multiple Sclerosis Society of Switzerland.

REFERENCES

Almazan G, Honegger P, Du Pasquier P, Matthieu J-M (1986) Dexamethasone stimulates the biochemical differentiation of fetal forebrain cells in reaggregating cultures. Dev Neurosci 8:14-23

Almazan G, Honegger P, Matthieu J-M (1985a) Triiodothyronine stimulation of oligodendroglial differentiation and myelination. A developmental study. Dev Neurosci 7:45-54

Almazan G, Honegger P, Matthieu J-M, Guentert-Lauber B (1985b) Epidermal growth factor and bovine growth hormone stimulate differentiation and myelination of brain cell aggregates in culture. Dev Brain Res 21:257-264

Bau D, Vernadakis A (1982) Effects of corticosterone on brain cholinergic enzymes in chick embryos. Neurochem Res 7: 821-829

Brunner C, Lassmann H, Waehneldt TV, Matthieu J-M, Linington C (1989) Differential ultrastructural localization of myelin basic protein, myelin/oligodendroglial glycoprotein, and 2',3'-cyclic nucleotide 3'-phosphodiesterase in the CNS of adult rats. J Neurochem 52:296-304

Cardwell MC, Rome LH (1988) RGD-containing peptides inhibit the synthesis of myelin-like membrane by cultured oligodendrocytes. J Cell Biol 107:1551-1559

Dorfman SH, Fry JM, Silberberg DH, Grose C, Manning MC (1978) Cerebroside antibody titers in antisera capable of myelination inhibition and demyelination. Brain Res 147:410-415

Dubois-Dalcq M, Niedieck B, Buyse M (1977) Action of anti-cerebroside sera on myelinated nervous tissue cultures. Path Europ 5:331-347

Eccleston PA, Silberberg DH (1985) Fibroblast growth factor is a mitogen for oligodendrocytes in vitro. Dev Brain Res 21: 315-318

Honegger P (1985) Biochemical differentiation in serum-free aggregating brain cell cultures. In: Bottenstein JE, Sato G (eds) Cell Culture in the Neurosciences. Plenum Publishing Corp, New York, p 223-243

Honegger P (1987) Oligodendrocyte development and myelination in serum-free aggregating brain cell cultures. In: Serlupi-Crescenzi G (ed) A Multidisciplinary Approach to Myelin Disease. NATO-ASI Series. Plenum Publishing Corp, New York, p 161-169

Honegger P, Guentert-Lauber B (1983) Epidermal growth factor (EGF) stimulation of cultured brain cells. I. Enhancement of the developmental increase in glial enzymatic activity. Dev Brain Res 11:245-251

Honegger P, Lenoir D (1980) Triiodothyronine enhancement of neuronal differentiation in aggregating fetal rat brain cells cultured in a chemically defined medium. Brain Res 199:425-434

Honegger P, Lenoir D, Favrod P (1979) Growth and differentiation of aggregating fetal brain cells in a serum-free defined medium. Nature 282:305-308

Honegger P, Matthieu J-M (1980) Myelination of aggregating fetal rat brain cell cultures grown in a chemically defined medium. In: Baumann N (ed) Neurological Mutations Affecting Myelination. INSERM Symposium No 14. Elsevier/North Holland Biomedical Press, Amsterdam, p 481-488

Honegger P, Matthieu J-M, Lassmann H (1989) Demyelination in brain cell aggregate cultures, induced by a monoclonal antibody against the myelin/oligodendrocyte glycoprotein (MOG). Schw Arch Neurol Psy 140:10-13

Honegger P, Richelson E (1976) Biochemical differentiation of mechanically dissociated mammalian brain in aggregating cell culture. Brain Res 109:335-354

Kerlero de Rosbo N, Honegger P, Lassmann H, Matthieu J-M Demyelination induced in aggregating brain cell cultures by a monoclonal antibody against myelin-oligodendroglial glycoprotein (MOG). Manuscript submitted.

Lassmann H, Linington C (1987) The role of antibodies against myelin surface antigens in demyelination in chronic EAE. In: Serlupi Crescenzi G (ed) A Multidisciplinary Approach to Myelin Disease. NATO-ASI Series. Plenum Publishing Corp, New York, p 219-225

Lelong GR, Robbins J, Condliffe PG (1989) Iodine and the Brain, Plenum Press, New York

Lenoir D, Honegger P (1983) Insulin-like growth factor I (IGF-I) stimulates DNA synthesis in fetal rat brain cell cultures. Dev Brain Res 7:205-213

Leutz A, Schachner M (1982) Cell type-specificity of epidermal growth factor (EGF) binding in primary cultures of early postnatal mouse cerebellum. Neurosci Lett 30:179-182

Matthieu J-M, Honegger P, Favrod P, Gautier E, Dolivo M (1979) Biochemical characterization of a myelin fraction isolated from rat brain aggregating cell cultures. J Neurochem 32: 869-881

Matthieu J-M, Honegger P, Trapp BD, Cohen SR, Webster HdeF (1978) Myelination in rat brain aggregating cell cultures. Neuroscience 3:565-572

Matthieu J-M, Roch J-M, Eng L, Honegger P (1987) Triiodothyronine controls myelin protein gene expression in aggregating brain cell cultures. Ann Endocrinol 48:164

Matthieu J-M, Roch J-M, Torch S, Tosic M, Carpano P, Insirello L, Giuffrida Stella AM, Honegger P Triiodothyronine increases the stability of myelin basic protein mRNA in aggregating brain cell cultures. In: Giuffrida Stella AM, deVellis J, Perez-Polo JR (eds) Regulation of Gene Expression in the Nervous System. AR Liss, Inc, New York, in press

McMorris FA, Dubois-Dalcq M (1988) Insulin-like growth factor I promotes cell proliferation and oligodendroglial commitment in rat glial progenitor cells developing in vitro. J Neurosci Res 21:199-209

McMorris FA, Smith TM, DeSalvo S, Furlanetto RW (1986) Insulin-like growth factor I/somatomedin-C: a potent inducer of oligodendrocyte development. Proc Natl Acad Sci USA 83: 822-826

Merrill JE, Kutsunai S, Mohlstrom C, Hofman F, Groopman J, Golde DW (1984) Proliferation of astroglia and oligodendroglia in response to human T cell-derived factors. Science 224:1428-1430

Meyer JS (1983) Early adrenalectomy stimulates subsequent growth and development of the rat brain. Exp Neurol 82: 432-446

Monnet-Tschudi F, Honegger P (1989) Influence of epidermal growth factor on the maturation of fetal rat brain cells in aggregate culture. Dev Neurosci 11:30-40

Moscona AA (1965) Recombination of dissociated cells and the development of cell aggregates. In: Willmer BM (ed) Cells and Tissues in Culture. Academic Press, New York, p 489-529

Noble M, Murray K, Stroobant P, Waterfield MD, Riddle P (1988) Platelet-derived growth factor promotes division and motility and inhibits premature differentiation of the oligodendrocyte/type-2 astrocyte progenitor cell. Nature 333: 560-562

Pieringer RA, Cabacungan E, Mittal R, Soprano D (1989) Regulation of proteolipid protein mRNA by T_3 in cultures of cells dissociated from mouse cerebra. J Neurochem 52:S194C

Preston SL, McMorris FA (1984) Adrenalectomy of rats results in hypomyelination of the central nervous system. J Neurochem 42:262-267

Raff MC, Lillien LE, Richardson WD, Burne JF, Noble MD (1988) Platelet-derived growth factor from astrocytes drives the clock that times oligodendrocyte development in culture. Nature 333:562-565

Raine CS (1984) Biology of disease. Analysis of autoimmune demyelination: its impact upon multiple sclerosis. Lab Invest 50:608-635

Raine CS, Johnson AB, Marcus DM, Suzuki A, Bornstein MB (1981) Demyelination in vitro. Absorption studies demonstrate that galactocerebroside is a major target. J Neurol Sci 52: 117-131

Roger JL, Fellows RE (1979) Evidence for thyroxine-growth hormone interaction during brain development. Nature 282: 414 415

Samuels HH, Forman BM, Horowitz ZD, Ye Z-S (1989) Regulation of gene expression by thyroid hormone. Annu Rev Physiol 51: 623-639

Saneto RP, Altman A, Knobler RL, Johnson HM, deVellis J (1986) Interleukin 2 mediates the inhibition of oligodendrocyte progenitor cell proliferation in vitro. Proc Natl Acad Sci USA 83:9221-9225

Saneto R, deVellis J (1985) Characterization of cultured rat oligodendrocytes proliferating in a serum free chemically defined medium. Proc Natl Acad Sci USA 82:3509-3515

Sarliève LL, Besnard F, Labourdette G, Yusta B, Pascual A, Aranda A, Luo M, Puymirat J, Dussault JH (1989) Investigation of myelinogenesis in vitro: transient expression of 3,5,3'-triiodothyronine nuclear receptors in secondary cultures of pure rat oligodendrocytes. In: LeLong GR, Robbins J, Condliffe PG (eds) Iodine and the Brain, 1st ed. Plenum Press, New York London, p 113-130

Seeds NW (1973) Differentiation of aggregating brain cell cultures. In: Sato G (ed) Tissue Culture of the Nervous System. Plenum Press, New York, p 35-53

Seeds NW, Haffke SC (1978) Cell junction and ultrastructural development of reaggregated mouse brain cultures. Dev Neurosci 1:69-79

Seeds NW, Vatter AE (1971) Synaptogenesis in reaggregating brain cell culture. Proc Natl Acad Sci USA 68:3219-3222

Shanker G, Campagnoni AT, Pieringer RA (1987) Investigation on myelinogenesis in vitro: developmental expression of myelin basic protein mRNA and its regulation by thyroid hormone in primary cerebral cell cultures from embryonic mice. J Neurosci Res 17:220-224

Shapiro DJ, Blume JE, Nielsen DA (1987) Regulation of messenger RNA stability in eukaryotic cells. BioEssays 6:221-226

Simpson DL, Morrison R, deVellis J, Herschman HR (1982) Epidermal growth factor binding and mitogenic activity of purified populations of cells from the central nervous system. J Neurosci Res 8:453-462

Trapp BD, Honegger P, Richelson E, Webster HdeF (1979) Morphological differentiation of mechanically dissociated fetal rat brain in aggregating cell cultures. Brain Res 160:117-130

Warringa RAJ, Hoeben RC, Koper JW, Sykes JEC, van Golde LMG, Lopez-Cardozo M (1987) Hydrocortisone stimulates the development of oligodendrocytes in primary glial cultures and affects glucose metabolism and lipid synthesis in these cultures. Dev Brain Res 34:79-86

Wilson SH, Shrier BK, Farber JL, Thompson EJ, Rosenberg RN, Blume AJ, Nirenberg MW (1972) Markers for gene expression in cultured cells from the nervous system. J Biol Chem 247:3159-3169

Yusta B, Besnard F, Ortiz-Caro J, Pascual A, Aranda A, Sarlième L (1988) Evidence for the presence of nuclear 3,5,3'-triiodothyronine receptors in secondary cultures of pure rat oligodendrocytes. Endocrinology 122:2278-2284

THE TRANSPLANTATION OF GLIAL CELLS INTO AREAS OF PRIMARY DEMYELINATION

W.F. Blakemore, A.J. Crang, R.J.M. Franklin
Department of Clinical Veterinary Medicine
University of Cambridge
Madingley Road
Cambridge, CB3 OES
England

Introduction

Injection of ethidium bromide into the white matter of the spinal cord produces a lesion in which demyelinated axons lie in a glial-free environment (Graça and Blakemore, 1986). This lesion therefore provides a system in which one can examine interactions between axons and glia which are involved in the reconstruction of a CNS environment around axons. By injecting glial cell cultures of differing composition it is possible to examine the ability of different cell types to influence repair and, in so doing, reveal aspects of their biology which may be difficult to document using other systems. It also provides a useful system in which to examine *in-vivo*, conclusions drawn from *in-vitro* investigations. For example, the importance of a stable extracellular matrix to the myelinating Schwann cell established by tissue culture studies (Bunge and Bunge, 1978; 1983), was well illustrated when Schwann cells were introduced into ethidium bromide lesions made in X-irradiated spinal cord of the cat (Blakemore and Crang, 1985). In these experiments it was found that myelin sheath formation following Schwann cell/axon contact only occurred in areas where there was collagen.

Ethidium bromide (EB) causes cell death by intercalating with nucleic acids to prevent nucleic acid and protein synthesis. Thus when injected into white matter of the rat spinal cord, EB selectively kills glial cells to produce areas of primary demyelination in which the demyelinated axons lie in an area devoid of both astrocytes and oligodendrocytes (Graça and Blakemore, 1986). Because the areas of demyelination contains neither astrocytes nor oligodendrocytes, the demyelinated axons are largely remyelinated by Schwann cells which migrate into the lesion from peripheral sites. Any oligodendrocyte remyelination that does occur is seen at the edges of the lesion, next to normal tissue.

When ethidium bromide is injected into areas of the spinal cord which have been exposed to 40 Grays of X-irradiation 3 days prior to the injection of ethidium bromide, there is no remyelination by either oligodendrocytes or Schwann cells (Blakemore and Patterson, 1978; Blakemore and Crang, 1985). Thus, any remyelination seen following the injection of cultures of glial cells can be considered to have been carried out by the transplanted cells and their progeny.

Using the X-irradiated lesion system we have made a series of observations which indicate that, following the injection of CNS cultures, demyelinated axons can be remyelinated by either oligodendrocytes or Schwann cells and the extent of Schwann cell remyelination is influenced by the number of cells of the oligodendrocyte lineage present in the injected cell suspension (Blakemore et al, 1987; Blakemore and Crang, 1988). Initially, we were surprised to find Schwann cell remyelination following the injection of cultures of CNS glial cells. However, immunostaining with anti-laminin and anti-S100 antibodies revealed that CNS mixed glial cell cultures routinely contain a small number of Schwann cells, less than 5%, despite careful attempts to minimize their presence. This low level of contamination only results in significant Schwann cell remyelination when the injected cultures contain low numbers of cells of the oligodendrocyte lineage. Thus, when we inject large numbers of oligodendrocytes and their precursors (Blakemore and Crang. 1988), the demyelinated axons are remyelinated mainly by oligodendrocytes at 18 days post transplantation. If fewer cells of the oligodendrocyte lineage are injected, either in terms of number of cells injected, or as a proportion of the cells injected, the extent of remyelination is reduced and the demyelinated axons are mainly remyelinated by Schwann cells (Crang and Blakemore, 1989).

From these observations it was concluded that, when large numbers of cells of the oligodendrocyte lineage are present, the extent of remyelination by low numbers of Schwann cells is limited. However, when the number of cells of the oligodendrocyte lineage is low, even small numbers of Schwann cells can become the predominant myelinating cell. These results are considered to reflect differences in the mitotic and migratory potential of the two types of myelinating cell when they are introduced into a glial-free area containing demyelinated axons. Myelin-forming oligodendrocytes are non-motile, non-mitotic cells derived from mitotically active, motile precursor cells (Kachner et al., 1986; Small et al., 1987; Noble et al., 1988; Raff et al., 1988); whereas Schwann cells are mitotically active and motile (Abercrombie and Johnson, 1942; Crang and Blakemore, 1987). Both Schwann cells and cells of the oligodendrocyte lineage can respond to axonal mitogens (Wood and Bunge, 1979; 1986). However, because not all the cells of the oligodendrocyte lineage can divide, the potential for this population to expand is more limited. Thus, one can understand why the extent of remyelination by oligodendrocytes reflects the number of cells of the oligodendrocyte lineage transplanted, while that achieved by Schwann cells does not.

When cultures of glial cells are injected into ethidium bromide lesions made in **non**-irradiated white matter, analysis of the effect of transplanting cells is more complex, as the injected cells will have to compete with, be influenced by, or influence local cells populations. The results of transplantation can be analysed in terms of alterations to the repair seen in non-transplanted lesions. This paper will examine the effect on repair of injecting glial cell cultures of differing composition into ethidium bromide lesions made in the **non**-irradiated spinal cord of the rat.

Methods

High-oligodendrocyte mixed glial cell cultures. These were prepared by mechanical dissociation of cerebral hemispheres from one-day-old inbred PVG/Ola rats (the preparation and immunocytochemical characterization of these cultures is described in Blakemore and Crang, 1989). Characterization of the cell suspension 4 days post transplantation indicated a composition of 7% O10+ve mature oligodendrocytes, 24% A2B5+ve progenitor cells, 52% GFAP+ve type-1 astrocytes, 5% A2B5/GFAP+ve type-2 astrocytes, the remaining cells being meningeal cells, Schwann cells and endothelial cells.

High-oligodendrocyte / low type-1 astrocyte cultures. These were prepared by shaking-off and sub-culturing top-dwelling cells from the mixed glial cell cultures (see Blakemore and Crang, 1989 for details). Characterization of the injected cell suspension 4 days post transplantation indicated a composition of 38% O10+ve mature oligodendrocytes, 22% A2B5+ve progenitor cells, 8% GFAP+ve type-1 astrocytes, 10% GFAP/A2B5+ve type-2 astrocytes, the remaining cells comprising mainly meningeal cells, Schwann cells and endothelial cells.

Low-oligodendrocyte mixed glial cell cultures. These were prepared by trypsinization of cerebral hemispheres from four-day-old PVG/Ola rats (see Crang and Blakemore, 1989 for details). Characterization of the injected cell suspension 4 days post transplantation indicated a composition of <3% galactocerebroside +ve oligodendrocytes, >85% GFAP +ve type-1 astrocytes, 5% co-expressed laminin and S100 (Schwann cells) and <5% were intensely fibronectin +ve (meningeal cells).

The ethidium bromide lesion system. One microlitre of a 0.1% solution of ethidium bromide was injected into the dorsal columns after performing a laminectomy on the first lumbar vertebra of inbred PVG/Ola adult rats. Three days later the rats was re-anaesthetized and 1 μl of a glial cell suspension was injected into the lesion area and the animals allowed to recover. Animals were killed at the end of the experiment by perfusion with 4% glutaraldehyde in phosphate buffer under general anaesthesia. The spinal cord was removed and 1 mm coronal slices made of the lesioned area. These were processed into TAAB resin and 1 μm sections cut from each block, stained with toluidine blue and examined by light microscopy.

Assessment of the effect of injecting the cultured cells. The relative proportions of Schwann cell remyelinated axons, oligodendrocyte remyelinated axons and demyelinated axons in each lesion-containing block was estimated on a five point scale. Thus, a lesion-containing block in which all the axons remain demyelinated, would have an oligodendrocyte remyelination score of 0, a Schwann cell remyelination score of 0, and a demyelinated axon score of 5. A lesion-containing block in which 20% of the axons were remyelinated by oligodendrocytes, 40% remyelinated by Schwann cells and 40% remain demyelinated would have an oligodendrocyte score of 1, a Schwann cell score of 2, and a demyelinated axon score of 2. To compare the nature of remyelination from different treatments, a plot of oligodendrocyte versus Schwann cell remyelination scores provides a convenient graphical representation of each treatment group. For each treatment, individual block scores were plotted for all lesion-containing blocks. The mean oligodendrocyte and Schwann cell remyelination scores are indicated ±2 SEM and the domains enclosed by these limits represent a frequency-weighted average for the type of repair in that treatment group. These plots give a semi-quantitative representation of the extent of remyelination as well as the balance between central- and peripheral-type remyelination. Non-overlapping domains in these representations indicate significantly different results; the direction of their displacement from each other indicates which parameters contribute to the difference.

Results

Non-transplanted lesions. The extent and type of remyelination achieved by 21 days and 41 days after the injection of 1.0 μl of 0.1% ethidium bromide is depicted in Figure 1.

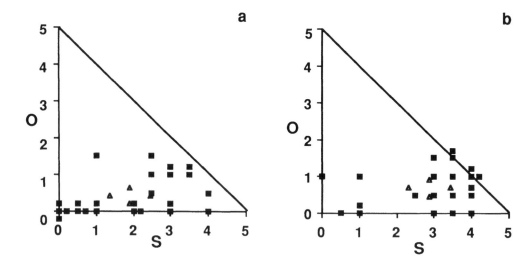

Figure 1 The nature of repair in non-transplanted animals at 21 days (a) and 42 days (b) after injection of 1 µl of 0.1% ethidium bromide. The squares represent the proportion of axons remyelinated by oligodendrocytes (O) plotted against the proportion of axons remyelinated by Schwann cells (S) for each lesion containing block from 5 animals (see text for details). The triangles indicate ±2 standard errors of the mean scores for each parameter and the area enclosed thus represents a frequency-weighted average for the nature of repair in this group of animals.

It can be seen that remyelination is carried out mainly by Schwann cells and that there is an increase in the number of remyelinated axons with time. The small areas of oligodendrocyte remyelination present in the lesions are located next to normal tissue at the edges of the lesion. The degree of astrocytosis in the areas of oligodendrocyte remyelination is minimal and astrocyte processes could not be detected in the areas of Schwann cell remyelination.

High-oligodendrocyte mixed glial transplants. When high-oligodendrocyte mixed glial cell cultures are introduced into the lesion three days after ethidium bromide demyelination has been induced, the type of remyelination observed in lesions from animals killed 30 days after transplantation is very different (compare Figure 2 with Figure 1).

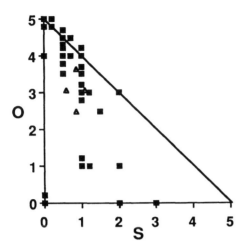

Figure 2 Diagram depicting the nature of repair in lesion containing blocks from 6 animals, 30 days after a high-oligodendrocyte mixed glial cell cultures were injected into ethidium bromide lesions.

In these lesions there was extensive oligodendrocyte remyelination and a reduction in the extent of Schwann cell remyelination from that seen in normally-repairing ethidium bromide lesions. The areas of Schwann cell remyelination were sharply demarcated from the areas of oligodendrocyte remyelination which showed a heavy astrocyte presence with individual axons often separated by a mat of astrocyte processes.

High-oligodendrocyte / low type-1 astrocyte transplants. When cultures enriched for oligodendrocytes, but depleted of type-1 astrocytes were injected into the lesions, the extent of remyelination by Schwann cells was greater than that seen in animals injected with the high-oligodendrocyte mixed glial cell cultures and the extent of oligodendrocyte remyelination was only slightly greater than that seen in non-transplanted animals (compare Figure 3 with Figure 2 and Figure 1). In some parts of the lesion, areas of oligodendrocyte remyelination were clearly separated from areas of Schwann cell remyelination. However, in other parts of the lesion, the separation of the two types of remyelination was not so clear-cut and there was intermingling of Schwann cells and oligodendrocyte-remyelinated axons.

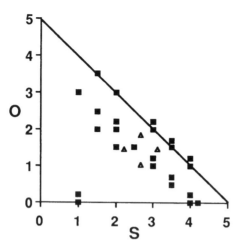

Figure 3 Diagram depicting the nature of repair in lesion containing blocks from 6 animals, 29 days after a high-oligodendrocyte/low type-1 astrocyte glial cultures were injected into an ethidium bromide lesions.

Low-oligodendrocyte mixed glial transplants. When low-oligodendrocyte mixed glial cell cultures were injected into the lesions, the extent of

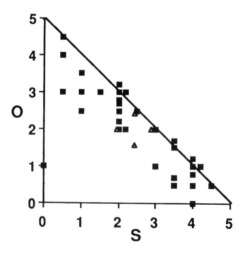

Figure 4 Diagram depicting the nature of repair in lesion containing blocks from 6 animals, 28 days after low-oligodendrocyte mixed glial cell culture were injected into ethidium bromide lesions.

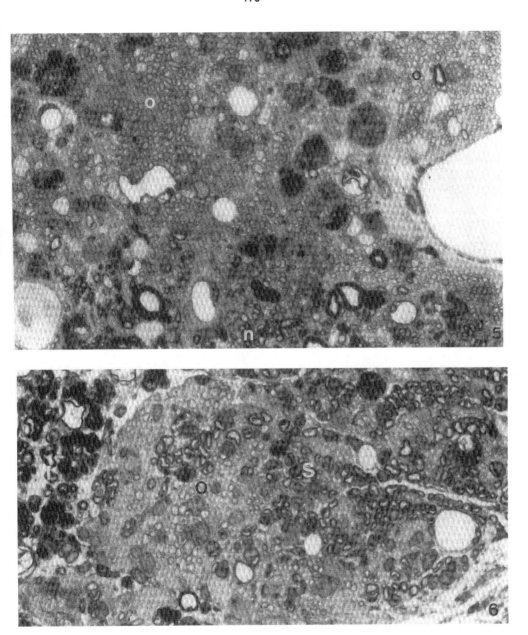

Figure 5 Area of lesion where demyelinated axons have been remyelinated by oligodendrocytes (o). At the bottom of the micrograph, normally-myelinated axons (n) can be seen. Toluidene blue, x900.

Figure 6 Area containing mixed Schwann cell and oligodendrocyte remyelination. The Schwann cell repair can be distinguished from oligodendrocyte repair by a thicker myelin sheath and/or a signet-ring appearance. There is no clear-cut boundary between central- and peripheral-types of repair. Toluidene blue, x900.

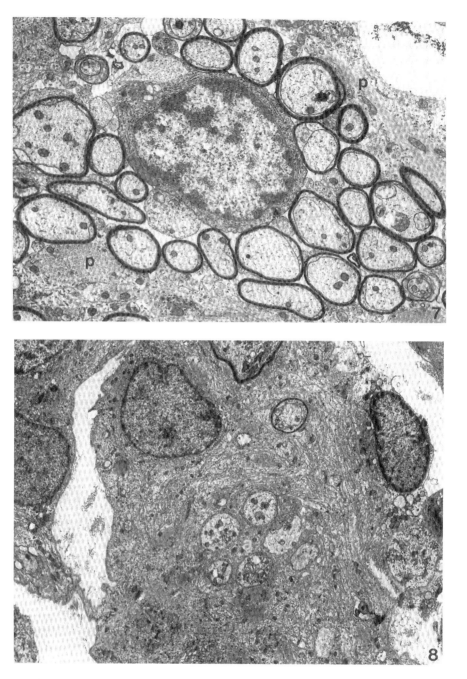

Figure 7 Oligodendrocyte-remyelinated axons in an area containing astrocyte processes (p), x8,400.

Figure 8 Group of demyelinated axons (a) surrounded by a dense network of astrocyte processes, x4,200.

oligodendrocyte-remyelination was increased over that seen in both the non-transplanted lesions and lesions injected with high-oligodendrocyte-low-type-1 astrocyte cultures, but less than that seen in the lesions injected high-oligodendrocyte mixed glial cell cultures (compare Figure 4 with Figures 1,3 and 2). In these animals the distribution of repair resembled that seen following oligodendrocyte-enriched/type-1 astrocyte-depleted transplants, in that clearly defined areas of oligodendrocyte (Figure 5) or Schwann cell remyelination were observed, along with less clearly delineated regions of mixed central- and peripheral-type repair (Figure 6). Oligodendrocyte remyelination occurred in areas containing large numbers of astrocyte processes (Figure 7); however, in contrast to other transplants, a small number of dense astrocytic processes were also observed in the areas of Schwann cell remyelination. In other areas of the lesion where there was a particularly dense astrocytosis there was a failure of remyelination (Figure 8).

Discussion

The irradiated and non-irradiated EB lesions constitute two distinct systems for investigating the ability of transplanted glia to influence remyelination. The results of these experiments using non-irradiated lesions differ from those carried out in irradiated lesions in that they identify the type-1 astrocyte component of the transplant as the factor which determines whether the lesion will be repaired by oligodendrocytes or Schwann cells. This result has to be considered in the context of the intrinsic repair capacity of the two systems.

Ethidium bromide causes the death of both astrocytes and oligodendrocytes so that axons lie in an area devoid of glia. In the irradiated lesion there is no intrinsic repair and thus the nature of the repair following transplantation is a direct reflection of the composition of the transplanted cell-suspension. In this system the number of oligodendrocytes transplanted appears to be a critical factor in establishing central-type repair in the face of competition from a small number of motile and mitotically active transplanted Schwann cells.

In contrast, the **non**-irradiated EB lesion has an inherent tendency to repair. The majority of the demyelinated axons are remyelinated by Schwann cells and only axons adjacent to normal tissue at the edge of the lesion are remyelinated by oligodendrocytes. Transplantation of glial cells into this lesion alters the nature and extent of the inherent repair process, the final repair achieved being a consequence of the interaction between transplant and host-derived cells. Since it has been observed that Schwann cell remyelination occurs following death of astrocytes (Blakemore, 1983), it has been proposed that astrocytes, rather than the oligodendrocytes, are the critical factor which determines the distribution of peripheral- and central-type remyelination. If this hypothesis is correct, it should follow that replacement of missing astrocytes in glial-free lesions will firstly, reduce the extent of Schwann cell remyelination, and secondly, allow more extensive oligodendrocyte remyelination. In the present experiments there is evidence for both effects. Thus, the extent of Schwann cell remyelination was only affected when the transplanted cultures contained a large number of type-1 astrocytes, indicating that this cell is important in containing Schwann cell invasion of the CNS. Furthermore, high type-1 astrocyte transplants increased the extent of oligodendrocyte repair irrespective of the oligodendrocyte content of the transplant. This last observation is at variance with results obtained from the irradiated system, where astrocytes alone are unable to control Schwann cell invasion unless accompanied by a minimum number of oligodendrocytes. Because the low oligodendrocyte-containing cultures behave differently in the two types of the lesion we may conclude that the different results can be related to the contribution made to the repair by host-derived oligodendrocytes. Oligodendrocytes may be contributing to the control of Schwann cells by masking axonal signals or by helping the astrocytes to become established in a stable manner which allows them to control Schwann cell access to demyelinated axons. Thus, the role of astrocytes in the **non**-irradiated lesion may be twofold. In addition to preventing Schwann cell invasion, astrocytes may also potentiate the recruitment of host oligodendrocytes, by the production of soluble migratogenic and mitogenic factors. The recent observation that type-1 astrocyte-derived "platelet derived growth factor" has a mitogenic effect on oligodendrocyte precursors may be of significance in this respect (Noble et al., 1988; Raff et al., 1988).

Acknowledgements

This work was supported by grants from the Multiple Sclerosis Society and the Wellcome Trust. R.J.M.Franklin holds a Wellcome Research Training Scholarship.

References

Abercrombie M & Johnson ML (1942) The outwandering of cells in tissue culture of nerves undergoing Wallerian degeneration. J. Exp. Biol. 19: 266-286.

Blakemore WF (1983) Remyelination of demyelinated spinal cord axons by Schwann cells. In: C.C.Kao R.P.Bunge P.J.Reier (eds) Spinal cord reconstruction. Ravan Press, New York, pp281-293.

Blakemore WF & Patterson RC (1978) Suppression of remyelination in the CNS by X-irradiation Acta Neuropathol. 42: 105-113.

Blakemore WF & Crang AJ (1985) The use of cultured autologous Schwann cells to remyelinate areas of persistent demyelination in the central nervous system. J. Neurol. Sci. 70: 207-223.

Blakemore WF Crang AJ & Patterson RC (1987) Schwann cell remyelination of CNS axons following injection of cultures of CNS cells into areas of persistent demyelination. Neurosci. Let. 77: 20-24.

Blakemore WF & Crang AJ (1988) Extensive oligodendrocyte remyelination following injection of cultured central nervous system cells into demyelinating lesions in the adult central nervous system. Dev. Neurosci. 10: 1-11.

Blakemore WF & Crang AJ (1989) The relationship between type-1 astrocytes, Schwann cells and oligodendrocytes following transplantation of glial cell cultures into demyelinating lesions in the adult rat spinal cord J. Neurocytol. in press.

Bunge RP & Bunge MB (1978) Evidence that contact with connective tissue matrix is required for normal interaction between Schwann cells and nerve fibres. J. Cell Biol. 78: 943-950.

Bunge RP & Bunge MB (1983) Interrelationship between Schwann cell function and extracellular matrix. T.I.N.S. 6: 499-505.

Crang AJ & Blakemore WF (1987) Observations on the migratory behaviour of Schwann cells from adult peripheral nerve explants. J. Neurocytol. 16: 423-431.

Crang AJ & Blakemore WF (1989) The effect of the number of oligodendrocytes transplanted into X-irradiated, glial free lesions on the extent of oligodendrocyte remyelination. Neurosci. Let. in press.

Griffin JW, Drucker N, Gold BG, Rosenfeld J, Benzaquen M, Charnas RC, Fahnstock KE & Stocks EA (1987) Schwann cell proliferation and migration during paranodal demyelination. J. Neurosci. 7: 682-699.

Graça DL & Blakemore WF (1986) Delayed remyelination in the rat spinal cord following ethidium bromide injection. Neuropathol. Appl. Neurobiol. 12: 593-605.

Kachar B, Behar T & Dubois-Dalcq M (1986) Cell shape and motility of oligodendrocytes cultured without neurons. Cell Tissue Res. 244: 27-38

Noble M, Murray K, Stoobant P, Waterfield MD & Riddle P (1988) Platelet-derived growth factor promotes division and motility and inhibits premature differentiation of oligodendrocyte/type-2 astrocyte progenitor cell. Nature 333: 560-562.

Raff MC, Lillien LE, Richardson WD, Burne JF & Noble M (1988) Platelet-derived growth factor from astrocytes drives the clock that times oligodendrocyte development in culture. Nature 333: 562-565.

Small RK, Riddle P & Noble M (1987) Evidence for the migration of oligodendrocyte-type-2 astrocyte progenitor cells in the developing rat optic nerve. Nature 328: 155-157.

Wood PM & Bunge RP (1979) Evidence that sensory axons are mitogens for Schwann cells. Nature 256: 283-308.

Wood PM & Bunge RP (1986) Evidence that axons are mitogenic for oligodendrocytes isolated from adult animals. Nature 320: 756-758.

REMYELINATION OF A CHEMICALLY INDUCED DEMYELINATED LESION IN THE SPINAL CORD OF THE ADULT SHIVERER MOUSE BY TRANSPLANTED OLIGODENDROCYTES.

O.Gout, A.Gansmuller, M.Gumpel.
Laboratoire de Neurobiologie Cellulaire, Moleculaire et Clinique
INSERM U 134
Hôpital de la Salpêtrière
47 Bd de l'Hôpital
75651 Paris Cedex 13 - France

INTRODUCTION

The idea of a spontaneous remyelination in the mammalian Central Nervous System (CNS) is now widely accepted. After the pionner works of Bunge et al (1961) and Bornstein et al (1962) a spontaneous remyelination process was described in numerous animal models of demyelination (reviews by Blakemore 1979-1981, Ludwin 1987). In Multiple Sclerosis (MS) a discrete remyelination was first described at the periphery of chronic lesions (review by Prineas and Connel, 1979). Lassmann (1983), after the observation of thinly myelinated fibers in fresh lesions, concluded that in acute forms of the disease, remyelination can begin within a short period of time after demyelination. This could result in the total remyelination of the plaque (formation of "shadow plaques") and contribute to clinical remission of the disease (Prineas, 1985). In animal models of demyelination as well as in MS, the newly formed myelin was identified on the criteria defined by Harrison and Mc Donald (1977) : abnormally thin myelin over the total internodal length, abnormally short internodes, presence in the remyelinating area of the different steps observed in the developmental myelination process (wrapping of the axon, compaction of the lamellae). The comparative study of animal models of chemical demyelination allowed Blakemore (1976-1978) to evidence that the remyelination process was very variable according to the species, the demyelinating agent being the same. In animal models, as well as in Multiple Sclerosis, the remyelination appears to be more efficient in acute lesions than in chronic ones (Ludwin, 1978 - Lassmann, 1983). In both cases Schwann cells and oligodendrocytes participate to the repair of the lesion. However, in animal models, the Schwann cells participation appears to be very variable according to the species : after lysolecithin injection in the spinal cord, the Schwann cells are very invasive in rat, less invasive in cat (Blakemore, 1976 - Blakemore et al, 1977). By contrast, in the mouse, after the same treatment, most of the remyelination observed (if not all) is carried out by oligodendrocytes (Duncan et al, 1981 - Aranella and Herndon, 1984 - Gout et al, 1988a, b).

If the process of remyelination in CNS is well documented in human diseases and in normal animal models, very few is known about the remyelination capabilities of the animal myelin deficient mutants. In our previous publications (Lachapelle et al, 1983-1984 - Gansmuller et al, 1986 - Gumpel et al, 1987) we focused our attention on criteria allowing to distinguish the myelin formed by intracerebral or intraspinal transplanted cells in an host environment. For this purpose, we used the "Shiverer model". The Shiverer mouse mutation is a recessive autosomal mutation characterized by a severe myelin deficit in the CNS. To utilize the shiverer mouse as a marker in transplantation experiments, we took advantage of two particularities of this mutant : biochemically (Dupouey et al, 1979), the CNS myelin is completely deprived of Myelin Basic Protein (MBP). Morphologically, at ultrastructural level (Privat et al, 1979), this myelin is abnormally thin, uncompacted and the Major Dense Line (MDL) is absent. Thus, when normal oligodendrocytes are transplanted in a Shiverer host CNS, the myelin formed by grafted oligodendrocytes could be detected by immunohistochemistry using an anti-MBP antiserum or by Electron Microscopy (EM).

Extensive ultrastructural analysis of the myelin sheaths formation in the shiverer mutant allowed Rosenbluth (1980) and Inoue et al (1981) to describe a number of abnormalities in the Shiverer oligodendrocyte function during the myelination process.

The aim of the present paper is to study the demyelination-remyelination process in the Shiverer adult spinal cord to understand how remyelination occurs in this mutant and to observe the function of CNS myelin-forming cells involved in the remyelination process. In experiments in which a graft containing normal myelin-forming cells was placed at a distance of a chemically induced lesion, we studied the role of grafted cells in remyelination and we compared the formation of the myelin sheath by normal and shiverer oligodendrocytes around shiverer demyelinated axons.

THE SHIVERER MYELINATION

As already reported by Rosenbluth (1980), Inoue et al (1981) and Billings-Gagliardi et al (1986) the shiverer myelination does not proceed normally. As noted above, this mutant is grossly myelin

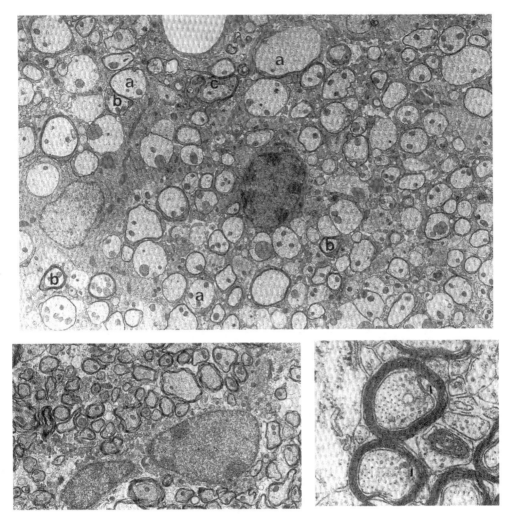

Fig.1 : Adult shiverer mouse dorsal funiculus. Numerous abnormalities can be observed : incomplete ensheathment of the axon (a), abnormal cytoplasmic loops (b), aberrant myelination (c) X 4200

Fig 2 : Adult normal mouse dorsal funiculus. Same magnification as fig 1 X 4200

Fig 3: Higher magnification of normal myelin sheaths showing the spiralling ensheathment (I = inner tongue) X 32000

deficient. Moreover, at EM analysis, the CNS shiverer myelinated fibers are markedly abnormal compared with those of normal mouse. First, and even if it is difficult to ascertain this idea without a carefull analysis of some defined fiber tracts, it was noted by Rosenbluth (1980) that some large axons which would be expected as ensheathed by myelin appeared to be completely unmyelinated. The thickness of myelin was very irregular and classical relationships between axonal diameter and number of myelin lamellae did not seem to be conserved. Frequently lamellae contain abnormal quantity of cytoplam (sometimes with the presence of numerous cytoplasmic organelles) and terminate in huge cytoplasmic loops. Moreover, lamellae frequently do not encircle completely the axons (fig.1). This abnormality can concern the external lamellae only and results in marked irregularity in the myelin thickness around the axonal circumference. But the lamellae can also be deposited on one part of the circle only, the axon being partially unmyelinated (fig.1) In this case, the myelin sheath is certainly formed non by spiralling as it is the case for normal myelin (fig. 2,3), but by folding of one oligodendrocyte process. Aberrant myelination can also occur, sometimes around the soma of an oligodendrocyte, without relation with any axon. Sporadic compaction with formation of a structure ressembling a MDL can be observed, but never on the whole circumference of the axon and never between all the lamellae of the myelin sheath.

DEMYELINATION IN THE SHIVERER ADULT SPINAL CORD AND SPONTANEOUS REMYELINATION.

The demyelination lesion was obtained by intramedullary injection of 2 µl of a solution of 1% lysolecithin (Lysophosphatidyl Choline (LPC) Sigma) in saline (Gout et al, 1988a, b - Gumpel et al 1989a, b). The demyelinated site is marked with charcoal powder. At the end of the experiment the mice were perfused intracardially, the spinal cord was dissected out and processed for standard EM (Gout et al, 1988b). The demyelinating effect of LPC in the CNS of the normal mouse was first studied by Hall (1972). The substance acts directly on the myelin sheath and provokes very quickly a focal area of demyelination. The aspect of the demyelinated lesion was similar in normal and shiverer spinal cords we studied and our observations were comparable to those of Hall (1972) and Aranella and Herndon (1984). Two days after the injection (fig.4,5) the axons appeared healthy. However an increase of the density of cytoplasmic organelles and an irregular outline of the

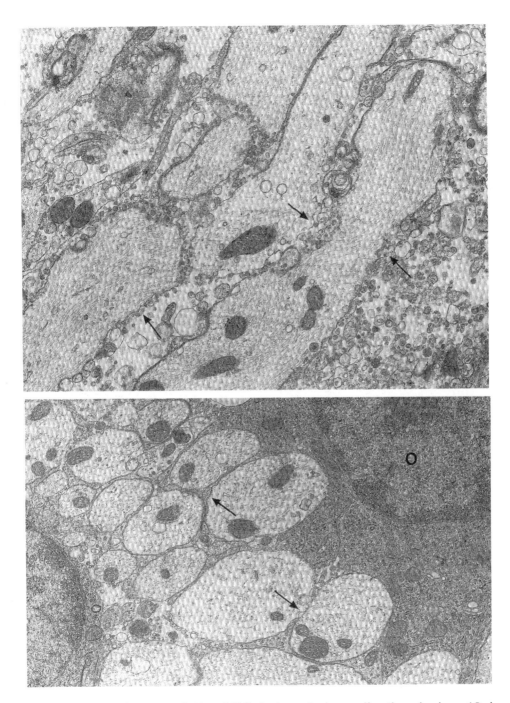

Fig.4 and 5 : Aspect of the LPC induced demyelination lesion 48 hrs after the injection. Longitudinal and transversal sections. Note the irregurarities of the axonal membrane (arrows) and the numerous cell debris around the axons O= healthy oligodendrocyte X 16400

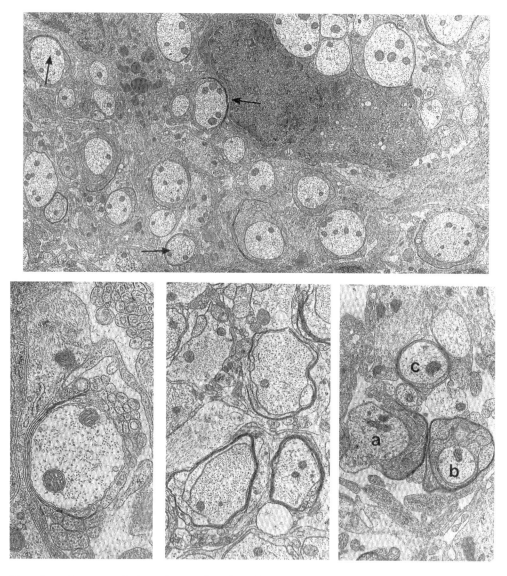

Fig. 6 : Aspect of the demyelination lesion in the shiverer mouse 7 days after LPC injection. During the first steps of the remyelination. Abnormal ensheathment by folding is clearly visible (arrows) X 5300

Fig 7 : Higher magnification of one of the ensheathments seen in fig.8. Incomplet ensheathment by folding and abnormally great number of oligodendrocytic microprocesses X 13000

Fig 8 : Remyelination in shiverer spinal cord 16 days after LPC injection. Numerous intramyelin loops of cytoplasm X 8200

Fig 9 : Same animal as fig lO. Incomplete ensheathment of the axon (a) abnormal cytoplasmic loops in the myelin (b) and complete ensheathment probably formed by folding (c) X 13000

axonal membrane indicates some degree of axonal shrinkage. Numerous membrane debris were visible around the axons, as well as macrophages filled with debris. Intact oligodendrocytes were visible. However very few damaged oligodendrocytes or oligodendroglial debris can also be observed. This suggests that the oligodendroglial population was partially destroyed in the myelinated area as concluded by Blakemore et al (1977), Blakemore (1983) . The debris disappeared very rapidly.

At D7, macrophages are still present in the lesions. But in the same time, oligodendrocytes start to ensheath actively the naked axons (fig.6). As during the developmental process of myelination the axons myelinated at the same time by the same oligodendrocytes have various diameters and are not at the same stage of ensheathment. At node of Ranvier, one axonal segment can be already covered by 4-5 lamellae while the adjacent segment is not yet covered by the first oligodendrocytic process. One-two months after the LPC injection the lesion can be considered as remyelinated in shiverer and in normal animals. The chronological process of remyelination appeared to be comparable in shiverer and normal spinal cord. Our observations are also comparable to those described in the litterature after injection of LPC in the normal adult mouse spinal cord (Hall, 1972 - Duncan et al, 1981). However in the normal mouse, even 3 months after the injection, the lesion was still very visible because the newly formed myelin was much thinner than the normal surrounding one. The remyelinated lesion appeared similar to the "shadow plaques" described in MS (Prineas, 1985). By contrast, at gross observation, the remyelinated shiverer lesion was more visible by the charcoal marking the point of LPC injection than by a specific morphological aspect. The newly formed myelin, was comparable to the surrounding myelin with a great irregularity in the number of lamellae and no clear relationships between the axonal diameter and the myelin thickness. The abnormalities observed in the adult shiverer myelin were however particularly frequent and severe in the remyelinated areas (fig. 6,7,8,9). From the first steps of remyelination, ensheathment by folding rather than spiralling can be observed. Numerous cytoplasmic loops thick enough to contain mitochondria were present between the lamellae. More frequently than in non demyelinated regions, largely incomplete myelin sheaths were clearly formed by folding of one oligodendrocyte process, sometimes covering less than one half of the axonal circumference. Schwann cell remyelination was never observed except in case of very traumatic superficial injection with an important destruction of the glia limiting membrane.

REMYELINATION IN THE ADULT SHIVERER SPINAL CORD BY GRAFTED OLIGODENDROCYTES

In these series of experiments the demyelination was performed in adult shiverer mice as described above. In the same time a graft was introduced in the spinal cord at a distance of 3-8mm from the demyelinated area, rostrally or caudally. The site of implantation was also marked with charcoal. The animal were killed between 9 days and 3 months after the operation and the spinal cords were processed for semi-thin sections and EM observation (Gout et al 1988a, b)

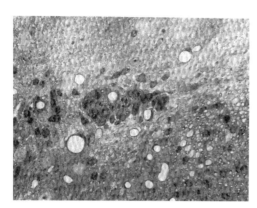 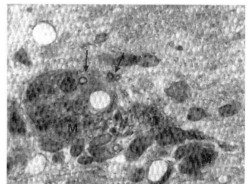

Aspect of the graft 3 months after the operation

Fig.10 : The graft is well delineated in the spinal cord parenchyma. X 130
Fig.11 : Higher magnification. Macrophage (M) Schwann cell myelination (arrows). X 1000

The graft was a fragment of olfactory bulb from a new-born normal mouse brain. It thus contained all the neural cell types at various stages of the development. Regarding the myelin forming cell lineage probably no oligodendrocytes characterized as Gal C (Galactosylceramide) positive cells (Raff et al, 1978) were present. Jacque et al (1985) assumed that the first GalC positive cells appear in the olfactory bulb of the normal mouse 3 days after birth. Their number increases slowly : at 7 days, the GalC positive cells represent only 8/1000 of the total cell population. Thus at the moment of the graft the myelinating cell population was completely represented by O2A progenitors (Raff et al, 1979, 1983) or earlier precursor cells.

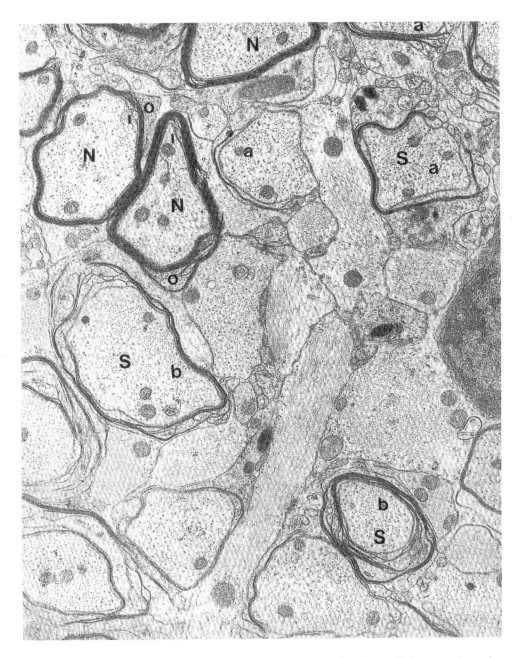

Fig.12 : Mixed remyelination in the shiverer spinal cord 3 months after LPC injection and graft Normal (N) and shiverer (S) myelin. The shiverer myelin presents all the abnormalities described in the text : uncomplete ensheathment by folding (a) internal loops of cytoplasm (b), irregularity of the myelin sheath. In the normal myelin, alternance of Major Dense and intraperiodic lines, and inner (I) and outer (O) cytoplasmic tongues evidence spiral wrapping. X 20000

Observed 2-3 months after implantation (fig. 10,11) the graft appeared most often well delineated in the spinal cord parenchyma.

However the basal lamina which develops and surrounds such CNS fragments when grafted into the newborn shiverer brain (Gansmuller et al, 1986) have never been observed in these series. Three months after transplantation, all types of neural cells could still be recognized inside the graft. Macrophages often containing charcoal debris were still present (fig.11) It was frequent to observe axons myelinated by Schwann cells at the periphery of the graft.

At the level of the lesion a spontaneous remyelination due to the shiverer oligodendrocytes and presenting the abnormalities already described occured in all the cases. Moreover axons in the previously demyelinated area were ensheated by normal myelin, MDL positive, thus formed by oligodendrocytes issued from the graft. This myelin did not present any of the defects characterictic of the shiverer myelin. The myelin was seen as a series of alternating dark (MDL) and less dark (intraperiod line) separated by unstained zone. This myelin surrounds the whole circumference of the axon, with an inner and outer cytoplasmic tongue. It is thus clearly formed by spiralling.

CONCLUSIONS AND DISCUSSION

A spontaneous remyelination occurs rapidly in the adult shiverer spinal cord following an acute demyelination by LPC injection. The chronology of the demyelination-remyelination process is comparable to that observed in the normal mouse (Hall, 1972). However, as it has been described in myelination (Rosenbluth, 1980-Inoue et al, 1981) the shiverer oligodendrocyte seems to function differently and to proceed abnormally to ensheath the axons. The normal steps of the myelination process are wrapping completely the axon, spiralling and compaction. In contrast, the shiverer mouse myelination and remyelination appears to proceed by folding rather than spiralling. This defective function of the shiverer oligodendrocyte leads to the abnormalities observed morphologically in the shiverer myelin : incomplete ensheathment of the axon, irregularities of the myelin thickness around the axonal circumference, presence of numerous cytoplasmic loops, absence of regular compaction of the lamellae, formation of myelin without relation with any axon. It worth noting that these abnormalities appears to be more severe in the remyelination than in the myelination process. However the shiverer remyelination myelin does not seem to be thinner than the adult shiverer myelin as it is the case in normal mouse.

When a graft containing normal oligodendrocyte precursor cells is placed at a distance of 3 up to 8 mm from the LPC induced demyelinated lesion in the shiverer spinal cord, cells were able to migrate from the graft to the lesion and participate to the remyelination (Gout et al, 1988a, b - Gumpel et al 1989a,b). In this case, the remyelination is mixed. Previously shiverer demyelinated axons are ensheathed either by normal myelin, or by shiverer myelin. Spiral wrapping of the axon by an oligodendrocytic process, followed by compaction and formation of the typical lamellae structure of normal myelin (i.e. alternance of MDL and interperiodic line all around the axon) evidenced that the "normal" myelin seen in the lesion was formed by transplanted normal oligodendrocytes. The mechanisms of remyelination by shiverer oligodendrocytes were similar to that observed in shiverer myelination or in case of remyelination without transplantation. Thus the process of myelin formation in the lesion is completely dependant of the origin (normal grafted or shiverer host) of the oligodendrocytes. This result reminds those obtained by Aguayo et al (1977) in the PNS : in cross transplantations between Trembler and normal mouse sciatic nerve, the myelin abnormalities depends upon the Schwann cells and are independant of the axon.

The shiverer mutation has been localized on chromosome 18 (Sidman et al, 1985) as well as the gene encoding for MBP in the mouse (Roach et al, 1985). In the shi/shi homozygous animals 5 out of 7 exons encoding for MBP are deleted (Roach et al, 1983 - Roach et al, 1985). Thus the deletion of the MBP gene leading to the absence of MBP in the CNS is considered to be responsible for the shiverer phenotype. This conclusion is emphasized by the fact that in transgenic mice, the expression of MBP gene corrects at least partially the dysmyelinating phenotype (Readhead et al, 1987). However, after a carefull observation of shiverer oligodendrocyte morphological abnormalities during the myelination and remyelination processes, the question raises up if the absence of MBP by itself can explain the mechanical defects of the myelin forming cells.

It is interesting to note that the remyelination in the adult shiverer spinal cord after LPC injection can be considered as completely due to oligodendrocytes. This fact was already reported in case of normal mouse (Duncan et al, 1981 - Aranella and Herndon, 1984). However some Schwann cell myelination is noted at the periphery of almost all the grafts. The presence of axons to myelinate at this place could correspond either to axonal projections from neurons present in the graft or to a sprouting from the host neurons damaged by the introduction of the graft Regarding the presence of Schwann cells, several hypotheses can be considered. Schwann cells are present in the CNS (Blakemore et al, 1987) Thus they could be present in the shiverer host environment as well as in the normal CNS grafted fragment. They could also be introduced by lesion of the roots during the transplantation procedure.

In mixed remyelination of a shiverer spinal cord lesion by transplantation of normal oligodendrocytes, the normal myelin appears rather thick, compared to normal remyelination myelin. This myelin is formed by normal immature grafted oligodendrocytes which myelinate for the first time. It would be interesting to know if the myelin formed by such cells around adult demyelinated axons is myelination or remyelination myelin. It should give some insights about the oligodendrocyte function at various steps of maturation.

ACKNOWLEDGEMENTS

This work was supported by INSERM, CNRS, ARSEP and Multiple Sclerosis Society (grant RG 1773A1) We thank M.Josien and F.Lachapelle for their help in the preparation of the manuscript.

REFERENCES

Aguayo AJ., Attiwell M., Trecarten J., Perkins J., Bray GM (1977) Abnormal myelination in transplanted Trembler mouse Schwann cells. Nature 265, 73-74.

Aranella LS., Herndon RM. (1984) Division following experimental demyelination in adult animals. Arch. Neurol. 41, 1162-1165

Billings-Gagliardi S., Wolf MK., Kirschner DA., Kerner AL. (1986) Shiverer-jimpy double mutant mice. II- Morphological evidence supports reciprocal intergenic suppression. Brain Res. 374, 54-62

Blakemore WF (1976) Invasion of Schwann cells into the spinal cord of the rat following local injections of lysolecithin. Neuropathol. Appl. Neurobiol. 2, 21-39

Blakemore WF. (1978) Observations on remyelination in the rabbit spinal cord following demyelination induced by lysolecithin. Neuropathol.Appl.Neurobiol. 4, 47-60

Blakemore WF (1979) Remyelination in the CNS. in "Progress in Neurological research" PO Behan, F.Clifford-Rose eds. Pitman Medical Bath, pp12-25

Blakemore WF (1981) Observations on myelination and remyelination in the central nervous system. in "Development in the nervous system" British Society for developmental biology. Symposium 5 D.R. Garrod and JD. Feldman eds. Cambridge University Press 289-308

Blakemore WF (1983) Remyelination of demyelinated spinal cord axons by Schwann cells. in "Spinal cord reconstruction. CC Kao, RP Bunge and PJ Reier eds. Raven Press, New-York, pp 281-291

Blakemore WF., Crang AJ. Evans RJ., Patterson RC.(1987) Rat Schwann cell remyelinisation of demyelinated cat CNS axons; evidence that injection of cell suspensions of CNS tissue results in Schwann cell remyelinisation. Neurosci. Lett. 77, 15-19

Blakemore WF., Eames RnA., Smith KJ., Mc Donald WI. (1977) Remyelination in the spinal cord of the cat following intraspinal injections of lysolecithin. J.Neurol.Sci. 33, 31-43

Bornstein MB., Appel SH., Murray MR. (1962). The application of tissue culture to the study of experimental allergic encephalomyelitis. Demyelination and remyelination. In H.Jacob ed. Proceedings IV International Congress of Neuropathology, 1961, Vol .2 Munich, Stuttgart, Georg Thieme Verlag, pp 279-282

Bunge MB., Bunge RP., Ris H. (1961) Ultrastructural study of remyelination in adult cat spinal cord. J.Biophys. Biochem. Cytol. 10, 67-94

Duncan ID., Aguayo AJ., Bunge RP., Wood PM. (1981) Transplantation of rat Schwann cells grown in tissue culture into the mouse spinal cord. J.Neurol.Sci. 49, 241-252

Dupouey P., Jacque C., Bourre JM, Cesselin F., Baumann N. (1979) Immunochemical studies of myelin basic protein in Shiverer mouse devoid of major dense line of myelin. Neurosci. Lett. 12, 113-118

Gansmuller A., Lachapelle F., Baron-Van Evercooeren A., Hauw JJ., Baumann N., Gumpel M. (1986) Transplantation of newborn CNS fragments into the brain of shiverer mutant mice. Extensive myelination by transplanted oligodendrocytes. II- Electron microscopic study Dev.Neurosci 8,197-207

Gout O., Gansmuller A., Baumann N., Gumpel M. (1988a) Remyelination by transplantated oligodendrocytes of a demyelinated lesion in the spinal cord of the adult shiverer mouse. Neurosci Lett. 87, 195-199

Gout O., Gansmuller A., Baumann N., Gumpel M. (1988b) Remyelination of a demyelinated lesion in the adult shiverer spinal cord by transplantation of normal neonatal cells. in Trends in European Multiple Sclerosis

Gumpel M., Gansmuller A., Lubetzki C., Lombrail P., Baron Van Evercooren A., Baulac M., Gout O., Baumann N., Lachapelle F. (1987) Transplantations intra-cérébrales d'oligodendrocytes chez la souris. Path. Biol. 35, 333-338

Gumpel M., Gout O., Gansmuller A.. (1989a) Spontaneous remyelination and intracerebral grafting of myelinating cells in the mammals. in "Neuronal Grafting and Alzheimer Disease" F.Gage, A.Privat and Y.Christen ed. Springer Verlag pp 43-53

Gumpel M., Gout O., Lubetzki C., Gansmuller A., Baumann N. (1989b) Myelination and remyelination in the central nervous system by transplanted oligodendrocytes using the shiverer model. Discussion on the remyelinating cell population in adult mammals. Dev. Neurosci 11, 132-139

Hall SM. (1972) The effect of injections of lypophosphatidyl choline into white matter of the adult mouse spinal cord. J.Cell Sci. 10, 535-546

Harrison BM, Mc Donald WI (1977) Remyelination after transient experimental compression of the spinal cord. Ann.Neurol. I, 542-551

Inoue Y., Nakamura R., Mikoshiba K., Tsukada Y. (1981) Fine structure of the central myelin sheath in the myelin deficient mutant shiverer mouse with special reference to the myelin formation by oligodendroglia. Brain Res. 219, 85-94

Jacque C., Collet A., Raoul M., Monge M., Gumpel M. (1985) Functional maturation of the oligodendrocytes and myelin basic protein expression in the olfactory bulb of the mouse. Dev. Brain. Res. 21, 277-282.

Lachapelle F., Gumpel M., Baulac M., Jacque C., Baumann N. (1983-84) Transplantation of fragments of CNS into the brain of shiverer mutant mice ; extensive myelination by implanted oligodendrocytes. I- Immuno- histochemical studies. Dev.Neurosci. 6, 326-33

Lassmann H. (1983) Comparative neuropathology of chronic experimental allergic encephalomyelitis and multiple sclerosis. Berlin-Heidelberg- New-York-Tokyo, Springer-Verlag.

Ludwin SK. (1978) Central nervous system demyelination and remyelination in the mouse. An ultrastructural study of Cuprizone toxicity. Lab. Invest. 39, 597-612

Ludwin SK. (1987) Remyelination in demyelinating diseases of the central nervous system. Critical review in Neurobiology 3, 1-28

Prineas J., Connell FW. (1979) Remyelination in multiple sclerosis. Ann. Neurol. 5, 22-31

Prineas JW. (1985). The neuropathology of multiple sclerosis. in "Handbook of clinical neurology, Vol 3 (47) : Demyelinating diseases. JC Koetsier ed. Elsevier Science Publishers, pp 213-225

Privat P., Jacque C., Bourre JM., Dupouey P., Baumann N. (1979) Absence of the major dense line in the mutant mouse Shiverer. Neurosci. Lett. 12, 107-112

Raff MC., Fields KL., Hakomori SI., Mirsky R.., Pruss RM., Winter J. (1979) Cell types specific markers for distinguishing and studying neurons and the major classes of glial cells in culture. Brain Res. 174, 283-308

Raff MC., Abney R., Cohen J., Lindsay R., Noble M. (1983) Two types of astrocytes in cultures of developing rat white matter : differences in morphology, surface gangliosides, and growth characteristics. J.Neurosci. 3, 1289-1300

Raff MC., Mirsky R.., Fields KL., Lisak PR., Dorfman SH., Silberberg DH., Gregson NA., Leibowitz S., Kennedy MC. (1978) Galactocerebroside as a specific cell-surface antigen marker for oligodendrocytes in culture. Nature, 274, 813-816

Readhead C., Popko B., Takahashi N., shine HD., Saavedra KA., Sidman RL., Hood L. (1987) Expression of a Myelin Basic Protein gene in transgenic shiverer mice : correction of the dysmyelinating phenotype. Cell, 48, 703-712

Roach A., Boylan K., Horvath S., Prusiner SB., Hood L. (1983) Characterization of cloned cDNA representing rat myelin basic protein absence of expression in brain of shiverer mutant mice. Cell. 34, 799-806

Roach A., Takahashi N., Pravtcheva D., Ruddle F., Hood L. (1985) Chromosomal mapping of mouse myelin basic protein gene and structure and transcription of the partially deleted gene in shiverer mutant mice. Cell. 42, 149-155

Rosenbluth J. (1980) Central myelin in the mouse mutant Shiverer. J.Comp.Neurol., 194, 639-648

Sidman RL., Conover CS., Carson JH. (1985) Shiverer gene maps near to the distal end of chromosome 18 in the house mouse. Cytogenet.cell genet. 39, 241-245

SIGNAL TRANSDUCTION AND REGULATORY EVENTS IN MYELIN-FORMING CELLS

cAMP-DEPENDENT PROTEIN KINASE: SUBUNIT DIVERSITY AND FUNCTIONAL ROLE IN GENE EXPRESSION

Matthias Meinecke, Wolfgang Büchler, Lilo Fischer, Suzanne M. Lohmann, and Ulrich Walter
Department of Internal Medicine
Laboratory of Clinical Biochemistry
University of Würzburg
Josef-Schneider-Str. 2
8700 Würzburg
West Germany

INTRODUCTION

The cAMP-dependent protein kinase (cAMP-PK) exists as an inactive tetramer of two regulatory (R) and two catalytic (C) subunits which are dissociated by cAMP to form an R dimer and two C monomers. The free C subunits are active for substrate phosphorylation. Several subunits have been purified and structurally defined by amino acid sequencing. More recently, heterogenous forms which appear to be products of different genes have been identified by cDNA sequencing. So far a greater number of R subunits than C subunits have been found. The spectrum of mammalian subunits is shown in Table I. From the four different R subunits and two different C subunits, at least 8 different tetrameric holoenzymes can be formed. This minimal number assumes only homodimers of identical R or C subunits. Heterodimers of types I and II R subunits are not found in a holoenzyme complex, however, it has not been demonstrated whether or not α and ß forms can combine to make heterodimers of R-I or R-II.

A functional correlate to subunit structural diversity has not yet been established. The cAMP-PK has important functions in several cellular compartments, such as regulation of intermediary metabolism in the cytosol (Beebe and Corbin, 1986; Edelman et al., 1987), ion channels in the cell membrane (Kaczmarek et al., 1980; Osterrieder et al., 1982; Levitan et al., 1983; and Schoumacher et al., 1987), and gene transcription in the nucleus (Comb et al., 1987; Roesler et al., 1988). Our functional studies reported here focus on the mechanism by which cAMP-PK subunits mediate cAMP-stimulated gene expression. A central question to be answered was whether the mechanism depends on a direct role of the R or the C subunit.

TABLE I.

mRNA Size (kb)**	Protein Mr (Kd)**		Subunit / Species		Sequence Reference	
	SDS-PAGE	Calculated			a.a.	nt
—	40	40.5	C	bovine heart	1	—
2.4	—	—	C_α	mouse lymphoma	—	2
4.3	—	—	C_β	mouse heart/brain	—	3
—	47	42.8	R-I	bov.skel. muscle	4	—
3.2(1.7)	—	55	$R-I_\alpha$	bovine testis	—	5
2.8	—	—	$R-I_\beta$	mouse brain	—	6
—	54	45	R-II	bovine heart	7	—
6.0 (2.4 testis)	—	51(brain)	$R-II_\alpha$	rat skel.muscle & mouse brain	—	8
3.2 (1.8)	52/51	—	$R-II_\beta$	rat ovary	—	9
—	74	76.3	cGMP-PK	bovine lung	10	—
6.2	—	76.4	$cGMP-PK_{\alpha,\beta}$	bovine lung	—	11
7.0 (4.2)	—	77.8	$cGMP-PK_\beta$	human placenta	—	12

**Sizes may vary somewhat with other species and tissues

1) Shoji et al., PNAS, 1981.
2) Uhler et al., PNAS, 1986.
3) Uhler et al., JBC, 1986.
4) Titani et al., Biochemistry, 1984.
5) Lee et al., PNAS, 1983.
6) Clegg et al., PNAS, 1988.
7) Takio et al., Biochemistry, 1984 a.
8) Scott et al., PNAS, 1987.
9) Jahnsen et al., JBC, 1986
10) Takio et al., Biochemistry, 1984 b.
11) Wernet et al., FEBS Lett., 1989
12) Sandberg et al., FEBS Lett., 1989

RESULTS

Regulatory Subunit Diversity

The cAMP-PK derives the greater part of its heterogeneity from its regulatory subunit (Lohmann et al., 1988 a). Sequence homologies of subdomains of R-II α and R-II$_\beta$ illustrate the specific nature of this heterogeneity (Fig. 1). The two cAMP binding sites are highly conserved, other parts of the surrounding carboxyl-terminal region somewhat less so. The amino-terminal region is the least constant and contains several functional domains which may be affected by this diversity. These domains were originally defined by biochemical functional analyses of bovine and porcine R-II. These include an R-R dimerization domain, corresponding approximately to amino acids 1-45 (Reimann, 1986), and the region of R-C interaction beginning with the two arginines preceding the ser-95 autophorphorylation site (Bechtel et al., 1977; Weber and Hilz, 1979; First et al., 1988) and extending as far as cysteine-97 (First et al., 1988), although amino acids 1-99 are apparently not sufficient (Edelman et al., 1987). The amino-terminal region of bovine heart R-II also contains phosphorylation sites for several different kinases. Autophosphorylation by C (Rangel-Aldao and Rosen, 1976) or phosphorylation by glycogen synthase kinase 3 (Hemmings et al., 1982) have some inhibitory effects on reassociation of R with C. These phosphorylation sites are not conserved among all the different R-II subunits shown in Fig. 1.

Thus a variable R amino-terminus could have several effects. It could possibly ensure against heterodimer formation between non-identical R forms. Interestingly, the kinase R subunit of Dictyostelium discoideum lacks amino acids prior to approximately number 66 of mouse brain R-II$_\alpha$, i.e. it lacks the dimerization domain, and it exists only as a monomer which can form a dimeric holoenzyme with C (Mutzel et al., 1987). A variant amino-terminus could cause holoenzymes composed of distinct R subunits to have different Ka's of activation (this has at least been demonstrated for R-I versus R-II in vitro, reviewed in Lohmann and Walter, 1984), or different kinetics of recombination with C due to their different extents of R subunit phosphorylation.

The antigenic site region borders the region of amino-terminal diversity. We have raised rabbit polyclonal antibodies that can selectively recognize R-II$_\alpha$ or R-II$_\beta$, but also other antibodies which can recognize both (DeCamilli et al., 1986, and unpublished data). Immunocytochemical studies in rat brain (DeCamilli, Lohmann, Walter, unpublished experiments) using two monospecific antibodies, each of which recognize a single R-II form, indicated that R-II$_\alpha$ and R-II$_\beta$ have some common

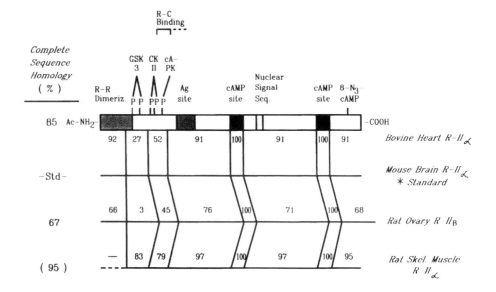

Fig. 1. **Homologies (%) between R subunits.** Complete protein sequences as well as certain subsites were analyzed using the GAP Wisconsin program with mouse brain R-IIα as the standard. GSK 3, glycogen synthase kinase 3; CK II, casein kinase II; cA-PK, cAMP-dependent protein kinase; Ag site, antigenic site; 8-N$_3$-cAMP, photoaffinity label of cAMP site.

localizations but also others which do not overlap. The distinct localizations of the R forms may be a result of specific R subunit interactions with other cellular proteins which have been demonstrated (Lohmann et al., 1988 b; Sarkar et al., 1984). More experiments are required to determine if the specificity of these interactions is encoded in the variable amino-terminal region of R.

Catalytic Subunit Diversity

The structure of C subunit and the mapping of important functional sites are shown in Fig. 2A. The catalytic subunit is much more highly conserved than is the regulatory subunit, in fact no distinct C subunit protein species have been distinguished. Only cDNA cloning has indicated that there are C$_\alpha$ and C$_\beta$ forms which are 91% identical, and even 95% identical in the large active site region (Fig. 2B). In contrast, the complete protein sequence homology between R-IIα and ß is only 67% (Fig. 1), and between R-IIα and R-I only 35% (Lohmann et al., 1988 a).

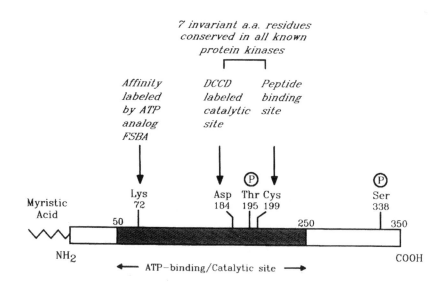

Fig. 2. **A. Defined regions of C subunit (Taylor, 1989).** FSBA, fluorosulfonylbenzoyl 5'-adenosine; DCCD, dicyclohexylcarbodiimide. **B. Homologies between C subunits.** Complete protein sequences as well as the active site region were compared using the GAP Wisconsin program with mouse lymphoma C_α as the standard.

Function of cAMP-Dependent Protein Kinase Subunits in Gene Expression

Cyclic AMP has been shown to stimulate an increase in the mRNA for a large number of mammalian proteins. Deletions in the promoter region of cAMP-regulated genes have defined a conserved cAMP response element (CRE) with enhancer characteristics, usually within 100-200 nucleotides upstream of the transcriptional start (Montminy et al., 1986; Comb et al., 1987; Roesler et al., 1988). The model shown in Fig. 3 indicates possible ways in which the cAMP-dependent protein kinase could mediate the effect of cAMP on gene expression.

It was considered that in mammals the kinase regulatory subunit might bind to DNA and activate transcription analogous to the way in which the cAMP-binding protein CAP, or CRP, in the prokaryote E-coli activates the lac operon (Nagamine and Reich, 1985). More recently, a high affinity cAMP-dependent binding of bovine heart R-II to the CRE consensus sequence was demonstrated (Wu and Wang, 1989). None of these studies however provided functional evidence for R-stimulated gene transcription. The cAMP-containing phosphoform of R-II from rat liver was reported to have intrinsic DNA topoisomerase activity (Constantinou et al., 1985), but another study refuted this, maintaining that the activity could be separated from R-II (Shabb

Fig. 3.

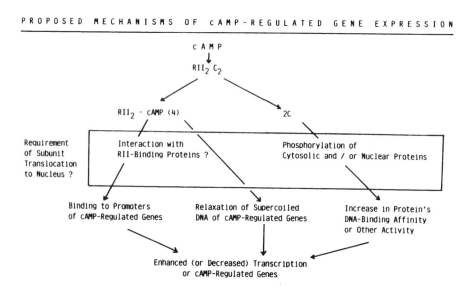

and Granner, 1988). Although R-II $_\alpha$ and ß contain what appears to be an acceptable nuclear signal sequence (see Fig. 1), based on known signal sequences and allowed substitutions (Dingwall et al., 1987), the use of this sequence or its significance in the mechanism of cAMP-regulated gene expression is not clear. C subunit has been shown to translocate from its Golgi area location into the nucleus upon stimulation by forskolin (Nigg et al., 1985) and it is small enough (40 Kd) not to need a signal sequence to enter the nucleus. These studies do not however established a functional link between C translocation and its being required for cAMP-stimulated gene expression. In Dictyostelium discoideum, a lower eukaryote, reports have indicated that both the the cell surface cAMP receptor (Oyama and Blumberg, 1986; Gomer et al., 1986), as well as the catalytic subunit of the intracellular cAMP-dependent protein kinase (Simon et al., 1989), mediate cAMP effects on gene expression. So far our results and those of others support a mechanism of C subunit phosphorylation of a substrate protein which participates in mammalian transcription activation.

Table II lists inhibitors of cAMP-dependent protein kinase and their effects on cAMP-stimulated mRNA levels in hepatocytes in primary cell culture. Our results indicated that C subunit activity is essential for the effect of cAMP on stimulation of

Table II.

EFFECT OF VARIOUS INHIBITORS OF cAMP-DEPENDENT PROTEIN KINASE ON GENE EXPRESSION IN HEPATOCYTES

Substance	Mechanism of Action	In vitro Inhibition Constant (K_i)	Intact Hepatocytes	
			Concentration for 1/2 max. Inhibition of $mRNA^{tat}$ or $mRNA^{pepck}$	Effect on $mRNA^{albumin}$
Rp-cAMP	Binds to cAMP-PK but inhibits holoenzyme dissociation	5 μM	40 - 50 μM	None
H8 (Isoquinoline-Sulfonamide)	Competitive inhibitor of ATP binding to catalytic subunit	3.5 μM	20 - 40 μM	None
PKI Peptide (Mr 2264)	High affinity competitive inhibitor (pseudosubstrate) of catalytic subunit	5 nM	500 μM	None
(ProProGly)$_{10}$ Peptide (Mr 2292)	Control peptide with no effect on cAMP-PK	——	No effect	None

phosphoenol pyruvate carboxykinase (pepck) and tyrosine aminotransferase (tat) mRNA levels (Büchler et al., 1987, 1988). Others have shown that transfection of a minigene for expression of the Walsh protein kinase inhibitor (PKI) fragment could inhibit cAMP-stimulated enkephalin gene expression in CV1 of COS-M6 cells (Grove et al., 1987). Likewise transfection of the cDNA for the entire PKI inhibited cAMP-stimulated prolactin gene transcription in rat GH_3 pituitary tumor cells, however it also inhibited the effects of several other agents such as epidermal growth factor, thyrotropin-releasing hormone, phorbol esters, and estrogen (Day et al., 1989).

To rule out possible ambiguities in the action of such inhibitors, we decided to test the action of R and C subunits directly by transfection and expression of the cloned cDNAs for these subunits in living cells. The protocol scheme for these experiments is shown in Fig. 4. The two transfected vectors were 1) pCEV, a C subunit expression vector (or an identically constructed R subunit expression vector, pREV), containing a Zn^{++}-regulated metallothionein promoter and a human growth hormone poly A signal, and 2) a reporter gene, pVIP-CAT, containing a CRE in -2 kb of the vasoactive intestinal peptide (VIP) promoter linked to the chloramphenicol acetyltransferase (CAT) coding cDNA and SV 40 poly A.

Fig. 4.

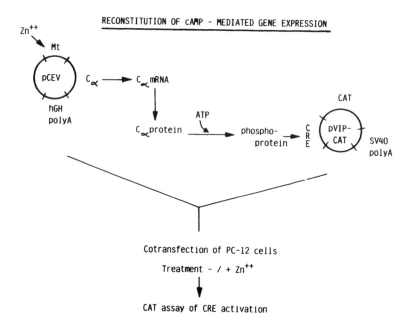

Cotransfection of pCEV and pVIP-CAT vectors into rat pheochromocytoma PC-12 cells, followed by CAT assay, showed a dose-dependent effect of C expression on CAT activity (Fig. 5A). Northern blot analysis demonstrated the presence of both endogenous C mRNA (2.4 kb) and the mRNA made from transfected pCEV cDNA (1.7 kb) which had a truncated 3'-untranslated region (Fig. 5B). The 1.7 kb mRNA was found only in the PC-12 cells transfected with pCEV, but not those transfected with pREV or untransfected. The maximal CAT activity observed with transfected pCEV was 50 times that of basal activity in untransfected cells or in cells transfected only with pVIP-CAT (Fig. 5A). This activity was 5 times greater than that caused by maximal 8-chlorophenylthio-cAMP (8CPT-cAMP) stimulation, reflecting that in transfected cells, transfected C was overexpressed beyond the level of endogenous C and that this was essentially free excess C that had no complementary R with which it could reassociate and be inactivated. This is not the case in Fig. 5B since there endogenous C is present in all cells whereas the transfected C is expressed in only the small per cent of transfected cells (Büchler et al., submitted for publication). Other experiments (not shown) demonstrated that pREV and pCEV cotransfection (ratio 2:1 and 4:1) together with pVIP-CAT caused 70 and 90% inhibition respectively of the pCEV stimulation of CAT activity, confirming the role of R subunit in this process to be limitation of the magnitude of C's effect on gene expression.

Recent results from other labs agree with our experiments and conclusions. The other studies were done by microinjection of C or R subunit protein into single rat C6 glioma cells which had been previously transfected with a pVIP-ßgal vector whose activation was assessed by expressed ß-galactosidase causing a substrate conversion-based cell color change (Riabowol et al., 1988). Other groups showed that expression of a C subunit transfection vector could stimulate transcription of the α-subunit of the glycoprotein hormones in JEG-3 cells (Mellon et al., 1989), and that either C$_\alpha$ or C$_\beta$ expression could stimulate prolactin transcription in GH$_3$ cells (Mauer, 1989).

DISCUSSION

Our results provide direct evidence that C and not R subunit can direct cAMP effects on gene expression. Since inhibitors of C subunit activity also blocked cAMP-stimulated mRNA levels, it would appear that C subunit phosphorylates a cellular protein which becomes the subsequent mediator of the cAMP signal (as shown in Figs. 3 and 4). Such a protein, designated CREB, has been purified from nuclear extracts using an affinity column containing a synthetic oligonucleotide CRE-sequence;

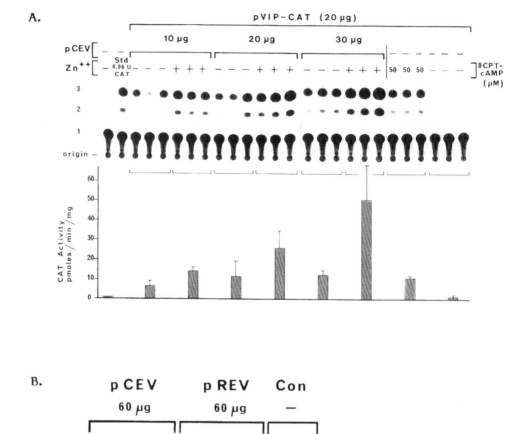

Fig. 5. A. Dependence of CAT activity stimulation on the amount of pCEV used to transfect PC-12 cells. B. Northern blot analysis of C subunit mRNAs in untransfected PC-12 cells (Con), or cells transfected with pCEV or pREV.

subsequently CREB was cloned (Hoeffler et al., 1988; Gonzalez et al., 1989). CREB has been shown to be an in vitro and in vivo substrate of the cAMP-dependent protein kinase (Montminy and Bilezikjian, 1987). Apparently in vitro transcription systems can be stimulated with either C subunit (Nakagawa et al., 1988) or CREB (Yamamoto et al., 1988). Since both C and CREB are cloned, they can now be transfected individually and together to test whether this can reconstitute the cAMP-dependent mechanism of gene regulation.

The CRE consensus sequence to which CREB binds is found in the promoter regions of a large number of cAMP-regulated genes (Montminy et al., 1986), some of which are shown in Fig. 6. In addition, VIP (Tsukada et al., 1987) and proenkephalin (Comb et al., 1988) have another site as shown. The site shown for TAT may not be functional, based on deletion studies from other labs. Promoters of c-fos and cytomegalovirus genes also contain CRE-like sequences (Montminy et al., 1986). A similar sequence is also found in the promoters of several adenovirus genes, the bovine leukemia virus long terminal repeat (LTR), and the human T-cell leukemia virus LTR, which bind transcription factors ATF (Hai et al., 1988), $p38^{tax}$ (Katoh et al., 1989), and NF-kB (Shirakawa et al., 1989), respectively. The various transcription factors have some ability to cross stimulate promoters other than their respective

Fig. 6.

	CRE	Transcription Factor
CONSENSUS :	T G A C G T C A	
Hormones / Transmitters / Enzymes		
• Somatostatin	-48 . . . . -41	CREB
• VIP (-86 CGTCA -82)	-76 . . . . T -69	
• Tyrosine hydroxylase	-45 . . . . -38	
• PEPCK	-90 . T . . . . -83	
• TAT	- 7 . . . G C 0	
Proenkephalin	-95 . . ▲ . . . . . . -89	
(-104 CGTAG -100)		
∝-Chorionic gonadotropin	-143 . . . . -136	
	-123 . . . . -116	
Oncogene		
c-fos	-66 . . . . . . ▲ . ˙ -60	—
Viruses		
Cytomegalovirus	-140 . . . . -133	—
Adenovirus	⌐ 10 sequences ¬	
(E1A, E2, E3, E4)	-438 . . . . . ⓉA Ⓐ -41	ATF
	A Ⓒ G	
BLV LTR	-160 A . . . . -153	$p38^{tax}$
HTLV II LTR	-178 . . . ⁀. . . . C -171	NF-kB
CONSENSUS PHORBOL ESTER :	. . . ▲ G . . .	AP1
cAMP (E.coli) :	T G . . . . ˙	CAP / CRP

ones. An important issue to resolve in the future will be whether multiple signal transduction pathways that regulate gene expression do so via separate or common transcription factors and DNA binding sites, and the nature of the specificities of these signals that enable the cell to distinguish them.

ACKNOWLEDGEMENTS

This work was supported by the Deutsche Forschungsgemeinschaft SFB 176 A5. M.M. was supported by a grant from the Wilhelm-Sander Stiftung.

REFERENCES

Bechtel PJ, Beavo JA, Krebs EG (1977) Purification and characterization of catalytic subunit of skeletal muscle adenosine 3':5'-monophosphate-dependent protein kinase. J. Biol. Chem. 252: 2691-2697

Beebe SJ, Corbin JD (1986) Cyclic nucleotide-dependent protein kinases. In: Boyer PD, Krebs EG (eds) The Enzymes. Control by phosphorylation, Part A, 3rd edn. Academic Press, London, pp 43-111

Büchler W, Meinecke M, Chakraborty T, Jahnsen T, Walter U, Lohmann SM (1989) Regulation of gene expression by transfected subunits of cAMP-dependent protein kinase. Submitted.

Büchler W, Walter U, Jastorff B, Lohmann SM (1988) Catalytic subunit of cAMP-dependent protein kinase is essential for cAMP-mediated mammalian gene expression. FEBS Lett.228: 27-32

Büchler W, Walter U, Lohmann SM (1987) Involvement of C subunit of cAMP-dependent protein kinase in mediation of gene expression in hepatocytes. J. Cell. Biol. 105: 65a

Clegg CH, Cadd GG, McKnight GS (1988) Genetic characterization of a brain-specific form of the type I regulatory subunit of cAMP-dependent protein kinase. Proc. Natl. Acad. Sci. USA 85: 3703-3707

Comb M, Hyman SE, Goodman HM (1987) Mechanisms of trans-synaptic regulation of gene expression. TINS 10: 473-478

Comb M, Mermod N, Hyman SE, Pearlberg J, Ross ME, Goodman HM (1988) Proteins bound at adjacent DNA elements act synergistically to regulate human proenkephalin cAMP inducible transcription. EMBO J. 7: 3793-3805

Constantinou AI, Squinto SP, Jungmann RA (1985) The phosphoform of the regulatory subunit R-II of cyclic AMP-dependent protein kinase possesses intrinsic topoisomerase activity. Cell 42: 429-437

Day RN, Walder JA, Mauer RA (1989) A protein kinase inhibitor gene reduces both basal and multihormone-stimulated prolactin gene transcription. J. Biol. Chem. 264: 431-436

DeCamilli P, Moretti M, Denis Donini S, Walter U, Lohmann SM (1986) Heterogeneous distribution of the cAMP receptor protein R-II in the nervous system: Evidence for its intracellular accumulation on microtubules, microtubule-organizing centers, and in the area of the Golgi complex. J. Cell Biol. 103: 189-203

Dingwall C, Dilworth SM, Black SJ, Kearsey SE, Cox LS, Laskey RA (1987) Nucleoplasmin cDNA sequence reveals polyglutamic acid tracts and a cluster of sequences homologous to putative nuclear localization signals. EMBO J. 69-74

Edelman AM, Blumenthal DK, Krebs EG (1987) Protein serine/threonine kinases. In: Richardson CC, et al. (eds) Annual Review of Biochemistry vol 56. Palo Alto, California, p 567-613

First EA, Bubis J, Taylor SS (1988) Subunit interaction sites between the regulatory and catalytic subunits of cAMP-dependent protein kinase. J.Biol.Chem. 263: 5176-5182

Gomer RH, Armstrong D, Leichtling BH, Firtel RA (1986) cAMP induction of prespore and prestalk gene expression in Dictyostelium is mediated by the cell-surface cAMP receptor. Proc. Natl. Acad. Sci. USA 83: 8624-8628

Gonzalez GA, Yamamoto KK, Fischer WH, Karr D, Menzel P, Biggs III W, Vale WW, Montminy MR (1989) A cluster of phosphorylation sites on the cyclic AMP-regulated nuclear factor CREB predicted by its sequence. Nature 337: 749-752

Grove JR, Price DJ, Goodman HM, Avruch J (1987) Recombinant fragment of protein kinase inhibitor blocks cyclic AMP-dependent gene transcription. Science 238: 530-533

Hai T, Liu F, Allegretto EA, Karin M, Green MR (1988) A family of immunologically related transcription factors that includes multiple forms of ATF and AP-1. Genes & Develop. 2: 1216-1226

Hemming BA, Aitkin A, Cohen P, Rymond M, Hofmann F (1982) Phosphorylation of the type-II regulatory subunit of cyclic AMP-dependent protein kinase by glycogen synthase kinase 3 and glycogen synthase kinase 5. Eur. J. Biochem. 127: 473-481

Hoeffler JP, Meyer TE, Yun Y, Jameson JL, Habener JF (1988) Cyclic AMP-responsive DNA-binding protein: structure based on a cloned placental cDNA. Science 242: 1430-1433

Jahnsen T, Hedin L, Kidd VJ,. Beattie WG, Lohmann SM, Walter U, Durica J, Schulz TZ, Schiltz E, Browner M, Lawrence CB, Goldman D, Ratoosh SL, Richards JS (1986) Molecular cloning, cDNA structure, and regulation of the regulatory subunit of type II cAMP-dependent protein kinase from rat ovarian granulosa cells. J. Biol. Chem. 261: 12352-12361

Kaczmarek LK, Jennings KR, Strumwasser F, Nairn AC, Walter U, Wilson FD, Greengard P (1980) Microinjection of catalytic subunit of cyclic AMP-dependent protein kinase enhances calcium action potentials of bag cell neurons in cell culture. Proc. Acad. Sci. USA 77: 7487-7491

Katoh I, Yoshinaka Y, Ikawa Y (1989) Bovine leukemia virus trans-activator p[38tax] activates heterologous promoters with a common sequence known as a cAMP-responsive element or the binding site of a cellular transcription factor ATF. EMBO J. 8: 497-503

Lee DC, Carmichael DF, Krebs EG, McKnight GS (1983) Isolation of a cDNA clone for the type I regulatory subunit of bovine cAMP-dependent protein kinase. Proc. Natl. Acad. Sci. USA 80: 3608-3612

Levitan IB, Lemos JR, Novak-Hofer I (1983) Protein phosphorylation and the regulation of ion channels. TINS 6: 496-499

Lohmann SM, Büchler W, Meinecke M, Walter U (1988) Multiple forms of cAMP-dependent protein kinase and their role in cAMP-mediated hormone regulation and gene expression. In: Imura H, et al. (eds) Progress in endocrinology. Elsevier, Amsterdam, p989-994

Lohmann SM, DeCamilli P, Walter U (1988) Type II cAMP-dependent protein kinase regulatory subunit-binding proteins. Methods in Enzymology 159: 183-193

Lohmann SM, Walter U (1984) Regulation of the cellular and subcellular concentrations and distribution of cyclic nucleotide-dependent protein kinases. In: Greengard, P. et al. (eds) Advances in Cyclic Nucleotide and Protein Phosphorylation Research vol 18. Raven Press, New York, p 63-117

Maurer RA (1989) Both isoforms of the cAMP-dependent protein kinase catalytic subunit can activate transcription of the prolactin gene. J. Biol. Chem. 264: 6870-6873

Mellon P, Clegg CH, Correll LA, McKnight GS (1989) Regulation of transcription by cyclic AMP-dependent protein kinase. Proc. Natl. Acad. Sci. USA 86: 4887-4891

Montminy MR, Bilezikjian LM (1987) Binding of a nuclear protein to the cyclic-AMP response element of the somatostatin gene. Nature 328: 175-178

Montminy MR, Sevarino KA, Wagner JA, Mandel G, Goodman RH (1986) Identification of a cAMP-responsive element within the rat somatostatin gene. Proc. Natl. Acad. Sci. USA 83: 6682-6686

Mutzel R, Lacombe M-L, Simon M-N, de Gunzberg J, Veron M (1987) Cloning and cDNA sequence of the regulatory subunit of cAMP-dependent protein kinase from Dictostelium discoideum. Proc. Natl. Acad. Sci. USA 84: 6-10

Nagamini Y, Reich E (1985) Gene expression and cAMP. Proc. Natl. Acad. Sci. USA 82: 4606-4610

Nakagawa J, von der Ahe D, Pearson D, Hemmings BA, Shibahara S, Nagamine Y (1988) Transcriptional regulation of a plasminogen activator gene by cyclic AMP in a homologous cell-free system. J. Biol. Chem. 263: 2460-2468

Nigg EA, Hilz H, Eppenberger HM, Dutly F (1985) Rapid and reversible translocation of the catalytic subunit of cAMP-dependent protein kinase type II from the Golgi complex to the nucleus. EMBO J. 11: 2801-2806

Osterrieder W, Brum G, Hescheler J, Trautwein W, Flockerzi V, Hofmann F (1982) Injection of subunits of cyclic AMP-dependent protein kinase into cardiac myocytes modulates Ca^{2+} current. Nature 298: 576-578

Oyama M, Blumberg DD (1986) Interaction of cAMP with the cell-surface receptor induces cell-type-specific mRNA accumulation in Dictyostelium discoideum. Proc. Natl. Acad. Aci. USA 83: 4819-4823

Rangel-Aldao R, Rosen OM (1976) Dissociation and reassociation of phosphorylated and nonphosphorylated forms of cAMP-dependent protein kinase from bovine cardiac muscle. J. Biol. Chem. 251: 3375-3380.

Reimann EM (1986) Conversion of bovine cardiac adenosine cyclic 3',5'-phosphate dependent protein kinase to a heterodimer by removal of 45 residues at the N-terminus of the regulatory subunit. Biochemistry 25: 119-125

Riabowol KT, Fink JS, Gilman MZ, Walsh DA, Goodman RH, Feramisco JR (1988) The catalytic subunit of cAMP-dependent protein kinase induces expression of genes containing cAMP-responsive enhancer elements. Nature 336: 83-86

Roesler WJ, Vandenbark GR, Hanson RW (1988) Cyclic AMP and the induction of eukaryotic gene transcription. J. Biol. Chem. 263: 9063-9066

Sandberg M, Natarajan V, Ronander I, Kalderon D, Walter U, Lohmann SM, Jahnsen T (1989) Molecular cloning and predicted full-length amino acid sequence of the type I_β isozyme of cGMP-dependent protein kinase from human placenta. Tissue distribution and developmental changes. FEBS Lett. 255: 321-329

Sarkar D, Erlichman J, Rubin CS (1984) Identification of a calmodulin-binding protein that co-purifies with the regulatory subunit of brain protein kinase II. J. Biol. Chem. 259: 9840-9846

Schoumacher RA, Shoemaker RL, Halm DR, Tallant EA, Wallace RW, Frizzell RA (1987) Phosphorylation fails to activate chloride channels from cystic fibrosis airway cells. Nature 330: 752-754

Scott JD, Glaccum MB, Zoller MJ, Uhler MD, Helfman DM, McKnight GS, Krebs EG (1987) The molecular cloning of a type II regulatory subunit of the cAMP-dependent protein kinase from rat skeletal muscle and mouse brain. Proc. Natl. Acad. Sci. USA 84: 5192-5196

Shabb JB, Granner DK (1988) Separation of topoisomerase I activity from the regulatory subunit of type II cyclic adenosine monophosphate-dependent protein kinase. Mol. Endocrinol. 2: 324-331

Shirakawa F, Chedid M, Suttles J, Pollok BA, Mizel SB (1989) Interleukin 1 and cyclic AMP induce k immunoglobin light-chain expression via activation of an NF-kB-like DNA- binding protein. Mol. Cell. Biol. 9: 959-964

Shoji S, Parmelee DC, Wade RD, Kumar S, Ericsson LH, Walsh KA, Neurath H, Long GL, Demaille JG, Fischer EH, Titani K (1981) Complete amino acid sequence of the catalytic subunit of bovine cardiac muscle cyclic AMP-dependent protein kinase. Proc. Natl. Acad. Sci. USA 78: 848-851

Simon M-N, Driscoll D, Mutzel R, Part D, Williams J, Veron M (1989) Overproduction of the regulatory subunit of the cAMP-dependent protein kinase blocks the differentiation of Dictyostelium discoideum. EMBO J. 8: 2039-2043

Takio K, Smith SB, Krebs EG, Walsh KA, Titani K (1984 a) Amino acid sequence of the regulatory subunit of bovine type II adenosine cyclic 3',5'-phosphate dependent protein kinase. Biochemistry 23: 4200-4206

Takio K, Wade, RD, Smith SB, Krebs EG, Titani K (1984 b) Guanosine cyclic 3', 5'-phosphate dependent protein kinase, a chimeric protein homologous with two separate protein families. Biochemistry 23: 4207-4218

Taylor SS (1989) cAMP-dependent protein kinase, model for an enzyme family. J. Biol. Chem. 264: 8443-8446

Titani K, Sasagawa T, Ericsson LH, Kumar S, Smith SB, Krebs EG, Walsh KA (1984) Amino acid sequence of the regulatory subunit of bovine type I adenosine cyclic 3', 5'-phosphate dependent protein kinase. Biochemistry 23: 4193-4199

Tsukada T, Fink JS, Mandel G, Goodman RH (1987) Identification of a region in the human vasoactive intestinal peptide gene responsible for regulation by cyclic AMP. J.Biol. Chem. 262: 8743-8747

Uhler MD, Carmichael DF, Lee DC, Chrivia JC, Krebs EG, McKnight GS (1986) Isolation of cDNA clones coding for the catalytic subunit of mouse cAMP-dependent protein kinase. Proc. Natl. Acad. Sci. USA 83: 1300-1304

Uhler MD, Chrivia JC, McKnight GS (1986) Evidence for a second isoform of the catalytic subunit of cAMP-dependent protein kinase. J. Biol. Chem. 261: 15360-15363

Weber W, Hilz H (1979) Stoichiometry of cAMP binding and limited proteolysis of protein kinase regulatory subunits R-I and R-II. Biochem. Biophys. Res. Commun.90: 1073-1081

Wernet W, Flockerzi V, Hofmann F (1989) The cDNA of the two isoforms of bovine cGMP-dependent protein kinase. FEBS Lett. 251: 191-196

Wu JC, Wang JH (1989) Sequence-selective DNA binding to the regulatory subunit of cAMP-dependent protein kinase. J. Biol. Chem. 264: 9989-9993

Yamamoto KK, Gonzalez GA, Biggs III WH, Montminy MR (1988) Phosphorylation-induced binding and transcriptional efficiency of nuclear factor CREB. Nature 334: 494-498

PROTEIN KINASE C IN NEURONAL CELL GROWTH AND DIFFERENTIATION

J.F. Kuo
Department of Pharmacology
Emory University School of Medicine
Atlanta, Georgia 30322, U.S.A.

INTRODUCTION

Protein kinase C (PKC) system is particularly predominant in nervous tissue compared to other tissues as indicated by a high level of the enzyme activity (Kuo et al., 1980) or amount (Girard et al., 1986; Yoshida et al., 1988) and by an abundant occurrence of its substrate proteins (Wrenn et al., 1980). It is likely, therefore, that PKC plays a pivotal role in neuronal function and regulation. In this chapter, I summarized some of our work on immunocytochemical localization of PKC and phosphorylation of endogenous proteins in brain and neuroblastoma cells as they are related to development and differentiation.

MATERIALS AND METHODS

These were indicated in the legends to the figures or detailed in the references cited.

RESULTS AND DISCUSSION

Immunocytochemical localization of PKC

The histogenesis and morphogenesis of rat cerebellum, including cell growth and migration, neuronal differentiation and synaptogenesis, occur mainly after birth (Altman, 1972). We reported previously that PKC catalytic activity (Turner et al., 1984) and amount (Girard et al., 1986) increased postnatally in this and other brain regions. More recently, Yoshida et al. (1988) further delineated the developmental increase in the three isoforms (types I, II and III) of PKC in rat brain. We investigated immunocytochemical localization of PKC, using rabbit polyclonal antisera raised against PKC (presumably a mixture of all isozymes) purified from pig brains (Girard et al., 1985), in certain regions of developing and adult rat brain (Girard et al., 1985, 1988; Wood et al, 1986). In low magnification images of 1- to 4-day postnatal rat cerebellum, prominent PKC immunoreactivity was observed in the region immediately adjacent to the early-forming internal granule cell layer occupied by fibers that would later become the white matter of the cerebellar cortex. At this level of magnification, a band of reaction product was also apparent at the Purkinje cell layer. At all stages, the PKC immunoreactivity in the fiber pathways entering and leaving the cerebellar cortex was present, but the intensity of staining diminished as development proceeded, resulting in a lower level of staining in axons of the white matter as observed previously for adult rat tissue. Thus, PKC immunoreactivity in the fiber tracts of the cerebellum was high in early development and decreased as development proceeded. In order to study the distribution of PKC immunoreactivity in forming synaptic contacts, we concentrated on an examination of the Purkinje cell, particularly its dendritic tree. In 2- and 4-day cerebellar cortex, PKC immunoreactivity was observed in the region of apical dendrites of Purkinje cells. Little staining was evident in the proliferative layer beneath the pial membrane. At approximately 4-5 days postnatal, there is a growth of Purkinje cell apical dendritic processes into the overlying molecular layer. A prominent immunolocalization of PKC was apparent at 4 and 6 days postnatal in the region of elongating processes containing extensive growth cones and early-forming synapses. In regard to the presence of immunolabel in extending axons and growth cones, it may be of some interest that a major substrate for PKC in brain appears to be B-50 protein (Aloyo et al., 1983) or growth associated protein (GAP-43) (Jacobson et al., 1986; Meiri et al., 1986). GAP-43 is an acidic M_r 43,000 protein that is part of a family of proteins whose levels are elevated during periods of axonal growth (Meiri et al. 1986). Although it is not yet possible to define an exact role for B-50/GAP-43, its increased level in development and regeneration, and its increased phosphorylation accompanying the induction of long-term potentiation in the rat hippocampal formation (Akers et al., 1986), are consistent with a major role of PKC in neuritic outgrowth of various types. Such a role is further strengthened by the findings showing high levels of PKC immunoreactivity is axons and forming synaptic terminals in developing brain.

PKC immunolabel within the nucleus of Purkinje neurons underwent an interesting developmentally regulated change in distribution. At early time points when there was very intense immunoreactivity in axons, growth cones, and early-forming synapses, there was no immunolabel in Purkinje cell nuclei. Beginning at approximately 10 days postnatal, and clearly apparent at 14 and 21 days, prominent immunoreactivity appeared in the Purkinje cell nuclei, predominantly in the nuclear envelope. Thus the antigenic sites on PKC recognized by our antisera appeared in neuritic processes of differentiating neurons prior to their appearance in neuronal nuclei.

Although distribution of anti-PKC immunoreactivity in adult rat brain was widespread, its concentration in the inner nuclear membrane (Wood et al., 1986), presynaptic terminals (Girard et al., 1985; Wood et al., 1986) and oligodendroglia (Girard et al., 1985) seems worth noting. An extensive analysis of neuronal somal labeling at the electron-microscopic level revealed a gradation of reaction product partially related to neuron type and partially to the depth of immunocytochemical analysis in tissue sections. Purkinje neurons exhibited a dense band of reaction product adjacent to the inner nuclear membrane and a sprinkling of reaction product throughout the remainder of the nucleus (Fig. 1, A and B). There were focal depositions of reaction product throughout

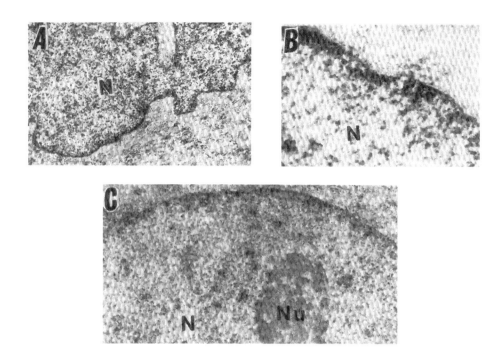

Figure 1. Electron micrographs of cerebellar cortex (A,B) and hippocampus (C) illustrating the fine structural distribution of PKC immunoreactivity. A striking distribution is seen at the nuclear envelope of Purkinje neurons (N) and sprinkling of reaction product is found throughout the nuclei. High-magnification views reveal that reaction product is concentrated against the inner nuclear membrane but nucleolus (Nu) appears unlabeled. (A) x 6,500; (B) x 37,000; (C) x 38,000. Adapted with permission from Wood et al. (1986).

the cytoplasm (Fig. 1A). Neurons in the dentate gyrus showed essentially the same staining pattern but the immunostain was frequently more restricted to the sites immediately adjacent to the inner nuclear membrane (Fig. 1C). It is attractive to speculate that some major substrate proteins for PKC might be located at the same sites and that they might be critical in nuclear function regulated by PKC.

We observed an uneven distribution of immunoreactivity within neuronal processes in rat brain. Electron micrographs revealed that it was highly concentrated in, for example, presynaptic terminals of cerebral cortex (Fig. 2A) and mossy fiber terminals of hippocampus (Fig. 2B), only rarely were labeled postsynaptic specialization elements seen. It is important to note, however, that postsynaptic structures could be labeled when high concentrations of antibodies were used. The immunostaining within presynaptic terminals tended to associate with various organelles, especially synaptic vesicles. However, it was not possible to delineate the exact organelles with which PKC was preferentially associated. Nonetheless, these results are consistent with a potential major role for PKC in presynaptic functions, perhaps related to phosphorylation of proteins such as synapsin (Ouimet et al., 1984).

A close inspection of white matter regions showed an interesting distribution of immunostain. The myelin itself was unstained with these experimental protocols, and the axons were moderately stained (Fig. 3A). There were labeled cells throughout the white matter that were characterized by a small oval soma from which fine labeled processes emanated. Although it was not possible to unequivocally identify these cells at the light-microscopic level, their morphological features and location strongly suggested that they were oligodendroglia. A major substrate for PKC in brain is myelin basic protein (Turner et al., 1982). Since oligodendroglia play a prominent role in the synthesis of myelin components, these localization results are consistent with a role for the enzyme in myelin function. To further assess the distribution of PKC in white matter (an area where penetration of antibodies is sometimes restricted), we treated some vibratome slices with 0.25% Triton X-100 in P_i/NaCl for 20 min before performing antibody cytochemistry. These slices revealed enhanced immunoreactivity in the axons and in the oligodendroglia (Fig. 3B) but no obvious increase in staining of the myelin sheath itself or of grey matter regions. These results suggest that the axons and oligodendroglia contain pools of anti-PKC immunoreactivity that exhibit restricted access to primary antibody or other reagents used in the protocols.

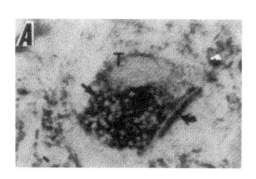
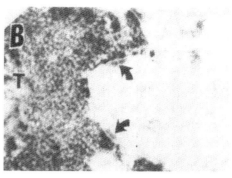

Figure 2. Electron micrographs showing PKC immunoreactivity in presynaptic terminals (T) of rat cerebral cortex (A) and mossy fiber terminals (T) of hippocampus (B). Reaction product is prominent around synaptic vesicles (straight arrows). The postsynaptic densities (curved arrows) are inherently electrondense but do not contain immunoreactivity. (A) x 44,800; (b) x 33,000. Adapted with permission from Girard et al. (1985) and Wood et al. (1986).

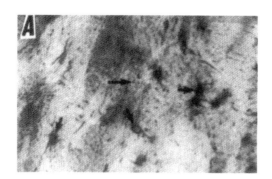
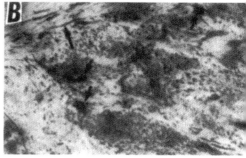

Figure 3. Light micrographs showing PKC immunoreactivity in cells closely resembling oligodendroglia (short arrows) and axons (long arrows) of rat brain white matter pretreated without (A) or with (B) Triton X-100. The myelin sheath is unstained and cannot be seen in these micrographs. (A) x 408; (B) x 240. Adapted with permission from Girard et al. (1985).

Protein phosphorylation

Because PKC immunoreactivity was heavily localized in oligodendroglia of white matter (Fig. 3), it is likely that myelin basic protein (MBP) in myelin membrane would be phosphorylated by the enzyme. Indeed, we found (Turner et al., 1982) that both large- and small-MBP from the soluble myelin fraction were phosphorylated exclusively by phosphatidylserine/Ca^{2+}-activated PKC endogenous to the membrane. We reported previously that MBP was an exceptionally effective substrate for PKC, with a low K_m and a high V_{max} and five sites of phosphorylation (Wise et al, 1982). Although the functional significance of MBP phosphorylation is still unclear, it is possible that the site-specific phosphorylation could modify its ability to induce or prevent experimental allergic encephalomyelitis (Chou et al., 1980).

We investigated ontogenetic changes of endogenous substrate proteins in different regions and fractions of developing rat brain (Turner et el., 1984). There was a major PKC substrate (M_r = 66,000) in the soluble fraction of white matter and, furthermore, its phosphorylation was highest in the neonate (1-day old) and progressively decreased with age (Fig. 4A). Other substrates (M_r 60,000 and 58-50,000) in which phosphorylation was stimulated to a similar extent by phospholipid/Ca^{2+} or calmodulin/Ca^{2+}, on the other hand, progressively increased as a function of age. In the particulate fraction of white matter, there were a number of substrates (e.g. M_r: 87,000, 66,000, 58-46,000, and MBP at gel front) were detected (Fig. 4B). It was worth noting that phosphorylation of the M_r 66,000 protein, for example, was maximal at days 13-18. Phosphorylation of MBP (M_r 18-14,000), in comparison, increased as a function of age. Similar ontogenetic changes in substrate proteins, some of which were not evident in white matter, were also noted in grey matter of rat brain. The observed changes in protein phosphorylation were likely due to altered levels of the substrate proteins, because addition of exogenous PKC,

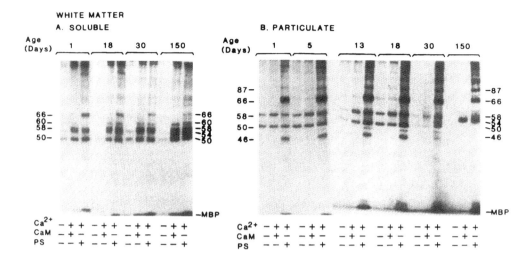

Figure 4. Autoradiograms showing phosphorylation of proteins from the soluble and particulate (solubilized) fractions of white matter of brain from rats of various ages. CaM, calmodulin (0.5 μg/0.2 ml); PS, phosphatidylserine (5 μg/0.2 ml). Numbers on right and left indicate $M_r \times 10^{-3}$. Taken with permission from Turner et al. (1984).

for example, to the soluble fraction of white matter from 150-day old rats did not increase phosphorylation of the M_r 66,000 protein. It seemed that PKC substrate proteins involved in the ontogenetic changes can be classified into two major categories. (i) Proteins that increased steadily and were present at the highest level in the adult. MBP is an example and they may be referred to as the "adult proteins" involved in maturation or aging of brain. (ii) Proteins that exhibited a biphasic developmental changes with a peak level closely following a peak growth rate of brain mass. These proteins, exemplified by the M_r 66,000 species in particulate fraction of white matter, may be termed "neonatal proteins" intimately involved in brain development. Investigations of the detailed subcellular and ultrastructural localization of these proteins related to PKC would undoubtedly yield new insights into the functional significance of the protein phosphorylation system in brain ontogeny.

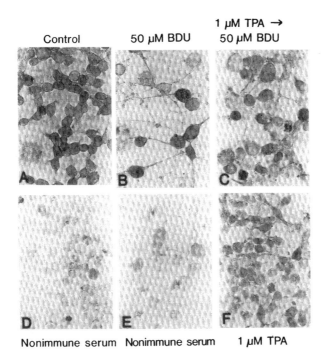

Figure 5. Immunocytochemical localization of PKC in Neuro 2a cells. Cells were plated for 1 day and cultured for 4 additional days without (A) or with (B) 50 μM BDU, or plated for 1 day, and then treated with 1 μM TPA for 15 h followed by a 4-day incubation without (F) or with (C) 50 μM BDU. The cells from (A) and (B) were also stained with nonimmune serum for comparison (D and E, respectively).

Differentiation of neuroblastoma cells

We employed several neuroblastoma cell lines as a model to examine the potential role of PKC in cell differentiation. 5-Bromo-2'-deoxyuridine (BDU) caused differentiation of Neuro 2a cells as indicated by an extensive network of neurite outgrowth, accompanied by an apparent decrease in cellular PKC immunostaining (Fig. 5). Cells pretreated with a high concentration (1 μM) of 12-0-tetradecanoylphorbol-13-acetate (TPA) exhibited marked reduction in immunoreactivity and responsiveness to BDU, suggesting depletion of cellular PKC has a negative effect on neuritogenesis. Immunoblotting indicated the PKC level in the BDU-induced differentiating cells had a reduced level of native PKC (M_r 80,000) in the soluble fraction (Fig. 6A). The differentiating cells also showed a decreased phosphorylation of two cellular proteins (M_r 96,000 and 80,000) (Fig. 6B). Similar decreases in cellular PKC in soluble fraction and phosphorylation of M_r 80,000 protein by PKC were also noted for SH-SY5Y cells induced to differentiate by TPA (date not shown), or in LA-N-5 cells induced to differentiate by TPA or retonoic acid (Girard and Kuo, 1989). As in Neuro 2a cells, pretreatment of SH-SY5Y and LA-N-5 cells with 1 μM TPA to deplete cellular PKC rendered these cells less responsive to the differentiating effect of the agents. Neither a reduced cellular PKC nor a decreased phosphorylation of M_r 80,000 protein was observed in CHP-100 cells treated with TPA, retinoic acid or nerve growth factor, a cell line shown to be resistant to differentiation induced by these agents (Spinell et al., 1982). The findings appeared to support an involvement of PKC in cell differentiation, consistent with a report showing inhibition of the nerve growth factor-induced neurite outgrowth in PC12 cells by PKC inhibitor sphingosine (Hall et al., 1988). A decreased phosphorylation of M_r 80,000 protein by PKC might represent a common pathway involved in differentiation of various neuroblastoma cell lines in response to various differentiation inducers. It is of interest that an enhanced phosphorylation of M_r 80,000-87,000 proteins in a variety of cell lines and tissues was stimulated by tumor promoters, mitogenic peptides and growth factors (Blackshear et al., 1986; Rozengurt et el., 1983), suggesting their potential role in cell growth. These proteins have been determined to be PKC substrates and comparison of their properties suggested that they may be an identical protein (Albert et al., 1986; Blackshear et al., 1986). It has been reported recently that the levels of the M_r 80,000 PKC substrate were lower in transformed cells than in untransformed controls (Smith et al., 1988; Wolfman et al., 1987), and were also lower in leukemia KG-1 cells responsive to the differentiating effect of TPA than in the unresponsive subline KG-1a cells (Kiss et al., 1987). The ubiquitous occurrence of

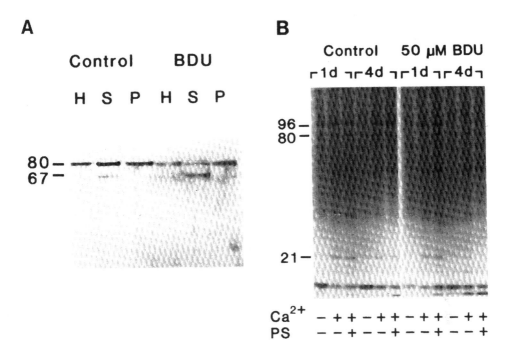

Figure 6. Levels of PKC and PKC-dependent phosphorylation during BDU-induced neurite outgrowth of Neuro 2a cells. (A) Immunoblots of PKC in subcellular fractions of Neuro 2a cells before (control) and after induction of morphological differentiation by BDU. Cells were plated 1 day prior to addition of BDU and subsequently incubated for 4 days. Homogenates (H) and soluble (S) and particulate (P) fractions were prepared and immunoblotted. (B) Autoradiograph of phosphorylated endogenous proteins in homogenates of Neuro 2a cells grown for 4 days in the absence or presence of 50 μM BDU.

this PKC substrate protein suggests its critical involvement in PKC-mediated signal transduction related to cell growth and differentiation. Further characterization of this phosphoprotein could shed light on the problem.

ACKNOWLEDGEMENTS

The original work from the author's laboratory was supported by USPHS Grants NS-17608, HL-15696 and CA-36777. We thank LaVonne Schoffner for her skillful preparation of the manuscript.

REFERENCES

Akers RF, Lovinger DM, Colley PA, Linden DJ, Routenberg A (1986) Translocation of protein kinase C activity may mediate hippocampal long-term potentiation. Science 132: 587-589

Albert KA, Walaas SI, Wang JK-T, Greengard P (1986) Widespread occurrence of "87 kDa" protein, a major specific substrate for protein kinase C. Proc Natl Acad Sci USA 83: 2822-2826

Aloyo VJ, Zwiers H, and Gispen WH (1983) Phosphorylation of B-50 protein by calcium-activated phospholipid-dependent protein kinase and B-50 protein kinase. J Neurochem 41: 649-653

Altman J. (1972) Postnatal development of cerebellar cortex in rat. II. Phases in the maturation of Purkinje cells and the molecular layer. J Comp Neurol 145: 399-464

Blackshear PJ, Wen L, Glynn BP, Witters LA (1986) Protein kinase C-stimulated phosphorylation in vitro of a M_r 80,000 protein phosphorylated in response to phorbol esters and growth factors in intact fibroblasts. Distinction from protein kinase C and prominence in brain. J Biol Chem 261: 1459-1469

Chou FC-H, Chou C-HJ, Fritz RB and Kibler RF (1980) Prevention of experimental allergic encephalomyelitis in Lewis rats with peptide 68-88 of guinea pig myelin basic protein. Ann Neurol 7: 336-339

Girard PR, Mazzei GJ, Kuo JF (1986) Immunoquantitation of Phospholipid/Ca^{2+}-dependent protein kinase and its fragments. Tissue levels, subcellular distribution and ontogenetic changes in brain. J Biol Chem 261: 370-375

Girard PR, Mazzei GJ, Wood JG, Kuo JF (1985) Polyclonal antibodies to phospholipid/Ca^{2+}-dependent protein kinase and immunocytochemical localization of the enzyme in rat brain. Proc Natl Acad Sci USA 82: 3030-3034.

Girard PR, Wood JG, Freschi JE, Kuo JF (1988) Immunocytochemical localization of protein kinase C in developing brain tissue and in primary neuronal culture. Devel Biol 126: 98-107

Girard PR, Kuo JF (1989) Protein kinase C and its 80-kDa substrate protein in neuroblastoma cell neurite outgrowth. J Neurochem, in press

Hall FL, Fernyhough P, Ishii DN, Vulliet PR (1988) Suppression of nerve growth factor-directed neurite outgrowth in PC12 cells by sphingosine, an inhibitor of protein kinase C. J Biol Chem 263: 4460-4466

Jacobson RD, Virag I, Skene JHP (1986) A protein associated with axon growth, GAP-43, is widely distributed and developmentally regulated in rat CNS. J Neurochem 6: 1843-1855

Kiss Z, Deli E, Shoji M, Koeffler HP, Pettit GR, Bogler WR, Kuo JF (1987) Differential effects of various protein kinase C activators on protein phosphorylation in human acute myeloblastic leukemia cell line KG-1 and its phorbol ester-resistant subline KG-1a. Cancer Res 47: 1302-1307

Kuo JF, Andersson RGG, Wise BC, Mackerlova L, Salomonsson I, Brackett NL, Shoji M, Wrenn RW (1980). Calcium-dependent protein kinases: Widespread occurrence in various tissues and phyla of animal kingdom and comparison of the effects of phospholipid, calmodulin and trifluoperazine. Proc Natl Acad Sci USA 77: 7039-7043

Meiri KF, Pfenninger KH, Willard MB (1986) Growth associated protein, GAP-43, a polypeptide that is induced when neurons extend axons, is a component of growth cones and corresponds to pp46, a major polypeptide of a subcellular fraction enriched in growth cones. Proc Natl Acad Sci USA 83: 3537-3541

Ouimet CC, McGuinness TL, Greengard P (1984) Immunocytochemisal localization of calcium/calmodulin-dependent protein kinase II in rat brain. Proc Nat Acad Sci USA 81: 5604-5608

Rozengurt E, Rodriguez-Pena M, Smith KA (1983) Phorbol esters, phospholipase C, and growth factors rapidly stimulate the phosphorylation of a M_r 80,000 protein in intact quiescent 3T3 cells. Proc Natl Acad Sci USA 80: 7244-7248

Smith BM, Colburn NH (1988) Protein kinase C and its substrates in tumor promoter-sensitive and resistant cells. J Biol Chem 263: 6424-6431

Spinell W, Sonnenfeld KH, Ishii DN (1982) Effects of phorbol ester tumor promoters and nerve growth factor on neurite outgrowth in cultured human neuroblastoma cells. Cancer Res 42: 5067-5073

Turner RS, Chou C-HJ, Kibler RF, Kuo JF (1982) Basic protein in brain myelin is phosphorylated by endogenous phospholipid-sensitive Ca^{2+}-dependent protein kinase. J Neurochem 39: 1397-1404

Turner RS, Raynor RL, Mazzei GJ, Girard PR, Kuo JF (1984) Developmental studies of phospholipid-sensitive Ca^{2+}-dependent protein kinase and its substrates and phophoprotein phosphatases in rat brain. Proc Natl Acad Sci USA 81: 3143-3147

Wise BC, Glass DB, Chou C-HJ, Raynor RL, Katoh N, Schtazman RC, Turner RS, Kibler RF, Kuo JF (1982) Phospholipid-sensitive Ca^{2+}-dependent protein kinase from heart. II. Substrate specificity and inhibition by various agents. J Biol Chem 257: 8489-8495

Wolfman A, Wingrove TG, Blackshear PJ, Macara IG (1987) Down-regulation of protein kinase C and of an endogenous 80-kDa substrate in transformed fibroblasts. J Biol Chem 262: 16546-16552

Wood JG, Girard PR, Mazzei GJ, Kuo JF (1986) Immuno-cytochemical localization of protein kinase C in identified neuronal compartments of rat brain. J Neurosci 6: 2571-2577

Wrenn RW, Katoh N, Wise BC, Kuo JF (1980) Stimulation of phosphtidylserine and calmodulin of calcium-dependent phos-phorylation of endogenous proteins from cerebral cortex. J Biol Chem 255: 12042-12046

Yoshida Y, Huang FL, Nakabayashi H, Huang K-P (1988) Tissue distribution and developmental expression of protein kinase C isozymes. J Biol Chem 263: 9868-9873

OLIGODENDROCYTE-SUBSTRATUM INTERACTION SIGNALS CELL POLARIZATION AND SECRETION

S. Szuchet and S.H. Yim
Department of Neurology and the Brain Research Institute
The University of Chicago
BH Box 425
5841 South Maryland Avenue
Chicago, IL 60637 (USA)

I. INTRODUCTION

The importance of cell-substratum interaction in determining morphogenetic events and the differentiation pathway a cell is to follow, has been amply documented (for reviews see, Hay, 1984; Ekblom et al., 1986; Grinnell, 1978). Examples abound, from extracellular matrix (ECM) induced changes in the shape of fibroblasts with the concomitant reorganization of their cytoskeleton to the influence of substratum on the differentiation of muscle cells or the modulation of secretion in the corneal epithelium by ECM components. Of particular relevance to the work to be discussed here is the role of ECM in peripheral nervous system (PNS) myelination.

The two myelinating cell types, the Schwann cell from PNS and the oligodendrocyte from the central nervous system (CNS) share many features but have also important differences. The Schwann cell may either ensheathe a group of axons or myelinate a single axon; in either case the unit is surrounded with a basal lamina and this triumvirate (as Bunge et al., 1986 refer to it) is suspended in an ECM. The oligodendrocyte does not ensheathe axons, and it will myelinate between 20 to 40 internodes. Thus, a one to one relationship between an oligodendrocyte and its axon is not the norm. This difference between the two myelinating cells in their interaction with axons has profound effects on their biology (Mirsky et al., 1980). Progress has been made in elucidating the factors that control and influence Schwann cell function (Bunge et al., 1986, and ref. therein; Eldridge et al., 1989). Schwann cells are largely responsible for the secretion of basal lamina components; to achieve this the cells must establish contact with an axon. *In vitro*, this dual contact with an axon and substratum leads to cell polarization. Only a polarized Schwann cell with its complete basal lamina is competent to myelinate or ensheathe (Bunge et al., 1986; Eldridge et al., 1989). There is no visible basal lamina in the CNS; little is known on the ECM in the CNS save for the fact that proteoglycans (PGs) have been isolated from brain (Margolis et al., 1986); likewise, nothing is known on what oligodendrocytes secrete to the medium, or, indeed, if they do.

In previous studies we have shown that attachment of OLGs to a substratum is required for the cells to express their myelinogenic properties (Szuchet et al., 1983; Yim et al., 1986; Szuchet et al., 1986; Vartanian et al., 1986, 1988). We are interested in defining the molecules that may mediate the substratum-induced signal transduction; we speculated that attachment may result in the spatial segregation of such molecules and in secretion. We have, therefore, undertaken to investigate if OLGs a) are polarized; b) have the equivalent of a basal lamina; and c) secrete components to the medium. Herein we present evidence that attachment of OLGs to a substratum leads to cell polarization and to *de novo* synthesis of components which may well serve a role of a rudimentary basal lamina. We also report that OLGs secrete a limited repertoire of PGs and glycoproteins (Gps) that are both sulfated and fucosylated. Finally, we describe the temporal modulation of OLG secretory products.

II. DESCRIPTION OF THE MODEL

All the work in our laboratory has been performed with ovine oligodendrocytes (OLGs) isolated from young but mature brains; mature in the sense that myelination has taken place. The procedures for cell isolation (Szuchet et al., 1980a), culture (Szuchet et al., 1980b), as well as the characterization of these cells (Mack and Szuchet, 1981; Mack et al., 1981; Szuchet et al., 1983; Yim et al., 1986; Szuchet et al., 1986; Vartanian et al., 1986, 1988; Soliven et al., 1988a,b; Szuchet, 1987a,b) have been described in detail.

We maintain OLGs under two sets of experimental conditions (Szuchet and Yim, 1984). We shall be referring to these cells as:

B3.f: these are OLGs from Band 3 (Szuchet et al., 1980a) plated into culture petri dishes and kept for 3 to 5 days in a floating state, since they do not adhere to plastic plates. Morphologically B3.f cells do not differ from freshly isolated cells; and

B3.fA: these are B3.f OLGs transferred to polylysine coated plates to which they adhere. After attachment, these cells undergo drastic changes in their morphology, ultrastructure, and metabolism (see ref. cited above).

III. OLIGODENDROCYTE SECRETORY PRODUCTS

Cells utilize two secretion pathways: regulated and constitutive (Burgess and Kelly, 1987). In the former mode, material to be secreted is concentrated in specialized granules and is only released to the medium upon receipt of a signal from a secretagogue. This type of secretion is utilized for components whose function requires them to be present at a relatively high concentration and at a precise time, e.g.; hormones, digestive enzymes, etc. In the constitutive form of secretion, components are discharged continuously, presumably, at the same rate as they are synthesized. Such components have functions that may demand steady state concentrations, e.g.; structural components, tropic and trophic factors. In what follows, we present evidence that OLGs secrete constitutively to the medium a limited repertoire of fucosylated and sulfated glycoproteins. They also secrete chondroitin sulfate PGs (CHSPGs). Both the Gps and PGs are temporally modulated.

Labeling OLGs cultures with $^{35}SO_4^{2-}$ in serum-free medium for 18-20 hours yields 2% of the label incorporated into cellular components and 98% into components secreted to the medium. With [^{3}H]fucose or [^{3}H]leucine as the label, the distribution is 78% and 93% for cellular components and 22% and 7%, respectively for secreted components. When the secreted proteins are resolved by sodium dodecylsulfate polyacrylamide gel electrophoresis (SDS-PAGE), they give patterns that resemble one another and are independent of the label used, indicating that the same proteins are both fucosylated and sulfated (Figs. 1a and b). In contrast, the pattern of cellular proteins is different for each label and is also distinct from the corresponding pattern of secreted proteins (Figs. 1a and b). This latter fact can be interpreted as showing that the secreted proteins are not mere products of surface shedding but are released into the medium for some functional purpose, even though presently we can only guess what this function might be.

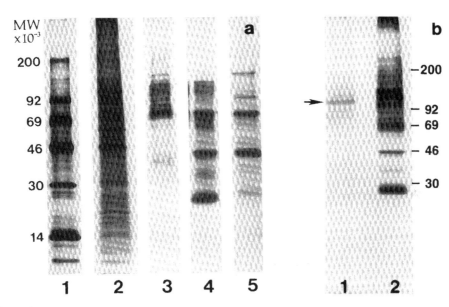

Figure 1
Incorporation of [^{3}H]leucine, [^{3}H]fucose, and $^{35}SO_4^{2-}$ by cultured OLGs. Comparison of the patterns of cellular and secreted components. Cultures were labeled and processed as described by Yim and Szuchet, (1989a). Radiolabeled components were resolved on SDS-PAGE and detected on x-ray film by fluorography. (From Yim and Szuchet, 1989a).

a. [^{3}H]leucine- and [^{3}H]fucose-labeled cellular and secreted components from a 20 day-old OLGculture. Lanes: 1, [^{14}C]mol.wt. standards; 2 and 3, respectively, [^{3}H]leucine-and [^{3}H]fucose-labeled cellular polypeptide chains; 4 and 5, respectively, [^{3}H]fucose- and [^{3}H]leucine-labeled polypeptide chains secreted to the medium. Note: 1) the difference in the patterns of lanes 2 and 3 showing that only a discrete number of cellular components are fucosylated; 2) the difference in the patterns of lanes 3 and 4 showing that cellular glycoproteins are distinct from those secreted, and 3) the similarity of patterns in Lanes 4 and 5 indicating that all major components found in the medium are fucosylated.

b. $^{35}SO_4^{2-}$-labeled polypeptides found in the cell homogenate and secreted to the medium by a 42 day-old OLG culture (see text). Lanes: 1, cell homogenate, notice that only one band is sulfated (small arrow). We have identified this band as MAG. 2, secreted polypeptides. This pattern is very different from that seen in lane 1 but resembles the polypeptide profiles shown in (a).

Characterization of secreted components. Analysis and quantitation of densitometric traces of [^{3}H]leucine-, [^{3}H]fucose-, and $^{35}SO_4^{2-}$- labeled fluorograms showed that OLGs secrete 4 major and a number of minor Gps. The apparent preponderance of these components changes somewhat depending on which label is used as the measuring stick (Yim and Szuchet, 1989a). The relative concentration of each component was calculated by taking the ratio of the area under the corresponding peak and dividing it by the total area. Average values for 20-60 day-old cultures are given in Table 1. It is seen in Table 1 that four components $M_r \times 10^{-3}$ =74, 70, 45, and 26 account for 73% of incorporated [³H]leucine. The same Gps account for also 60% of incorporated [³H]fucose. However, there are four other Gps that are highly fucosylated namely, Gp36, Gp93, Gp66, and Gp105. The ratio of [³H]fucose/[³H]leucine appears highest for Gp36 followed by Gp26, Gp93, and Gp105, but see below. The $M_r \times 10^{-3}$ of these Gps, as assessed from SDS-PAGE, range from 147±10 to 26±1 (Table I). Note that the data from $^{35}SO_4^{2-}$-label correspond to proteins

Table I

Relative concentrations of secreted glycoproteins[a]

Components[b]	[³H]Leucine	[³H]Fucose	$\dfrac{[\text{Fucose}]}{[\text{Leucine}]}$	$^{35}SO_4^{2-}$	$\dfrac{^{35}SO_4^{-2}}{\text{Leucine}}$
Gp147[c]	2.1 ± 0.7 (5)[d]	1.4 ± 0.5 (8)	0.7	1.8 ± 0.5 (2)[e]	0.85
Gp125	3.8 ± 0.6 (6)	3.8 ± 0.3 (8)	1.0	7.0 ± 0.6 (4)	1.8
Gp105	3.4 ± 0.7 (5)	5.5 ± 0.3 (2)[e]	1.6	6.4 ± 1.5 (3)	1.9
Gp93	4.0 ± 0.3 (5)	6.9 ± 1.6 (3)	1.7	9.3 ± 2.4 (3)	2.3
Gp74	19.9 ± 1.9 (6)	11.1 ± 0.7 (8)	0.6	0	0
Gp70	6.5 ± 1.3 (6)	5.5 ± 0.4 (4)	0.8	10.0 ± 1.6 (5)	1.5
Gp66	4.8 ± 0.6 (6)	6.4 ± 0.7 (5)	1.3	2.4 ± 0.3 (5)	0.5
Gp51	2.2 ± 0.4 (5)	1.2 ± 0.2 (8)	0.5	1.7 ± 0.2 (5)	0.8
Gp45	25.2 ± 1.9 (6)	10.3 ± 0.6 (8)	0.4	4.1 ± 0.4 (5)	0.2
Gp36	3.2 ± 0.5 (5)	10.0 ± 0.5 (8)	3.1	6.8 ± 0.8 (5)	2.1
Gp26	11.0 ± 0.3 (6)	27.0 ± 0.8 (8)	2.4	17.0 ± 1.3 (5)	1.5

a. Relative concentrations are defined as $A_i / \Sigma A_i$ where A_i is the area under a given peak and ΣA_i is the total area under the densitometric trace of a fluorogram. (From Yim and Szuchet, 1989a)

b. Components are identified by their $M_r \times 10^{-3}$

c. Gp = glycoprotein

d. Mean values ± SEM; number of experiments are given in parenthesis

e. Mean values ± SD

secreted by 40-60 day-old cultures (Table 1). Because in cultures younger than 40 days, there is secretion of sulfated PGs (see below) and because the latter do not resolve well on SDS-PAGE but produce a smear, the densitometric traces of such gels tend to be distorted and are, therefore, of little value for comparison with [³H]fucose-labeled or [³H]leucine-labeled gels. Even at late time points, it is still difficult to obtain good quantitative estimates of the high molecular weight components. All in all, there is reasonable correlation between the extent of sulfation and fucosylation (Table 1), suggesting that the sulfate groups may be localized on the carbohydrate moiety. However, not all of the secreted proteins seem to be sulfated (Table 1); a notable example is Gp74 (Fig 2).

In Fig.2 we have compared the relative concentration of each secreted component as obtained from the incorporation of each of the three labels examined. Two components, Gp45 and Gp74, stand out for their high leucine content; all others have more fucose and/or sulfate than leucine. The

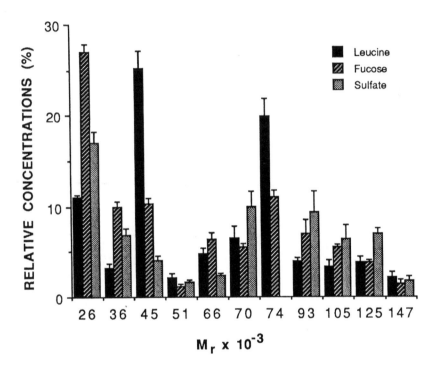

Figure 2
Comparison of the relative concentration of each secreted component, as assessed from the incorporation of each of the labels. Concentrations were calculated from densitometric traces of the corresponding fluorograms as indicated in Table 1. Values given are mean ± SD. (From Yim and Szuchet, (1989a).

trend seems to be that of sulfate prevailing over fucose for components with $M_r \geq 74,000$ and fucose prevailing over sulfate for components with $M_r \leq 70,000$. However, when the ratios [^{3}H]fucose/[^{3}H]leucine and $^{35}SO_4^{2-}$/[^{3}H]leucine for each component are compared (Table 1), they are, with a few exceptions, amazingly close. These results in conjunction with evidence that most of these polypeptides chains are subunits of complex molecules (see below) are the first indication that the native Gps may be assemblies of heterologous subunits that differ in both their mass and carbohydrate content.

Two-dimensional profile of secreted glycoproteins. When secreted Gps that had been labeled with [³H]fucose are subjected to isoelectric focusing in one dimension followed by SDS-PAGE in the other, the pattern shown in Fig. 3, is obtained (Yim and Szuchet, 1989a). It is seen that each component has given rise to two or more "spots" when separated on the basis of charge (Fig. 3); Gp26 stands out among them for having resolved into six polypeptides that cover the full spectra of pI. These may be isoforms that have arisen via differential splicing of a single gene, not unlike fibronectin; the 20 isoforms of fibronectin are the products of differential splicing of pre-mRNA (reviewed by McDonald, 1988). Alternatively, some or all of them may be products of different genes; preliminary data indicate the later may be the case for Gp36 and Gp26. In contrast, there seems to be only one Gp45. The range of pI covered by the secreted proteins is rather narrow; most of them fall within an apparent pI of 6.2 to 7.0. But three polypeptides are truly alkaline with pI of 8.2 or above.

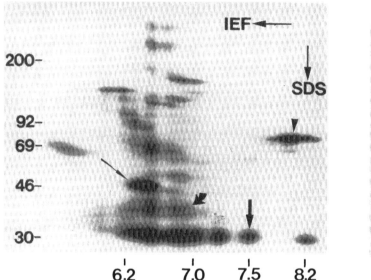
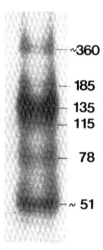

Figure 3
Fluorogram of [³H]fucose-labeled secreted components resolved by two-dimensional gel electrophoresis IEF, marks the direction of isoelectric focusing; SDS, shows the direction of SDS-PAGE. Numbers on the ordinate are $M_r \times 10^{-3}$ for [¹⁴C] standards; numbers on the abscissa denote apparent pI. Note that the majority of polypeptides have pI between 6.2 and 7.0. Many components are present as multiple isotypes: Gp26 (thick arrow); Gp36 (curved arrow); Gp74 (arrowhead). In contrast, Gp45 (thin arrow) may be a single polypeptide chain. (From Yim and Szuchet, 1989a).

Figure 4
Protein profile of [³H]fucose-labeled secreted proteins under native conditions. 3-25% gradient gel in a 0.09M Tris, 0.08M borate, 25mM EDTA buffer pH 8.2. Numbers give $M_r \times 10^{-3}$ Compare range of mol.wt with those of Figs. 1a,b. (From Yim and Szuchet, 1989a).

The native proteins. Given the complexity of the two-dimensional pattern of secreted components, it was of interest to determine which of the components were single polypeptide chains, which were subunits of large proteins. Since the secreted proteins are soluble in aqueous solvents, it was possible to measure the molecular masses of the native proteins, compare the results with those obtained from the denatured proteins and draw inferences thereon. Figure 4 gives the profile of the native proteins. Six proteins are seen; their apparent molecular weights ranging from 51,000 to 360,000. Taken together, these results (see Fig. 1a and Table 1) allowed us to conclude that Gp26 and Gp36 are subunits of one or more of the larger proteins (Fig. 4). Even though no protein with a M_r=45,000 could be seen in the native pattern (Fig. 4), considering that measurement of molecular weights of glycoproteins by SDS-PAGE is not very accurate (Tanford and Reynolds, 1976), the 12% difference between Gp45 and Gp51 (native protein) may not be significant, i.e., they may be one and the same protein.

Native to denatured pattern. In an attempt to establish a correspondence between components in the native and denatured states, we performed the following experiment. [^{3}H]fucose-labeled secreted components were resolved by PAGE under native conditions, the gel was cut so as to separate individual components, gel pieces were thoroughly minced, equilibrated with SDS sample buffer, boiled for 1 min to dissociate the proteins and applied to a 7%-25% gradient SDS-PAGE under standard conditions (Yim and Szuchet, 1989a). The results (not shown) are intriguing for they suggest that all components with M_r >100,00 are complexes of four to five subunits having a considerable span in their apparent molecular weights. In contrast, components with M_r < 100,000 seem to have a simple composition of only two to three subunits. The results also indicated that each of the three Gp36 and each of the six Gp26 (Fig. 3) may be subunits of different Gps; this could be taken as an indication that each of them is a distinct molecular species unrelated to the others save for having the same molecular mass. While this may be the case, it is intriguing that all of the secreted proteins should have either a 36,000 or 26,000 subunit (Yim and Szuchet, 1989a), this may be suggestive of a common function.

The carbohydrate moiety. Our work indicates that all fucose residues are attached to N-linked sugars; we also have evidence that most of the sulfate is on sugar chains (data not shown). But there are at least three polypeptides with sulfate on them. Preliminary results suggest that the carbohydrate chains, on the average, contribute to the total mass of the molecules with 1000-2000 Da. Though we could not detect O-linked sugars, we cannot exclude the possibility that minor components may contain such moieties (Yim and Szuchet, 1989a).

Temporal modulation of secreted components. It was apparent from the data given in Table 1 that, within the limits of resolution of our system (i.e., one dimensional gel electrophoresis), there were no changes in the pattern of secretion by 20 to 65 day-old cultures (last time point studied). The first inkling that this may not be so for early cultures (i.e.; $\leq$20 days) came from experiments using $^{35}SO_4^{2-}$ as the label. Figure 5 illustrates the patterns of secretion of $^{35}SO_4^{2-}$-labeled components from 10 day-old and 63 day-old cultures. Note that what distinguishes the younger of the two cultures, is a marked smear at the top of the gel; the smear is much attenuated in the 63 day-old culture. There are also differences in the banding patterns of the two gels: a component (M_rx10^{-3}=49) is strongly expressed in the 10 day-old culture but not in the 63 day-old, whereas the apposite is the case for two other components ($M_r \times 10^{-3} = 36$ and 45; see Fig. 1b). It is these observations that prompted us to undertake longitudinal studies. Figure 6 depicts densitometric traces of $^{35}SO_4^{2-}$-labeled fluorograms obtained at various time points from 5 day-old to 63 day-old cultures. The striking features of these patterns is the presence of high molecular weight components in the early cultures; these components loose conspicuity in old cultures; for example, there is a 56% drop in the relative concentration of these components in a 26 day-old culture as compared to a 5 day-old culture. In lieu, a new set of components, $M_r \times 10^{-3}$ of 26 to 125, become prominent, though not at the same time (Szuchet and Yim, 1989). When a similar study was performed (Szuchet and Yim 1989) using [^{3}H]fucose or [^{3}H]leucine as the label, the high molecular weight components were not detected. However, the pattern of secretion from, say, a 10 day-old culture was still simpler than that of older cultures. Of particular note was the absence of Gp45. In general, there is one pattern of secretion that characterizes the early cultures (up to 15 days) and another one for the late cultures, i.e., >20 days. Therefore, all changes occur within a very narrow window (~ 5 days). Figure 7 demonstrates the modulation of Gp45 over time in culture, for the three labels used. It is evident that the largest increase in Gp45 takes place between the second and third week in culture, after which time a plateau is reached (Fig 7a, b). The sulfate data seem to suggest a gradual change in the concentration of Gp45 (Fig. 7c), but given the difficulties in quantifying such data (see above), this point has to be further substantiated.

<u>High molecular weight sulfated components</u>. The physical appearance of $^{35}SO_4^{2-}$-labeled fluorograms was reminiscent of patterns obtained with PGs. Hence, we investigated if, indeed, OLGs secrete PGs. After collecting the secreted products as described (Yim and Szuchet, 1988, 1989b); the strategy followed was to digest the proteins with papain and fractionate the glycosaminoglycans (GAGs) by chromatography on Sepharose CL-6B. Elution from this column gave the profile shown in Fig. 8a. Fractions were pooled as indicated in Fig. 8a, desalted by dialysis and concentrated by lyophilization. Each pool (designated I; II; III) was then divided into three aliquots; one of which was left as control, the other two were used for carbohydrate identification. Figure 8b shows the results of treating aliquots of fraction I with either chondroitinase ABC or HNO_2. 75% of material from fraction I was degraded by the enzyme whereas none was affected by HNO_2. Taken together, the data indicate that: a) 75% of fraction I is chondroitin sulfate GAGs; and b) there is no heparin sulfate PGs (HSPGs) in this fraction. We have not, as yet, identified the material that was resistant to both treatments. Material from fraction II was also resilient to the action of either reagent (Fig. 8c). We suspect that this fraction contains sulfated glycopeptides. Most of the $^{35}SO_4^{2-}$-label in fraction III was lost during dialysis, suggesting that these were low molecular weight components. In conclusion, the majority of the high molecular weight sulfated components secreted by OLGs are CHSPGs (Szuchet and Yim, 1989).

IV. SUBSTRATUM-INDUCED OLIGODENDROCYTE POLARIZATION

In preliminary work, we have shown that PGs and sulfated Gps were peripherally located on the surface of cultured OLGs and could be displaced with heparin, a sulfated glycosaminoglycan (Yim and Szuchet, 1989c). In light of our previous findings on the effect of adhesion on OLG function (Szuchet et al., 1983; Szuchet et al., 1986; Yim et al., 1986; Vartanian et al., 1986, 1988), we investigated if adhesion affected the expression of these components (Yim and Szuchet, 1989b).

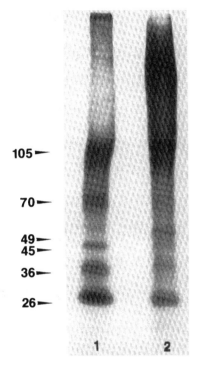

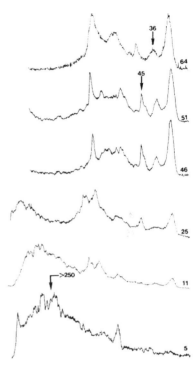

Figure 5
Fluorograms of $^{35}SO_4^{2-}$-labeled secreted components resolved by SDS-PAGE. Numbers on the left are $M_r \times 10^{-3}$ of standards. Note the smear on top of the gels. This is characteristic of PGs. Lanes: 1, 63-day old culture; 2, 5-day old culture. Note the difference in the banding patterns of the two gels. (From Yim and Szuchet, 1989a).

Figure 6
Temporal modulation of $^{35}SO_4^{2-}$-labeled secreted ocmponents. The media from $^{35}SO_4^{2-}$-labeled cells were collected, dialyzed, lypphilized and resolved on SDS-PAGE, 7-25% gradient gels. The dried gels were exposed to Kodak XAR-2f film. Densitometric traces of flyorograms. Numbers on the right are days in culture. Notice the change in the patterns: most of the high molecular weight components that predominate at early time points are chondroitin sulfate proteoglycans. Note that the increase in this expression of Gp45 is after two weeks in culture. (From Yim and Szuchet, 1989).

We compared the synthesis of heparin-susceptible sulfated components by B3.f and B3,fA OLGs after 5 days *in vitro*; 24h of attachment for B3.fA OLGs. Note that the entire surface of B3,f OLGs is exposed to heparin extraction, whereas in B3.fA OLGs only the surface facing the medium, apical domain, is available for heparin extraction (OLGs do not spontaneously detach during this step). Harvesting the cells uncovers the entire cell perimeter. B3.fA OLGs are, therefore, treated with heparin twice: first, while the cells are adhered to the substratum; the second time, after they have been detached. As mentioned above, now all the plasmalemma is accessible to heparin, however because the apical components have already been removed, the second heparin extraction yields components that had been in contact with substratum, i.e.; the basal domain.

In the first set of experiments, we compared the size distributions, as assessed from the elution profiles of a (1x100) cm Sepharose CL-6B chromatographic column, of sulfated components that were removed with heparin from B3.f and B3.fA OLGs. Figures 9a, b and c show the three elution profiles. The patterns are clearly different. The high counts associated with fractions F3 (Fig. 9a) and A3 (Fig. 9b) may be due to free label; these fractions will not be considered further. Notice that what distinguishes the material removed from the apical surface is fraction A2, eluted between tube No 35-55 (Fig. 9b); A2 is approximately 4 fold larger than F2 (cf. Figs. 9a and b). In contrast, the material that was removed from the basal surface has a main fraction, B1, eluted after the void volume ($M_r \geq 10^6$). There are several implications to these results: first, they indicate an adherence-induced expression of selective components; second, they denote a segregation of peripheral components within the plasma membrane with a concomitant creation of two distinct domains, apical and basal, i.e.; and adherence-induced cell polarization has taken place; third, they suggest an adherence-induced development of what could be considered a "rudimentary basal lamina" or ECM (Yim and Szuchet, 1989b).

All this occurred within the first 24h of adherence of OLGs to a substratum. In epithelial cells, prototypes of polarized cells, compartmentalization is achieved through the formation of tight junctions between cells; however, adhesion to a solid substratum will also lead to cell polarization, even in the absence of tight junction formation (Sabatini et al., 1983). Oligodendrocytes also make tight junction contacts, but the junctions are only assembled in adherent cells, furthermore, they do not polarize OLGs in the same manner as they do epithelial cells (Massa et al., 1984; Mugnaini et al; 1987). It is doubtful that the number of tight junctions available in a 1 day adherent culture could be solely responsible for the observations reported here. Hence, we postulate that it is adherence, per se, that induces OLG polarization and the events that follow (Yim and Szuchet, 1989b).

<u>Initial characterization of peripheral components</u>. It could be argued that attachment-induced polarization entails a quantitative redistribution of components without any qualitative differences between the two domains, i.e.; apical and basal. To address this issue, fractions from apical and basal domains obtained after chromatography on Sepharose CL-6B (see Figs 9a, b and c), were examined by SDS-PAGE directly and after pretreatment with enzymes that specifically degrade different classes of PGs or remove glycans. The outcome (Yim and Szuchet, 1989b) was as follows: first, corresponding fractions from apical and basal domains were clearly distinct; second, all the PGs were degraded by heparitinase, and therefore, belong to the heparan sulfate class, none of the PGs was degraded by chondroitin ABC indicating that there is very little or no CHSPGs or dermatan sulfate PGs; third, in addition to PGs, sulfated Gps were also found.

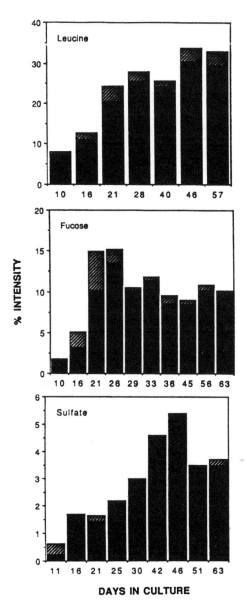

Figure 7
Changes in the expression of Gp45 overtime in culture. OLG cultures at different time points were labeled with either [^{3}]fucose, or [^{3}]leucine or $^{35}SO_4^{2-}$, media were collected and handled as described by Yim and Szuchet, 1989a. Flyorograms were traced with a laser densitometer and the % of Gp45 calculated as described in Table I. Numbers in the abscissa indicate time in culture. Ordinates are expressed in arbitrary units. Values given are mean ± SD. (From Szuchet and Yim, 1989).

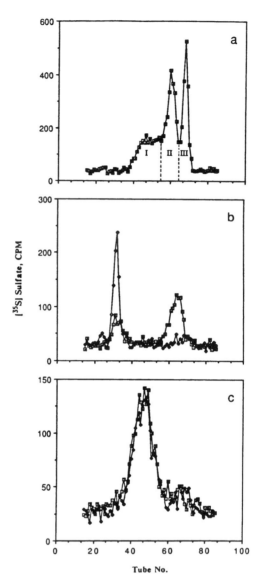

Figure 8
Characterization of sulfate-labeled secreted components. The medium of sulfate labeled OLGs was collected, dialyzed, lyophilized, digested with papain and resolved by chromatography on Sepharose CL-6B. Fractions were pooled as indicated in (a). Each fraction was divided into three aliquots; one was left as control, the other was treated with chondroitinase ABC and the third one, with nitrous acid. (From Yim and Szuchet, 1989b).

a. Elution profile from Sepharose CL6B.

b. Elution profiles from a Sephadex G-50 column of fraction I after treatment with □ choindroitinase ABC; ■ HNO_2. Material from fraction I was resistant to HNO_2.

c. Elution profiles from a Sephadex G-50 column of material from fraction II, treated as in (b). This fraction was resistant to both □ choindroitinase ABC and ■ HNO_2.

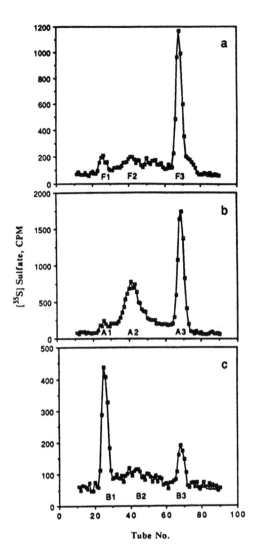

Figure 9
Comparison of heparin-extracted, $^{35}SO_4^{2-}$-labeled material from non-adherent (a) and adherent (b,c) OLGs. Elution profiles from Sepharose CL-6B chromatographic columns. B3.f and B3.fa OLGs (differing by 24h adherence) were labeled with carrier-free $^{35}SO_4^{2-}$ for 18-20h. After removing media and free-label, cells were extracted with 0.1 mg/ml of heparin in Tris-buffered saline for 45min at 37°C. B3.fa are extracted twice, first to remove material from the upper surface while cells are still attached; the second time after detaching cells. (From Yim and Szuchet, 1989b).

a. Components removed from B3.f OLGs

b. Material extracted from B3.fA OLGs upper surface

c. Material extracted from the basal surface

Note that the elution profiles in (b) and (c) are different indicating that these components are spatially segregated.

V. CONCLUSIONS

We have presented evidence that upon adherence of OLGs to a polylysine substratum, there is spatial segregation of plasmalemma peripheral components in such a way that two distinct domains are created: one, apical, is exposed to the medium; the other, basal, is in contact with substratum. We have also shown that OLG polarization is accompanied by an attachment-induced synthesis of HSPGs and sulfated Gps that are susceptible to heparin extraction. The fact that the HSPGs and Gps from the apical and basal surfaces are distinct components, allow us to postulate that: a) specific receptors must exist on OLG plasmalemma that bind these components; b) these receptors have to be segregated in the two domains, i.e., there must, be an adhesion-induced polarization of membrane receptors; and c) adhesion has to initiate vectorial transport.

We have demonstrated that OLGs constitutively secrete to the medium a discrete number of fucosylated and sulfated Gps. These are complex molecules composed of heterologous polypeptide chains with the sugar moiety in an N-linked bond. OLGs also secrete CHSPGs. Secretory products are temporally modulated, but the modulation is different for the CHSPGs and Gps. CHSPGs seem to be under negative control by adhesion. Thus, B3.f OLGs have the highest secretion of CHSPGs; upon attachment and over time in culture there is a diminution in CHSPG discharge, i.e.; B3,fA OLGs secrete less CHSPGs than B3.f OLGs. In line with this observation, it is interesting that Margolis et al., (1986) found no extracellular localization of CHSPGs in the adult rat brain but did find them in the developing brain. Secretion of Gps appears to be independent of adhesion, per se, but is influenced by time in culture; specifically one component, Gp45, is induced 4-fold between the second and third week in culture, when it reaches plateau levels.

Presently, we have no clues as to the functional role of neither the peripheral nor the secreted components, so we can only speculate as to what their function might be. We hypothesize that peripherally disposed PGs and Gps in conjunction with some but not all secreted components may constitute the scaffolding on which cell adhesion occurs. In this regard, these components may effectively act as a rudimentary basal lamina/extracellular matrix that OLGs utilize for cell-substratum interaction and/or cell-cell interaction. That this may be the case can be surmised from our finding that after removal of peripheral components, OLGs no longer attach to the substratum. Thus, as other cell types do, e.g.; Schwann cells, OLGs too may be secreting the molecules they require to accomplish their biological functions. Finally, we speculate that some of the secreted components, e.g.; Gp45 may have trophic/tropic functions either on OLGs themselves, i.e., autocrine or on other cells, i.e.; paracrine. We believe that this work opens up a new avenue in our quest for understanding the molecular and cellular mechanisms of myelin formation in the CNS.

ACKNOWLEDGEMENTS

We are grateful to our colleagues for help in many aspects of this work. In particular, we would like to thank Mr. John Kazwell for the steady supply of healthy oligodendrocytes. This work was supported by grants GR 1223-D5 from the National Multiple Sclerosis Society and P01-NS 24575 from the National Institute of Health.

REFERENCES

Bunge RP, Bunge MB, Eldridge CF (1986) Linkage between axonal ensheathment and basal limina production by Schwann cells. Ann Rev Neurosci 9:305-328

Burgess TL, Kelly RB (1987) Constitutive and regulated secretion of proteins. Ann Rev Cell Biol 3:243-293

Ekblom P, Vestweber D, Kemler R (1986) Cell-matrix interactions and cell adhesion during development. Ann Rev Cell Biol 2:27-47

Eldridge CF, Bunge MB, Bunge RP (1989) Differentiation of axon-related Schwann cells in vitro:II. Control of myelin formation by basal lamina. J Neurosci 9:625-638

Grinnell F (1978) Cellular adhesiveness and extracellular substrata. Intl Rev of Cyt 54:67-144.

Hay ED (1984) Cell-matrix interaction in the embryo:cell shape, cell surface, cell skeletons, and their role in differentiation. In: Trelstad RL (ed) The role of extracellular matrix in development. Liss, New York, 1-31

McDonald JA (1988) Extracellular matrix assembly. Ann Rev Cell Biol 4:183-208

Mack SR, Szuchet S, Dawson G (1981) Synthesis of Gangliosides by cultured oligodendrocytes. J Neurosci 6:361-367

Mack SR, Szuchet S (1981) Synthesis of myelin glycosphingolipids by isolated oligodendrocytes in tissue culture. Brain Res 214:180-185

Margolis RU, Aquino DA, Klinger MM, Ripellino JA, Margolis RK (1986) Structure and localization of nervous tissue proteoglycans. Annals NY Acad of Sci 481:46-54

Massa P, Szuchet S, Mugnaini E (1984) Cell-cell interactions of isolated and cultured oligodendrocytes: formation of linear occluding junctions and expression of peculiar intramembrane particles. J Neurosci 4:3128-3139

Mirsky R, Winter J, Abney ER, Pruss RM, Gavrilovic J, Raff M (1980) Myelin-spedific proteins and glycolipids in rat Schwannn cells and oligodendrocytes in culture. J Cell Biol 84:483-494

Sabatini DD, Griepp EB, Rodriquez-Boulan EJ, Dolan WJ, Robbins ES, Papadopoulos S, Ivanov IE and Rindler MJ (1983) Biogenesis of epithelial cell polarity. In: McIntosh, JR (ed) Spatial organization of eukaryotic cells, Vol 2:419-450 Modern Cell Biol, Alan R Liss Inc. NY

Salminen A, Novick PJ (1987) A ras-like protein is required for a post-Golgi event in yeast secretion. Cell 49:527-538

Soliven B, Szuchet S, Arnason BGW, Nelson D (1988) Voltage-gated potassium currents in cultured ovine oligodendrocytes. J Neurosci 8:2131-2141

Soliven B, Szuchet S, Arnason BGW, Nelson DJ (1988) Forskolin and phorbol esters modulate the same K^+ conductance in cultured oligodendrocytes. J Membrane Biol 105:177-186

Szuchet S, Stefansson K, Wollmann RL, Dawson G, Arnason BGW (1980) Maintenance of isolated oligodendrocytes in long-term culture. Brain Res 200:151-164

Szuchet S, Arnason BGW, Polak PE (1980) Separation of ovine oligodendrocytes into two distinct bands on a linear sucrose gradient. J Neurosci Methods 3:7-19

Szuchet S, Yim SH, Monsma S (1983) Lipid metabolism of isolated oligodendrocytes maintained in long-term culture mimics events associated with myelinogenesis. Proc Natl Acad Sci (USA) 80:7019-7023

Szuchet S, Yim SH (1984) Characterization of a subset of oligodendrocytes separated on the basis of selective adherence properties. J Neurosci Res 11:131-144

Szuchet S, Polak PE, Yim SH (1986) Mature oligodendrocytes cultured in the absence of neurons recapitulate the ontogenic development of myelin. Dev Neurosci 8:208-221

Szuchet S (1987a) The plasticity of mature ologodendrocytes: a role for substratum in phenotype expression. In: Serlupi Crescenzi G. (ed) A multidisciplinary approach to myelin diseases. Plenum Press, New York London, 143-160

Szuchet S (1987b) Myelin palingenesis: the reformation of myelin by mature oligodendrocytes in the absence of neurons. In: Althaus HH, Seifert W (eds) Glial-neuronal communication in development and regeneration. Springer-Verlag, Berlin Heidelberg New York London Paris Tokyo, 755-778

Szuchet S, Polak PE, Yim SH, Lange Y (1988) The plasma membrane of cultured oligodendrocytes. II. Possible structural and functional domains. Glia 1:54-63

Szuchet S, Yim SH (1989) Oligodendrycyte products of secretion: II. Temporal modulation. Submitted

Tanford C, Reynolds JA (1976) Characterization of membrane proteins in detergent solutions. Biochem Biophy Acta 457:133-170

Vartanian T, Szuchet S, Dawson G, Campagoni AT (1986) Oligodendrocyte adhesion activates protein kinase C-mediated phosphorylation of myelin basic protein. Science 234:1395-1397

Vartanian T, Sprinkle TS, Dawson G, Szuchet S (1988) Oligodendrocyte-substratum adhesion modulates expression of adenylate cyclase linked receptors. Proc Natl Acad Sci (USA) 85:939-943

Vartanian T, Dawson G, Szuchet S (1988) Intracellular messengers: influence of oligodendrocyte substratum adhesion. Annals N.Y. Acad Sci 540:433-436

Yim SH, Szuchet S, Polak PE (1986) Cultured oligodendrocytes: a role for cell-substratum interaction in phenotypic expression. J Biol Chem 261:11808-11815

Yim SH, Szuchet S (1988) Proteoglycans associated with oligodendrocyte membrane. Trans Am Soc Neurochem 19:260

Yim SH, Szuchet S (1989a) Oligodendrocyte products of secretion: I. Isolation and initial charactarizaiton. Submitted

Yim SH, Szuchet S (1989b) Myelin palingenesis involves oligodendrocyte polarization and *de novo* synthesis of proteoglycans. Submitted

Yim SH, Szuchet S (1989c) Heparan Tethered Oligodendrocyte surface proteoglycans. Trans Am Soc Neurochem 20:335

PROTEIN KINASES A AND C ARE INVOLVED IN OLIGODENDROGLIAL PROCESS FORMATION

H.H. Althaus, P. Schwartz[1], S. Klöppner, J. Schröter, V. Neuhoff
Max–Planck–Institut für Experimentelle Medizin
Forschungsstelle Neurochemie
Hermann–Rein–Str. 3
3400 Göttingen, FRG

INTRODUCTION

Oligodendrocytes (Ol), shorn of their processes during the isolation procedure, regenerate their fibres under appropriate culture conditions (Althaus et al., 1984). various diffusible and non–diffusible compounds contribute to this regenerative process. Some of the signaling factors seem to be receptor mediated and transduced via the protein kinases. Activation of protein kinase C by phorbol esters such as tetradecanoylphorbolacetate (TPA) or $4-\beta$ phorboldibutyrate ($4-\beta$ PDB) enhanced the oligodendroglial fibre formation appreciably, as revealed by phase contrast microscopy (Schröter and Althaus, 1987). However, a closer insight into the sequence of oligodendroglial process regeneration was obtained by using scanning electron microscopy (SEM). This showed that activation of protein kinase A (Pk–A) and protein kinase C (Pk–C) supports different steps of the initial remyelination process.

MATERIALS AND METHODS

Reagents: All chemicals were of analytical grade if possible, culture media and fetal calf serum (Biochem.–Berlin); $N^6-2'-O-$Dibutyryladenosine $3',5'$cyclic monophosphate (Bt_2cAMP), Forskolin, IBMX (Sigma–Weinheim); Phorbol,12–myristate, 13–acetate (TPA) (Calbiochem–Frankfurt); $4-\beta$–phorbol–12,13–dibutyrate ($4-\beta$–PDB) (Serva–Heidelberg); 2–methylpiperazine dihydrochloride (H–7) (Seikagaku–USA)

Cell culture: Oligodendrocytes were isolated from adult pig brain as previously described (Gebicke–Härter et al., 1984; Althaus et al. 1987) and cultured on poly–D–lysine

---

[1]Dept. of Anatomy, University of Göttingen, 3400 Göttingen, FRG

coated glass plates in Petri dishes (Althaus et al., 1984); the culture medium consisted of MEM/HAM's F 10 (1:1, v/v) containing fetal calf serum (10%), transferrin 10 μg/ml, insulin 5 μg/ml, mezlocillin 40 μg/ml, cytosin–β–D–arabinoside (16 μM) was present for the first 5 days in vitro (DIV). On day 6 in vitro the culture medium was removed and replaced by fresh culture medium to which TPA or 4–β PDB was added; the phorbol esters were dissolved in methanol, final methanol concentration did not exceed 0,1%. H–7 (50 μM) or forskolin (100μM) or Bt_2cAMP (1 mM) plus IBMX (100 μM) was added after removal of the phorbol ester at day 8 in vitro.

Scanning electron microscopy: The cells were fixed with glutaraldehyde 2.5% in 0.1 M cacodylate buffer (2 h) and postfixed in 1% O_3O_4/0.1 M cacodylate buffer (1 h) after washing in 0.2 M cacodylate buffer; the next steps included dehydration, critical point drying, and sputtering with gold palladium. The cells were viewed with a Zeiss Novascan 30.

RESULTS

Pk–C /oligodendroglial process formation. Activation of Pk–C with TPA (10 nM) or 4–β PDB (50 nM) resulted in an enhanced fibre growth. After 24 h the phorbol ester treated cultures developed a dense network of fibres. The fibres reached a length of several times the body diameter and were highly branched (Fig. 1).

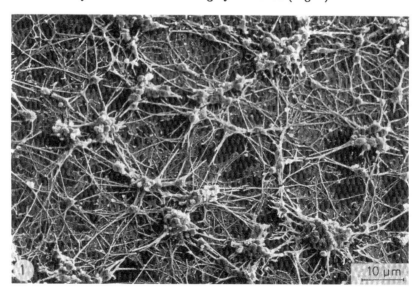

Figure 1: Oligodendrocytes (6 DIV) treated with TPA (10 nM) for 24 h; a feltlike network of fibres is furnished; some lamellae are still present.

In contrast, control cultures produced fewer less branched thin fibres with a length 3–4 the body diameter. Instead, a carpet like layer of lamellae or flattened membranes was predominant (Fig. 2).

Figure 2: Oligodendrocytes cultured under normal conditions (7 DIV); at this time in culture a carpet like layer of membranes is predominantly produced.

These lamellae were still present after 24 h of phorbol ester treatment although they were less apparent than in the controls. They disappeared completely after another 24 h (Fig. 3). In control cultures (8 DIV), these lamellae, which extend from the OL processes (Fig. 4), remained present. When H–7 (50 μM) was added to the culture medium after removal of 4–β PDB the lamellae reappear (Fig. 5).

For these experiments 4–β PDB was used because it can be more easily washed off than TPA.

Pk–A/oligodendroglial process formation. Bt$_2$cAMP/IBMX (1 mM/100 μM) delayed the production of oligodendroglial processes, compared with controls. A 24 h pretreatment of the cells with Bt$_2$cAMP/IBMX (1 mM/100 μM) followed by the addition of TPA (1 nM) reduced the amount of elongated and branched processes after 24 h: a 'less densely woven'network of fibres was apparent; with 10 nM TPA the fibre production was similar to that without pretreatment. When 4–β PDB was removed after 48 h and replaced by Bt$_2$cAMP/IBMX (1 mM/100 μM) or forskolin /IBMX (100 μM/100 μM), lamellae reappear (Fig. 6).

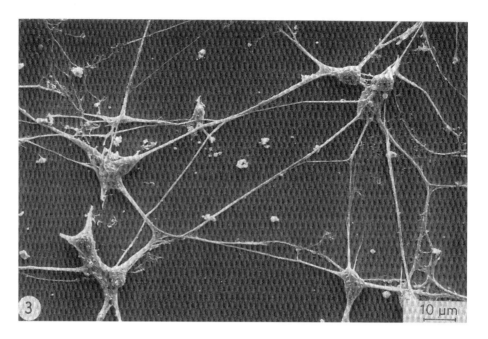

Figure 3: Flattened membranes extend from processes which are only a few microns long (7 DIV).

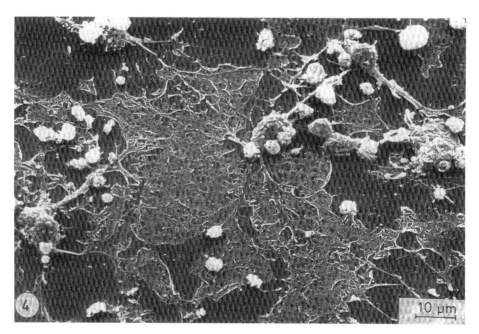

Figure 4: Oligodendrocytes (6 DIV) treated with 4–β PDB (100 nM) for 48 h; note the absence ot flattenes membranes.

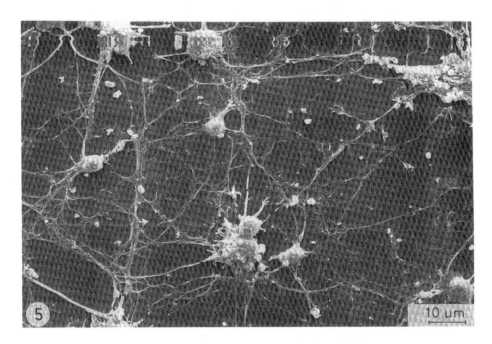

Figure 5: 4–β PDB, present for 48 h, was removed and H–7 (50 μM) was added; lamellae can again be observed.

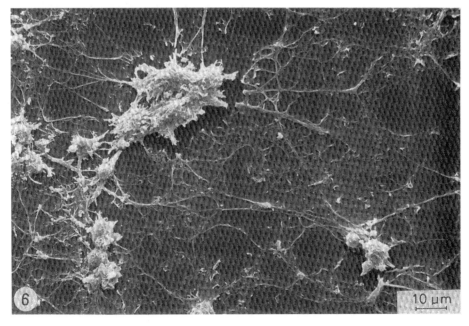

Figure 6: Oligodencrocytes (6 DIV) treated with 4–β PDB (100 nM) for 48 h, removal of 4–β PDB, and addition of Bt$_2$cAMP/IBMX (1 mM/100μM): after 24 h some lamellae reappear.

DISCUSSION

Oligodendroglial processes have to fulfill at least two tasks: 1) they have to explore the environmental terrain as well as 2) to recognize the axon. Depending on the distance to the axon − longer or shorter processes have to be furnished − the one or the other function might have priority. Subsequent myelination is characterized by the enwrapping of the axon and lateral broadening. By and large, these key events occur similarly during perinatal myelogenesis and remyelination in adulthood. In cell culture glial cells from newborn mouse forebrain produce flattened membranes which can be positively labelled for MBP and GC (Carey et al., 1988). Oligodendrocytes isolated from adult pig brain regenerate their fibres and also form shovel like GC⁺ lamellae. Fibre formation and production of flattened membranes seem to be under the control of different kinases. Activation of Pk−C favours the formation of fibres whereas the production of flattened membranes ceases.

Stimulation of Pk−A initiates the production of flattened membranes but not that of processes. Hence, concerning remyelination, the balanced interaction of Pk−C and Pk−A seems to be very important. For example, a preponderance of Pk−A activity upon Pk−C at the beginning of a regenerating − remyelinating process in adulthood would lead to the formation of flattened membranes which probably cannot overcome the distances to the axons. Furthermore, in the case of astrocytic fibre formation Pk−A and Pk−C do not seem to act antagonistically (Kimelberg et al., 1978; Mobley et al., 1986) which again would handicap an oligodendrocyte to remyelinate. A predominance of Pk−A activity could occur, for example, because of an endogenous decrease of the Pk−C activity which has been found in freshly isolated oligodendrocytes. On the other hand, exogenous factors present in the serum (as under normal culture conditions which include 10% fetal calf serum) could favour Pk−A activation. Whether other kinases contribute an important part to this Pk−A/Pk−C duo is as yet unknown.

Acknowledgement. This work was supported in part by the DFG (SFB 236).

REFERENCES

Althaus HH, Montz H, Neuhoff V, Schwartz D (1984) Isolation and cultivation of mature oligodendroglial cells. Naturwissenschaften 71: 309–315.

Althaus HH, Burgisser P, Klöppner S, Rohmann A, Schröter J, Schwartz P, Siepl C, Neuhoff V (1987) Oligodendrocytes ensheath carbon fibres and produce myelin in vitro. In: Althaus HH, Seifert W (eds.), NATO ASJ Series H Vol. 2: Glial–Neuronal Communication in Development and Regenertion, 779–798, Springer, Berlin, Heidelberg, New York.

Carey EM, Reynolds R, Herschkowitz N (1988) Expression of 2',3'–cyclic nucleotide 3'–phosphohydrolase and myelin basic protein, and formation of oligodendrocytes in mixed glial cultures maintained in a defined medium. Biochem. Soc. Trans., 900–901.

Gebicke–Härter PJ, Althaus HH, Rittner I, Neuhoff V (1984) Bulk separation and long–term culture of oligodendrocytes from adult pig brain. 1. Morphological studies. J. Neurochem 42: 357–368.

Kimelberg HK, Narumi S, Bourke RS (1978) Enzymatic and morphological properties of primary rat brain astrocyte cultures, and enzyme development in vivo. Brain Res. 153: 55–77.

Mobley PL, Scott SL, Cruz EG (1986) Protein kinase C in astrocytes: a determinant of cell morphology. Brain Res. 398: 369.

Schröter J, Althaus HH (1987) The phorbolester TPA dramatically accelerates oligodendroglial process regeneration. Naturwissenschaften 74: 393–394.

GROWTH FACTORS IN HUMAN GLIAL CELLS IN CULTURE

S.U. Kim and V.W. Yong
Division of Neurology, Department of Medicine
University of British Columbia
2211 Wesbrook Mall
Vancouver, B.C. V6T 1W5
Canada

INTRODUCTION

In the central nervous system (CNS), astrocytes and oligodendrocytes are the major glial cell types, while in the peripheral nervous system, Schwann cells, the equivalent of CNS oligodendrocytes, form the only glial cell type. During the normal development of the nervous system, glial cells or their precursors undergo proliferation at certain time points, possibly influenced by the schedule of an internal time clock and just before the onset of myelination. The glial cells eventually become post-mitotic and normally do not divide in the adult nervous system except in response to injury. The latter is well documented for astrocytes during insults of various kinds (Cavanagh 1970, Skoff 1975, Latov et al. 1979) and for Schwann cells during Wallerian degeneration (Abercrombie and Johnson 1946, Bradley and Asbury 1970, Salzer and Bunge 1980) and paranodal demyelination (Griffin et al. 1987). In the case of oligodendrocytes, however, the literature has been controversial. While some reports have shown that oligodendrocytes cannot proliferate after injury (Cavanagh 1970, Skoff 1975), others have suggested that doubling of oligodendrocytes can occcur after physical brain trauma (Ludwin 1984, 1985), following recovery from demyelinating viruses (Herndon et al. 1977, Rodriguez et al. 1987) and in an experimental allergic encephalomyelitis (EAE) (Raine and Tragott 1983).

Given that glial cells may undergo cell division under certain conditions, it is beneficial to identify and understand the factors that control their proliferation, be these endogenous growth factors, contact with axons or exogenous agents. In disease states where cells are lost, such as for oligodendrocytes in multiple sclerosis, induction of mitosis may achieve a critical number of the same cell type to enable restoration of function. This aim is not without clinical support. In multiple sclerosis, proliferation of oligodendrocytes, albeit limited in its extent, has already been reported in the relatively normal white matter adjacent to disease plaques (Raine et al. 1981). What is therefore required are stimuli to enhance this limited regenerative capacity already observed.

In vivo studies of glial cell proliferation are intrinsically difficult. Dividing cells may not be reliably identified as having undergone mitosis, and morphological characterization of the cell type may also present problems. The difficulties existing in the identification of dividing oligodendrocytes were underlined by Cavanagh (1970) who remarked that even when he could not find definite evidence for oligodendrocyte proliferation, it was conceivable that they had divided, but that in so doing, they had become morphologically indistinguishable from a normal oligodendrocyte.

For such reasons, many studies of glial cell proliferation have resorted to tissue culture techniques. Enriched population of a particular cell type can be attained (Wood, 1976, McCarthy and deVellis 1980, Kim et al. 1983a, 1983b, Kim 1985), and cells can be also readily identified by immunocytochemistry for the presence of their cell type specific markers, such as galactocerebroside for oligodendrocytes and glial fibrillary acidic protein (GFAP) for astrocytes (Bignami et al. 1972, Raff et al. 1979, Kim 1985).

Using cultured cells, many mitogens have been reported for glial cells, and the results are summarized on Tables I and II. For astrocytes, these include fibroblast growth factor (FGF) (Pruss et al. 1982), platelet-derived growth factor (PDGF) (Heldin et al. 1977), epidermal growth factor (EGF) (Westermark 1976, Leutz and Schachner 1981, Simpson et al. 1982), glial growth factor from the bovine pituitary (GGF-BP) (Brockes et al. 1980, Pruss et al. 1982, Kim et al. 1983a), myelin basic protein (Sheffield and Kim 1977, Bologa et al. 1985) and interleukin-1 (Giulian and Lachman 1985). For Schwann cells, among the reported mitogens are axolemma fragments derived from either the peripheral or central nervous system (Salzer and Bunge 1980, DeVries et al. 1982, Sobue et al. 1984, Pleasure et al. 1985), a myelin-enriched fraction (Yoshino et al. 1987), GGF-BP (Raff et al. 1978, Brockes et al. 1980, Pleasure et al. 1986), cAMP (Raff et al. 1978, Sobue et al. 1986), and forskolin, an adenyl cyclase activator that results in cellular accumulation of cAMP (Porter et al. 1986).

Mitogens for Schwann cells have recently been comprehensively summarized by Ratner et al. (1986). In the case of oligodendrocytes, proliferative factors are reported to be 'glia promoting factors' (Giulian et al. 1985, 1986), contact with neuronal axons (Wood and Williams 1984, Wood and Bunge 1986), PDGF (Besnard et al. 1987), FGF (Eccleston and Silberberg 1985), T-cell derived factors (Merrill et al. 1984) and interleukin-2 (Benveniste and Merrill 1987).

To identify factors that can induce cell proliferation, it is necessary to have a technique that is reproducible and unambiguous. Three methods are commonly used, but each has its own limitations. The first relies on the counting of cell numbers in a given area (Morrison et al. 1982, Giulian et al. 1985, Hatten 1985). The second utilizes the incorporation of ^{3}H-thymidine or ^{125}I-iododeoxyuridine into the DNA of dividing cells followed by quantification of the amount of radioactivity in a scintillation counter (Raff et al. 1978, Pettman et al. 1982, Simpson et al. 1982, Merrill et al. 1984). For these two methods, unless the cell population is homogeneous, results can be misleading. Type I astrocytes have a fibroblast-like morphology (Raff et al. 1983), and it is conceivable that when a positive result for astrocytes is reported by these techniques, the cells that were proliferating were mainly fibroblasts, which normally thrive in culture. Another potential source of error is that the original density of cells may vary from one sample to the next such that comparisons of test samples from controls become flawed.

The third method also employs ^{3}H-thymidine incorporation into the DNA of dividing cells but then uses autoradiography to determine the number of cells with silver grains on their nuclei (Besnard et al. 1987, Kim et al. 1983a, Westermark 1976, Yoshino et al. 1987). The cell type undergoing mitosis can be positively identified if such cells have been additionally immunostained with their cell type specific markers. Although this method is commonly used, the disadvantages include the high level of background that is frequently encountered with autoradiography (Skoff 1975) and the long period (days or even weeks) that is required for the development of the radiolabelling. Subjectivity associated with discrimination between unlabelled and weakly labelled cells can also constitute a problem.

Due to such difficulties, we have developed a double labelling immunofluorescence technique to assess cell proliferation (Yong and Kim 1987). The method uses nuclear incorporation of bromodeoxyuridine (BrdU), an analog of thymidine, as an index of DNA replication (Gratzner 1982,

Dolbeare et al. 1983, Morstyn et al. 1983). By using a specific antibody directed against BrdU and another antibody directed against the cell type specific marker (for example, GFAP for astrocytes), the proliferating cell type can be readily identified. This technique is simple, rapid, reproducible and unambiguous.

As shown on Tables I and II, most studies of mitogens for glial cells have relied on cells isolated from the rodent nervous system. Results using human glial cells are scarce, and it cannot be assumed that mitogens for rodent cells will necessarily have similar effects on human cells. For this reason, we have embarked on an extensive study of factors that might induce proliferation of cultured human glial cells. This chapter, using the BrdU method, summarizes our findings with fetal astrocytes, fetal Schwann cells, adult astrocytes and adult oligodendrocytes. Part of the results have been published elsewhere (Yong et al. 1988a, 1988b).

MATERIALS AND METHODS

Cell culture

Human adult astrocytes and oligodendrocytes were obtained by previously described methods (Kim et al. 1983b, 1985a) from the autopsied corpus callosum of subjects of ages between 60 and 90 years old. In brief, the corpus callosum was cut into fragments of 3^3 mm and incubated with 0.25% trypsin and 0.002% DNAse in calcium- and magnesium-free Hanks' balanced salt solution (CMF-HBSS) for 1 hour at 37°. Dissociated cells were passaged through a nylon mesh of 100 μm pore size, mixed with Percoll (Pharmacia) (final concentration of 30% Percoll) and centrifuged at 15,000 rpm for 25 minutes. Oligodendrocytes and astrocytes, floating between an upper myelin layer and a lower erythrocyte layer, were diluted with 3 volumes of HBSS and harvested by centrifugation at 1,500 rpm for 10 minutes. The cells were washed twice in HBSS, suspended in feeding medium and plated onto 9 mm Aclar plastic coverslips coated with 10 μg/ml polylysine (Sigma, molecular weight of 400,000) at a density of 10^4 cells/coverslip. Feeding medium consisted of 5% fetal calf serum, 5 mg/ml glucose and 20 μg/ml gentamicin in Eagle's minimum essential medium. The cells were used for the present proliferation studies after 5-7 weeks in culture (Figure 1).

TABLE I. REPORTED MITOGENS FOR ASTROCYTES AND OLIGODENDROCYTES IN CULTURE

Reported mitogen	Sources of cells	Method*	Ref.
ASTROCYTES			
Glial growth factor from bovine pituitary	Neonatal rat brain Neonatal rat brain	C D	6 29.54
Fibroblast growth factor	Neonatal rat brain	D	54
Epidermal growth factor	Neonatal mouse brain Neonatal rat brain Human glial cell line	D B C	37 66 71
Platelet-derived growth factor	Human glial cell line	C	25
Myelin basic protein	Mouse embryonic brain Adult mouse brain or C6TK cell line	D B	7 65
Interleukin-1	Neonatal rat brain	B	17
OLIGODENDROCYTES			
'Glia promoting factors'	Neonatal rat brain	A	16,18,19
Fibroblast growth factor	Neonatal rat brain	D	13
Platelet-derived factor	Neonatal rat brain	C	3
Axonal contact	Adult rat spinal cord	D	73,74
T-cell derived factors	Neonatal rat brain	B	44
Interleukin-2	Neonatal rat brain	B	2

*Refers to the method of assessing cell proliferation: A—Counting of cells not identified by immunostainings, B—^{3}H-thymidine or ^{125}I-iododeoxyuridine uptake followed by scintillation counting, C—^{3}H-thymidine incorporation followed by autoradiography with cell-type specific markers not used, D—^{3}H-thymidine uptake followed by autoradiography, cell-type specific markers used.

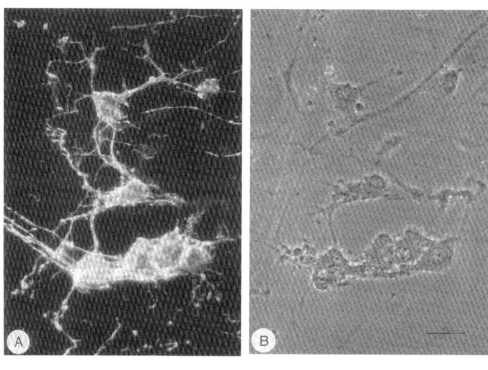

Figure 1. Human adult oligodendrocytes grown in culture for more than 2 months in vitro. A: galactocerebroside immunostaining, B: phase contrast microscopy. Bar indicates 20 µm.

TABLE II. REPORTED MITOGENS FOR SCHWANN CELLS IN CULTURE

Reported mitogen	Sources of cells	Method*	Ref.
Glial growth factor from bovine pituitary	Neonatal rat sciatic	B	56
	Neonatal rat sciatic	C	52,53
	Adult human sural	C	52
	Adult human trigeminal	C	47
Axolemma	Neonatal rat sciatic	C	52,64,69
	Adult human sural	C	52
	Neonatal rat dorsal root ganglion	C	11
cAMP	Neonatal rat sciatic	B	56
	Neonatal rat sciatic	C	70
Myelin	Neonatal rat sciatic	C	78
Forskolin	Neonatal rat sciatic	C	53

*Refer Table I for legend

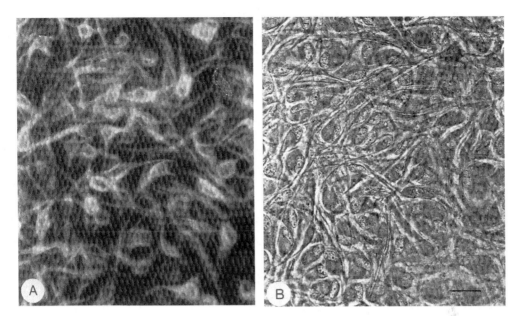

Figure 2. Human fetal astrocytes grown in culture for 2 weeks in vitro. A: GFAP immunostaining, B: phase contrast microscopy. Bar indicates 20 μm.

Fetal astrocytes were isolated from human fetuses, legally and therapeutically aborted, of 8-10 weeks' gestation. Whole brain was cut into fragments of 2^3 mm and incubated with 0.25% trypsin and 0.002% DNase in CMF-HBSS for 30 minutes at 37°. The fragments were then dissociated into single cells by gentle pipetting. The suspension was centrifuged at 1,500 rpm to collect the dissociated cells. Two washings in HBSS ensued, and the cells were finally suspended in feeding medium identical to that for adult astrocyte/oligodendrocyte. The cells were seeded onto polylysine-coated Aclar coverslips at a density of 10^4 cells/coverslip. Cell types include neurons, fibroblasts and astrocytes. Oligodendrocytes could only be detected occasionally using immunostainings of antibodies directed against galactocerebroside. The proliferation experiments were performed after the cells had been in culture for at least 11 days at the time GFAP-positive astrocytes occupy more than 90% of the cell population (Figure 2).

For the isolation of Schwann cells, dorsal root ganglia were obtained from fetuses of 8-10 weeks' gestation. Dissociated cultures were established as previously described (Kim et al. 1985b) in a manner similar to that for fetal astrocytes above. Feeding medium was Eagle's minimum essential medium supplemented with 10% fetal calf serum, 5 mg/ml glucose and 20 μg/ml gentamicin. No nerve growth factor was added, and this resulted in depletion of neuronal populations (Levi-Montalcini et al. 1954, 1968, Mobley et al. 1977) such that within 2 weeks in vitro, neurons comprised less than 1% of the total cell population - Schwann cells (30%) and fibroblasts (70%) being predominant. These neuron-deficient cultures were used for the present proliferation studies.

Incubation of cells with growth factors and BrdU

For fetal astrocytes, the test growth factors were added to the culture medium for 3 days prior to immunofluorescence studies, while BrdU (Sigma, 10 μM) was introduced for the last 2.5 hours. Preliminary experiments had indicated that for control fetal astrocytes, 2.5 hours of incubation with BrdU-containing medium was adequate to achieve BrdU nuclear labelling of 10-30% of cells. For adult astrocytes and oligodendrocytes, test growth factors and BrdU were added for 3 days before immunostainings were performed. BrdU was added for 3 days because preliminary studies had shown that these adult cells were slowly dividing cells, if at all. In addition, the concentration (10 μM) of BrdU added for 3 days was found not to be toxic to the cells as assessed by cell morphology and survival. In the case of fetal Schwann cells, test factors and 10 μM BrdU were applied to the culture medium simultaneously for 2 days prior to immunostainings.

Test growth factors were recombinant Interleukin-2 (Genzyme Corporation, 5 and 50 U/ml), EGF and FGF (Bethesda Research Laboratories, 0.5 and 5 μg/ml), PDGF (Collaborative Research, 5 and 50 mU/ml), dibutyryl cAMP (DBcAMP, Sigma, 100 μM and 1 mM), nerve growth factor (NGF, Sigma, 100 ng/ml and 1 μg/ml), forskolin (Sigma, 10 and 100 μM), a tumor-promoting agent 4B-phorbol 12, 13-dibutyrate (PDB, Sigma, 10 and 100 nM) (Boutwell 1974, Nishizuka 1986) and GGF-BP (1 μg/ml). GGF-BP was prepared from adult bovine pituitary homogenates by ammonium sulfate fractionation, carboxymethylcellulose and phosphocellulose column chromatographies (Brockes et al. 1980) and was a gift of Dr. D. Pleasure. In addition, a combination of GGF-BP (1 μg/ml) and forskolin (10 μM) was tested.

Immunocytochemistry

The general process has been described in detail for astrocytes (Yong and Kim 1987). After removal of BrdU-containing medium and washing the cells thoroughly with phosphate-buffered saline (PBS) to ensure complete removal of unbound BrdU, astrocytes and Schwann cells on coverslips were fixed with 70% ethanol at -20° for 30 minutes. Rat monoclonal antibody to glial fibrillary acidic protein (GFAP) (1:2) or polyclonal rabbit anti-GFAP was applied to astrocytes for 30 minutes, followed by goat anti-rat immunoglobulin conjugated to rhodamine (1:40). This enabled subsequent identification of astrocytes. For oligodendrocytes, prior to the ethanol fixation process, cells were incubated with rabbit antiserum against galactocerebroside (1:20, 30 minutes), followed by goat anti-rabbit immunoglobulin conjugated to rhodamine (1:40) for another 30 minutes. Oligodendrocytes were then fixed with 4% paraformaldehyde at 4° for 15 minutes, followed by ethanol fixation. Without the paraformaldehyde pretreatment, galactocerebroside immunostainings tended to be obliterated by the acid treatment described below. After the ethanol fixation step, cells were incubated with 0.25% Triton-X in PBS for 30 minutes. This improvisation was necessary because the paraformaldehyde treatment resulted in the nuclear membrane being less permeable to the BrdU antibody.

Hydrochloric acid at 2 M concentration was introduced to the cells for 10 minutes to denature DNA. This step is essential because the antibody to BrdU cannot bind to BrdU of intact DNA. Following a wash in PBS, sodium borate at pH 9 and 0.1 M was added for 10 minutes to neutralize the HCl. After rinsing off the borate with PBS, mouse monoclonal antibody to BrdU was applied for 30 minutes at 1:10 dilution. This was followed by goat anti-mouse immunoglobulin conjugated to fluorescein (1:40) for 30 minutes. The coverslips were then mounted on glass slides with glycerol-PBS and examined on a Zeiss Universal fluorescence microscope equipped with phase contrast, fluorescein and rhodamine optics. Except for the cell fixation processes, the entire staining and acid denaturation procedure was carried out at room temperature.

Thus, proliferating cells could be detected by their nuclear incorporation of BrdU. Astrocytes could be simultaneously recognized by the presence of GFAP while oligodendrocytes were identified by galactocerebroside. No double stainings were performed on Schwann cells, but these cells could be readily identified by their bipolar, spindle-shaped morphology.

Sources of antibodies were as follows: mouse anti-BrdU monoclonal from Becton Dickinson, rat anti-GFAP monoclonal from Dr. V. Lee (Lee et al. 1984), polyclonal rabbit anti-GFAP from Dako Corporation, and all secondary antibodies (e.g., goat anti-mouse immunoglobulin-fluorescein) were from Cappel Laboratories. Antiserum to galactocerebroside purified from bovine brain was raised in rabbits in this laboratory. Dilutions of antibodies were made in PBS.

Tabulation of data

Under the immunofluorescence microscope, GFAP or galactocerebroside positive cells (astrocytes and oligodendrocytes respectively) were identified and counted. Of these, the number of identified cells with BrdU labelling in their nuclei was assessed to obtain the percentage of astrocytes or oligodendrocytes that were proliferating on the particular coverslip. Cell countings were done at x200 magnification in 10-20 fields such that between 100-400 cells were tabulated. The result was divided by the average of similar results obtained from control coverslips to give the proliferation index (PI) of the test factor. Thus, a PI value of 1 represents no mitotic capability while a PI value of 2 indicates a two-fold mitotic response. Value of 1.5 or greater was taken as a positive proliferative result.

For Schwann cells, the percentage of bipolar, spindle-shaped cells with BrdU-nuclear labelling was similarly tabulated and the PI value obtained.

In all cases, controls were sister cultures that were exposed to BrdU but not to any of the test growth factor.

RESULTS

For the convenience of the reader this section has been sub-divided into the three types of glial cells.

Oligodendrocytes

Adult oligodendrocytes did not incorporate BrdU into their nuclei even in the presence of any of the test growth factors. This inability was not due to the stage of differentiation of oligodendrocytes in culture since undifferentiated oligodendrocytes, as well as their differentiated counterparts (Figure 3), did not show positive labellings with BrdU. That the BrdU method works for oligodendrocytes is shown by galactocerebroside

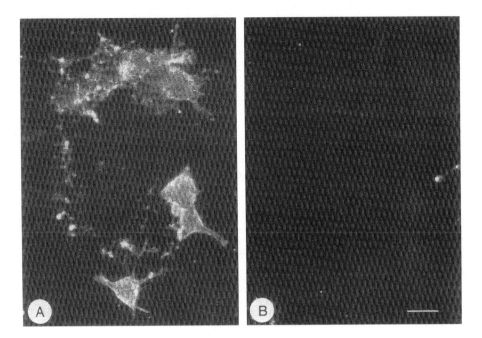

Figure 3. Differentiated human adult oligodendrocytes do not incorporate bromodeoxyuridine into their nuclei. A: galactocerebroside immunostaining, B: bromodeoxyuridine labelling. Bar indicates 20 μm.

and BrdU stainings of rat oligodendroccyte isolated on postnatal Day 1 and incubated with BrdU for 48 hours (Figure 4).

Astrocytes

Figure 5 shows positive immunolabellings of a control fetal astrocyte culture with the antibodies directed against GFAP and BrdU. The GFAP staining was cytoplasmic while BrdU labelling was nuclear. Different degrees of nuclear labelling by BrdU could be seen, presumably representing different stages of the mitotic cycle of the cell. Although there was occasional cross-reactivity between the two secondary antibodies when mouse anti-BdrU and rat anti-GFAP were used simultaneously, both being directed against rodent immunoglobulins, the distinct spatial labellings of cytoplasmic GFAP on the one hand and nuclear BrdU on the other, makes the results clear, interpretable and valid. However, to avoid confusion, we have recently begun to use polyclonal rabbit anti-GFAP and mouse anti-BdrU antibodies.

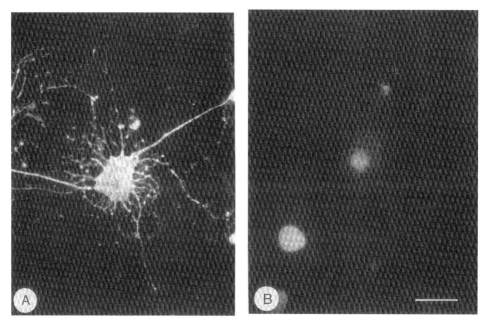

Figure 4. Postnatal day 1 rat brain at 3 days in vitro was incubated with 10 μm BrdU for 48 hours. A: galactocerebroside immunostaining, B: BrdU labelling. The positive BrdU labelling is evidence that the method can deter mitosis of oligodendrocytes. Bar indciates 20 μm.

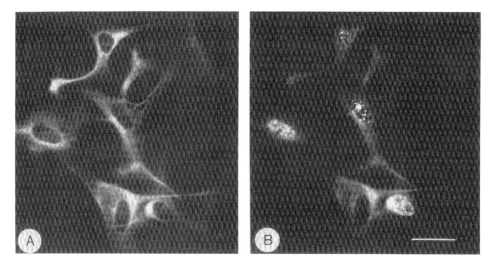

Figure 5. Control fetal astrocytes culture showing GFAP (A) and bromodeoxyuridine (B) immunostaining. Bar indicates 20 μm.

As shown on Table III, four mitogens were identified for fetal astrocytes. These were GGF-BP at 1 μg/ml, PDGF at 5 and 50 mU/ml, FGF at the higher concentration tested (5 μg/ml), and PDB at 100 nM but not 10 nM. All other agents tested produced a PI value of about 1 and thus were ineffective as mitogens. Figure 3 shows a higher density of BrdU labelling for fetal astrocytes exposed to GGF-BP when compared to controls.

The proliferative ability of fetal astrocytes appeared to be dependent on the period of time in culture. Figure 7 represents the percent of control astrocytes (those that were not treated with growth factors) that incorporated BrdU as a function of time in vitro. At 11 days of culture, 57% of astrocytes had positive labelling of BrdU in their nuclei. This gradually declined such that after 112 days in vitro, only 11% of fetal astrocytes incorporated BrdU.

In the case of adult astrocytes, no BrdU labelling could be observed for control astrocytes or those exposed to the various growth factors. It appeared that these adult cells had lost the ability to undergo mitosis. Figure 5 shows the absence of BrdU incorporation in the nucleus of adult astrocytes.

It should be noted that most fetal astrocytes were flat and fibroblast-like (Figures 5 and 6) while most of their adult counterparts were protoplasmic and process-bearing (Figure 8). The fibroblast-like astrocytic morphology has been designated Type I while protoplasmic and process-bearing astrocytes have been described as Type II (Raff et al. 1978). The failure of adult astrocytes to incorporate BrdU did not seem to be due to their Type II morphology since similar cells in fetal astrocyte cultures were found to incorporate BrdU.

Schwann cells

In Table III, the PI values of various growth factors on human fetal Schwann cells are presented. Mitogenic factors were GGF-BP (1 μg/ml), NGF (100 ng/ml and 1 ug/ml), PDB (10 and 100 nM), and the combination of GGF (1 μg/ml) and forskolin (10 μM). PDGF produced a positive proliferative response only at the higher concentration tested (50 mU/ml). On the average (n of 13 cultures containing between 100-200 Schwann cells each), 34% of control Schwann cells incorporated BrdU into their nuclei over a 48 hour period. Figure 9 compares BrdU immunolabelling of control Schwann cells with those exposed to GGF-BP. As noted, more Schwann cells were labelled in cultures treated with GGF-BP.

TABLE III.   PROLIFERATION INDEX (PI) OF VARIOUS GROWTH FACTORS ON HUMAN
FETAL ASTROCYTES AND SCHWANN CELLS IN CULTURE

Growth factor	Concentration		PI astrocyte	PI Schwann
Interleukin-2	5	U/ml	1.2 ± 0.2	1.2 ± 0.1
	50	U/ml	0.9 ± 0.1	1.0 ± 0.1
Glial growth factor from bovine pituitary	1	μg/ml	3.6 ± 0.1*	1.7 ± 0.1*
Platelet-derived growth factor	5	mU/ml	2.1 ± 0.1*	1.0 ± 0.1
	50	mU/ml	1.8 ± 0.2*	1.7 ± 0.2*
Fibroblast growth factor	0.5	μg/ml	1.0 ± 0.2	1.0 ± 0.1
	5	μg/ml	1.5 ± 0.1*	1.2 ± 0.2
Epidermal growth factor	0.5	μg/ml	0.5 ± 0.1	0.9 ± 0.1
	5	μg/ml	1.2 ± 0.1	1.2 ± 0.1
Nerve growth factor	100	ng/ml	1.1 ± 0.1	1.5 ± 0.1*
	1	μg/ml	1.4 ± 0.2	1.7 ± 0.1*
Phorbol-12, 13-dibutyrate	10	nM	1.1 ± 0.1	1.7 ± 0.2*
	100	nM	3.1 ± 0.4*	1.6 ± 0.1*
Dibutyryl cAMP	100	μM	1.1 ± 0.0	1.1 ± 0.1
	1	mM	0.9 ± 0.0	0.8 ± 0.0
Forskolin	10	μM	not tested	0.9 ± 0.2
	100	μM	not tested	1.1 ± 0.1
Glial growth factor and forskolin	1	μg/ml	not tested	1.8 ± 0.1*
	10	μM		

PI values are mean ± SEM, of between 3 and 6 coverslips each.   PI for
control is 1.   *Denotes effective mitogens.   SEM of less than 0.05 is
shown as 0.0.   The average % of control fetal astrocytes that
incorporated BrdU was 16 ± 1.2 (n=12) while that of control fetal Schwann
cells was 34 ± 2.1 (n=13) (mean ± SEM).

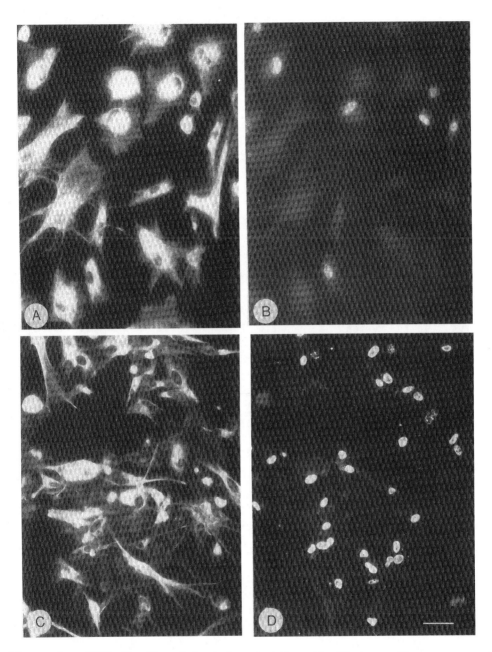

Figure 6. GFAP (A, C) and bromodeoxyuridine (B, D) immunofluorescence of control fetal astrocytes (A, B) or astrocytes exposed to GGF-BP (C, D). Note the higher density of bromodeoxyuridine labelling in culture exposed to GGF-BP (D). Bar indicates 20 μm.

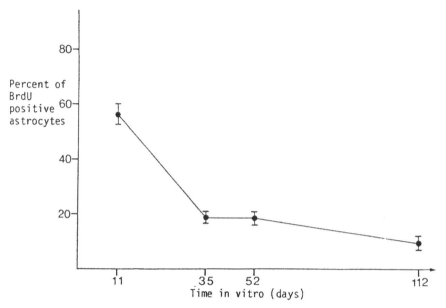

Figure 7. Percent of control human fetal astrocytes undergoing proliferation as a function of time in culture. Each point is the mean ± SEM of 3-6 determinations.

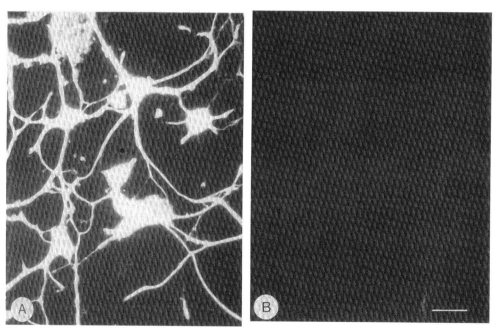

Figure 8. Human adult astrocytes do not incorporate bromodeoxyuridine into their nuclei. A: GFAP immunofluorescence, B: bromodeoxyuridine labelling. Bar indicates 20 μm.

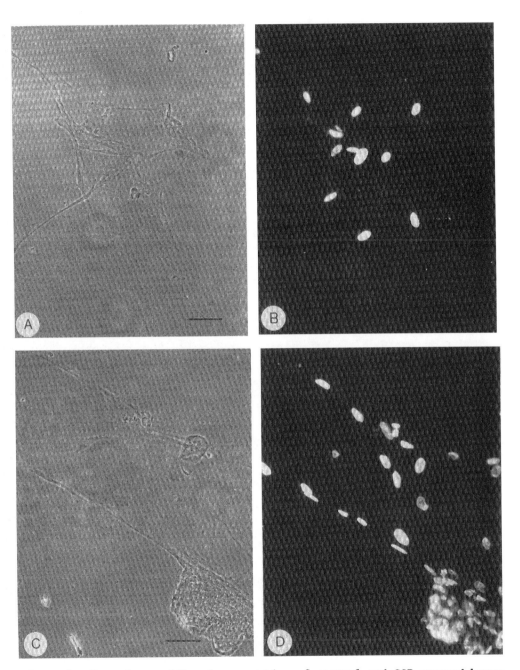

Figure 9. Bromodeoxyuridine incorporation of control and GGP exposed human fetal Schwann cells. Bar indicates 20 μm.
  A. Control, phase contrast microscopy
  B. Control, Bromodeoxyuridine immunofluorescence
  C. GGF, phase contrast microscopy
  D. GGF, Bromodeoxyuridine immunofluorescence

DISCUSSION

The search for factors that can promote proliferation of glial cells has generated many studies, some with contradictory results. For instance, while EGF was found to be mitogenic for neonatal rodent astrocytes in the studies of Simpson et al. (1982) and Leutz and Schachner (1981), negative result was obtained by Pruss et al. (1982). A probable reason for the discrepancies in the literature lies in the methods commonly used to detect cell proliferation; namely ^{3}H-thymidine uptake followed by scintillation counting or autoradiography (Tables I and II). As aforementioned, these techniques can have potential flaws such as cell identification, the lack of a homogeneous cell population in the case of scintillation counting, and the subjectivity that may become important in autoradiography when differentiating between weakly labelled and unlabelled cells.

In the present series of experiments, a less ambiguous method (immunofluorescence of BrdU) was used to assess whether or not several soluble factors have the potential to produce proliferation of glial cells derived from human fetuses or adults. In addition, since astrocytes and oligodendrocytes were simultaneously labelled with their cell type specific markers, positive identification of the proliferating cell type becomes possible.

Our results for fetal astrocytes show that four agents were capable of producing a mitogenic response: GGF-BP, PDGF, FGF and phorbol ester. The effects of FGF and phorbol ester were dose-dependent in that only the higher concentrations tested resulted in proliferation. The findings with GGF-BP confirm the reports of others for rat astrocytes (Pruss et al. 1982, Brockes et al. 1982, Kim et al. 1983a). On the other hand, the negative actions of EGF contrast the findings of Simpson et al. (1982) and Leutz and Schachner (1981). Besides the difference in the technique used to assess cell proliferation, a possible explanation for the discrepancy is that the astrocytes in the present study are derived from humans and not rodents. Another possibility is that the present experiment was conducted in the presence of serum-containing medium while many previous studies were performed in serum-free medium. Indeed, Morrison et al. (1982) showed that the proliferation of astrocytes in 10% serum-containing medium was about 3 times greater than those maintained in serum-free medium supplemented with a mitogen, FGF. Thus, the presence of serum in the present experiment, itself capable of promoting mitosis, might have masked the proliferative ability of EGF. However, serum-containing medium presents a more physiological environment to the cells than that of serum-free medium, and the present series of experiments was performed in serum-containing medium for that reason. The capability of phorbol ester to stimulate division of fetal astrocytes, as well as fetal Schwann cells (see below), has never been reported and may represent a widespread tumor-inducing property of phorbol esters (for review, see Boutwell 1974).

Figure 6 shows that the potential of control astrocytes to undergo proliferation decreases with the length of time that they have been maintained in culture. Many cell types lose their proliferative ability after repeated mitosis and passaging (Hayflick and Moorhead 1961, Hay and Strechler 1967) and in the present experiment, human fetal astrocytes appear to fall into this category.

That adult astrocytes in culture were incapable of mitosis confirms our earlier observations (Kim et al. 1985a) using ^{3}H-thymidine radiolabelling to detect cell division. This failure did not appear to be due to their Type II morphology (Raff et al. 1983) since some of the Type II astrocytes in the fetal cultures incorporated BrdU. In addition, Miller et al. (1986) showed that both Type I and Type II astrocytes could undergo mitosis in response to injury. From the results in Figure 6, where fetal astrocytes lose their proliferative capability with aging in vitro, it appears that the most likely reason for the failure of adult astrocytes to undergo mitosis is due to the advanced age of the adult astrocytes, being derived from humans of between ages 60-90. It is important to elucidate the mechanism by which proliferation of astrocytes (gliosis) is initiated. Gliosis is a central problem in CNS regeneration in which uncontrolled astrocytic multiplication blocks paths of regrowing nerves.

As shown on Table I, factors such as 'glial promoting factors', FGF, PDGF, axonal contact and Interleukin-2 have been reported to produce mitosis of oligodendrocytes isolated mostly from neonatal rat brain. It is important to confirm these findings since such results have important implications for diseases such as multiple sclerosis where oligodendrocytes are lost. Promoting remaining oligodendrocytes in such diseases to proliferate, and then to differentiate to produce remyelination, may eventually result in restoration of lost functions. Since multiple sclerosis is a disease of adults, it is especially necessary to test such agents on human adult oligodendrocytes. We attempted such an experiment and found that adult human oligodendrocytes in culture do not undergo mitosis. In addition, none of the agents tested, such as interleukin-2, could promote proliferation. The search for effective agents continues.

Several studies have examined the factors that might induce the mitosis of neonatal rat Schwann cells in vitro (Table II). The results have led Raff et al. (1978) to propose that there might exist two pathways that led to the proliferation of Schwann cells: one involved cAMP and the other did not. This was supported by the observation that the combination of forskolin (an adenyl cyclase activator with resultant accumulation of cAMP) and GGF-BP (which did not increase cAMP content) produced a greater proliferation index on rat Schwann cells than those obtained when either was used individually (Porter et al. 1986). We repeated these experiments using human fetal Schwann cells. In agreement with the literature (Raff et al. 1978, Brockes et al. 1980, Pleasure et al. 1986), GGF-BP was found to increase the proliferation rate of human fetal Schwann cells (Table III). Contrasting the results of others (Raff et al. 1978, Salzer and Bunge 1980, Pruss et al. 1982), NGF (at 100 ng/ml and 1 μg/ml) and PDGF (at the higher concentration of 50 mU/ml) were mitogenic for human fetal Schwann cells. The mitogenic property of NGF is interesting in view of the presence of nerve growth factor receptors on these cells (Kim et al. 1989).

Unlike the hypothesis of Raff et al. (1978), we found no evidence for the requirement of a cAMP-dependent mechanism in increasing the mitotic response of human fetal Schwann cells. This was shown by the lack of effects of two concentrations of dibutyryl cAMP or forskolin (Table II). The combination of forskolin and GGF-BP did not result in a higher response than that of GGF-BP alone, leading to the conclusion that the positive effect of the combination was due to GGF only.

In conclusion, we have assessed the ability of various agents to promote proliferation of human glial cells in culture. We began with the hypothesis that growth factors described for rodents cannot be assumed to be similarly effective in humans and, indeed, this may be the main reason why some of the results presented here contradict those in the literature. The previous observation that interleukin-2 promoted proliferation of oligodendrocytes (Merrill et al. 1984) was of special interest because interactions between the immune system and the nervous system are increasingly being recognized to be important in the pathogenesis of several nervous disorders. However, our results with human glial cells could not replicate that finding. The observations that phorbol ester, thought to activate directly protein kinase C (Kikkawa et al. 1983, Burgess et al. 1986, Nishizuka 1986), stimulated the mitosis of human fetal astrocytes and Schwann cells, suggest that activation of this important ubiquitous enzyme may trigger key events that lead to cell proliferation. Human adult astrocytes and oligodendrocytes could not be induced to proliferate in the present series of experiments, and it is important to continue the search for factors that might induce their mitosis in culture. As stated, such experiments have strong implications for diseases such as multiple sclerosis and CNS injury.

ACKNOWLEDGEMENTS

This study was supported by grants from the Medical Research Council of Canada, the Multiple Sclerosis Society of Canada, and the Jacob Cohen Fund for Research into Multiple Sclerosis. The skilled technical assistance of M. Kim and D. Osborne is acknowledged.

REFERENCES

Abercrombie M, Johnson ML (1946) Quantitative histology of Wallerian degeneration. I. Nuclear population in rabbit sciatic nerve. J Anat 80:37-50

Benveniste EN, Merrill JE (1986) Stimulation of oligodendroglial proliferation and maturation by interleukin-2. Nature 321:610-613

Besnard F, Perraud F, Sensenbrenner M, Labourdette G (1987) Platelet-derived growth factor is a mitogen for glial but not for neuronal rat brain cells in vitro. Neurosci Letts 73:287-292

Bignami A, Eng L, Dahl D, Uyeda C (1972) Localization of the glia fibrillary acidic protein in astrocytes by immunofluorescence. Brain Res 43:429-435

Bologa L, Deugnier MA, Joubert R, Bisconte JC (1985) Myelin basic protein stimulates the proliferation of astrocytes: Possible explanation for multiple sclerosis plaque formation. Brain Res 346:199-203

Bradley WG, Asbury AK (1970) Duration of synthesis phase in neurilemma cells in mouse sciatic nerve during degeneration. Exp Neurol 26:275-282

Brockes JP, Lemke GE, Balzer DR Jr (1980) Purification and preliminary characterization of a glial growth factor from the bovine pituitary. J Biol Chem 255:8374-8377

Boutwell RK (1974) The function and mechanism of promoters of carcinogenesis. Crit Rev Toxicol 2:419-443

Burgess SK, Sahyoun N, Blanchard SG, LeVine H III, Chang KJ, Cuatrecasas P (1986) Phorbol ester receptors and protein kinase C in primary neuronal cultures: Development and stimulation of endogenous phosphorylation. J Cell Biol 102:312-319

Cavanagh JB (1970) The proliferation of astrocytes around a needle wound in the rat brain. J Anat 106:471-487

DeVries GH, Salzer JL, Bunge RP (1982) Axolemma-enriched fractions isolated from PNS and CNS are mitogenic for cultured Schwann cells. Dev Brain Res 3:295-299

Dolbeare F, Gratzner H, Pallavicini MG, Gray JW (1983) Flow cytometric measurement of total DNA content and incorporated bromodeoxyuridine. Proc Natl Acad Sci (USA) 80:5573-5577

Eccleston PA, Silberberg DH (1985) Fibroblast growth factor is a mitogen for oligodendrocytes in vitro. Dev Brain Res 21:315-318

Giulian D, Tomozawa Y, Hindman H, Allen RL (1985) Peptides from regenerating central nervous system promote specific populations of macroglia. Proc Natl Acad Sci (USA) 82:4287-4290

Giulian D, Lachman LB (1985) Interleukin-1 stimulation of astroglial proliferation after brain injury. Science 228:497-499

Giulian D, Allen RL, Baker TJ, Tomozawa Y (1986) Brain peptides and glial growth. I. Glia-promoting factors as regulators of gliogenesis in the developing and injured central nervous system. J Cell Biol 102:803-811

Giulian D, Young DG (1986) Brain peptides and glial growth. II. Identification of cells that secrete glia-promoting factors. J Cell Biol 102:812-820

Gratzner HG (1982) Monoclonal antibody to 5-bromo- and 5-iododeoxyuridine: A new reagent for detection of DNA replication. Science 218:474-475

Griffin JW, Drucker N, Gold BG, Rosenfeld J, Benzaquen M, Charnas LR, Fahnestock KE, Stocks EA (1987) Schwann cell proliferation and migration during paranodal demyelination. J Neurosci 7:682-699

Hatten ME (1985) Neuronal regulation of astroglial morphology and proliferation in vitro. J Cell Biol 100:384-396

Hay RJ, Strechler BL (1967) The limited growth span of cell strains isolated from the chick embryo. Exp Gerontol 2:123-135

Hayflick L, Moorhead PS (1961) The serial cultivation of human diploid cell strains. Exp Cell Res 25:585-621

Heldin CH, Wasteson A, Westermark B (1977) Partial purification and characterization of platelet factors stimulating the multiplication of normal human glial cells. Exp Cell Res 109:429-437

Herndon RM, Pricce DL, Weiner LP (1977) Regeneration of oligodendroglia during recovery from demyelinating disease. Science 195:693-694

Immamoto K, Paterson J, Leblond CP (1968) Radioautographic investigation of gliogenesis in the corpus callosum of young rats. I. Sequential changes in oligodendrocytes. J Comp Neurol 180:115-138

Kikkawa U, Takai Y, Tanaka Y, Miyake R, Nishizuka Y (1983) Protein kinase C as a possible receptor protein of tumor-promoting phorbol esters. J Bil Chem 258:11442-11445

Kim SU, Stern J, Kim MW, Pleasure DE (1983a) Culture of purified rat astrocytes in serum-free medium supplemented with mitogen. Brian Res 274:79-81

Kim SU, Sato Y, Silberberg DH, Pleasure DE, Rorke L (1983b) Long-term culture of human oligodendrocytes. Isolation, growth and identification. J Neuro Sci 62:295-301

Kim SU, Moretto G, Shin DH, Lee VM (1985a) Modulation of antigenic expression in cultured adult human oligodendrocytes by derivatives of adenosine 3', 5'-cyclic monophosphate. J Neurol Sci 69:81-91

Kim SU, Kim KM, Moretto G, Kim JH (1985b) The growth of fetal human sensory ganglion neurons in culture: A scanning electron microscopic study. Scan Electron Microscopy II:843-848

Kim SU (1985) Antigen expression by glial cells grown in culture. J Neuroimmunol 8:255-282

Kreider BQ, Messing A, Doan H, Kim SU, Lisak RP, Pleasure DE (1981) Enrichment of Schwann cell cultures from neonatal rat sciatic nerve by differential adhesion. Brain Res 207:433-444

Latov N, Nilayer G, Zimmerman EA, Johnson WG, Silverman JA, Detenionini R, Cote L (1979) Fibrillary astrocytes proliferate in response to brain injury. Dev Biol 72:381-384

Lee V, Page K, Wu H, Schlaepfer WW (1984) Monoclonal antibodies to gel excised glial filament protein and their reactivity with other intermediate filament proteins. J Neurochem 42:25-32

Leutz A, Schachner M (1981) Epidermal growth factor stimulates DNA-synthesis of astrocytes in primary cerebellar cultures. Cell Tissue Res 220:393-404

Levi-Montalcini R, Meyer H, Hamburger V (1954) In vitro experiments as the effects of mouse sarcomas 180 and 37 on spinal and sympathetic ganglia of the chick embryo. Cancer Res 14:49-57

Levi-Montalcini R, Angeletti PU (1968) Nerve growth factor. Physiol Rev 48:534-569

Ludwin SK (1984) Proliferation of mature oligodendrocytes after trauma to the central nervous system. Nature 308:274-275

Ludwin SK (1985) Reaction of oligodendrocytes and astrocytes to trauma and implantation: A combined autoradiographic and ummunohi..ochemical study. Lab Invest 52:20-30

McCarthy KD, de Vellis J (1980) Preparation of separate astroglial and oligodendroglial cell cultures from rat cerebral tissue. J Cell Biol 85:890-902

Merrill JE, Kutsunai S, Mohlstrom C, Hofman F, Groopman J, Colde DW (1984) Proliferation of astroglia and oligodendroglia in response to human T cell-derived factors. Science 224:1428-1430

Miller RH, Abney ER, David S, French-Constant C, Lindsay R, Patel R, Stone J, Raff MC (1986) Is reactive gliosis a property of a distinct subpopulation of astrocytes? J Neurosci 6:22-29

Mobley WC, Server AC, Ishii DN, Riopelle RP, Shooter EM (1977) Nerve growth factor. New Engl J Med 297:1096-1104

Moretto G,Kim SU, Shin DH, Pleasure DE, Rizzuro N (1984) Long-term cultures of human adult Schwann cells isolated from autopsied materials. Acta Neruopathol 64:15-21

Morrison RS, Saneto RP, de Vellis J (1982) Developmental expression of rat brain mitogens for cultured astrocytes. J Neurosci Res 8:435-442

Morstyn G, Hsu SM, Kinsella T, Gratzner H, Russo A, Mitchell JB (1983) Bromodeoxyuridine in tumors and chromosomes detected with a monoclonal antibody. J Clin Invest 72:1844-1850

Nishizuka Y (1986) Studies and perspectives of protein kinase C. Science 233:305-311

Pettman B, Weibel M, Daune M, Sensenbrenner M, Labourdette G (1982) Stimulation of proliferation and maturation of rat astroblasts in serum-free culture by an astroglial growth factor. J Neurosci Res 8:463-476

Pleasure DE, Kreider B, Shuman S, Sobue G (1985) Tissue culture studies of Schwann cell proliferation and differentiation. Dev Neurosci 7:364-

373

Porter S, Clark MB, Glaser L, Bunge RP (1986) Schwann cells stimulated to proliferate in the absence of neurons retain full functional capability. J Neurosci 6:3070-3078

Pruss RM, Bartlett PF, Gavrilovic J, Lisak RP, Rattray S (1982) Mitogens for glial cells: A comparison of the response of cultured astrocytes, oligodendrocytes and Schwann cells. Dev Brain Res 2:19-35

Raff MC, Mirsky R, Fields KL, Lisak RP, Dorfman SH, Silberberg DH, Gregson NA, Leibowitz S, Kennedy MC (1978) Galactocerebroside is a specific cell-surface antigenic marker for oligodendrocytes in culture. Nature 274:813-816

Raff MC, Abney E, Brockes JP, Hornby-Smith A (1978) Schwann cell growth factors. Cell 15:813-822

Raff MC, Fields KL, Hakomori SI, Mirsky R, Pruss RM, Winter J (1979) Cell type-specific markers for distinguishing and studying neurons and the major classes of glial cells in culture. Brain Res 174:283-308

Raff MC, Abney ER, Cohen J, Lindsay R, Noble M (1983) Two types of astrocytes in cultures of developing rat white matter: differences in morphology, surface gangliosides, and growth characteristics. J Neurosci 3:1289-1300

Raine CS, Scheinberg L, Waltz JM (1981) Multiple sclerosis: Oligodendrocyte survival and proliferation in an active established lesion. Lab Invest 45:534-546

Raine CS, Traugott U (1983) Chronic relapsing experimental autoimmune encephalomyelitis: Ultrastructure of the central nervous system of animals treated with combinations of myelin components. Lab Invest 48:275-284

Ratner N, Bunge RP, Glaser L (1986) Schwann cell proliferation in vitro: An overview. Ann NY Acad Sci 486:170-181

Rodriguez M, Lennon VA, Benveniste EN, Merrill JE (1987) Remyelination of oligodendrocytes stimulated by antiserum to spinal cord. J Neuropathol Experimental Neurol 46:84-95

Rogers AW (1979) Techniques of autoradiography. Elsevier, Amsterdam

Salzer JL, Bunge RP (1980) Studies of Schwann cell proliferation. I. An analysis in tissue culture of proliferation during development, Wallerian degeneration, and direct injury. J Cell Biol 84:739-752

Sheffield WD, Kim SU (1977) Myelin basic protein causes proliferation of lymphocytes and astrocytes in vitro. Brain Res 132:580-584

Simpson DL, Morrison R, de Vellis J, Herschman HR (1982) Epidermal growth factor binding and mitogenic activity on purified populations of cells from the central nervous system. J Neurosci Res 8:453-462

Skoff RP, Vaughn JE (1971) An autoradiographic study of cellular proliferation in degenerating rat optic nerve. J Comp Neurol 141:133-156

Skoff RP (1975) The fine structure of pulse labelled (^{3}H-thymidine) cells in degenerating rat optic nerve. J Comp Neurol 161:595-612

Sobue G, Brown MJ, Kim SU, Pleasure D (1984) Axolemma is a mitogen for human Schwann cells. Ann Neurol 15:449-452

Sobue G, Shuman S, Pleasure D (1986) Schwann cell responses to cyclic AMP: Proliferation, change in shape, and appearance of surface galactocerebroside. Brain Res 362:233-32

Westermark B (1976) Density dependent proliferation of human glia cells stimulated by epidermal growth factor. Biochem Biophy Res Comm 69:304-310

Wood PM (1976) Separation of functional Schwann cells and neurons from normal peripheral nerve tissue. Brain Res 115:361-375

Wood PM, Williams AK (1984) Oligodendrocyte proliferation and CNS myelination in cultures containing dissociated embryonic neuroglia and dorsal root ganglion neurons. Dev Brain Res 12:225-241

Wood PM, Bunge RP (1986) Evidence that axons are mitogenic for oligodendrocytes isolated from adult animals. Nature 320:756-758

Yong VW, Kim SU (1987) A new double labelling immunofluorescence technique for the determination of proliferation of human astrocytes in culture. J Neurosci Methods 21:9-16

Yong VW, Kim SU, Pleasure DE (1988a) Growth factorss for fetal and adult human astrocytes in culture. Brain Res 444:59-66

Yong VW, Kim SU, Kim MW, Shin DH (1988b) Growth factors for human glial cells in culture. Glia 1:113-123

Yoshino JE, Mason PW, DeVries GH (1987) Developmental changes in myelin-induced proliferation of cultured Schwann cells. J Cell Biol 104:655-660

REGULATION OF OLIGODENDROCYTE DEVELOPMENT BY INSULIN-LIKE GROWTH FACTORS AND CYCLIC AMP

F. Arthur McMorris, Richard W. Furlanetto[1,2], Robin L. Mozell[3],
Monica J. Carson and David W. Raible
The Wistar Institute
36th Street at Spruce
Philadelphia, PA 19104
USA

INTRODUCTION

The mechanisms that regulate the development of oligodendrocytes and the synthesis of myelin are very incompletely understood. Clarification of these processes would lead to a better
understanding of nervous system development and function in
health and disease, and may lead to the development of therapies
to promote remyelination in multiple sclerosis and other
demyelinating diseases.

We have used tissue cultures of cells explanted from cerebrum of 1-day-old rats to investigate the regulation of oligodendrocyte development. Few, if any, oligodendrocytes are present
in cerebrum at the time of explantation, but oligodendrocytes
develop during the first few weeks in these cultures, following a
timetable very similar to that observed *in vivo* (Abney et al.,
1981; McMorris, 1983). We found that insulin increased the number of oligodendrocytes that developed in these cultures, but to
be effective, insulin must be present at a concentration higher
than that needed to saturate insulin receptors (McMorris, 1983;

---

[1]Division of Endocrinology/Diabetes, Department of Pediatrics,
Children's Hospital of Philadelphia, Philadelphia, PA 19104, USA
[2]Present address: Division of Endocrinology, Department of
Pediatrics, University of Rochester School of Medicine,
Rochester, NY 14642, USA
[3]Present address: Department of Neuroscience, Children's
Hospital Medical Center, Boston, MA 02115, USA

McMorris et al., 1986). Therefore, we suspected that insulin might be cross-reacting with receptors for other growth factors, such as the insulin-like growth factors (IGFs) IGF-I and IGF-II. The IGFs, also known as somatomedins, are polypeptide hormones with approximately 50% amino acid sequence homology with insulin. Micromolar concentrations of insulin are known to bind to type I IGF receptors and mimic the action of physiological concentrations of IGF-I and IGF-II (reviewed by Froesch et al., 1985). The IGFs, particularly IGF-I, mediate many of the growth effects of growth hormone (GH), which is known to affect CNS myelination *in vivo* (Pelton et al., 1977; Noguchi et al., 1982, 1985). Therefore, we have tested the IGFs *in vitro* and *in vivo*, and have found that both IGF-I and IGF-II are potent inducers of oligodendrocyte development and myelination.

We have also found that 3',5'-cyclic AMP (cAMP) plays an important role in the regulation of oligodendrocyte development. We previously reported that cAMP induces the activity of the oligodendrocyte marker enzyme, 2',3'-cyclic nucleotide 3'-phosphohydrolase (CNP) in cloned C6 rat glioma cells, and that treatment of rat oligodendrocytes with cAMP analogs *in vitro* increases the amount per cell of three myelin components: CNP, myelin basic protein (MBP) and galactocerebroside (GC) (McMorris, 1977, 1983, 1985; McMorris et al., 1985). We have now investigated earlier events during the development of oligodendrocytes *in vitro* and have found that cAMP analogs, or agents that activate cellular adenylate cyclase, accelerate the development of oligodendrocytes.

MATERIALS AND METHODS

Surface cultures of glial cells were established by mechanical dissociation of cerebra of 1-day-old rats and inoculation into culture vessels coated with polylysine or with polylysine and fibronectin as described previously (McMorris, 1983; McMorris et al., 1986). Cultures were maintained in serum-free medium or in medium containing 1% or 10% fetal bovine serum (FBS) as described (McMorris et al., 1986; McMorris and Dubois-Dalcq, 1988; Raible and McMorris, 1989). Aggregate cultures were established from brains of 16-day fetal rats and maintained in rotating micro-Fernbach flasks as described (Matthieu et al., 1979; Mozell and McMorris, 1988a,b). At the appropriate times, cultures were fixed in 4% paraformaldehyde, permeabilized by freezing and thawing in 60-96% glycerol, and stained by the indirect immunofluorescence method or the avidin-biotin complex (ABC) immunoperoxidase method using antibodies against CNP (Raible and McMorris, 1989), MBP (McMorris et al., 1981), GC (Ranscht et al., 1982), A2B5 antigen (Eisenbarth et al., 1979) or glial fibrillary acidic protein (GFAP; Lee et al., 1984) as described (McMorris et al., 1986; Raible and McMorris, 1989). For fluorescence-activated cell sorting (FACS), live cells were dissociated with trypsin-EDTA solution, immunostained with fluorescein-conjugated antibodies, and sorted on a Becton-Dickinson FACS-IV cell sorter as described (McMorris et al., 1986). Myelin was isolated from cultures or brain tissue by the method of Norton and Poduslo (1973).

Transgenic mice of line Tg(Mt-1,IGF-I)Bri45, also known as line 1219-6, containing a mouse metallothionein promoter/human IGF-I cDNA fusion construct, were produced by Mathews et al. (1988), and were propagated by breeding hemizygous transgenic males with normal females. Transgenic mice were distinguished from non-transgenic littermates by dot hybridization of tail DNA as described (Brinster et al., 1985).

Recombinant human IGF-I was obtained from Imcell, Terre Haute, IN or from Amgen, Thousand Oaks, CA. IGF-II was purified to homogeneity from human plasma by fractionation and HPLC. Bovine insulin, dibutyryl cAMP (dbcAMP), 8-bromo cAMP and other biochemicals were purchased from Sigma Chemical Corp., St. Louis, MO, and second antibodies and other reagents for immunostaining were purchased from Cappel Labs, West Chester, PA, Vector Labs, Burlingame, CA, or Southern Biotechnology Associates, Birmingham, AL.

RESULTS AND DISCUSSION

Physiological concentrations of IGF-I or IGF-II dramatically increased the number of oligodendrocytes that developed in glial cell cultures. Fig. 1 shows that IGF-I induced up to a 60-fold increase in the number of oligodendrocytes in cultures grown for 2 weeks in serum-free medium. Non-oligodendroglial cells (cells negative for GC, CNP or MBP) increased less than 2-fold under the same conditions (data not shown) (see McMorris et al., 1986). Insulin was effective only at concentrations high enough to cross-react with IGF receptors. IGF-I also increased the number of oligodendrocytes that developed in the presence of 10% FBS, but the effect was not as dramatic as in serum-free medium, possibly because of the presence in serum of IGFs and other growth factors which support oligodendrocyte development in control cultures, and of IGF binding proteins which blunt the action of exogenous IGFs (data not shown; see McMorris et al., 1986). IGF-II, which is more abundant in brain than IGF-I both pre- and

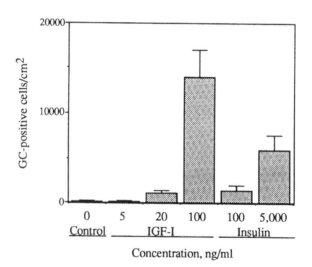

Figure 1. Stimulation of oligodendrocyte numbers in serum-free culture medium. Insulin or IGF-I were added 18 hours after explantation. Cells were fixed and immunostained 14 days later, and GC-positive cells (oligodendrocytes) were counted under the microscope. Error bars represent SEM.

postnatally, also dramatically increased oligodendrocyte numbers, but was 2- to 4-fold less potent than IGF-I (data not shown; see McMorris and Furlanetto, 1989).

IGFs might increase oligodendrocyte numbers by acting as mitogens or as differentiation factors, or both. To test the former possibility, we incubated cells in serum-free medium in the presence or absence of 100 ng/ml IGF-I, and then pulsed them with ^{3}H-thymidine for 5-24 hours prior to fixation and immunostaining. Table 1 shows that IGF-I substantially increased the number of CNP-positive oligodendrocytes that incorporated ^{3}H-thymidine during a 24-hour pulse, indicating that IGF-I stimulated DNA synthesis in oligodendrocytes or in precursors within 24 hours of when they began to express CNP. Table 1 also shows that bipotential glial progenitors, defined by the presence of A2B5 antigen and the absence of GC and GFAP, respond to IGF-I by an increase in DNA synthesis. Cells in cultures from optic nerve (Raff et al., 1983) or cerebrum (Behar et al., 1988) with this pattern of antigenic expression have been shown to be bipotential progenitors that can develop into either oligodendrocytes or astrocytes. Therefore, IGF-I is a mitogen both for CNP-positive oligodendrocytes and for A2B5-positive glial progenitors (McMorris, unpublished data; McMorris and Dubois-Dalcq, 1988).

Table 1. IGF-I is a mitogen for CNP-positive oligodendrocytes and A2B5-positive precursors. Cells growing in serum-free medium were exposed to 0.5 μCi/ml ^{3}H-thymidine for the last 24 hr (expt. 1) or the last 5 hr (expt. 2) before fixation on day 6 *in vitro*. IGF-I was added on day 1 (expt. 1) or day 4 (expt. 2) after explantation.

Experiment	Cell type scored	Condition	% ^{3}H-thymidine labeled ± SEM
1	Oligodendrocytes (CNP$^+$)	Control IGF-I, 100 ng/ml	1.4 ± 1.3 11.7 ± 1.5
2	Precursors (A2B5$^+$,GC$^-$,GFAP$^-$)	Control IGF-I, 100 ng/ml	34.7 ± 2.5 61.0 ± 2.4

To determine whether IGF-I acts as a differentiation factor, we used FACS to isolate 97%-pure populations of A2B5-positive cells from 4-day-old cultures, and maintained the cells for an additional 3-9 days in serum-free culture medium in the presence or absence of 100 ng/ml IGF-I. Very few oligodendrocytes were present in the original sorted populations, but when IGF-I was added, 30-40% of the cells rapidly differentiated into CNP-positive oligodendrocytes (Table 2). However, in the absence of IGF-I, the same proportion of the cells, 30-40%, continued to express A2B5 antigen and remained undifferentiated, as indicated by the absence of staining with antibodies against CNP, GC, MBP and GFAP (data not shown). Therefore, we conclude that IGF-I acts as a differentiation factor, and induces bipotential precursors to differentiate into oligodendrocytes (McMorris, unpublished data; McMorris and Dubois-Dalcq, 1988). Our interpretation differs from the model of Raff et al. (1983), who have proposed that A2B5-positive glial progenitors differentiate autonomously into oligodendrocytes in serum-free medium in the absence of any inducing factors. Our data show that an inducing factor is required, and that IGF-I acts as an inducer. The difference in our interpretations is probably due to the fact that the serum-free medium used by Raff et al. (1983) contained insulin at a concentration high enough to cross-react with IGF receptors and act as an IGF analog.

Table 2. IGF-I induces A2B5-positive progenitors to develop into oligodendrocytes in serum-free medium. A2B5-positive cells were isolated by FACS from 4-day-old cerebral cell cultures and inoculated into culture wells. One day later (day 5), 100 ng/ml IGF-I was added to some of the wells. On day 8 or 14, cells were fixed, immunostained, and scored.

Days after explantation	Condition	% CNP-positive $\pm$ SEM
8	Control	3.7 $\pm$ 0.9
8	IGF-I	17.3 $\pm$ 1.7
14	Control	0.0 $\pm$ 0.0
14	IGF-I	31.3 $\pm$ 2.0

Next, we tested whether IGF-I increases the synthesis and accumulation of myelin. For these studies, we used aggregate cultures established from 16-day fetal rat brain (Mozell and

McMorris, 1988a,b). In this culture system, the dissociated brain cells form floating 3-dimensional aggregates approximately 1/3 - 1/2 mm in diameter; neurons, oligodendrocytes and astrocytes all develop; and myelination occurs (Matthieu et al., 1979). Myelin was isolated from the aggregate cultures by density gradient centrifugation and quantitated. Addition of IGF-I to the culture medium for 18-25 days approximately doubled the amount of myelin in the cultures (Table 3), and doubled the number of oligodendrocytes (data not shown; see Mozell and McMorris, 1988a,b). Therefore, we conclude that IGF-I induces the synthesis and accumulation of myelin as well as the development of oligodendrocytes (Mozell and McMorris, 1988a,b, and unpublished data).

Table 3. IGF-I increases the myelin content of aggregate cultures. IGF-I (100 ng/ml) was added at day 2 *in vitro*. Myelin was isolated and quantitated at the times indicated.

		μg myelin protein	
Days *in vitro*	Condition	per culture	per mg total protein
20	Control	2.61	2.30
	IGF-I	8.07	3.96
27	Control	18.09	16.70
	IGF-I	34.09	22.50

Our ultimate aim is to understand the regulation of oligodendrocyte development and myelination *in vivo*. Therefore, we tested whether IGF-I affects myelination in brain *in vivo*, using a transgenic mouse line that over-expresses IGF-I in many tissues and has approximately 2-fold elevated levels in brain (Mathews et al., 1988). While body weight was increased approximately 30% in these mice, their brains were 50-60% larger than those of their non-transgenic littermates, and contained twice as much myelin (Table 4) (Carson et al., 1988a,b, 1989). Thus, brain growth is disproportionately sensitive to elevated IGF-I, and myelination is relatively more affected than total brain growth (Carson et al., 1988a,b, 1989). These data show that IGF-I is a potent inducer of myelination *in vivo* as well as *in vitro*, and suggest that the effect of GH on CNS myelination *in vivo* is mediated by IGFs.

Table 4. Brain weight and myelin content of IGF-I transgenic (TG) mice and their littermate controls. Data ± SEM are from a representative litter of 55-day-old mice.

Measurement	Control (N = 3)	IGF-I TG (N = 3)	Ratio, TG/control
Brain wet weight (grams)	0.47 ± 0.01	0.73 ± 0.01	1.55 ± 0.03
mg myelin/g brain weight	19.50 ± 1.40	29.00 ± 2.10	1.48 ± 0.15
mg myelin/g total brain protein	176.5 ± 15.2	241.8 ± 53.6	1.37 ± 0.33
Total mg myelin per brain	9.18 ± 0.67	21.31 ± 1.80	2.32 ± 0.26

Taken together, our data show that IGF-I and IGF-II are potent inducers of oligodendrocyte development; that IGFs act both by promoting cell proliferation and by directing lineage decisions of uncommitted precursors; that they increase the synthesis and accumulation of myelin; and that they act *in vivo* as well as *in vitro*.

Cyclic AMP is another agent that regulates oligodendrocyte development and the expression of myelin components. Cyclic AMP is not involved in IGF receptor signal transduction, and thus represents a separate regulatory pathway or pathways. Our previous work showed that cAMP analogs or agents that elevated intracellular cAMP increased CNP activity in C6 rat glioma cells, and that this was due to a cAMP-dependent increase in the synthesis rate of CNP (McMorris, 1977; McMorris et al., 1985). Treatment of oligodendrocytes with cAMP analogs increased the amount per cell of three myelin components: CNP, MBP and GC (McMorris, 1983, 1985). We have now tested the effect of cAMP analogs at earlier stages of development, when few oligodendrocytes are present in the cultures. Treatment of cultures with 1 mM dbcAMP or 1 mM 8-bromo cAMP during the first postnatal week resulted in a rapid, 2- to 3-fold increase in the number of cells that expressed CNP, GC or MBP (Figure 2) (Raible and McMorris, 1989). Subsequently, oligodendrocyte numbers began to increase rapidly even in control cultures (Figure 2). However, we found that cAMP acts quite differently than IGF-I to increase oligodendrocyte numbers. ^{3}H-thymidine incorporation and autoradiography experiments showed that dbcAMP had no mitogenic effect on oligodendrocytes or precursors. Moreover, experiments with FACS-purified A2B5-positive cells showed that dbcAMP did not increase the number of precursors that ultimately developed into oligodendrocytes. Rather, dbcAMP shortened the time interval between the last precursor S-phase and the onset of expression of CNP. Thus, cAMP acts by accelerating the rate at which oligodendrocytes progress through their developmental program without affecting lineage decisions (Raible and McMorris, 1989). This acceleration is consistent with the acceleration of CNP synthesis that we previously observed at later stages of oligodendrocyte development (McMorris et al., 1985). Quantitative dot blot analysis showed that a 2- to 3-fold elevation in CNP mRNA content preceded the increase in CNP-positive cell number by approximately one

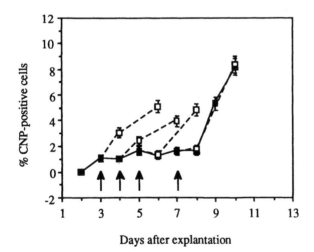

Figure 2. dbcAMP increases the number of CNP-positive oligodendrocytes in culture. Cultures were grown in the presence of 10% FBS and 5 µg/ml insulin. dbcAMP (1 mM) was added at the times indicated by the arrows, and slides were fixed 1 or 3 days later, immunostained, and scored. Control (■); dbcAMP (□).

day, suggesting that cAMP may be acting at the transcriptional level (data not shown; Raible and McMorris, unpublished), consistent with our earlier observation that induction of CNP activity by cAMP analogs could be suppressed by inhibitors of mRNA synthesis (McMorris et al., 1985).

We also tested whether agents that activate adenylate cyclase would promote oligodendrocyte development. Cholera toxin (250 ng/ml), which constitutively activates adenylate cyclase by covalently modifying the regulatory G_S protein, and forskolin (5 µM), which activates adenylate cyclase by interacting directly with the catalytic subunit, both increased the number of oligo-dendrocytes that developed in glial cell cultures in the same manner as exogenous dbcAMP or 8-bromo cAMP (Raible and McMorris, submitted). Thus, our findings show that oligodendrocyte precursors have functional adenylate cyclase and regulatory G proteins, suggesting that cholera toxin, forskolin and cAMP analogs mimic the action of a hormone or other factor that binds to cyclase-coupled receptors on the cell surface and activates adenylate cyclase in developing glial cells (Raible and McMorris, unpublished data). These results support the hypothesis that oligo-dendrocyte development is regulated in part by factors that use cAMP as a second messenger (McMorris, 1977, 1983; Raible and McMorris, 1989).

Thus, both IGFs and cAMP are potent inducers of oligoden-drocyte development. Other hormones and growth factors, such as thyroid hormones, glucocorticoids, PDGF, EGF, FGF, and lympho-kines have been reported to affect oligodendrocyte development and the expression of differentiated properties of oligodendro-cytes. A better understanding of CNS development will require more information on what factors affect oligodendrocyte develop-ment and biochemical expression, how the factors act, and how they interact with each other to result in the timely and coordi-nated production of myelin in the CNS.

ACKNOWLEDGMENTS

Supported by National Multiple Sclerosis Society grant RG 1767-B-2; NIH NS 26119, NS 11036, CA 09171 and CA 38981; and NSF BNS 8518023.

REFERENCES

Abney ER, Bartlett PP, Raff MC (1981) Astrocytes, ependymal cells and oligodendrocytes develop on schedule in dissociated cell culture of embryonic rat brain. Dev Biol 83:301-310

Behar T, McMorris FA, Novotny EA, Barker JL, Dubois-Dalcq M (1988) Growth and differentiation properties of O-2A progenitors purified from rat cerebral hemispheres. J Neurosci Res 21:168-180

Brinster RL, Chen HY, Trumbauer ME, Yagle MK, Palmiter RD (1985) Factors affecting the efficiency of introducing foreign DNA into mice by microinjecting eggs. Proc Natl Acad Sci USA 82:4438-4442

Carson M, Behringer RR, Mathews LS, Palmiter RD, Brinster RL, McMorris FA (1988a) Myelin and 2',3'-cyclic nucleotide 3'-phosphohydrolase levels are elevated in transgenic mice producing increased levels of insulin-like growth factor-1 (IGF-I). Trans Am Soc Neurochem 19:82

Carson MJ, Behringer RR, Mathews LS, Palmiter RD, Brinster RL, McMorris FA (1988b) Myelin content increased in transgenic mice producing elevated levels of insulin-like growth factor-I (IGF-I). Neurosci Absts 14:119

Carson MJ, Behringer RR, Mathews LS, Palmiter RD, Brinster RL, McMorris FA (1989) Hypomyelination caused by growth hormone deficiency is reversed by insulin-like growth factor I in transgenic mice. Trans Am Soc Neurochem 20:286

Eisenbarth GS, Walsh FS, Nirenberg M (1979) Monoclonal antibody to a plasma membrane antigen of neurons. Proc Natl Acad Sci USA 76:4913-4917

Froesch ER, Schmid C, Schwander J, Zapf J (1985) Actions of insulin-like growth factors. Ann Rev Physiol 47:443-467

Lee VM, Page CD, Wu HL, Schlaepfer WW (1984) Monoclonal antibodies to gel-excised glial filament protein and their reactivities with other intermediate filament proteins. J Neurochem 42:25-32

Mathews LS, Hammer RE, Behringer RE, D'Ercole AJ, Bell GI, Brinster RL, Palmiter RD (1988) Growth enhancement of transgenic mice expressing human insulin-like growth factor-I. Endocrinology 123:2827-2833

Matthieu JM, Honegger P, Favrod P, Gautier E, Dolivo M (1979) Biochemical characterization of a myelin fraction isolated from rat brain aggregating cell cultures. J Neurochem 32:869-881

McMorris FA (1977) Norepinephrine induces glial-specific enzyme activity in cultured glioma cells. Proc Natl Acad Sci USA 74:4501-4504

McMorris FA (1983) Cyclic AMP induction of the myelin enzyme 2′, 3′-cyclic nucleotide 3′-phosphohydrolase in rat oligodendrocytes. J Neurochem 41:506-515

McMorris FA (1985) Cyclic AMP induces oligodendroglial cerebroside and myelin basic protein. Trans Am Soc Neurochem 16:285

McMorris FA, Dubois-Dalcq M (1988) Insulin-like growth factor I promotes cell proliferation and oligodendroglial commitment in rat glial progenitor cells developing in vitro. J Neurosci Res 21:199-209

McMorris FA, Furlanetto RW (1989) Insulin-like growth factor II induces development of oligodendrocytes from rat brain. The Endocrine Society, 71st Annual Meeting, Abstract #603

McMorris FA, Miller SL, Pleasure D, Abramsky O (1981) Expression of biochemical properties of oligodendrocytes in oligodendrocyte x glioma cell hybrids proliferating *in vitro*. Exp Cell Res 133:395-404

McMorris FA, Smith TM, DeSalvo S, Furlanetto RW (1986) Insulin-like growth factor I/somatomedin C: a potent inducer of oligodendrocyte development. Proc Natl Acad Sci USA 83:822-826

McMorris FA, Smith TM, Sprinkle TJ, and Auszmann JM (1985) Induction of myelin components: cyclic AMP increases the synthesis rate of 2′,3′-cyclic nucleotide 3′-phosphohydrolase in C6 glioma cells. J Neurochem 44:1242-1251

Mozell RL, McMorris FA (1988a) Insulin-like growth factor I stimulates regeneration of oligodendrocytes in vitro. Ann NY Acad Sci 540:430-432

Mozell RL, McMorris FA (1988b) Insulin-like growth factor I increases myelin synthesis in rat brain aggregate cultures. Trans Am Soc Neurochem 19:83

Noguchi T, Sugisaki T, Tsukada Y (1982) Postnatal action of growth and thyroid hormones on the retarded cerebral myelinogenesis of Snell dwarf mice (dw). J Neurochem 38:257-263

Noguchi T, Sugisaki T, Tsukada Y (1985) Microcephalic cerebrum with hypomyelination in the growth hormone-deficient mouse (lit). Neurochem Res 10:1097-1106

Norton WT, Poduslo SE (1973) Myelination in rat brain: method of myelin isolation. J Neurochem 21:749-757

Pelton EW, Grindeland RE, Young E, Bass NH (1977) Effects of immunologically induced growth hormone deficiency on myelinogenesis in developing rat cerebrum. Neurology 27:282-288

Raff MC, Miller RH, Noble M (1983) A glial progenitor cell that develops in vitro into an astrocyte or an oligodendrocyte depending on culture medium. Nature (London) 303:390-396

Raible DW, McMorris FA (1989) Cyclic AMP regulates the rate of differentiation of oligodendrocytes without changing the lineage commitment of their progenitors. Dev Biol 133:437-446

Ranscht B, Clapshaw PA, Price J, Noble M, Seifert W (1982) Development of oligodendrocytes and Schwann cells studied with a monoclonal antibody against galactocerebroside. Proc Natl Acad Sci USA 79:2709-2713

PLATELET-DERIVED GROWTH FACTOR AND ITS RECEPTORS IN CENTRAL NERVOUS SYSTEM GLIOGENESIS

I.K. Hart, E.J. Collarini, S.R. Bolsover[1], M.C. Raff and
W.D. Richardson.

Department of Biology (Medawar Building) and
[1]Department of Pathology
University College London
Gower Street
London WC1E 6BT

INTRODUCTION

The optic nerve contains three types of post-mitotic glial cells - oligodendrocytes and two types of astrocytes - but no neuronal cell bodies. These cells can be distinguished *in vitro* by their morphology and reactivity with certain antibodies. Oligodendrocytes are multipolar cells *in vitro* which stain with antibodies against galactocerebroside (GC, Ranscht et al., 1982). Both types of astrocytes stain for glial fibrillary acidic protein (GFAP); type-2 astrocytes are process-bearing cells which label in addition with monoclonal antibody A2B5 (Eisenbarth et al., 1979) while type-1 astrocytes are flat fibroblast-like cells which are A2B5-negative. Oligodendrocytes are the myelin-forming cells of the CNS, while type-2 astrocytes apparently extend processes to the gaps (nodes of Ranvier) between adjacent myelinated regions (internodes)(ffrench-Constant and Raff, 1986). The function of type-2 astrocytes is unknown, but it seems probable that they somehow assist saltatory propagation of action potentials. Type-1 astrocytes extend processes to the surface of the nerve and to blood vessels where they induce the capillary endothelial cells to form tight junctions, resulting in the formation of a blood-brain barrier (Janzer and Raff, 1987).

During development, oligodendrocytes and type-2 astrocytes are derived from common, bipotential progenitor cells (A2B5^{+}GC^{-}GFAP^{-}) known as O-2A progenitors, while type-1 astrocytes are derived from a different, A2B5^{-} precursor cell (Raff et al., 1984). O-2A progenitors are highly motile cells *in vitro* (Small et al., 1987) and there is evidence that they are migratory cells *in vivo*, moving into the developing optic nerve (Small et al., 1987) and cerebellum (Reynolds and Wilkin, 1988) from a germinal zone(s) elsewhere in the brain. In contrast, type-1 astrocytes in the optic nerve probably differentiate from the neuroepithelial cells that form the optic stalk.

The glial cells of the optic nerve appear during development on a predictable schedule. Type-1 astrocytes first appear around embryonic day 16 (E16) (Miller et al., 1985). Small numbers of 0-2A progenitors appear in the nerve around the same time and proliferate for several weeks after this (Skoff et al., 1976a,b), some differentiating into oligodendrocytes starting on the day of birth (E21/P0) and others into type-2 astrocytes from the second postnatal week (P7-P10) (Miller et al., 1985). This timing breaks down, however, when embryonic optic nerve cells are dissociated and grown in monolayer culture. Then all the 0-2A progenitors prematurely stop dividing and spontaneously differentiate within a day or two, giving rise to oligodendrocytes if maintained in low ($\leq 0.5\%$) fetal calf serum (FCS) or to type-2 astrocytes in 10% FCS (Raff et al., 1983). The same behaviour is observed even if a single 0-2A progenitor cell is cultured on its own in a microwell (Temple and Raff, 1985). These observations demonstrate that the behaviour of 0-2A progenitor cells is profoundly influenced by the extracellular environment, and suggest that oligodendrocyte differentiation is the constitutive or "default" pathway which is automatically triggered when 0-2A progenitors are deprived of signals from other cells. Type-2 astrocyte differentiation on the other hand seems to be an induced pathway which is activated by an inducing agent(s) present in the optic nerve, that is mimicked by a component of FCS.

An important advance in understanding the development of the 0-2A cell lineage was made by the discovery that the normal timing of oligodendrocyte differentiation can be reconstituted *in vitro* by culturing embryonic optic nerve cells on a monolayer of astrocytes from rat cerebral cortex, or in astrocyte-conditioned medium (ACM). Now instead of differentiating prematurely the 0-2A progenitors continue to divide and differentiate for several weeks (Noble and Murray, 1984), and the first oligodendrocytes appear on the *in vitro* equivalent of the day of birth (Raff et al., 1985). This finding implies that cortical astrocytes (which resemble type-1 astrocytes in the optic nerve) secrete a growth factor which keeps 0-2A progenitors dividing and prevents premature differentiation *in vivo*. We now have compelling evidence that this astrocyte-derived growth factor (ADGF) is a form of PDGF. The evidence is as follows. 1) Pure human or porcine PDGF is strongly mitogenic for 0-2A progenitor cells *in vitro* (Noble et al., 1988; Richardson et al., 1988), and like ACM can recreate the normal timing of oligodendrocyte differentiation *in vitro* (Raff et al., 1988). 2) ADGF comigrates with pure

PDGF on a size-exclusion column (Richardson et al., 1988). 3) Anti-PDGF immunoglobulin can neutralise more than 90% of the activity of ADGF (Richardson et al., 1988; Noble et al., 1988). 4) Metabolically labelled PDGF dimers can be immunoprecipitated from ACM with antibodies against human PDGF (Richardson et al., 1988). There is also evidence that PDGF is present in the brain and optic nerve at the time that O-2A progenitor cells are proliferating *in vivo*. First, mRNA encoding PDGF can be detected both in rat brain by Northern blot analysis (Richardson et al., 1988) and in rat optic nerves by *in situ* hybridisation (Pringle et al., 1989), from before birth to adulthood. Second, protein extracts of three-week-old rat optic nerves contain mitogenic activity for O-2A progenitors, the majority of which can be neutralised by anti-PDGF antibodies (Raff et al., 1988). The source of this PDGF in the nerve is probably type-1 astrocytes, because the spatial distribution of PDGF mRNA in the optic nerve resembles that of mRNA encoding GFAP, an astrocyte-specific marker (Pringle et al., 1989). The combined evidence therefore strongly supports the idea that PDGF, secreted by type-1 astrocytes, controls the proliferation and differentiation of O-2A progenitor cells during development.

PDGF is a disulphide-linked dimer of A and B chains, with the structure AA, AB or BB depending on its source (see Heldin and Westermark, 1989 for review). PDGF from human platelets is mainly PDGF-AB (Hammacher et al., 1988), PDGF from porcine platelets resembles PDGF-BB (Stroobant and Waterfield, 1984), and several human tumour cell lines secrete PDGF-AA (Heldin et al., 1986; Nistér et al., 1988). The major isoform of PDGF in the CNS is probably PDGF-AA, because although we were able to detect readily mRNA encoding the A chain in the brain and optic nerve, we have been unable to detect significant amounts of mRNA encoding the B chain (Richardson et al., 1988; Pringle et al., 1989). There are at least two species of PDGF receptors with different specificities for the three dimeric isoforms of PDGF; type-A receptors bind all three PDGF isoforms while type-B receptors bind PDGF-BB and PDGF-AB but not PDGF-AA (references in Heldin and Westermark 1988). Type-A and type-B receptors are closely related transmembrane proteins with immunoglobulin-like extracellular domains and intracellular "split" tyrosine kinase domains (Yarden et al., 1986; Matsui et al., 1989). ^{125}I-PDGF binding studies demonstrate that human fibroblasts possess mainly type-B receptors (Heldin et al., 1988; Hart et al., 1988) which explains why PDGF-AA has a low mitogenic effect on these cells (Nistér et al., 1988; Heldin et al., 1988). The same is true of rat fibroblasts (Pringle et al., 1989). In contrast, O-2A progenitor cells seem to possess only type-A PDGF receptors (Hart et al., 1989) and respond better to PDGF-AA than to PDGF-BB (Pringle et al., 1989).

O-2A progenitors in cultures of embryonic rat optic nerve cells do not proliferate indefinitely when maintained with PDGF *in vitro*, but stop responding to PDGF after a number of cell divisions, and spontaneously differentiate into oligodendrocytes. The first oligodendrocytes appear in these cultures at the same time as they would have done so *in vivo*. Thus, pre-programmed loss of responsiveness to PDGF may determine when O-2A progenitors stop dividing and differentiate *in vivo*. Where does the block to PDGF responsiveness lie? One possibility could be that the PDGF receptors on O-2A progenitors are diluted with each cell division, or are turned over more rapidly than they are replenished, until the remaining receptors are too few to stimulate mitosis. Alternatively, the secondary messenger systems which transduce the mitotic signal from the PDGF receptors may be blocked at some point downstream of the receptors. In this article we describe experiments we have performed which indicate that the latter explanation is the correct one, and go some way towards determining where in the intracellular signalling cascades the block to PDGF responsiveness might occur.

MATERIALS AND METHODS

Cell cultures

Optic nerves were dissected from Sprague-Dawley rats of various ages and dissociated into individual cells using trypsin, EDTA and collagenase, as previously described (Miller et al., 1985). About 5000-20000 cells were cultured at 37°C in 5% CO_2 on poly-D-lysine (PDL) coated glass coverslips in 0.5 ml supplemented DMEM (Bottenstein, 1986), with or without 0.5% FCS.

^{125}I-PDGF binding and autoradiography

Cultures of optic nerve cells on coverslips were washed once in DMEM containing 10% FCS, 0.1% BSA and made up to pH 7.4 (binding buffer). The cells were then incubated for 1 hour at room temperature in 30 μl binding buffer containing ^{125}I-PDGF (human) at a concentration of 33 ng/ml. After radiolabelling the cells were fixed with 4% paraformaldehyde, washed, immunostained and subjected to autoradiography as described before (Hart et al., 1989).

For cross-competition binding studies, cultures of P0 optic nerve cells were cultured for 1 day, then incubated with ^{125}I-PDGF-AA or ^{125}I-PDGF-BB (kindly supplied by C.-H. Heldin, Ludwig Institute for Cancer Research, Uppsala, Sweden) with or without a 100-fold excess of unlabelled PDGF-AA or PDGF-BB (both recombinant material produced in yeast). After binding, the cells were washed, solubilised and counted in a gamma counter as described before (Hart et al., 1989). Competitive binding studies in cultured rat fibroblasts (NRK cells) were performed as described in Pringle et al., (1988).

BrdU incorporation and immunofluorescence staining

Bromodeoxyuridine (BrdU, Boehringer) was added to cultures at a final concentration of 5×10^{-5} M for the last 24 hours of the culture period. After fixation with 4% paraformaldehyde at room temperature, cells were surface-stained with monoclonal anti-GC and/or A2B5 antibodies, post-fixed and subsequently stained with monoclonal anti-BrdU (Magaud et al., 1988) as described previously (Hart et al., 1989; Pringle et al., 1989).

Measurement of cytosolic Ca^{2+}

P2 optic nerve cells were cultured for one day in supplemented DMEM (without PDGF) on PDL-coated coverslips on which a grid had been drawn in indelible ink. The cells were incubated with the tetra-acetoxymethyl ester of the fluorescent Ca^{2+} indicator fura-2 (fura-2M; 4 μM) in supplemented DMEM for 30 minutes at room temperature. The fura-2M was aspirated and the cells incubated in L15 medium for a further 90 minutes at 37°C in air, before they were examined in a fluorescence microscope adapted to alternate rapidly between 350nm and 380nm excitation wavelengths. A field of cells was visualised via a television camera connected to an image processor. and the fluorescence was recorded before and after the addition of PDGF (2 ng/ml) or the vehicle used to dissolve the PDGF. Free Ca^{2+} levels were calculated for individual cells within the field as described previously (Silver et al.,1989). Polaroid photographs were taken of the fields of cells, the cells were immunolabelled as described above and the field of interest relocated using the photograph and the grid on the coverslip.

RESULTS

0-2A progenitor cells possess type-A PDGF receptors

Human fibroblasts possess two types of high-affinity PDGF receptors. Type-A receptors bind all three dimeric isoforms of PDGF while type-B receptors bind PDGF-BB and PDGF-AB but not PDGF-AA (Heldin et al., 1988; Hart et al., 1988). The receptor types are distinguished by competitive binding experiments: an excess of PDGF-AA will effectively compete with ^{125}I-PDGF-BB for binding to type-A but not type-B receptors. Table 1 shows representative

Table 1. Binding of ^{125}I-PDGF to rat 0-2A progenitors and rat fibroblasts

unlabelled competitor	^{125}I-PDGF bound, % of maximum	
	^{125}I-PDGF-AA	^{125}I-PDGF-BB
(0-2A progenitors)		
none	100 (890)	100 (600)
PDGF-AA	8	15
PDGF-BB	13	16
(NRK fibroblasts)		
none	100 (539)	100 (1696)
PDGF-AA	40	105
PDGF-BB	36	31

^{125}I-PDGF-AA or ^{125}I-PDGF-BB were allowed to bind to the surface of cells in the presence or absence of a 100-fold excess of unlabelled PDGF-AA or PDGF-BB. Bound ^{125}I-PDGF was determined by gamma counting, and is expressed as a percentage of the binding in the absence of competitor. The actual cpm bound are in parentheses. The means of at least two independent experiments are tabulated. The non-specific binding depends on the batch of ^{125}I-PDGF and the cell type, and is different in these two sets of experiments. Data from Hart et al. (1989) and Pringle et al. (1989).

binding data of this sort for NRK fibroblasts and 0-2A progenitor cells (both rat). Whereas a 100-fold excess of recombinant PDGF-AA strongly diminishes binding of both ^{125}I-PDGF-AA and ^{125}I-PDGF-BB to 0-2A progenitors, the same excess of PDGF-AA competes with ^{125}I-PDGF-AA but not ^{125}I-PDGF-BB for binding to NRK fibroblasts. Thus rat O-2A progenitors appear to possess mainly or exclusively type-A receptors unlike rat fibroblasts which possess mainly type-B receptors.

Newly differentiated oligodendrocytes possess functional PDGF receptors

One possible reason that O-2A progenitors drop out of division and differentiate even in the presence of PDGF might be that they progressively lose their PDGF receptors with each cell division, eventually reaching a point where the remaining receptors are unable to elicit a mitogenic response. This hypothesis predicts that newly differentiated oligodendrocytes should have relatively few PDGF receptors compared to actively dividing O-2A progenitors. We therefore performed ^{125}I-PDGF binding and autoradiography on cultures of P0 (newborn) optic nerve cells grown for one day in the presence of PDGF. The

Table 2. Binding of ^{125}I-PDGF to cells in optic nerve cultures, determined by autoradiography

cell type	Percentage of cells labelled	
	in absence of unlabelled PDGF	in presence of unlabelled PDGF
type-1 astrocytes	0	0
O-2A progenitors	61 ± 2	7 ± 2
oligodendrocytes	50 ± 1	8 ± 2
non-macroglial cells	<5	<1

The optic nerve cells were incubated in ^{125}I-PDGF (human), with or without a 100-fold excess of unlabelled human PDGF. They were then immunolabelled and subjected to autoradiography as described in the text. The means and standard errors of at least three independent cultures are tabulated (data from Hart et al., 1989)

rationale for this experimental design is that there are very few oligodendrocytes in the P0 optic nerve, so the majority of oligodendrocytes present at the end of the culture period will have been born *in vitro*, presumably as a result of O-2A progenitor cells becoming unresponsive to PDGF and differentiating. The numbers of A2B5$^+$GC$^-$ O-2A progenitor cells and GC$^+$

oligodendrocytes which bound ^{125}I-PDGF in this experiment are listed in Table 2, and fluorescence micrographs of typical positive cells are shown in Figure 1. Similar proportions of O-2A progenitors and oligodendrocytes bound PDGF and there was no discernible difference in the numbers of silver grains

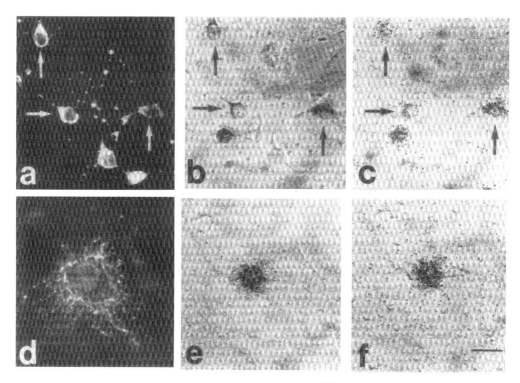

Figure 1. Immunofluorescence autoradiographs of ^{125}I-PDGF binding to cells in 1-day-old cultures of P0 optic nerve. Cells were incubated with ^{125}I-PDGF, immunostained and subjected to autoradiography as described in the text. They were photographed using phase-contrast (B,C,E,F), rhodamine (A) or fluorescein (D) optics. Three of the four A2B5$^+$ O-2A progenitor cells shown in (A) are seen to be radiolabelled when viewed in different planes of focus in (B) and (C) (arrows). The GC$^+$ oligodendrocyte in (D-F) is also radiolabelled, while other GC$^-$ cells in the same field are not radiolabelled. Scale bar, 20 μm. Taken from Hart et al., 1989.

over each cell type, indicating that the number of receptors on O-2A progenitors and oligodendrocytes are similar. This makes it rather unlikely that PDGF receptor loss is the reason why O-2A progenitor cells initially lose responsiveness to PDGF.

To determine if the PDGF receptors on newly-formed oligodendrocytes are functional, we studied the effect of PDGF on cytosolic Ca^{2+} levels in these cells. P2 optic nerve cells were cultured for one day and then loaded with the Ca^{2+} indicator fura-2M. The cytosolic free Ca^{2+} concentration was then determined in individual cells before and after addition of PDGF to the cultures. The cells were first identified by morphology, and the assignments later checked by labelling with anti-GC and A2B5 antibodies. In 12 experiments, 7/60 oligodendrocytes showed at least a 100% rise in cytosolic Ca^{2+} within 5 minutes after addition of PDGF. For comparison, 15/88 O-2A progenitor cells showed at least a 100% rise in cytosolic Ca^{2+} within 5 minutes. Addition of the vehicle in which the PDGF was dissolved had no effect on Ca^{2+} levels in 20/20 oligodendrocytes and 25/25 O-2A progenitors examined. Nor did PDGF cause an elevation in cytosolic Ca^{2+} in 50/50 type-1 astrocytes examined, in keeping with our previous finding that these cells do not possess PDGF receptors (Hart et al., 1989). Thus it appears that PDGF receptors on at least some newly-formed oligodendrocytes are coupled to the intracellular biochemical events linking receptor occupation to cytosolic Ca^{2+} elevation.

PDGF induces Fos and Myc protein expression in O-2A progenitor cells and oligodendrocytes

Growth stimulation of quiescent fibroblasts by serum or PDGF leads to transcriptional activation of several cellular genes including the proto-oncogenes c-*fos* and c-*myc*. The proteins encoded by c-*fos* and c-*myc*, Fos and Myc respectively, are found in the cell nucleus and are thought to play essential roles in signal transduction. In order to examine the expression of Fos in PDGF-stimulated O-2A progenitor cells and newly-formed oligodendrocytes, we cultured P0 optic nerve cells for one day in defined medium containing 0.5% FCS, then added PDGF (4 ng/ml) for one hour before fixing the cells and staining with a monoclonal anti-Fos antibody. In the final 30 minutes of the culture period, i.e. before fixation, we added either anti-GC or A2B5 antibody to assist subsequent identification of oligodendrocytes or O-2A progenitors. In these cultures, approximately 50% of the O-2A progenitors are still able to incorporate BrdU following addition of PDGF after being deprived of PDGF for 24h. We found that approximately 20% of the $A2B5^+$ O-2A progenitors gave clear nuclear staining with the anti-Fos antibody, compared to less than 5% in control cultures to which we added only the vehicle used to dissolve the PDGF. These figures are probably minimum estimates because our staining procedure, which is designed for optimal Fos staining, results in poor A2B5 staining so that not all O-2A progenitors can be positively identified. In parallel P0 optic nerve cultures, a proportion of GC⁺ oligodendrocytes were also stimulated to express Fos, but the numbers of oligodendrocytes that we have observed in these cultures is presently too small to provide reliable statistics. In order to obtain larger numbers of oligodendrocytes, we switched to using P7 optic nerve cultures. After depriving P7 cultures of PDGF for one day in and then adding PDGF for one hour, approximately 20% of GC⁺ oligodendrocytes displayed strong nuclear fluorescence with the anti-Fos antibody (see Figure 2), compared to less than 5% in parallel control cultures which received no PDGF.

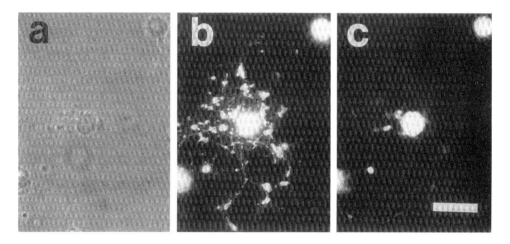

Figure 2. Double immunofluorescence micrographs of an cells in a P7 optic nerve culture. Cells were deprived of growth factors for 24 hours and then stimulated with PDGF for 1 hour, before fixing and staining with anti-GC and anti-Fos antibodies as described in the text. The cells were photographed with phase-contrast (a), rhodamine (b) or fluorescein (c) optics. The GC⁺ oligodendrocyte (b) stains brightly in the nucleus with anti-Fos antibodies (c), as do other GC⁻ cells in the field. Scale bar 25 μm.

To examine the expression of Myc protein in O-2A lineage cells after PDGF stimulation, we followed protocols similar to the above, except that PDGF was added for three hours instead of one hour, and an anti-Myc monoclonal antibody was used. Myc staining was not as distinct as Fos staining in any cell type, but was nevertheless present above background levels in similar proportions of O-2A progenitors and oligodendrocytes (not shown). Thus it appears that the mitotic signalling pathways in oligodendrocytes are intact from the cell surface PDGF receptors to the points of activation of Fos and Myc.

DISCUSSION

We have provided evidence that homodimers of PDGF A chains (PDGF-AA) play an essential role in stimulating the proliferation of 0-2A progenitor cells during perinatal and postnatal development. In low serum and in the absence of PDGF, 0-2A progenitors in cultures of optic nerve cells or in single cell microcultures prematurely stop dividing and differentiate into oligodendrocytes, whereas in the presence of PDGF they continue to divide a number of times before differentiating on the same time schedule as they would have done so *in vivo*. PDGF therefore seems to be important not only for expanding the pool of 0-2A progenitors, but also for controlling the time and rate at which they differentiate to form oligodendrocytes. Oligodendrocyte differentiation appears to be automatically triggered when 0-2A progenitor cells lose the ability to divide in response to PDGF. If we could understand why 0-2A progenitors become incapable of responding to PDGF in the first place, this would be an important step towards understanding not only oligodendrocyte differentiation but also how the multitude of other specialised cell types of an animal are produced at defined and predictable times during development.

In 3T3 cells, PDGF activates multiple intracellular signalling pathways (see Rozengurt, 1986; Williams, 1989 for reviews). These include the inositol phospholipid pathway that leads to protein kinase C (C-kinase) activation and an elevation in cytosolic Ca^{2+}, and the cAMP pathway that leads to activation of the cAMP-dependent protein kinase (A-kinase). In addition, PDGF receptors may be physically associated with a phosphatidylinositol kinase (PI kinase) (Coughlin et al., 1989), that may generate a novel set of intracellular signalling molecules (Williams, 1989). On top of this, the intrinsic protein tyrosine kinase activity of the PDGF receptors generates secondary messages by a mysterious mechanism involving receptor autophosphorylation and phosphorylation of other unidentified target molecules. These parallel signalling pathways ultimately converge to produce an altered pattern of gene expression which leads to cell division. Thus, PDGF and other stimuli rapidly (usually within 5 minutes) induce the transcription of several cellular genes including the proto-oncogene c-*fos* (Greenberg and Ziff, 1984; Cohen and Curran, 1988). Other genes, including the proto-oncogene c-*myc* (Kelly et al., 1983) are induced less rapidly. It is not certain how the Fos or Myc proteins lead to the initiation of DNA replication and mitosis, but it is known that Fos (and possibly Myc) is a transcription factor which helps to activate a further set of genes (for a review see Curran and Franza, 1988). Some of these, such as the collagenase gene (Setoyama et al., 1986) may be needed to effect the complex cellular rearrangements required to enter mitosis while others may participate directly or indirectly in DNA replication.

Despite being unable to divide or synthesise DNA in response to PDGF, newly-produced oligodendrocytes seem to possess PDGF receptors which are indistinguishable from those on dividing O-2A progenitor cells by all the criteria we have investigated. Thus, the receptors are present at similar levels on both types of cell, and can stimulate a similar increase in cytosolic Ca^{2+} following PDGF binding. Furthermore, the receptors on both O-2A progenitors and oligodendrocytes can activate the expression of Fos and Myc proteins. These observations suggest that the block to PDGF-stimulated mitosis in newly-formed oligodendrocytes, and most likely in the O-2A progenitors which gave rise to them, lies somewhere downstream of the points at which Fos and Myc act in the mitotic signal transduction network. Further experiments are now needed to see if, for example, proteins which are absolutely required for DNA replication are present in newly-formed oligodendrocytes as well as in O-2A progenitor cells.

O-2A progenitors are thought to migrate into the optic nerve (Small et al., 1987) and cerebellum (Reynolds and Wilkin, 1988) from germinal zone(s) elsewhere in the brain, but their sites of origin and migration pathways are unknown. Progress in studying O-2A progenitors *in vivo* has been hampered by the lack of specific markers which distinguish these cells from immature or mature neurons. O-2A progenitors and their newly differentiated progeny are the only glial cells in the optic nerve to possess PDGF receptors (Table 2), so it will be worth investigating whether antibodies against the type-A PDGF receptor may be useful reagents to study O-2A progenitor cells *in situ*.

ACKNOWLEDGEMENTS

We are most grateful to Carl-Henrik Heldin and Bengt Westermark for invaluable advice and for providing us with PDGF and ^{125}I-PDGF. We thank Gerard Evan for antibodies to Fos and Myc. We also thank the other members of our laboratories for help and discussions. This work was supported by U.K. Medical Research Council grants to M.C.R. and W.D.R.

REFERENCES

Claesson-Welsh L, Eriksson A, Morn A, Severinsson L, Ek B, Östman A, Betsholtz C, Heldin C-H (1988) cDNA cloning and expression of a human platelet-derived growth factor (PDGF) receptor specific for B-chain-containing PDGF molecules. Mol. Cell. Biol. 8: 3476-3486

Cohen DR, Curran T (1988) *fra*-1: a serum-inducible cellular immediate-early gene that encodes a Fos-related antigen. Mol. Cell. Biol. 8: 2063-2069

Coughlin SR, Escobedo JA, Williams LT (1989) Role of phosphatidylinositol kinase in PDGF receptor signal transduction. Science 243: 1191-1194

Curran T, Franza BR (1988) Fos and Jun: the AP-1 connection. Cell 55: 395-397

Eisenbarth GS, Walsh FS, Nirenberg M (1979) Monoclonal antibody to a plasma membrane antigen of neurons. Proc. Natl. Acad. Sci. USA 76: 4913-4917

ffrench-Constant C, Raff MC (1986) The oligodendrocyte-type-2 astrocyte cell lineage is specialized for myelination. Nature 223: 335-338

Greenberg ME, Ziff EB (1984) Stimulation of 3T3 cells induces transcription of the c-*fos* proto-oncogene. Nature 311: 433-438

Gronwald RGK, Grant FJ, Haldeman BA, Hart CE, O'Hara PJ, Hagen FS, Ross R, Bowen-Pope DF, Murray MJ (1988) Cloning and expression of a cDNA coding for the human platelet-derived growth factor receptor; evidence for more than one receptor class. Proc. Natl. Acad. Sci. USA 85: 3435-3439

Hammacher A, Hellman U, Johnsson A, Östman A, Gunnarsson K, Westermark B, Wasteson Å, Heldin C-H (1988) The major part of PDGF purified from human platelets is a heterodimer of one A and one B chain. J. Biol. Chem. 263: 16493-16498

Hart CE, Forstrom JW, Kelly JD, Seifert RA, Smith RA, Ross R, Murray MJ, Bowen-Pope DF (1988) Two classes of PDGF receptor recognize different isoforms of PDGF. Science 240: 1529-1531

Hart IK, Richardson WD, Heldin C-H, Westermark B, Raff MC (1989) PDGF receptors on cells of the oligodendrocyte-type-2 astrocyte (0-2A) cell lineage. Development 105: 595-603

Heldin C-H, Bäckström G, Östman A, Hammacher A, Rönnstrand L, Rubin K, Nistér M, Westermark B (1988) Binding of different dimeric forms of PDGF to human fibroblasts: evidence for two separate receptor types. EMBO J. 7: 1387-1393

Heldin C-H, Westermark B (1989) Platelet-derived growth factor: three isoforms and two receptor types. Trends Genetics 5: 108-111

Janzer RC, Raff MC (1987) Astrocytes induce blood-brain barrier properties in endothelial cells. Nature 325: 253-257

Kelly K, Cochran BH, Stiles CD, Leder P (1983) Cell-specific regulation of the c-*myc* gene by lymphocyte mitogens and platelet-derived growth factor. Cell 35: 603-610

Magaud JP, Sargent I, Mason DY (1988) Detection of human white cell proliferative responses by immunoenzymatic measurement of bromodeoxyuridine uptake. J. Immunol. Meth. 106: 95-100

Matsui T, Heidaran M, Miki T, Popescu N, LaRochelle W, Kraus M, Pierce J, Aaronson S (1989) Isolation of a novel receptor cDNA establishes the existence of two PDGF receptor genes. Science 243: 800-804

Miller RH, David S, Patel R, Abney ER, Raff M (1985) A quantitative immunohistochemical study of macroglial cell development in the rat optic nerve: in vivo evidence for two distinct astrocyte lineages. Dev. Biol. 111: 35-41

Nistér M, Hammacher A, Mellström K, Siegbahn A, Rönnstrand L, Westermark B, Heldin C-H (1988) A glioma-derived PDGF A chain homodimer has different functional properties than a PDGF AB heterodimer purified from human platelets. Cell 52: 791-799

Noble M, Murray K (1984) Purified astrocytes promote the in vitro division of a bipotential glial progenitor cell. EMBO J. 3: 2243-2247

Noble M, Murray K, Stroobant P, Waterfield MD, Riddle P (1988) Platelet-derived growth factor promotes division and motility and inhibits premature differentiation of the oligodendrocyte/type-2 astrocyte progenitor cell. Nature 333: 560-562

Pringle N, Collarini EJ, Mosley MJ, Heldin C-H, Westermark B, Richardson WD (1989) PDGF A chain homodimers drive proliferation of bipotential (O-2A) glial progenitor cells in the developing rat optic nerve. EMBO J. 8: 1049-1056

Raff MC, (1989) Glial cell diversification in the rat optic nerve. Science 243: 1450-1455

Raff MC, Abney ER, Fok-Seang J (1985) Reconstitution of a developmental clock in vitro: a critical role for astrocytes in the timing of oligodendrocyte differentiation. Cell 42: 61-69

Raff MC, Abney ER, Miller RH (1984) Two glial cell lineages diverge prenatally in rat optic nerve. Dev. Biol. 106: 53-60

Raff MC, Lillien LE, Richardson WD, Burne JF, Noble MD (1988) Platelet-derived growth factor from astrocytes drives the clock that times oligodendrocyte development in culture. Nature 333: 562-565

Raff MC, Miller RH, Noble M (1983) A glial progenitor cell that develops in vitro into an astrocyte or an oligodendrocyte depending on the culture medium. Nature 303: 390-396

Ranscht B, Clapshaw PA, Price J, Noble M, Seifert W (1982) Development of oligodendrocytes and Schwann cells studied with a monoclonal antibody against galactocerebroside. Proc. Nat. Acad. Sci. USA 79: 2709-2713

Reynolds R, Wilkin GP (1988) Development of macroglial cells in rat cerebellum II. An in situ immunohistochemical study of oligodendroglial lineage from precursor to mature myelinating cell. Development 102: 409-425

Richardson WD, Pringle N, Mosley MJ, Westermark B, Dubois-Dalcq M (1988) A role for platelet-derived growth factor in normal gliogenesis in the central nervous system. Cell 53: 309-319

Rozengurt E, (1986) Early signals in the mitogenic response. Science 234: 161-166

Setoyama C, Frunzio R, Liau G, Mudryj M, deCrombrugghe B (1986) Transcriptional activation encoded by the v-*fos* gene. Proc. Natl. Acad. Sci. USA 83: 3213-3217

Silver RA, Lamb AG, Bolsover SR (1989) Elevated cytosolic calcium in the growth cone inhibits neurite elongation in neuroblastoma cells: correlation of behavioural states with cytosolic calcium concentration. J. Neurosci. in press

Skoff R, Price D, Stocks A (1976a) Electron microscopic autoradiographic studies of gliogenesis in rat optic nerve. I. Cell proliferation. J. Comp. Neurol. 169: 291-312

Skoff R, Price D, Stocks A (1976b) Electron microscopic autoradiographic studies of gliogenesis in rat optic nerve. II. Time of origin. J. Comp. Neurol. 169: 313-333

Small RK, Riddle P, Noble M (1987) Evidence for migration of oligodendrocyte-type-2 astrocyte progenitor cells into the developing rat optic nerve. Nature 328: 155-157

Stroobant P, Waterfield MD (1984) Purification and properties of porcine platelet-derived growth factor. EMBO J. 3: 2963-2967

Temple S, Raff MC (1985) Differentiation of a bipotential glial progenitor cell in single cell microculture. Nature 313: 223-225

Williams LT, (1989) Signal transduction by the platelet-derived growth factor receptor. Science 243: 1564-1570

Yarden Y, Escobedo JA, Kuang W-J, Yang-Feng TL, Daniel TO, Tremble PM, Chen EY, Ando ME, Harkins RN, Francke U, Fried VA, Ullrich A, Williams LT (1986) Structure of the receptor for platelet-derived growth factor helps define a family of closely related growth factor receptors. Nature 323: 226-232

TRANSFECTED CELLS AS A TOOL IN MYELIN RESEARCH

GENE TRANSFER OF RAT MATURE OLIGODENDROCYTES AND O-2A PROGENITOR CELLS WITH THE ß-GALACTOSIDASE GENE.

C. GOUJET-ZALC*, C. LUBETZKI*, C. EVRARD°, P. ROUGET° AND B. ZALC*.
*Laboratoire de Neurochimie,
INSERM U-134,
Hôpital de la Salpêtrière
°Laboratoire de Biochimie Cellulaire,
Collège de France,
Paris, France.

Introduction

In normal conditions, central nervous system cells are devoid of species specific antigens. In grafting experiments it is therefore difficult or even impossible to distinguish host from donor cells. With murine species, one way to circumvent this problem is to use mutant models such as the shiverer (*shi/shi*) dysmyelinating mutant (Gumpel et al. 1983). As shiverer myelin contains no myelin basic protein (MBP) (Dupouey et al., 1979) or major dense line (Privat et al., 1979), it is possible to distinguish between host and donor cells, by either immunocytochemistry (using an anti-MBP antibody) or electron microscopy (presence or absence of the major dense line in the myelin sheath). Recently, using the *shi/shi* model, we demonstrated that adult rat oligodendrocytes, when grafted into a newborn shiverer brain were able to survive, migrate and myelinate (Lubetzki et al., 1988). However one wonders whether grafted normal oligodendrocytes would behave in the same way if transplanted into a normal host.

In order to avoid the limitations inherent in the shiverer transplantation model, we decided to introduce the E. Coli LacZ gene (coding for the enzyme ß-galactosidase) into mature oligodendrocytes or bipotential glial progenitor cells (O2A cells) to facilitate the detection of the grafted cells in a normal host. This technic has been reported to be a reliable labeling system (Sanes et al., 1986, Price et al., 1987, Turner et al., 1987). It allows the easy detection of the transfected (or infected) cells, with a histoenzymatic reaction using 5-bromo-4-chloro-3-indolyl-ß-D-galactopyranoside (X-Gal) as the chromogenic substrate. We tested both tranfections and retroviral infections to determine the best method of ß-galactosidase gene transfer in these cells.

Cell cultures

Mature rat oligodendrocytes were obtained from 4-6 week old rat brains, using a Percoll density gradient as previously described (Lubetzki et al., 1986). The preparation contained 90% oligodendrocytes, as shown by immunocytochemistry, using galactosylceramide (GalC) as a marker, and by electron microscopy, which showed the typical morphology of mature oligodendrocytes. The cells survived several weeks in culture and continued to express their specific markers, GalC and MBP. We recently developped a similar procedure for 1 day old rat brains, yielding a 70% pure O2A cell populations. The O2A progenitor cells were identified in our neuron-free cultures by the co-expression of both the A2B5 and GD3 markers (Eisenbarth et al., 1979, LeVine and Goldman, 1988). When maintained in culture, these progenitor cells developped a typical bipolar morphology and proliferated in a medium supplemented with PDGF, as demonstrated by the incorporation of 5-bromo-desoxy-uridine (BrDU). During the first week, about 30% of the cells showed BrDU incorporation. This percentage decreased during the following weeks, independently of the culture conditions (i.e. in the presence of 10% or 1% fetal calf serum (FCS)). The proportion of A2B5 positive cells also decreased as a function of time in culture, while an increasing number of cells started to express either GFAP or GalC , depending on the culture conditions.

ß-galactosidase gene transfer

Transfection

Several methods of transfection were tested using the pCH110 (Pharmacia) construction which contains the E. Coli LacZ gene inserted into pBR 322 plasmid between the enhancer-promoter sequences of the SV40 virus and the SV40 polyadenylation sequences. The best results were obtained using the phosphocalcium precipitate method described by Graham and Van der Erb (1973) modified by Chen and Okayama (1987). With this method, 1 to 3% of the mature oligodendrocytes and 3 to 5% of the O2A cells were transfected and appeared dark blue after histoenzymatic revelation of the expression of the ß-galactosidase gene with X-Gal.

The cells were able to survive and express their specific markers after the transfection. Three weeks after transfection, we were still able to detect a few blue cells in the mature oligodendrocytes population, suggesting that some of these supposedly non dividing cells had integrated the ß-galactosidase gene into their genome. One of the limitations of this technic was the false positive staining observed in microglial cells, due to high amounts of endogenous ß-galactosidase, indicating that immunocytological characterisation of the LacZ positive cells is extremely important for a confident interpretation of the data. The main problem of this technic, however, is the low yield of transfection and the cytotoxicity of the phospho-calcium precipitate, which causes a high level of cell death.

Other transfection methods were tested on mature oligodendrocytes. Electroporation led to drastic cell death and was thus rapidly abandonned. In an other series of experiments, pCH110 was introduced into liposomes. The liposomes, coated or not with anti-GalC antibodies were then added to the cell preparations for various periods of time. No evidence of transfection was ever observed with this method. Electroporation and fusion with DNA-carrying liposomes were not tested on O2A progenitors.

Retroviral infections

For these infections, we used a culture medium from a helper cell line secreting a recombinant Moloney murine leukemia virus (MoMuLV) in which the LacZ gene replaces the viral structural genes *gag, pol, env*. The secreted provirus, after infecting a cell, cannot replicate and is only transmitted to the progeny of the infected cell. It can thus be used as a lineage marker (Sanes et al., 1986, Price et al., 1987, Turner et al., 1987).

In a first series of experiments, we tried to infect glial progenitors *in vivo* by intracerebral infection of newborn rat or mice. Since glial cells are actively proliferating at birth, we expected a high rate of infection and the subsequent appearance of oligodendrocytes carrying the LacZ gene. After isolation of these oligodendrocytes, it would have been possible to graft them and follow their migration. In these experiments, newborn rat and mouse brains were injected near the thalamus with 10ul of the helper cells supernatant mixed with Polybrene (at a final concentration of 8ug/ml). Three weeks later, the animals were sacrificed, and the brains removed, fixed, sectionned and processed for the X-Gal reaction. The result were disappointing. False positive staining was observed in microglial cells, particularly in proximity to meningeal vessels and the yield of infection was extremely low. Only a few LacZ positive cells were detected in each brain.

We then studied the retroviral gene transfer system in vitro, either on mature oligodendrocytes or on O2A progenitor cells. Two hundred ul of unconcentrated or concentrated infecting medium, mixed with 8ug/ml of Polybrene was added to each well (containing about 100.000 cells) for 2 hours, then removed and replaced by fresh culture medium.

With adult oligodendrocytes, we were unable to detect retro-viral infected cell. These negative results could in fact have been predicted as genomic incorporation of an exogenous gene requires, theoretically, at least one cellular division, and in our culture conditions oligodendrocytes were not dividing.

More encouraging results were obtained when O2A progenitor cells were subjected to retroviral infection. The percentage of infection was higher when concentrated infecting medium was used: about 1% of the cells were blue 48 hours after infection, most of which were A2B5 positive cells. Surprinsingly, when the culture took place in 1% FCS, the percentage of infected cells increased as a function of time in culture, as the cells proliferated. When tested at 7 and 21 days post infection, 2% and 7% of the cells were LacZ positive respectively. This suggested that i) gene dilution did not occured during cell divisions and ii) when cultivated in the presence of 1% FCS, the relative abundancy of O2A progenitor cells increased as a function of time in the culture. The infected cells had a normal O2A morphology. In view of cell lineage study, the most promising data was our observation that, as expected, the LacZ positive cells appeared in clusters of increasing number of cells. The number of cells in a given cluster was always very close from n^2

where n represents the number of cell divisions after the infection. The largest clone observed comprised nearly 64 blue cells, suggesting that 6 divisions had occured after infection 2 weeks previously. For the moment, we have been able to study only a few ß-galactosidase positive clones by immunocitochemistry. These were either 100% A2B5 positive or 100% GFAP positive. We have not yet observed clusters of LacZ positive differenciated oligodendrocytes

This technic should provide a promising tool to study migration and myelination potency of these cells when transplanted into newborn normal brain. Furthermore, it should also be a reliable system to study O2A lineage in vitro.

Acknowledgments   We thank Dr. M. Ruberg for helpful suggestions regarding the manuscript. This investigation was partly supported by grants from La Fondation de France (88-1800), Association pour la Recherche sur la Sclérose en Plaques and Fidia-France to B.Z.

References
-Chen C, Okayama H (1987) High efficiency transformation of
    mammalian cells by plasmid DNA. Mol. and Cell. Biology, 7:2745-
    2752.
-Einsenbarth GS, Walsh FS, Niremberg M (1979) Monoclonal antibody to
    a plasma membrane antigen of neurons. Proc. Natl. Acad. Sci., 4913-
    4917
-Graham FL, Van der Erb AJ (1973) A new technic for the assay of infec-
    tivity of human adenovirus 5DNA. Virology, 52: 456-467.
-Gumpel M, Baumann N, Raoul M, Jacque C (1983) Survival and differen-
    tiation of oligodendrocytes from neural tissue transplanted into
    new-born mice brain. Neurosci. Lett. 37: 307-311.
-Dupouey P, Jacque C, Bourre JM, Cesselin F, Privat A, Baumann N.
    (1979) Immunochemical studies of myelin basic protein in shiverer
    mouse devoid of major dense line of myelin. Neurosci. Lett. 12: 113-
    118.
-LeVine, S.M. and Goldman J.E. (1988) Ultrastructural characteristic of
    GD3 ganglioside-positive immature glia in rat forebrain white
    matter J. Comp. Neur., 277: 456-464.

-Lubetzki C, Lombrail P, Hauw JJ, Zalc B (1986) Multiple sclerosis: rat and human oligodendrocytes are not the target for cerebrospinal fluid immunoglobulins. Neurology, 36: 524-528

-Lubetzki C, Gansmüller A , Lachapelle F, Lombrail P, Gumpel M. (1988) Myelination by oligodendrocytes isolated from 4-6week old rat central nervous system and tranplanted into newborn shiverer brain J. Neurol. Sci., 88: 161-175

-Price J, Turner D, Cepko C (1987) Lineage analysis in the vertebrate nervous system by retrovirus mediated gene transfer. Proc. Natl. Acad. Sci., 84: 156-160.

-Privat A, Jacque C, Bourre JM, Dupouey P, Baumann N (1979). Absence of the major dense line in the mutant mouse shiverer. Neurosci. Lett., 12: 107-112.

-Sanes JR, Rubenstein JLR, Nicolas JF (1986) Use of a recombinant retrovirus to study post implantation cell lineage in mouse embryos. Embo J., 5: 3133-3142.

-Turner D, Cepko C, (1987) A common progenitor for neurons and glia persists in rat retina late in development. Nature, 328: 131-136.

IMMORTALIZATION OF OLIGODENDROCYTE PRECURSORS FROM THE OPTIC NERVE OF THE RAT WITH A TEMPERATURE-SENSITIVE FORM OF THE SV40 T ANTIGEN USING A RETROVIRUS VECTOR

Guillermina Almazan
Department of Pharmacology
McGill University
3655 Drummond St.
Montreal, Quebec H3G-1Y6
Canada

INTRODUCTION

Considerable amount of information regarding the cell of origin and the mode of differentiation of oligodendroglial cells has been derived from studies using primary culture. In the optic nerve of the rat, oligodendrocytes differentiate from a bipotential progenitor cell, the O2A precursor, which also gives rise to type 2 astrocyte. The O2A lineage is characterized by the expression on its plasma membrane of gangliosides recognized by a monoclonal antibody A2B5 (Raff et al. 1983). The differentiation step from the progenitor to the astroglial or oligodendroglial lineage can be influenced by growth factors present in the medium. Thus, in the presence of fetal calf serum most cells differentiate into type 2 astrocyte expressing glial fibrillary acidic protein (GFAP), whereas in serum-free medium they will develop into oligodendrocytes expressing galactocerebroside (GC, Hughes &Raff 1987). PDGF and CNTF (ciliary neurotrophic factor) are implicated as growth factor signals from type 1 astrocytes which influence the proliferation and differentiation of O2A progenitors (Richardson et al. 1988; Noble et al. 1988, Lillien et al., 1988). An oligodendrocyte progenitor cell with character-istics similar to the O2A optic lineage has also been obtained from cerebellum (Levi et al. 1986) and cerebral hemispheres (Behar et al. 1988) in rats. These two regions of the central nervous system are the primary source for the study of glial differentiation in tissue culture.

Oligodendrocytes development in culture proceed through various maturation steps generating several myelin specific components and synthesizing a myelin-like membrane (Bradel & Prince 1983, Szuchet al. 1986, Rome et al., Nussbaum et al. 1988). From GC positive cells (Sarlieve et al. 1980, Bologa-Sandru et al. 1980, Poduslo et al. 1985), oligodendrocytes differentiate into cells that also express the myelin specific enzyme 2',3'-cyclic-nucleotide 3'-phosphodiesterase (CNP)

(Pfeiffer et al. 1981, Bansal & Pfeiffer 1985, Reynolds et al. 1989), the main myelin associated glycoprotein (MAG) (Dubois-Dalcq et al. 1986), myelin basic proteins (MBP) (Barbarese & Pfeiffer 1981, Knapp et al. 1987) and the proteolipid protein (PLP) (Dubois-Dalcq et al. 1986). These differentiation steps are also dependent on specific growth factors and signals (Almazan et al. 1985, Ecleston & Silberberg 1985, McMorris et al. 1986, Benveniste & Merrill 1986, McMorris & Dubois-Dalcq 1988, Saneto et al. 1988.

We are interested to study the molecular and cellular mechanisms controlling the growth of oligodendrocyte or its putatite precursor _in vitro._ To approach this problem we have immortalized optic nerve cells by infection with a retroviral vector expressing a temperature sensitive transforming protein, the SV40 T antigen. One would expect that at the permissive temperature for the oncogene (33⁰), the cells infected with the virus will be transformed and grow indefinitely. Shifting to the non-permissive temperature (39⁰) will result in inactivation of the oncogene and return of the cells to their normal course of development.

In this study, we report the use of this method to establish cell lines from the optic nerve of the rat. One of the clonal cell lines, tsU19-5, can be differentiated morphologically and immunocytochemically to express some characteristics of oligodendrocyte cells. Preliminary results suggest that PDGF, FGF, EGF and cyclic AMP analogues regulate the growth of these cells.

MATERIALS AND METHODS

Construction of recombinant virus

The construction of recombinant plasmid pZipNeoSV40U19tsA58 from plasmids containing the tsA58 (Tegtmeyer & Ozer 1971) and U19 (Paucha et al. 1986) mutants of the SV40 T antigen have been described elsewhere (Jat et al. 1986, Almazan & McKay submitted). Psi2, viral producer, cell lines were established after plasmid transfection by the calcium phosphate methods. Cell lines with a titer of 10^5 cfu/ml were selected for infection of primary cultures.

Establishment of cell lines

Optic nerves were dissected out of 2 day-old Sprague Dawley rats (Taconic, N.Y.). Primary culture was prepared according to the technique of Raff et al. (1983). The cell suspension corresponding to 4 optic nerves was plated on poly-D-ornithine-coated dishes in DMEM, 10% FCS and incubated at 37⁰. One day after seeding, the cells were infected for 2h with the recombinant retrovirus in the presence of 8 ug/ml polybrene (Aldrich) for 2h at 33⁰. Selection for the cells integrating the viral DNA was initiated 48h later in 200ug/ml G418 (Geneticin,

Gibco). Of 12 selected colonies, 5 gave stable cell lines, one of which is described here: tsU19-5.

Characterization of ts19-5 cell line

Cells were passaged in DMEM, 10% FCS or 5% FCS/5%CS at 33°. Several antibodies were used to characterize tsU19-5 cell line growing on coverslips at 33° or 39° according to standard techniques. Staining for surface antigens GC (Ranscht et al. 1982) and A2B5 (Eisenbarth et al. 1979, ATCC) was carried out on live cells or pre-fixed for 5 min in 4% paraformaldehyde/PBS. Staining for intracellular proteins was performed on cultures fixed in 2% formaldehyde or 4% paraformaldehyde for 10 min at RT. Cells were permeabilized with 0.1% Triton X-100 before the application of anti-MBP (Fritz and Chou 1983 or Colman et al. 1982), anti-CNP (Bernier et al. 1987) or anti-GFAP (ICN). The second antibodies were fluorescein conjugated (Cappel). Non-specific binding was determined by omitting the primary antibody in PBS/serum.

To determine the effects of various growth factors on the growth of tsU19-5 line, the cells were grown in a serum free medium (SFM) with the following composition: DMEM, 15 mM Hepes; 50 ug/ml transferrin, 5 ug/ml insulin, 300 nM triiodothyronine, 30 nM sodium selenite, 20 nM progesterone, 0.1 nM putrescine and 20 nM hydrocortisone. The concentration for growth factors were: 10 ng/ml FGF (ICN), 100 ng/ml EGF (Collaborative Research), 26 ng/ml recombinant v-sis (PDGF, AmGen) or 1 mM dbcAMP (Sigma). Cells which had been grown under the different experimental conditions for 48 h were labeled with ^{3}H-thymidine (1uCi/ml) for 2 h. For determination of TCA-insoluble material, cells were rinsed once with PBS, 3X with 5% ice-cold TCA and solubilized in 0.2 N NaOH. The amount of radioactivity was determined by liquid scintillation counting. For autoradiography, cells were immunolabeled with anti-GC, fixed in 5% acetic/ethanol, coated with NTB-2 (Kodak) and developed 12 d later. Coverslips mounted with Immuno-mount were examined under a Zeiss microscope. The photographs were taken using Kodak film.

RESULTS

Growth and morphological characterization

The cell line tsU19-5 was isolated 4 weeks after viral infection of optic nerve primary culture. Southern blot analysis of the high molecular weight DNA extracted from the cells demonstrated a single viral insertion suggesting that the cell line is clonal (results not shown).

The growth of the cell line was dependent on the concentration of serum in the medium. In the presence of 10% FCS their doubling time was 24-36 h. The cells were also grown in serum free medium in the presence of growth factors. PDGF,

EGF or FGF stimulated the proliferation of the cells at 33°. This effect was determined by the incorporation of ^{3}H-thymidine into TCA-precipitated DNA and by counting the number of labeled nuclei by autoradiography (Table I). dbcAMP had no significant effect on DNA synthesis.

Table I. <u>Effect of growth factors on ^{3}H-thymidine incorporation.</u> Results are expressed as mean +/- standard deviation for 3 determinations. The numbers of cells growing on coverslips were counted in 25 different fields.

	cpm+/-SD		labeled nuclei(100%)	GC+ (100%)	#cells
SFM	1481 +/-	114	32	30	136
PDGF	3969	632	52	42	218
FGF	5214	1345	55	62	296
EGF	4459	956	59	62	200
dbcAMP	1530	71	31	72	170

At 33°, tsU19-5 cells were elongated and bipolar, but as the number of cells increased in the dish they became polymorphic and flat with 2 or more cellular processes. Many of the processes terminated in enlarged endfeet. Switching the temperature to 39° caused dramatic changes in morphology. A few days after, large membrane sheets were extended. The morphologies were very diversified as illustrated in figures 1, 2 and 3. Many cells had dense an vesiculated cytoplasm, while few isolated cells acquired oligodendrocyte-like morphology with smaller cell bodies and thin processes.

<u>Antigenic properties</u>

Cells growing at 33° were positive for the nuclear T antigen (Figure 2A). The expression was lost 2-3 d after switching at 39°. This cell line was selected from other clones for its immunoreactivity to the monoclonal antibody A2B5 which recognizes a surface ganglioside (Fig.2B). A2B5 expression was lost after 1 week at 39°.

We also examined the cells for the presence of several oligodendrocyte antigens: GC, CNP, MBP and PLP. At 33°, the cell line was always negative for all the markers except for GC (Fig. 1C), and CNP (results not shown) which were expressed in few more differentiated cells. The percentage of cells expressing GC was increased in serum-free medium and all the growth factors enhanced differentiation. The most dramatic effect was observed after the addition of dbcAMP in the culture medium as more than 70% of the cells were GC positive (Table I and Figure 1C). MBP and PLP were expressed in a few cells in cultures after 4-5 d at 39°. Figure 3C shows that PLP was enriched in the perinuclear region while MBP was found more homogeneously distributed in cell bodies (results not shown). Both antibodies stained thin processes. The distribution of the two antigens

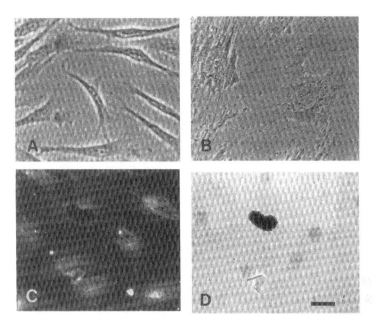

Figure 1. Phase-contrast micrograph of live tsU19-5 cells growing at 33° (A) or 21 d ar 39° (B). Cells labeled with ^{3}H-thymidine and anti-GC antibody growing for 48 h in dbcAMP: GC (C) and phase contrast (D).

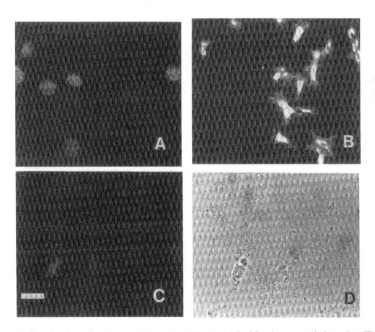

Figure 2. Antigenic properties of cells at 33°. Nuclear staining for T antigen (A), surface staining for A2B5, and negative reaction for GFAP (C) with its corresponding phase contrast. Scale bars, 20 um for A,C,D and 32 um for B.

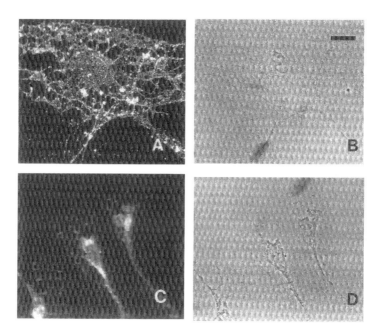

Figure 3. Antigenic properties of differentiated cells for 4-5 d at 39°. Surface staining for GC (A) and intracellular staining for PLP (C) with corresponding phase contrast micrographs (B) and (D), Scale bars, 20 um.

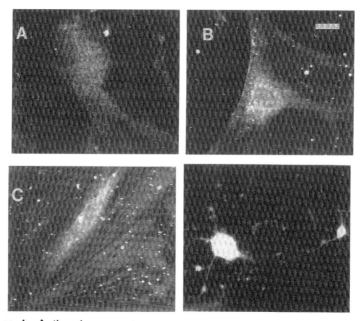

Figure 4. Antigenic properties of cells differentiated at 39° for 21 d. Staining for GC (A), CNP (B), MBP (C) and PLP (D). Scale bars, 20 um for A, B, C and 16 um for D.

corresponded to the description in primary culture of oligodendrocytes (Barbarese & Pfeiffer 1981, Dubois-Dalcq et al. 1986, Zeller et al. 1985, Knapp et al. 1987). After 21 days in culture many cells became positive for the four myelin components, although GC expression was less intense than in younger cultures (Fig. 4A vs. 3A).

Because astrocyte are also derived from A2B5 positive cells, the astrocytic marker GFAP was also included here. Staining for GFAP was always negative at 33° (Figure 2C). Although, FCS is known to promote the differentiation of the O2A precursor cell to the type 2 astrocyte (Hughes & Raff 1978) our cell line showed only once a very small group of cells to be positive for GFAP at 39°.

DISCUSSION

A cell line, tsU19-5, was established after infection of P3 optic nerve cultures with a retrovirus vector carrying the conditional oncogene SV40U19tsA58. At this early stage of development optic nerve cultures are composed of O2A progenitors and type 2 astrocytes (Raff et al. 1983). Our aim was to immortalize the O2A progenitor to study the early cellular and molecular signals controlling oligodendrocyte development. The clonal cell line tsU19-5 was selected from others for its immunoreactivity to A2B5 at 33° which is the permissive temperature for the oncogene. This monoclonal antibody recognizes a surface ganglioside present on the O2A precursor. To differentiate the tsU19-5, the temperature was raised to 39° (the non-permissive temperature for the oncogene) and after 2 days no T antigen expression was observed and proliferation was slowed down several fold. The morphology of the cells changed progressively from elongated to very flat with elaboration of extensive membranes. After 5 days the cells lost the expression of A2B5, and GC expression became very pronounced. Staining for the myelin components CNP, MBP and PLP revealed varied morphologies. Vesicular structures around the nucleus and thick processes stained intensively for all myelin markers. Occasionally, there were cells resembling primary oligodendrocytes in culture with small cell bodies and thin cellular processes. Expression of MBP in primary cultures from rat optic nerve has been reported (Dubois-Dalcq et al. 1986) and we have also detected the presence of CNP (unpublished observation).

A complex series of signals regulate the timing of glial cell differentiation as well as the distribution and numbers of glial cells in the central nervous system. Type 1 astrocytes first appear at embryonic day 16, oligodendrocytes on the day of birth and type 2 astrocytes around postnatal day 10 (Abney et al., 1981; Miller et al. 1985). In vitro different culture conditions influence the choice of developmental pathway taken by the bipotential O2A cell (Hughes & Raff 1987). PDGF and CNTF are implicated as growth factor signals from type 1 astrocytes which influence the proliferation and differentiation of O2A progenitors (Richardson et al. 1988, Noble et al. 1988, Lillien et al., 1988). Several growth factors other than PDGF are reported to stimulate proliferation and differentiation of oligodendrocytes or their precursors in vitro, including FGF (Ecleston & Silberberg 1985), EGF (Almazan et al. 1985) and dbcAMP (McMorris 1983).

EGF, FGF and PDGF increased proliferation of tsU19-5 at 33° as demonstrated both by the incorporation of ^{3}H-thymidine into TCA-precipitated DNA and by autoradiography. The percentage of GC-positive cells increased after treatment with all the growth factors reflecting an effect on differentiation, survival or both. After growing the cells for 48 h in the presence of dbcAMP, the number of GC positive cells was more than doubled (70% vs. 30% in controls) but the number of labeled nuclei remained the same. This results confirmed a role for cyclic AMP in the differentiation of oligodendrocytes as has been previously suggested (McMorris 1983). tsU19-5 cells may be useful in further defining receptors and the intracellular signals responsible for the effects of growth factors on oligodendrocytic differentiation and proliferation. We are presently characterizing the receptors for the growth factors in tsU19-5 cultured at both 33° and 39°.

By inactivating the immortalizing oncogene the tsU19-5 cell differentiates in large numbers to an oligodendrocytic features. The differentiation of this cell line to a GFAP positive state is not a major response to elevated temperature even in the presence of fetal calf serum. Two interpretations of this observation seem likely. First, that a specific extracellular signal is required to generate astrocytes. CNTF may cause this differentiation as it has recently been shown to cause primary O2A cells to become Type 2 astrocytes (Lillien et al., 1988). A second possibility is that some O2A cells can not be switched to give GFAP positive cells. This possibility may be supported by the observation that a maximum of 25% of A2B5-positive cells from optic nerve become GFAP-positive in the presence of a CNTF like factor (Lillien et al., 1988). It will be important to test the effect of CNTF on tsU19-5 cells.

In conclusion, tsu19-5 cell line has some properties resembling oligo-dendrocyte precursors and can differentiate morphologically and biochemically to express properties of primary oligodendroglial cells in vitro. This cell line may be suitable for the study of the biochemistry and pharmacology of oligodendrocyte differentiation. Our future goal is to use this cell line to identify genetic signals involved in myelination.

ACKNOWLEDGEMENTS

I am very grateful to Ron McKay at MIT where most of the work was carried out. I like to thank D. Colman, B. Ranscht and M. Dubois-Dalcq for gifts of antibodies, P.J. Jat for donations of pZipNeoSV40tsA58 and pZipNeoSV40U19, D. Wu for the cloning of the psi2 cell lines and K. Csonka for technical assistance. This work was supported by grants from the National Institute of Health and the Rita Allen Foundation to R. McKay, and a grant from the Multiple Sclerosis Society of Canada to G. Almazan.

REFERENCES

Abney E.R., Bartlett P.P. and Raff M.C. (1981) Astrocytes, ependymal cells, and oligodendrocytes develop on schedule in dissociated cell cultures of embryonic rat brain. Develop. Biol. 83:301-310.

Almazan G., Honegger P., Matthieu J.-M. and Guentert-Lauber B. (1985) Epidermal growth factor and bovine growth hormone stimulate differentiation and myelination of brain cell aggregates in culture. Dev. Brain Res. 21: 257-264.

Almazan G. and McKay (1989) Oligodendrocyte precursor cell lines from the rat optic nerve (submitted for publication).

Bansal R. and Pfeiffer S.E. (1985) Developmental expression of 2',3'-cyclic nucleotide 3'-phosphohydrolase in dissociated fetal rat brain cultures and rat brain. J. Neurosci. Res. 14:21-34.

Barbarese E and Pfeiffer S.E. (1981) Developmental regulation of myelin basic protein in dispersed cultures. Proc. Natl. Acad. Sci. 78:1953-1975.

Behar T., McMorris F.A., Novotny E.A., Barker J.L. and Dubois-Dalcq M. (1988) Growth and differentiation properties of O-2A progenitors purified from rat cerebral hemispheres. J. Neurosci. Res. 21:168-180.

Bernier L., Alvarez F., Norgard E.M., Raible D.W., Mentaberry A., Schembri J.G., Sabatini D.D. and Colman D.R. (1987) Molecular cloning of a 2',3'-cyclic necleotide 3'-phosphodiesterase: mRNAs with different 5' ends encode the same set of proteins in nervous and lymphoid tissues. J. Neurosci. 7: 2703-2710.

Benveniste E.N. and Merrill J.E. (1988) Stimulation of oligodendroglial proliferation and maturation by interleukin-2. Nature 321:610-613.

Bologa-Sandru J.C., Joubert R., Marangos P.J., Derbin C., Rioux F. and Herschkowitz N. (1982) Accelerated differentiation of oligodendrocytes in neuronal rich embryonic mouse brain cell cultures. Brain Res. 252:129-136.

Bradel E.J. and Prince F.P. (1983) Cultured neonatal rat oligodendrocytes elaborate myelin membrane in the absence of neurons. J. Neurosci. Res. 9:381-392.

Colman D.R., Kreibich G., Frey A.B. and Sabatini D.D. (1982) Synthesis and incorporation of myelin polypeptides into CNS myelin. J. Cell Biology 95:598-608.

Dubois-Dalcq M., Behar T., Hudson L. and Lazzarini R.A. (1986) Timely emergence of three myelin proteins in oligodendrocytes. J. Cell Biology 102:384-392.

Eccleston P.A. and Silberberg D.H. (1985) Fibroblast growth factor is a mitogen for oligodendrocytes in vitro. Dev. Brain Res. 21:315-318.

Eisenbarth G.S., Walsh F.S. and Nirenberg M. (1979) Monoclonal antibody to a plasma membrane antigen of neurons. Proc. Nat. Acad. Sci. 76:1286.

Fritz R.B. and Chou C.H. (1983) Epitopes of peptide 43-88 of guinea pig myelin basic protein: localization with monoclonal antibodies. J. Immunology 130:2180-2183.

Hughes S.H. and Raff M.C. (1987) An inducer protein may control the timing of fate switching in a bipotential glial progenitor cell in rat optic nerve. Development 101:157-167.

Jat P.S. and Sharp P.A. (1986) Large T-antigens of simian virus 40 and polyoma virus efficiently establish primary fibroblasts. J. Virol. 59:746-750.

Knapp P.E., Bartlett W.P. and Skoff R.P. (1987) Cultured oligodendrocytes mimic in vivo phenotypic characteristics: cell shape, expression of myelin-specific antigens, and membrane production. Dev. Biol. 120:356-365.

Levi G., Gallo V. and Ciotti M.T. (1986) Bipotential precursors of putative fibrous astrocytes and oligodendrocytes in rat cerebellar cultures express distinct surface features and "neuron-like" gamma-aminobutyric acid transport. Proc. Natl. Acad. Sci. USA 83:1504-1508.

Lillien L.E., Sendtner M., Rohrer H., Hughes S.M. and Raff M.C. (1988) Type-2 astrocyte development in rat brain cultures is initiated by a CNTF-like protein produced by type-1 astrocytes. Neuron 1: 485-594.

McMorris F.A. (1983) Cyclic AMP induction of the myelin enzyme 2',3'-cyclic nucleotide 3'-phosphohydrolase in rat oligodendrocytes. J. Neurochem. 41:506-515.

McMorris F.A. and Dubois-Dalcq M.(1988) Insulin-like growth factor I promotes cell proliferation and oligodendroglial commitment in rat glial progenitor cells developing in vitro. J. Neurosci. Res. 21:199-209.

McMorris F.A., Smith T.M., De Salvo S. and Furlanetto R. W. (1986) Insulin-like growth factor I/somatomedin C: A potent inducer of oligodendrocyte development. Proc. Nat. Acad. Sci. 83:822-826.

Miller R.H., David S., Patel P., Abney E.R. and Raff M.C. (1985) A quantitative immunohistochemical study of microglial cell development in the rat optic nerve: in vivo evidence for two distinct astrocyte lineages. Dev. Biol. 111:35-41.

Noble M., Murray K., Stroobant P., Waterfield M.D. and Riddle P. (1988) Platelet-derived growth factor promotes division and motility and inhibits premature differentiation of the oligodendrocyte/type-2 astrocyte progenitor cell. Nature 333:560-565.

Nussbaum J.L., Espinosa de los Monteros A., Pari F.M., Doerr-Schott J., Roussel G and Neskovic N.M. (1988) A morphological and biochemical study of the myelin-like membrane structures formed in cultures of pure oligodendrocytes. Int. J. Dev. Neuroscience 6:395-408.

Paucha E., Kalderon K., Harvey R.W. and Smith A.E. (1986) Simian virus 40 origin DNA-binding domain on large T antigen. J. Virol. 57:50-54.

Pfeiffer S.E., Barbarese E. and Bhat S. (1981) Noncoordinate regulation of myelinogenic parameters in primary cultures of dissociated fetal rat brain. J. Neurosci. Res. 6:369-380.

Poduslo S.E., Curbeam R. , Miller K. and Reier P. (1985) Purification and characterization of cultures of oligodendroglia from rat brain. J. Neurosci. Res. 14:433-447.

Raff M.C., Miller R.H. and Noble M. (1983) A glial progenitor cell that develops in vitro into an astrocyte or an oligodendrocyte depending on culture medium. Nature 303:390-396.

Ranscht B., Clapshaw P.A., Price J., Noble M. and Seifert W. (1982) Development of oligodendrocytes and Schwann cells studied with a monoclonal antibody against galactocerebroside. Proc. Natl. Acad. Sci. USA 79:2709-2713.

Reynolds R., Carey E.M. and Herschkowitz N. (1989) Immunohistochemical localization of myelin basic protein and 2',3'-cyclic nucleotide 3'-phosphohydrolase in flattened membrane expansions produced by cultured oligodendrocytes. Neuroscience 28:181-188.

Richardson W.D., Pringle N., Mosley M.J., Westermark B. and Dubois-Dalcq M. (1988) A role for platelet-derived growth factor in normal gliogenesis in the central nervous system. Cell 53:309-319.

Rome L.H., Bullock P.N., Chiappelli F., Cardwell, Adinolfi A.M. and Swanson D. (1986) Synthesis of a myelin-like membrane by oligodendrocytes in culture. J. Neurosci. Res. 15:49-65.

Saneto R.P., Low K.G., Melner M.H. and de Vellis J. (1988) Insulin/insulin-like growth factor I and other epigenetic modulators of myelin basic protein expression in isolated oligodendrocyte progenitor cells. J. Neurosci. Res. 21:210-219.

Sarlieve L.L., Rao G.S., Campbell G. L. and Pieringer R.A. (1980) Investigations on myelination in vitro: Biochemical and morphological changes in cultures of dissociated brain cells from embryonic mice. Brain Res. 189:70-90.

Szuchet S., Polak P.E. and Yim S.H. (1986) Mature oligodendrocytes cultured in the absence of neurons recapitulate the ontogenic development of myelin membranes. Dev. Neurosci. 8:208-221.

Tegtmeyer P. and Ozer H.L. (1971) Temperature-sensitive mutants of simian virus 40: Infection of permissible cells. J. Virol. 8:516-520.

Watanabe T. and Raff M.C. (1988) Retinal astrocytes are immigrants from the optic nerve. Nature 332:834-836.

Zeller N.K., Behar T.N., Dubois-Dalcq M.E. and Lazzarini R.A. (1985) The timely expression of myelin basic protein gene in cultured rat brain oligodendrocytes is independent of continuous neuronal influences. J. Neurosci. 5:2955-2962.

EXPRESSION OF NERVOUS SYSTEM cDNAS IN GLIAL AND NON-GLIAL CELL LINES

*#B. Allinquant, *S.M. Staugaitis, *D. D'Urso, ° G. Almazan, *S. Chin, *P.J. Brophy, and *D.R. Colman
*Departments of Anatomy and Cell Biology, and Pathology
College of Physicians & Surgeons
Columbia University
New York, NY 10032

INTRODUCTION:

Advances in recombinant DNA technologies have given us the capacity to introduce (by transfection), with relative ease, cDNAs encoding expressible polypeptides into host cells. In these novel intracellular environments, properties of the encoded proteins may be revealed that are not easily studied when the same proteins are expressed in their natural surroundings. Using the transfection paradigm, information about the mechanism of polymerization of the individual neurofilament proteins can be obtained in cells such as Cos and L tk⁻ cells that express a variety of other cytoskeletal proteins (Chin and Liem, 1989). The proteins involved in myelin formation may also be studied in this way. In myelinogenesis, the interactions between membrane lipids and proteins that lead to the assembly of the multilamellar myelin sheath are extremely complex. However, by transfecting cDNAs encoding the individual myelin proteins into non-glial cells that act as "foreign" hosts, the intracellular behavior and fate of each myelin protein (in the absence of the others) can be

---

INSERM U 134
Hôpital de la Salpétrière
75651 Paris Cedex 13
France
° Department of Pharmacology R171321
Mc Intyre Med. Building
3655 Drummond Street
Montreal
Quebec H3G 1Y6

monitored, and data interpreted in light of what is known about myelin assembly from other studies. Moreover, the existence of rodent myelin mutants that express at null or very low levels certain structural genes of myelin (Campagnoni and Macklin, 1988) potentially allows us to introduce into the mutant myelin-forming cells "minigenes" encoding the missing complete polypeptides or only small segments thereof. By mapping the distribution of the encoded polypeptides using antibodies as probes, functional domains in these proteins can perhaps be identified. The development of stable lines of glial cells that under certain circumstances express highly differentiated properties would provide a significant advantage for these kinds of studies, since an unlimited supply of cells would be available.

In this article, we show how the transfection paradigm can be employed to gain information about some of the physiological properties of the small myelin basic protein (SMBP), the 2'3' cyclic nucleotide 3' phosphodiesterase (CNP_l), protein zero (P_0) of peripheral nerve myelin, and the middle molecular weight neurofilament (NF-M), when these are expressed in a "foreign" cytoplasm. Further, we report our progress in obtaining glial cell lines using a transforming retroviral agent (see article by Almazan et al., this volume). Lastly, we show that some of the lines that we are developing may be transfected with expressible cDNAs, and the polypeptide products detected with antibodies.

MATERIALS AND METHODS:

Preparation of cDNAs:

The cDNAs encoding the small MBP (Mentaberry et al., 1986), CNP_l (Bernier et al., 1987), and P_0 (Lemke and Axel, 1985) were subcloned into the pECE vector (Ellis et al., 1986) that contains the SV40 promoter. The cDNA coding for NF-M (Chin and Liem, 1989) was subcloned into a vector containing the Rous Sarcoma virus (RSV) promoter (Forman et al., 1988). The plasmids were purified by two cycles of CsCl equilibrium density gradient centrifugation, followed by dialysis and precipitation.

Transfection of cells:

Cells (HeLa and shiverer glial cell lines) were transfected with lipofectin (Bethesda Research Laboratories) according to the BRL protocol. 2 µg of plasmid DNA (in lipofectin) per 12 mm diameter coverslip was used (5 h at 37°C). The cells were then rinsed twice and fed. Immunostaining for the proteins encoded by the cDNAs was performed 48 h later, using affinity purified antibodies to these proteins followed by second antibodies labelled with fluorochromes.

Immortalization of glial cell lines:

Cortices from neonatal shiverer mice were dissected, trypsinized, filtered through a 63 µm nytex filter, and plated either in defined medium with 3% fetal bovine serum or in DMEM with 10% fetal bovine serum. In either case, PDGF was included at 10 ng/ml for 48 h at 37°C. Incubation temperature was dropped to 33°C, and the cells were infected for 2 h with a thermosensitive retroviral agent containing the neomycin resistance gene (see G. Almazan et al., this volume) in the presence of polybrene. After infection, the cells were put back into medium with PDGF for 48 h at 33°C. Selection of the neomycin-resistant cell clones was performed after 3 weeks of growth in the presence of G418. These neomycin resistant cells were maintained at 33°C where their division is induced by the T antigen. In order to promote differentiation, the temperature was raised to 39°C, at which point cell division ceased.

RESULTS AND DISCUSSION:

The P_0 glycoprotein of peripheral nerve myelin is detectable at the plasma membrane of transfected cells:

P_0 is a transmembrane protein whose N-terminus is exposed at the extracellular surface (D'Urso et al, submitted). This extracellular segment consists for the most part of a single Ig-like domain (Lai et al., 1987; Lemke et al., 1988) that is believed to mediate, homotypically, the binding interactions between the adjacent membrane surfaces of the Schwann cell plasma membrane that lead to the formation of compact myelin (Lemke and Axel, 1985). A single membrane spanning domain separates the N-terminal segment from a positively charged C-terminal domain that must function to bring about adhesion of the cytoplasmic aspects of apposed membrane bilayers (Ganser and Kirschner, 1980).

P_0 is thus a type I membrane protein, like the myelin-associated glycoprotein, and presumably several other members of the Ig superfamily of proteins (Salzer and Colman, 1989). In HeLa cells transfected with a P_0 cDNA, P_0 can be detected by immunofluorescence at 24 h post-transfection in the perinuclear regions (D'Urso et al., submitted), consistent with its site of synthesis on rough endoplasmic reticulum (RER), and passage through the Golgi apparatus (Poduslo et al., 1985; Trapp et al., 1981). Weak surface staining was also detectable. These data demonstrate that the P_0 molecule contains all the information necessary to be incorporated into the plasma membrane of even non-glial host cells. Interestingly, by 48 hours post-transfection, defined regions of contact between adjacent labelled cells developed in which P_0 was concentrated (D'Urso et al., submitted). The appearance of these regions (Fig. 1A) was highly suggestive of adhesive zones, presumably induced by the interaction of P_0 molecules on the apposed surfaces. This may be an extremely useful model to directly examine the molecular basis for the assembly of the intraperiod line of peripheral nerve myelin.

The 14 Kd MBP and CNP$_I$, when expressed in host cells, reveal very different membrane-associative properties:

The MBPs are a set of positively charged peripheral membrane proteins that are found exclusively in compact myelin in adult animals and mediate the association of membrane bilayers to form the major dense line in the CNS. These are very reactive molecules, which will readily associate with membrane vesicles *in vitro*. The targeting mechanism that ensures that the MBPs associate specifically with forming myelin lies in the ability of the oligodendrocyte to transport selectively MBP mRNAs in "free" polysomes to myelinating zones. Thus, in oligodendrocytes *in situ* with well established cytoplasmic processes that are elaborating myelin, neither MBP mRNAs nor the polypeptides are detectable in the oligodendrocyte cell bodies. The insertion of the most abundant forms of MBP (14 and 18.5 Kd) into the nascent myelin bilayer must be virtually instantaneous (Colman et al., 1982), and therefore in this system it is difficult to study how the association of MBP with membranes is achieved.

In HeLa cells that express the 14 Kd rat MBP cDNA (SMBP) striking perinuclear staining, highly reminiscent of the pattern obtained with P_0 transfectants, is obtained (Fig. 1B). Since SMBP is not a transmembrane protein, this perinuclear pattern most probably represents the association of MBP with cytoplasmic aspects of the bilayer of perinuclear membranes. It is likely that SMBP mRNA, on emerging from the HeLa cell nucleus, is first rapidly incorporated into polysomes. It is not expected that HeLa cells possess the subcellular mechanism that evolved in the oligodendrocyte to distribute these mRNAs, and so translation proceeds in the perinuclear areas, where abundant membranes (RER, Golgi) that could act as acceptors for the newly synthesized SMBP, are located. Our results demonstrate the strong affinity that SMBP has for intracellular membranes under physiological conditions.

By contrast, CNP_I, also synthesized on free polysomes in the oligodendrocytes, for the most part behaves as a soluble protein in the transfected HeLa cells, seeming to fill the cytoplasm, even into the long microvilli that are present on these cells (Fig. 1C). Interestingly, in many cells the plasma membrane was somewhat more heavily stained than the cytoplasm, analogous to observations on CNP distribution in oligodendrocytes *in situ* (Trapp et al., 1988).

The HeLa cell model is useful for studying certain fundamental properties of these myelin proteins, which will enable us in the near future to express simultaneously two or more of these proteins and observe how they interact with one another in the host cells. However, for studying MBP function, shiverer mouse cell lines may present the most productive avenue for exploration.

The existence of the shiverer mutant mouse in which the MBP gene is truncated and therefore inoperative (Roach et al., 1985), offers a useful model for testing upon the transfection of shiverer oligodendrocytes, the intracellular behavior of the MBPs in the presence of all the other normal myelin components. We have tried to develop glial cell lines from this mutant with a thermosensitive retroviral agent. Two types of cell lines have been observed: one is astrocyte-like, and the other is oligodendrocyte-like. Most of the cells of the astrocyte-like type can express the glial fibrillary acidic protein (GFAP) at 33°C (Fig. 2A, 2B), after 6 passages, while no appropriate marker has been detected at 33°C in the oligodendrocyte-like cells. Gal C and CNP (Fig. 3A) expression appear to be the first markers detectable after several days at 39°C in the oligodendrocyte-like line.

Most recently, we have been able to transfect (at 33°C) both cell lines with a test cDNA construct expressing NF-M (Chin and Liem, 1989), driven off the strong Rous sarcoma virus promoter. This construction, which expresses at very high levels, readily allows detection of the expressed product (Fig. 2C, 3B). In the astrocyte-like cells, NF-M distributes in a pattern highly similar if not identical to GFAP, suggesting that the 2 cytoskeletal proteins may associate in these cells. In the oligodendrocyte-like cells, filamentous staining was frequently seen emanating from a juxtanuclear zone, which may act as a "nucleation" site for NF-M polymerization.

Ultimately, the introduction of cloned "minigenes" encoding truncated or altered nervous system proteins into host cells such as those described may allow us to precisely define the functional domains within the normal polypeptides that mediate the variety of intracellular behaviors we observe.

Figure 1. Expression of myelin proteins in HeLa cells after transfection with expressible cDNAs. Examples of the typical patterns of immunofluorescence obtained in HeLa cells 48 hours after transfection with (A) P_0, (B) SMBP and (C) CNP_I cDNAs. In (A) note the perinuclear fluorescence, and the strong fluorescence at the plasma membranes of apposed cells. Perinuclear fluorescence is apparent in (B), and CNP_I distributes throughout the cell (C).

Figure 1

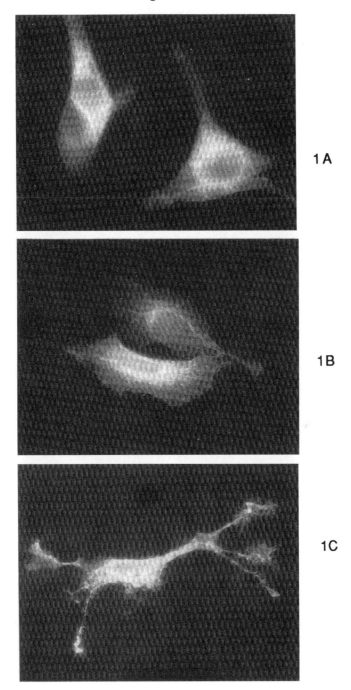

1A

1B

1C

Figure 2

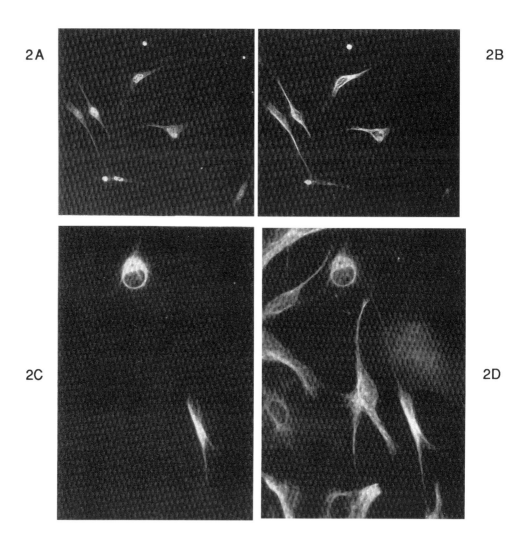

Figure 2. Immortalized shiverer astrocyte-like cells express the SV40 T antigen and are transfectable at 33°C. Shiverer glial cells were immortalized with a thermosensitive retrovirus. An astrocyte-like line obtained in this way expresses SV40 T antigen (A) and GFAP (B) at 33°C. After transfection (at 37°C) with pRSVi-NF-M, and incubation at 39°C for 2 days, cells were stained for NF-M (C), and GFAP (D).

Figure 3

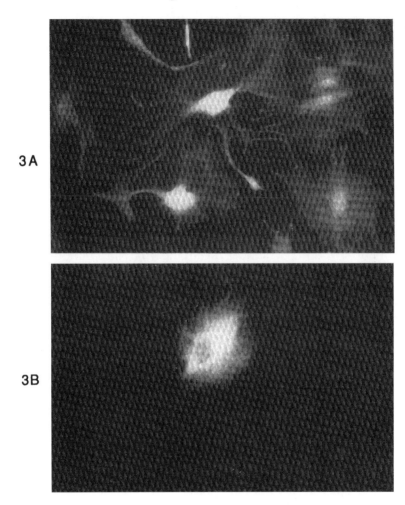

Figure 3. Immortalized shiverer oligodendrocyte-like cells begin to differentiate at 39°C and are transfectable at 33°C. Most of the shiverer cell lines obtained by immortalization with a thermosensitive retrovirus do not express any oligodendrocyte markers at 33°C. After 6 days at 39°C, a cell line showing CNP expression is observed (A). This cell line is also transfectable at 33°C with the pRSVi-NF-M, and expression of the NF-M can be observed after 48 hours at 33°C (B).

REFERENCES:

Bernier L, Alvarez F, Norgard EM, Raible DW, Mentaberry A, Schembri JG, Sabatini DD, Colman DR (1987) Molecular cloning of a 2'3' cyclic nucleotide 3' phosphodiesterase: mRNAs with different 5' ends encode the same set of proteins in nervous and lymphoid tissues. J Neurosci 7:2703-2710.

Campagnoni AT, Macklin WB (1988) Cellular and molecular aspects of myelin protein gene expression. Mol Neurobiol 2:41-89.

Chin SSM, Liem RKH (1989) Expression of rat neurofilament proteins NF-L and NF-M in transfected non-neuronal cells. Submitted.

Colman DR, Kreibich G, Frey AB, Sabatini DD (1982) Synthesis and incorporation of myelin polypeptides into CNS myelin. J Cell Biol 95:598-608.

D'Urso D, Brophy PJ, Staugaitis SM, Gillespie CS, Frey AB, Colman DR. Protein zero of peripheral nerve myelin: biosynthesis, membrane insertion and evidence for homotypic interactions. Submitted.

Ellis L, Clauser E, Morgan DO, Edery M, Roth RA, Rutter WJ (1986) Replacement of insulin receptor tyrosine residues 1162 and 1163 compromises insulin stimulated kinase activity and uptake of 2 deoxyglucose. Cell 45:721-732.

Forman BM, Yang C, Stanley F, Casanova J, Samuels HH (1988) c-erb-A protooncogenes mediate thyroid hormone-dependent and independent regulation of the rat growth hormone and prolactin genes. Mol Endocrinol 2:902-911.

Ganser AL, Kirschner DA (1980) Myelin structure in the absence of basic protein in the shiverer mouse. In Baumann N (ed) Neurological mutations affecting myelination. Elsevier.

Lai C, Watson JB, Bloom FE, Sutcliffe JG, Milner RJ (1987) Neural protein IB2361/myelin associate glycoprotein (MAG) defines a subgroup of the immunoglobulin superfamily. Immunol Rev 100:129-151.

Lemke G, Axel R (1985) Isolation and sequence of a cDNA encoding the major structural protein of peripheral meylin. Cell 40:501-508.

Lemke G, Lamar E, Patterson J (1988) Isolation and analysis of the gene encoding peripheral myelin protein zero. Neuron 1:73-83.

Mentaberry A, Adesnik M, Atchison M, Norgard EM, Alvarez F, Sabatini DD, Colman DR (1986) Small basic proteins of myelin from central and peripheral nervous systems are encoded by the same gene. Proc Natl Acad Sci USA 83:1111-1114.

Poduslo J, Dyck PJ, Berg CT (1985) Regulation of myelination: Schwann cell transition from a myelin-maintaining state to a quiescent state after permanent nerve transfection. J Neurochem 44:388-400.

Roach A, Takahashi N, Pravtcheva D, Ruddle F, Hood L (1985) Chromosomal mapping of mouse myelin basic protein gene and structure and transcription of the partially deleted gene in shiverer mutant mice. Cell 42:149-155.

Salzer JL, Colman DR (1989) Mechanisms of cell adhesion in the nervous system: role of the immunoglobulin gene superfamily. Dev Neurosci, in press.

Trapp BD, Bernier L, Andrews SB, Colman DR (1988) Cellular and subcellular distribution of 2'3-cyclic nucleotide 3'-phosphodiesterase and its mRNA in the rat central nervous system. J Neurochem 51:859-868.

Trapp BD, Itoyama Y, Sternberger NH, Quarles RH, Webster H deF (1981) Immunocytochemical localization of P_0 protein in Golgi membranes and myelin of developing rat Schwann cells. J Cell Biol 90:1-6.

ACKNOWLEDGEMENTS:

Supported by NIH grant NS 20147, and a grant from the National Multiple Sclerosis Society (RG 2090) to DRC. The full-length P_0 cDNA insert used in these studies was a gift from G. Lemke (see article, this volume).

PHYLOGENETIC ASPECTS
OF MYELINATION

MYELIN AND MYELIN-FORMING CELLS IN THE BRAIN OF FISH - A CELL CULTURE APPROACH

G. Jeserich, T. Rauen and A. Stratmann
Abt. Zoophysiologie
University of Osnabrück
Barbarastr. 11
4500 Osnabrück, FRG

INTRODUCTION

Myelin is an extension of the glial plasma membrane which is wrapped around nerve fibers in a typical multilamellar fashion. The occurence of densely compacted myelin around axons is a characteristic feature common to almost all vertebrate species, while it is only rarely observed in non-vertebrates (Bullock and Horridge 1965, Heuser and Doggenweiler 1966). The invention of the myelin sheath, which probably dates back to the Devonian period, several hundred million years of evolutionary history ago, can be regarded as a key step during evolution of the vertebrate brain, since it provided the basis for a most efficient mode of impulse propagation (saltatory conduction) by acting as an insulator against transmembrane ion currents and thus helped to save time, space and energy required for information processing in the brain (Rogart and Ritchie 1976).

Biochemically myelin is characterized by a very specialized protein composition comprising only a limited number of dominating components which are unique to this membrane and not found elsewhere. In the mammalian CNS a number of basic proteins (MBP) and a highly hydrophobic proteolipid protein (PLP) are the major components conveying adhesion of adjacent membrane lamellae, while in the PNS PLP is replaced by the hydrophobic glycoprotein Po which constitutes nearly 60% of the total protein content (for review see Lees and Brostoff 1984).

Immunologically related polypeptides in molecular weight

closely matching with their mammlian counterparts have been identified in all higher vertebrate classes, indicating that these components have been well conserved during vertebrate evolution (Waehneldt et al. 1985, 1986). When turning to phylogenetically older classes, like fishes, however, marked differences in the myelin protein composition are encountered. Thus in the CNS myelin of bony fishes a salient component of 36,000 dalton MW occurs for which as yet no immunologically related counterpart has been identified in other vertebrate classes (Jeserich 1983, Jeserich and Waehneldt 1986a). Furthermore in the brain of bony and cartilaginous fishes Po-like glycoproteins occur as major constituents of the myelin sheath functionally replacing PLP (Tai and Smith 1983, Waehneldt and Jeserich 1984, Jeserich and Waehneldt 1987). Since in higher vertebrates Po is a product unique to Schwann cells (Mirsky et al. 1980) whereas PLP is a molecular marker specific to oligodendrocytes (for review see Sternberger 1984) the question arises as to the nature of the myelin-forming cells in the brain of fish. The intriguing fact that glial cells in the CNS of fish have a marked capacity for remyelination comparing with those of Schwann cells (Murray 1976, Wolburg 1981) adds further interest to this question.

Electronmicroscopical analyses (Kruger and Maxwell 1967, Jeserich and Waehneldt 1986b) as well as immunocytochemical studies (Jeserich and Waehneldt 1986a) on glial cells in the CNS of teleosts have revealed close structural relationships with mammalian oligodendrocytes but not with Schwann cells, in that the cells extended multiple slender processes to myelinating fibers in their vicinity. In the present study a detailed characterization of the physiological properties of this particular type of oligodendrocytes was performed in a cell culture approach. Thereby the attention was focused on the regulation of myelinogenic expression, since mammalian oligodendrocytes and Schwann cells are known to basically differ in their dependence upon axonal signals to maintain their myelin-related phenotype (Brockes et al. 1979, Szuchet 1987).

RESULTS

Antibodies Against Myelin Proteins of Trout CNS

The protein pattern of myelin isolated from the CNS of trout after Coomassie Blue staining is shown by Fig. 1. Apart from a fastly migrating component, which was previously identified as a basic protein (Waehneldt and Jeserich 1984) two glycoproteins, designated intermediate proteins (IP1 23,000 dalton mol. wt; IP2 26,200 dalton mol. wt.), and a polypeptide of about 36,000 dalton (termed 36K) are the dominating components. To allow an immunohistochemical localization of these protein compounds in dissociated cell culture of trout CNS they were isolated by preparative electrophoresis and polyclonal antibodies were raised in rabbits against them. The specificity of the antisera was tested by immunoblotting of myelin proteins and membrane proteins of dissociated cells derived from trout CNS. The anti-36K antiserum in both cases selectively recognized a single band which was closely corresponding to 36K in electrophoretic mobility. The antiserum against IP2 specifically reacted with its respective antigen, too, but in addition it weakly stained the IP1 component (Fig. 1). This was not unexpected, since in previous biochemical studies significant sequence homologies had been revealed between the two IP-components of trout CNS myelin (Jeserich and Waehneldt 1986b). The antiserum against IP1 cross-reacted equally well with both intermediate glycoproteins and hence was not suitable for immunocytochemical work. Therefore monoclonal antibodies were raised in mice against this component. After fusion of NS1 myeloma with spleen cells from BALB/c mice that had been immunized with isolated IP1 approximately 40 antibody secreting hybridoma clones could be identified in an ELISA-test with isolated IP1. The supernatants of these immunopositive clones were further analyzed by Westernblotting of trout CNS myelin proteins (Fig. 2). Using this screening strategy two different types of antibodies were selected for immunocytochemical purposes: (1) monoclonal antibody 7C4 (mAb 7C4), which was of the IgG_{2a} subtype, selectively reacted with

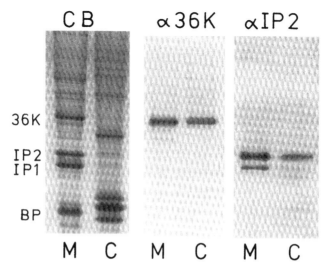

Fig. 1: SDS-PAGE of myelin proteins (M) and total particulate material of dissociated cells (C) from trout CNS. CB Coomassie blue staining; α 36K, α IP2 immunoblot analyses employing rabbit anti-36K and anti-IP2 antiserum, respectively. Myelin proteins were diluted 20 fold for immunoblotting to avoid excessive staining of bands. Immunoreactive bands were visualized using biotinylated anti-rabbit IgG in combination with alkaline phosphatase-conjugated streptavidin. 36K bony fish CNS myelin protein of 36,000 dalton MW; IP1, IP2 intermediate glycoproteins; BP basic protein.

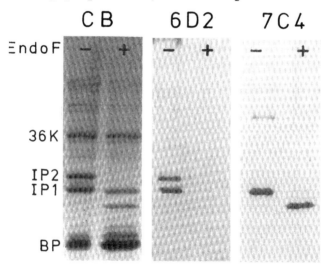

Fig. 2: SDS-PAGE of myelin proteins from the CNS of trout with (+) and without (-) endoglycosidase F treatment. CB Coomassie blue staining; 6D2, 7C4 immunoblot analyses using the monoclonal antibodies mAb 6D2 and mAb 7C4, respectively. For further details see Fig. 1.

the glycoprotein IP1 but not with IP2 on immunoblots. To more closely define the antigenic site, recognized by this antibody myelin proteins of trout CNS were enzymatically deglycosylated by Endoglycosidase F prior to immunoblotting. As a result both IP components underwent a reduction in molecular size of about 3,000 daltons. The immunoreactivity of mAb 7C4 with the cleavage product of IP1 was fully retained, however, indicating that the epitope was located in the protein portion of the IP1 molecule and not in the oligosaccharide chain. (2) monoclonal antibody 6D2 (mAb 6D2), which was an IgG_1, reacted with an antigenic site shared by both intermediate glycoproteins of trout CNS myelin. Obviously the epitope was residing in the carbohydrate moiety since any 6D2 immunoreactivity was eliminated after Endoglycosidase F treatment (Fig. 2).

Thus a panel of antibodies against myelin-specific proteins of trout CNS was available, to be used as specific molecular probes for the immunocytochemical identification and phenotypic characterization of fish oligodendrocytes in a cell culture approach.

Studies on Cell Cultures

Primary cultures of glial cells were prepared from brainstems and spinal cords of young trout by trypsinisation and mechanical dispersion of the tissue according to established procedures. The final cell supension was purified from cellular debris and myelin fragments by centrifugation through a Percoll density gradient using a similar protocoll as was suggested by Gebicke-Härter et al. (1984). During centrifugation at 10,000 g for 30 min. three bands (A,B,C) and a pellet were forming. Microscopical inspection revealed that band A on top of the gradient largely contained myelin fragments and cellular debris, whereas the remaining two bands mainly consisted of small round cell bodies. For identification of oligodendrocytes the cells from band B and C were immunohistochemically examined directly after isolation. By double-labeling immunostaining using anti-36K antiserum in

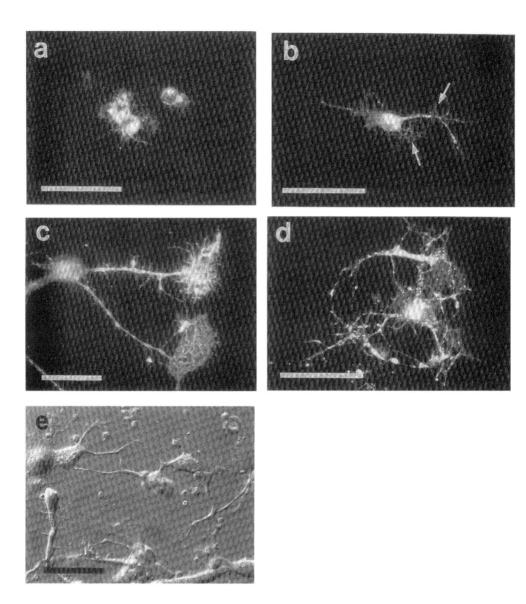

Fig. 3: Immunostaining of cultured trout oligodendrocytes using mAb 6D2 after 24 hours (a), 5 days (b) and two weeks (c,d) of culturing. Arrows in (b) indicate small membrane patches emanating from the cell processes. e differential interference contrast microscopy of a ten days old culture. Bars indicate 50 µm in a, b, d, e and 25 µm in c.

conjunction with mAb 6D2 72% of all cells in band C were strongly labeled for both these myelin markers, whereas in band B the proportion of immunolabeled cells was considerably lower (34% of all cells). The isolated cells were suspended in nutrient medium consisting of DMEM supplemented with 5% fetal calf serum and seeded on poly-D-lysine coated petri dishes. Soon after plating the cells attached to the substrate and started to regenerate their processes. Immunostaining with mAb 6D2 allowed to unambiguously identify oligodendrocytes in culture and proved instrumental to visualize the more delicate cell structures. After one day in vitro most of the oligodendrocytes were surrounded by a dense coat of short hairlike extensions (Fig. 3a). During the following days the cells elaborated larger sized cytoplasmic processes which were extensively ramifying into numerous fine branches and sometimes spread out into small membrane patches (Fig. 3b). By ten days in culture oligodendrocytes were often arranged in groups of cells forming loose networks of interacting fibers (Fig. 4a, c). Occasionally flat membrane sheets were seen to emanate from the cell bodies or from the tips of processes (Fig. 3c, d). Interestingly the 6D2-immunostaining was not evenly distributed within these membrane expansions but was concentrated in an irregularly arranged network-like pattern. In general, these membrane sheets were significantly smaller sized than those described for mammalian oligodendrocytes in culture. On account of their highly branched morphology cultured trout oligodendrocytes nevertheless closely resembled their mammalian counterparts in terms of morphology.

In addition a second type of glial cells with a flattened polygonal shape occurred in the cultures, which did not stain for any of the myelin markers but were strongly labeled with anti-GFAP antibodies instead.

Regulation of myelinogenic expression

To further analyze the ability of fish brain oligodendrocytes to maintain their myelin-related cell functions in the

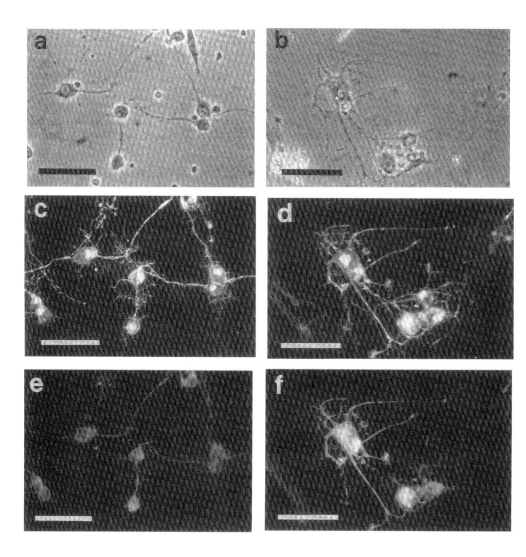

Fig. 4: Double-labeling immunostaining of cultured trout oligodendrocytes using anti-36K antiserum (e,f) in conjuction with mAb 6D2 (c,d) after ten days (a,c,e) and four weeks (b,d,f) of culturing. a and b are phase contrast images of the cells shown in fields c,e and d,f, respectively. Bars indicate 50 µm.

absence of axonal contacts a series of double-labeling experiments were performed. At first the expression of 36K and intermediate proteins was monitored by double-labeling immunostaining using anti-36K antiserum in conjunction with mAb 6D2. As it is shown by Fig. 4 and Fig. 5 the level of expression of 36K was significantly changing during a four weeks period in culture, whereas the intensity of staining with mAb 6D2 remained fairly constant in parallel. Until 3 days after plating oligodendrocytes were brightly immunostaining for both 36K and the intermediate proteins. By one week in culture, however, most of the 6D2-staining oligodendrocytes were only weakly labeled for 36K or even 36K negative. During the second week in vitro 36K started to reappear, however, and after four weeks in culture more than 80% of the 6D2-positive cells exhibited a strong 36K-labeling again (Fig. 4 and 5).

Interestingly the overall pattern of immunostaining was quite different for both these myelin markers. Whereas 36K appeared uniformly distributed in the cytoplasm, the staining for the intermediate glycoproteins was typically arranged in small dots, being heavily clustered in the Golgi region opposite to the nucleus (Fig. 3 and 4).

Since the epitope recognized by mAb 6D2 is shared by both IP

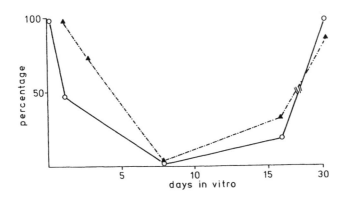

Fig. 5: Time course of expression of myelin proteins 36K (▲--▲) and IP1 (o——o) by cultured trout oligodendrocytes. The percentage of immunolabeled oligodendrocytes was determined after double-labeling immunostaining as described in the text.

glycoproteins, in a subsequent approach mAb 7C4 was used in conjunction with anti-IP2 antiserum to separately analyse the fate of IP1 and IP2 in cultured trout oligodendrocytes. As shown by Fig. 5 the time course of expression was quite different for both these glycoproteins. Whereas IP2 was continuously expressed by the oligodendrocytes during four weeks in culture, IP1 was progressively lost from the cells within the first week after plating and by ten days in vitro was no longer immunocytochemically detectable (Fig. 5 and 7). Similar as it was observed in the case of 36K, the expression of IP1 was reinduced during the second week in culture and by three weeks after plating the majority of IP2-positive oligodendrocytes were brightly staining for IP1 again. Finally the cells were examined for the presence of galactocerebroside (Fig. 6), which in the mammalian nervous system is an established molecular marker for both oligodendrocytes and Schwann cells. After double-labeling immunostaining using anti-IP2 antiserum in conjunction with monoclonal anti-GalC antibodies (Ranscht et al. 1982) until three days after plating

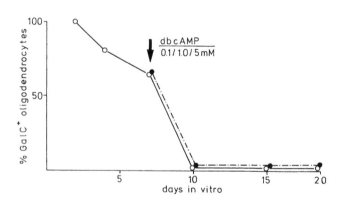

Fig. 6: Effect of dbcAMP on Galc-expression by cultured trout oligodendrocytes. On day seven after plating (arrow) 0.1, 1.0 or 5 mM dbcAMP were added to the cultures and the percentage of oligodendrocytes expressing Galc was determined after various intervals by double-labeling immunostaining using rabbit anti-IP2 antiserum in conjuction with mouse monoclonal anti-Galc antibodies. o—o control cultures, •—• dbcAMP treated cultures.

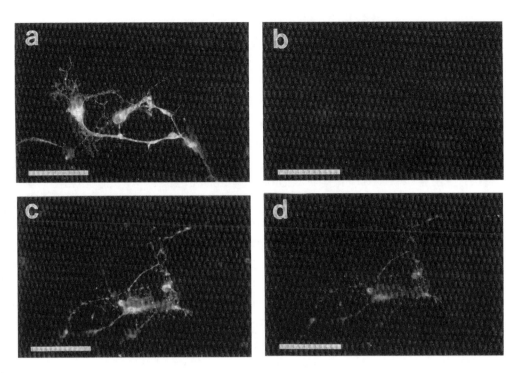

Fig. 7: Oligodendrocytes of trout brain in ten days old cultures double-labeled using rabbit anti-IP2 antiserum (a,c) and mouse mAb 7C4 (b,d). In untreated cultures (a,b) oligodendrocytes exhibit strong IP2-staining but are still negative for IP1. In cultures treated for six days with 200 µM forskolin (c,d) oligodendrocytes start to re-express IP1 already. Bars indicate 50 µm.

virtually all of the IP2 positive oligodendrocytes were intensely labeled for GalC, too. Thereafter GalC was gradually lost from the surface of the cells and after ten days in vitro staining for this glycolipid was no longer detectable. Unlike the myelin proteins IP1 and 36K, which used to reappear in the oligodendrocytes after a certain delay, GalC was not reexpressed even after three weeks in culture. This finding was in sharp contrast to the well known in vitro behavior of mammalian oligodendrocytes which continuously synthesize this glycolipid over extended periods in vitro (Szuchet et al. 1983). It was, however, highly reminiscent to the pattern of antigenic expression observed in mammalian Schwann cells, which rapidly fail to express GalC after axon withdrawal (Mirsky et al. 1980, Ranscht et al. 1982). Since for mammalian Schwann cells it has been shown that GalC expression can be experimentally reinduced by agents elevating intracellular cAMP (Sobue and Pleasure 1984) the effect of dbcAMP and forskolin on the molecular phenotype of cultured trout oligodendrocytes was examined. As shown by Fig. 6 treatment of the cultures with various concentrations of dbcAMP starting on day seven after plating did not affect the disappearance of GalC from the surface of the cells and neither elicited a reinduction of this glycolipid within three weeks of culturing. By contrast, for the myelin glycoprotein IP1 a significant stimulation of antigenic expression was revealed as a results of dbcAMP treatment (Fig. 7). In cultures continuosly exposed for six days to either dbcAMP or 200 µM forskolin, a potent activator of adenylate cyclase, numerous IP1-staining oligodendrocytes already occured by 10 days in vitro, whereas in untreated control cultures staining for this glycoprotein was barely detectable at that time. In addition, distinct morphological changes were elicited by the cAMP derivative since oligodendrocytes grown in the presence of 1 mM dbcAMP for six days elaborated a much denser network of fine processes than their untreated counterparts.

DISCUSSION

Myelin in the CNS of bony fish is exceptional in its protein composition, since it contains two major glycoproteins which are immunologically related to the mammalian PNS myelin glycoprotein Po, the major gene product of Schwann cells. This prompted us to investigate in detail the expression and maintenance of myelinogenic properties of the myelin-forming cells of trout CNS in a cell culture approach. Thereby antibodies against the myelin proteins 36K and IP2, which had been previously characterized immunochemically (Jeserich and Waehneldt 1986a), proved invaluable for the unambiguous identification and phenotypic characterization of oligodendrocytes by indirect immunofluorescence staining. Attempts to raise a polyclonal antiserum selectively recognizing IP1 failed, however, owing to the high degree of structural homology between the two intermediate glycoproteins (Jeserich and Waehneldt 1986b). This problem could be surmounted, however, by raising monoclonal antibodies against this component: mAb 7C4, which was directed against an epitope in the amino acid chain, indeed selectively labeled IP1 but not IP2; mAb 6D2, on the other side, which recognized an epitope in the oligosaccharide chain cross-reacted with both myelin glycoproteins, indicating structural homologies in the carbohydrate moiety of these two compounds.

In cell cultures prepared from the brain of mature trout two morphologically distinct cell populations developed: (1) flattened polygonal cells, morphologically resembling Type 1 astrocytes of mammalian glial cultures, which stained intensively for GFAP but were negative for the myelin proteins 36K, IP1 or IP2, and (2) cells with an irregularly branched morphology reminiscent of mammalian oligodendrocytes, which brightly immunostained for all three myelin proteins. A striking morphological feature of these cells was the presence of thin membrane sheets emanating either from the cell bodies or from the tips of processes, which was clearly visible only after immunostaining. These sheets were conspicuously smaller sized than those previously described for mammalian oligodendrocytes in culture (Knapp et al. 1987), suggesting that oligodendrocytes of fish require further exogenous stimulation for proper membrane production. In situ this could be provided for instance by neuronal elements that were absent from our cultures. Characteristic differences were observed concerning the pattern of immunolocalization for 36K on the one side and the intermediate glycoproteins on the other. The uniform intracellular distribution of 36K closely resembled the pattern of immunostaining observed for MBP in mammalian oligo-dendrocytes, suggesting a similar cytoplasmic site of synthesis for both these proteins. The dotted, granular appearance of 6D2-immunolabeling, on the other side, which was concentrated in a juxtanuclear position presumably reflects processing of the intermediate glycoproteins in the Golgi-apparatus and transport into the periphery by membrane vesicles.

Regarding the time course of antigenic expression by cultured trout oligodendrocytes marked differences were revealed for the various myelin antigens. The myelin glycoprotein IP2, for instance, proved as a most stable antigen which continued expression for the least 30 days in vitro. This shows that an ongoing axonal contact is not required to maintain expression of this myelin component. This finding was remarkable in so far, as in Schwann cells the homologous glycoprotein Po is rapidly down-regulated after axon withdrawal (Poduslo 1984, Mirksy et al. 1980). The myelin proteins IP1 and 36K, on the other side, gradually decreased in most of the cells during culturing. This was a transient phenomenon, yet, since after two weeks in vitro both myelin markers reappeared. According to recent developmental studies (Jeserich et al. 1989) IP2 is an early marker of oligodendroglial differentiation while the presence of IP1 and 36K defines a more mature state of the cells. Hence the loss of these myelin antigens during culturing means a partial dedifferentiation of the cells, possibly due to the lack of appropriate environmental signals or hormones which in situ might act to stabilize the differentiated phenotype of oligodendrocytes. The finding that both 36K and IP1 spontaneously reappeared in the cells after a certain delay was puzzling and is difficult to explain at present. Assuming, however, that the reappearance of these proteins was triggered by signals from other cells, astrocytes would be most likely candidates, since they represented the only other cell type in the cultures. In general it would be interesting to know, if the antigenic plasticity of trout oligodendrocytes observed in cell culture is in some way correlated with the regenerative capacity of these cells in vivo. If this is the case similar changes in the expression of IP1 and 36K should occur in the tissue after crush injury and/or during remyelination of CNS fiber tracts.

Concerning the expression of Galc cultured trout oligodendrocytes surprisingly mimicked the in vitro behaviour of mammalian Schwann cells, in that they ceased to express this glycolipid after a few days in culture. This was in sharp contrast to the well known in vitro behaviour of mammalian oligodendrocytes, which continuously synthesize these molecules indepedently from neuronal influences (Szuchet et al. 1983). For mammalian nervous tissue it has been demonstrated that by restoration of axon-Schwann cell contacts in vitro a reappearance of major myelin constituents can be induced, including basic proteins, Po and Galc (Politis et al. 1982). Therefore it will be important to establish, if in trout oligodendrocytes a similar modulation of antigenic expression is effected by coculturing with neurons. For such an approach retinal explants of the regenerating fish optic system should provide a most convenient source of target axons (Landreth and Agranoff 1979). The inductive effect of axonal membranes on mammalian Schwann cell cultures can be mimicked experimentally by raising the level of intracellular cyclic AMP which alone provokes a reexpression of certain myelin markers. In trout oligodendrocytes at least in the case of Galc a different mode of regulation seems to operate since the expression of this glycolipid was not affected by agents elevating intracellular cAMP. By contrast, concerning the regulation of IP1, good evidence for an involvement of cAMP was obtained, since the reappearance of this component was significantly accelerated by dbcAMP or forskolin treatment. Similar as it has been reported for other systems cAMP did not elicit an immediate response of the cells but required several days of continuous exposure to evoke a visible effect on IP1 expression (Sobue and Pleasure 1984, Raible, and McMorris 1989).

ACKNOWLEDGEMENTS

This study was supported by the Deutsche Forschungsgemeinschaft (grant Je 115/2-5). The authors would like to thank Dr. B. Ranscht (La Jolla, CA) and Dr. E. Bock (Copenhagen) for providing antibodies. The secreterial help of Mrs. H. Knehans is gratefully acknowledged. We thank also Prof. Lueken for general support and encouragement.

REFERENCES

Brockes IP, Raff MC, Nishiguchi DH, Winter J (1979) Studies on cultured rat Schwann cells. III. Assays for peripheral myelin protein. J Neurocytol 9:66-77.

Bullock TH, Horridge GA (1965) Structure and function in the nervous system of invertebrates, Freemann, San Francisco.

Gebicke-Härter PJ, Althaus HH, Ritter J, Neuhoff V (1984) Bulk separation and long-term culture of olgidendrocytes from adult pig brain. I. Morphological studies. J Neurochem 42:357-368.

Heuser JE, Doggenweiler CF (1966) The fine structural organization of nerve fibers, sheaths, and glial cells in the prawn, Palaemonetes vulgaris. J Cell Biol 30:381-403.

Jeserich G (1983) Protein analysis of myelin isolated from the CNS of fish: Developmental and species comparison. Neurochem Res 8:957-969.

Jeserich G, Müller A, Jacque C (1989) Developmental expression of myelin protein by oligodendrocytes in the CNS of trout. Dev Brain Res (in press).

Jeserich G, Waehneldt TV (1986a) Characterization of antibodies against major fish CNS myelin proteins: Immunoblot analysis and immunohistochemical localization of 36K and IP2 proteins in trout nerve tissue. J Neurosci Res 15:147-158

Jeserich G, Waehneldt TV (1986b) Bony fish myelin: evidence for common major structural glycoproteins in central and peripheral myelin of trout. J Neurochem 46:525-533.

Jeserich G, Waehneldt TV (1987) Antigenic sites common to major fish myelin glycoproteins (IP) and to major tetrapod PNS myelin glcyoprotein (Po) reside in the amino acid chains. Neurochem Res 12:821-825.

Knapp PE, Bartlett WP, Skoff RP (1987) Cultured oligodendrocytes mimic in vivo phenotypic characteristics: cell shape, expression of myelin-specific antigens, and membrane production. Dev Biol 120:356-365.

Kruger L, Maxwell DS (1967) Comparative fine structure of vertebrate neuroglia: Teleosts and reptiles. J comp Neurol 129:155-142.

Lees MB, Brostoff SW (1984) Proteins of myelin. In: Myelin, Morell P (ed) Plenum Press New York pp 197-224.

Landreth GE, Agranoff BW (1979) Explant culture of adult goldfish retina: a model for the study of CNS regeneration. Brain Res 161:39-53.

Mirsky E, Winter J, Abney ER, Pruss RM, Gavrilovic J, Raff MC (1980) Myelin specific proteins and glycolipids in rat Schwann cells and oligodendrocytes in cultures. J Cell Biol 84:483-393.

Murray M (1976) Regenerating retinal fibers into the goldfish optic tectum. J comp Neurol 168:175-196.

Poduslo JF (1984) Regulation of myelination: biosynthesis of the major myelin glycoprotein by Schwann cells in the presence and absence of myelin assembly. J Neurochem 42:493-503.

Politis JM, Sternberger N, Ederle K, Spencer PS (1982) Studies on the control of myelinogenesis. IV. Neuronal induction of Schwann cell myelin-specific proteins synthesis during nerve fiber regeneration. J Neurosci 2:1252-1266.

Raible DW, McMorris FA (1989) Cyclic AMP regulates the rate of differentiation of oligodendrocytes without changing the lineage commitment of their progenitors. Dev Biol 133:437-446.

Rauscht B, Clapshaw PA, Price J, Noble M, Seifert W (1982) Development of oligodendrocytes and Schwann cells studied with a monoclonal antibody against galactocerebroside. Proc Natl Acad Sci USA 79:2709-2713.

Rogart RB, Ritchie JM (1976) Physiological basis of conduction in myelinated nerve fibers. In: Myelin Morell P (ed) Plenum Press New York pp 117-159.

Sobue G, Pleasure D (1984) Schwann cell galactocerebroside induced by derivatives of adenosine 3',5'-monophosphate. Science 224:72-74.

Sternberger N (1984) Patterns of oligodendrocyte function seen by oligodendroglia immunocytochemisty. In: Adv Neurochem Vol 5 Norton WT (ed) Plenum Press New York pp 125-173.

Szuchet S (1987) Myelin palingenesis: the reformation of myelin by mature oligodendrocytes in the absence of neurons. In: Glial-neuronal communication in development and regeneration, Althaus HH, Seifert W (eds) Springer Verlag Heidelberg pp 756-777.

Szuchet S, Yim SH, Monsma S (1983) Lipid metabolism of isolated oligodendrocytes maintained in long-term culture mimics events associated with mylinogenesis. Proc Natl Acad Sci USA 80:7019-7023.

Tai FL, Smith R (1983) Shark CNS myelin contains four polypeptides related to the PNS protein Po of higher classes. Brain Res 278:350-353.

Waehneldt TV, Jeserich G (1984) Biochemical characterization of the central nervous system myelin proteins of the rainbow trout, Salmo gairdneri. Brain Res 309:127-134.

Waehneldt TV, Malotka J, Karin NJ, Matthieu JM (1985) Phylogenetic examination of vertebrate CNS myelin proteins by electro-immunoblotting. Neurosci Lett 57: 97-102.

Waehneldt TV, Matthieu JM, Jeserich G (1986) Appearance of myelin proteins during vertebrate evolution. Neurochem Int 9:463-474.

Wolburg H (1981) Myelination and remyelination in the regenerating visual system of the goldfish. Exptl Brain Res 43:199-206.

MYELIN PROTEOLIPID PROTEIN:

CLADISTIC TOOL TO STUDY VERTEBRATE PHYLOGENY

T.V. Waehneldt, J. Malotka, C.A. Gunn* and C. Linington*
Max-Planck-Institut für experimentelle Medizin
Forschungsstelle Neurochemie
D-3400 Göttingen, FRG

*Department of Medicine, University of Wales
Heath Park, Cardiff CF 4 4XN, UK

An anthropocentric view of myelin proteins

In view of the pressing problems of demyelinating diseases such as multiple sclerosis the main focus of myelin research is the human situation. However, it should be born in mind that the entire mammalian class constitutes less than 10 per cent of myelin-synthesizing vertebrates. Thus, a wealth of information as regards the structural/functional relationships of myelin components may be lost if gnathostome classes other than mammals are neglected.

Before examining the myelin proteins of lower vertebrate classes the protein composition of mammalian CNS and PNS myelin should be briefly reviewed, placing particular emphasis on their major hydrophobic constituents. Aside from the myelin-associated glycoprotein MAG (Quarles 1988), Wolfgram protein WP and/or 2',3'-cyclic nucleotide 3'-phosphodiesterase CNPase (Vogel and Thompson 1988; Sprinkle 1989) and myelin basic protein MBP (Lees and Brostoff 1984), which are present in both CNS and PNS myelin, albeit in different proportions, mammalian CNS myelin is characterized by the exclusive presence of unglycosylated proteolipid protein as major component (PLP; apparent mol. wt.$\sim$25,000 Daltons) (Lees and Macklin 1988). In contrast, the major component of PNS myelin is the glycosylated P_0 protein (P_0; apparent mol. wt.$\sim$30,000 Daltons) (Lemke and Axel 1985). PLP and P_0 are synthesized by oligodendrocytes and Schwann cells, respectively; both proteins are hydrophobic and partly embedded in the lipid matrix of the compact myelin membrane (Laursen et al 1984; Stoffel et al 1984; Lemke and Axel 1985) where they function presumably in an analogous way, maintaining the tight appositions in compact myelin (Braun 1984). However, despite these gross similarities PLP and P_0 have no significant sequence homology and their genomic arrangements are entirely different (Diehl et al 1986; Lemke et al 1988). Therefore, it seems justified to establish the distinct triads "CNS - oligodendrocyte - PLP" on the one hand, and "PNS - Schwann cell - P_0" on the other, even taking the presence of low levels of PLP in Schwann cell cytoplasm into consideration (Puckett et al 1987).

Fishes differ from tetrapods

Is this description of mammalian myelin proteins valid for all the other myelin-producing vertebrate classes, birds, reptiles, amphibians and bony and cartilaginous fishes ? There have been early observations to indicate that this may not be the case. Elam (1974), analyzing the proteins of goldfish brain myelin, noted the presence of two equally staining components in the typical size range of major hydrophobic proteins, instead of only one dominant protein, PLP. Franz et al (1981), in a systematic study comprising CNS myelin of all vertebrate classes, extended this observation to a larger number of elasmobranch and teleostean fish species and also found that these CNS protein pairs were glycosylated, as is P_0 and quite in contrast to PLP.

A brief overview of the major hydrophobic CNS myelin proteins of representative vertebrate classes shows the presence of two components in approximately equal proportions in the case of the chondrichthyan and actinopterygian fishes (Fig. 1A, T1 and T2, IP1 and IP2). These are glycosylated, as documented in Fig. 1B. By contrast, only one dominant hydrophobic protein is found in the amphibian and mammalian tetrapods (Fig. 1A, PLP), which lacks glycosylation. The relationship of these components to mammalian PLP and P_0 was determined by immunoblotting using appropriate antisera. Anti-mammalian PLP antisera clearly identified PLP as the major component of amphibian as well as mammalian CNS myelin, together with putative dimeric aggregates (Fig. 1C). In addition, the DM-20 isoform of PLP is observed in mammalian but not amphibian myelin. In contrast, the pairs of hydrophobic CNS myelin proteins seen in the fish are not recognized by anti-PLP antibodies; instead they react strongly with anti-mammalian P_0 antibodies (Fig. 1D).

Thus, there exists a break or discontinuity as regards major hydrophobic proteins of CNS myelin when stepping from the fish to land-living

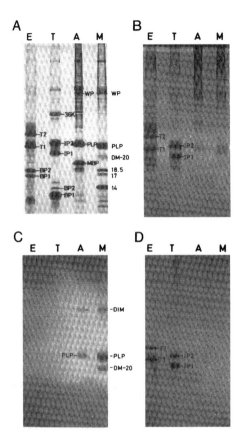

Fig. 1. Electrophoretic separation of CNS myelin proteins from different vertebrate classes on 14% polyacrylamide gel in the presence of sodium dodecylsulfate. Panel A, Coomassie Blue staining for protein; panel B, Concanavalin A-horseradish peroxidase staining for glycoprotein; panel C, immunoblotting with anti-rat PLP antiserum (dilution 1 : 150); panel D, anti-bovine P_0 antiserum (dilution 1 : 500). Individual lanes are as follows: E, elasmobranch (electric ray, Torpedo marmorata); T, teleost (trout, Salmo gairdneri); A, amphibian (frog, Rana pipiens); M, mammal (rat, Rattus rattus). T1, T2 and IP1, IP2, glycosylated hydrophobic myelin proteins of elasmobranchs and teleosts, respectively; BP1 and BP2, fish myelin basic proteins; 36K, teleost CNS myelin-specific intrinsic protein; WP, Wolfgram protein; PLP, proteolipid protein; DM-20, intermediate protein, homologous with PLP; MBP, myelin basic protein; 18.5, 17, 14, myelin basic proteins of 18.5, 17 and 14 KDaltons, respectively.

tetrapods. This has been confirmed by additional observations on chondrichthyan and teleostean fishes (Waehneldt et al 1984, 1985, 1986b). Therefore, at least in the case of trout (Jeserich and Waehneldt 1986) "CNS - oligodendrocyte - PLP" has to be replaced by "CNS - oligodendrocyte - P_0". Whether this applies also to other actinopterygian and even chondrichthyan species is presently not known since morphological and biochemical data on myelin-synthesizing cells in these classes are completely lacking.

A short look at some vertebrate living fossils

Was this "switch" in CNS myelin from P_0 to PLP exactly paralleled by the appearance of the first tetrapods, the amphibians ? Or - in view of the enormous number of adaptations that must be involved during the conquest of land - were there pre-tetrapodal fish-like stages that incorporated PLP instead of the ubiquitous P_0-like material into their CNS myelin ? Analysis of "living fossils" which may be closely related to such a putative transitional organism may resolve this question.

When the living lungfishes (Dipnoi) were first discovered in South America (<u>Lepidosiren</u>) and Africa (<u>Protopterus</u>) during the first half of the last century, they were grouped either with amphibians (Fitzinger 1837; Natterer 1837; Bischoff 1840) or with fishes (Owen 1839, 1841; Hyrtl 1845). This ambivalent taxonomic positioning prevailed even after the discovery of the more archetypal Australian lungfish (<u>Neoceratodus</u>) (Krefft 1870; Günther 1871) (Fig. 2), although this species was then taken to reinforce the views of Darwinian evolution.

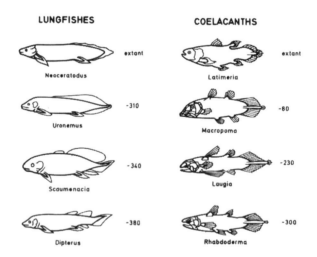

Fig. 2. Lungfishes and coelacanths in various geological times from Devonian to present. Not drawn to scale. Figures to the right are in million years. Adapted from Müller (1985), Forey (1988) and Balon et al (1988).

Similar to lungfishes (Rosen et al 1981; Forey 1986), the position of the living coelacanth (<u>Latimeria</u>) (Fig. 2) is also highly disputed among palaeontologists and phylogeneticists (Forey 1988, Balon et al 1988). Discovered only in 1938 (Smith 1939) it is grouped together with the crossopterygian fishes, among which the extinct rhipidistians are placed on the branch leading to tetrapods (Schultze 1986). Lungfishes and crossopterygian fishes in turn belong to the sarcopterygian or "lobe-finned" fishes - in contrast to the actinoperygian or "ray-finned" fishes, such as the teleosts. Moreover, lungfishes and coelacanths show amazingly few morphological changes since lower Devonian times, a period of nearly 400 million years (Fig. 2). Therefore, it is these few extant sarcopterygian species that may represent descendants of those extinct species that bridged the fish/tetrapod transition and whose CNS myelin may contain an indication of the evolutionary history of the PLP.

Lobe-finned fishes carry PLP in their CNS myelin

Inspection of CNS myelin proteins from four tetrapod classes and the two sarcopterygian fishes shows the following features (Fig. 3, left panel). In each species a single dominant band is migrating in the molecular size range of rat PLP. Only lungfish gPLP is retarded in migration due to its glycosylation. Myelin basic protein (MBP) is seen only clearly in tetrapods. However, its apparent absence in coelacanth and

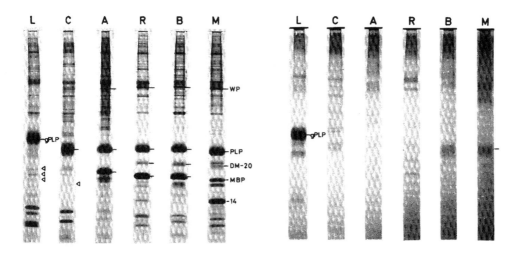

Fig. 3. Electrophoretic separation of CNS myelin proteins from sarcopterygian vertebrates. Left panel, protein staining; right panel, glycoprotein staining. L, lungfish (Neoceratodus forsteri); C, coelacanth (Latimeria chalumnae); A, amphibian (Rana pipiens); R, reptile (Ameiva ameiva); B, bird (Gallus gallus); M, mammal (Rattus rattus). Triangles point at the positions of myelin basic proteins in lungfish and coelacanth as revealed by immunblotting; gPLP, glycosylated proteolipid protein of lungfish. For further abbreviations, see Fig. 1.

lungfish can be ascribed to lengthy tissue handling and ensuing proteolysis during transportation and storage periods, as the presence of MBP-reactive proteins in the typical molecular weight range (14-18 KDaltons) has been confirmed by immunoblotting with anti-human MBP antibodies (Waehneldt et al 1987, 1989; Waehneldt and Malotka 1989). With respect to glycosylation only lungfish gPLP reacts intensely with the lectin Concanavalin A (Fig. 3, right panel), weak staining of material in the region of PLP in other species may be due to underlying traces of myelin/oligodendroglial glycoprotein (MOG; Linington et al 1984).

Fig. 4 shows the immunoblot results applying four anti-PLP antisera. First (A), rabbit antibodies against the entire rat PLP molecule react with PLP of all six species; the staining is strongest with the original rat

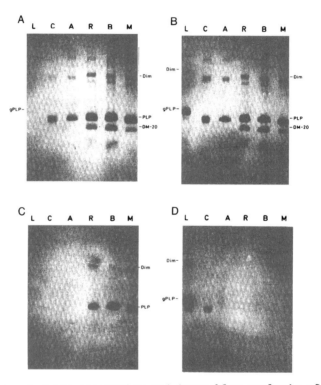

Fig. 4. Electrophoretic separation and immunoblot analysis of CNS myelin proteins from sarcopterygian vertebrates. Panel A, anti-rat PLP antiserum (dilution 1 : 150); panel B, anti-P6 antiserum (residues 257-276) (dilution 1 : 100); panel C, anti-P-1 antiserum (residues 109-128) (dilution 1 : 100); panel D, anti-lungfish gPLP antiserum (dilution 1 : 1000). Dim, putative dimeric form of PLP and gPLP.

The presence in amniotes of an additional band of slightly higher mobility than that of the PLP dimer is intriguing (panels A and B, pronounced in reptile and bird). Presumably this band consists of a DM-20 dimer, since it is not seen in the amphibian tetrapod and in the sarcopterygian fishes. A conceivable PLP - DM-20 heterodimer of intermediate mobility is not observed.

antigen and with PLP from bird and reptile and decreases substantially in amphibian and coelacanth, to be barely visible in lungfish. A similar trend is seen in the putative dimeric PLP aggregate (Dim) and higher oligomers as well as in DM-20. However, DM-20, defined as PLP having an internal deletion of 35 residues (Macklin et al 1987; Nave et al 1987a) leading to a characteristic reduction of approximately 4000 Daltons in apparent molecular weight (Agrawal et al 1972; Nussbaum and Mandel 1973), is absent in amphibian, coelacanth and lungfish, in agreement with earlier results (Franz et al 1981; Waehneldt et al 1985, 1986a, 1986c, 1989; Waehneldt and Malotka 1989). Second (B), rabbit antisera against the carboxyl terminal segment of bovine PLP, P6 (residues 257-276), react almost equally with PLP from all classes, including the dimeric form of lungfish. This points to an important region of PLP, conserved in all species, which is presumed to be exposed either at the cytoplasmic (Laursen et al 1984; Diehl et al 1986) or the extracellular side of the membrane (Hudson et al 1989). Third (C), in contrast to the staining patterns seen in (A) and (B), antiserum P-1, specific for an internal domain of PLP subtended by residues 109-128, only recognizes the PLPs of the amniotic classes (mammal, bird, reptile). Furthermore, this antiserum fails to detect the DM-20 isoform of these species indicating the epitope(s) recognized by this antiserum lie within the overlap of the DM-20 deletion (116-150) with the sequence used for immunization (109-128), i.e. within residues 116-128. The lack of immuno-reactivity of this antiserum with amphibian, coelacanthidan and dipnoan PLP is most probably due to alteration of one or a few amino acid residues within the sequence 116-128. In a speculative vein, the amphibian and fish PLP genes may not possess the alternative splice site for the synthesis of DM-20 mRNA that has been shown to exist in mammals (Nave et al 1987a; Macklin et al 1987; Simons et al 1987). The change of a single nucleotide would suffice to introduce the alternative splice site between residues 115 and 116, leading to the DM-20 isoform in reptile, bird and mammal with concomitant substitution of amino acid residue 116 in PLP, as follows:

Amphibia		Reptilia, Aves, Mammalia	
115	116	115	116
Thr	Leu	Thr	Val
ACG	CTA*	ACG	GTA

*or TTA (Leu), or ATA (Ile)

The altered amino acid residue, in turn, may represent the epitope (or part of the epitope) recognized only in amniotic PLP by the mammalian derived antibody P-1. Therefore, this putative point mutation would entail important alterations during the evolution from amphibians to reptiles, birds and mammals; in its far-reaching consequences this is reminiscent of the single base change A $\longrightarrow$ G in the splice site of intron 4 of the jimpy PLP gene (Nave et al 1987b). An alternative explanation for the absence of the DM-20 band in amphibians has been given by Kirschner et al (1989). Finally (D), anti-lungfish gPLP antibodies react strongly only with the lungfish component and its higher aggregates, while the reaction of coelacanth is very weak and that of tetrapods is virtually undetectable. This result, obtained by immunochemically approaching the vertebrates from the "other side", attests to substantial phylogenetic distance of lungfish gPLP.

With the exception of asparagine-linked glycosylation of lungfish gPLP - a unique dipnoan feature leading to an elevation of about 3000 Daltons (Waehneldt et al 1986c, 1987) - the apparent molecular weight of unglycosylated monomeric PLP is rather constant and ranges from 24,000 to 26,000 Daltons in the PLP-carrying species examined (Fig. 3). This contrasts with the variations seen in the P_0-like fish proteins, which range from 23,000 Daltons to about 30,000 Daltons (Franz et al 1981; Waehneldt et al 1986b). Since bovine PLP consists of 276 amino acid residues of presicely 29,869 Daltons (Lees and Brostoff 1984) it can be assumed that the number of residues in PLPs other than bovine may vary by only a few, unless particular amino acid residues and/or posttranslational modifications contribute to the minor variations seen in SDS gel electrophoresis.

Phylogenetic considerations

A consideration of the immunochemical data leads to the cladogram presented in Fig. 5. The presence of PLP in CNS myelin of lungfish (Dipnoi), coelacanth (Actinistia) and tetrapods, the sarcopterygians in a wide sense (Schultze 1986), clearly separates this group from actinopterygian and chondrichthyan fishes. Within the sarcopterygians, the observed immunochemical crossreactivities of PLP attest to extremely conserved structures of the molecule. They go far beyond those similarities

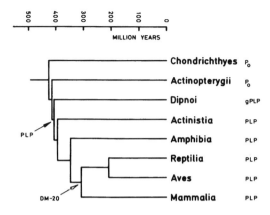

Fig. 5. Phylogeny of gnathostomates based on the presence of major hydrophobic proteins in CNS myelin, P_0 and PLP. The initial appearance of PLP and DM-20 is marked by arrows. Also included are the geological times at which the dichotomies occurred (adapted from Løvtrup 1977). It should be noted that the diversion into the marsupial-mammalian sister group dates back to about 120 million years. The marsupials are not depicted herein; however, they are known to possess PLP in their CNS myelin (Kerlero de Rosbo, Stuart and Bernard, personal communication).

noted in the mammalian class alone (Lin and Bartlett 1988). Nevertheless, clear-cut trends are seen. Amniotic vertebrates (mammal, bird, reptile) display fairly equal immunoreactivities, which extend also to the DM-20 component and higher aggregates (Fig. 4A), indicating close similarities of PLP among these classes. Another group is formed by the amphibian and the coelacanth which immunostain to a lesser extent than the amniotes, presumably due to sequence alterations. A third member of this grouping is the lungfish gPLP which barely reacts with anti-mammalian PLP antibodies. This graded response does not apply to the carboxyl terminal region which is highly conserved throughout all classes pointing to important structural/functional aspects of this portion of PLP. Similar results have been obtained for the carboxyl terminus of mammals alone (Potter and Lees 1988). The appearance of DM-20 from reptiles onward is a more recent development. DM-20 represents only a minor constituent in myelin of mature amniotes (however, see Nave et al 1987a, Kronquist et al 1987, Gardinier and Macklin 1988 for DM-20 in immature mammals). To date, it is not known whether DM-20 serves a special role; possibly it aids in the deposition of PLP, because jimpy mice transfected with the PLP gene were not able to incorporate PLP into the myelin membrane despite normal levels of PLP mRNA (L. Hudson, oral communication). Our results (Fig. 4), as well as those of

Kirschner et al (1989), tend to indicate alterations in PLP sequence when progressing from amphibia to reptiles; the ancestral pre-reptilian PLP component apparently suffices to both initiate and maintain myelination, whatever the putative role of DM-20 during later stages of evolution.

In contrast to chondrichthyan and actinopterygian fishes - together about 25,000 species - which carry at least two P_0-like glycosylated proteins as major CNS myelin components, the expression of PLP in CNS myelin of the extremely rare dipnoan and actinistian species represents a novel pre-tetrapodal acquisition which occurred some 400 million years ago (Fig. 5). Compared to P_0, the advantage of the new and entirely different PLP molecule is obscure; both components are presumed to serve similar adhesive functions in the compaction of myelin lamellae. The origin of PLP can momentarily only be the subject of hypotheses (Laursen et al 1983); however, the availability of PLP of largely identical size from widely separated vertebrate classes offers the unique possibility to draw comparative conclusions about sequence characteristics that are of crucial structural and functional importance.

Acknowledgements

The authors wish to thank Dr. C.V. Sullivan and Dr. J.A. Musick of the Explorer's Club and the Virginia Institute of Marine Science Coelacanth Program for brain material of a coelacanth (VIMS 8118). Dr. N.K. Zeller kindly provided antiserum P-1.

REFERENCES

Agrawal HC, Burton RM, Fishman AM, Mitchell RF, Prensky AL (1972) Partial characterization of a new myelin protein component. J Neurochem 19:2083-2089
Balon EK, Bruton MN, Fricke H (1988) A fiftieth anniversary reflection on the living coelacanth, *Latimeria chalumnae*: some new interpretation of its natural history and conservation status. Environ Biol Fishes 23:241-280
Bischoff TLW von (1840) *Lepidosiren paradoxa*. Anatomisch untersucht und beschrieben. Leipzig 34p
Braun PE (1984) Molecular organization of myelin. In: Myelin, Morell P (ed) Plenum New York p 97-116
Diehl H-J, Schaich M, Budzinski R-M, Stoffel W (1986) Individual exons encode the integral membrane domains of human myelin proteolipid protein. Proc Natl Acad Sci USA 83:9807-9811
Elam JS (1974) Association of axonally transported proteins in goldfish brain myelin fractions. J Neurochem 23:345-354

Fitzinger LJFJ (1837) Vorläufiger Bericht über eine höchst interessante Entdeckung Dr. Natterers in Brasil. Isis Jena p 379-380

Forey PL (1986) Relationships of lungfishes. J Morph Supp 1:75-91

Forey PL (1988) Golden jubilee for the coelacanth *Latimeria* *chalumnae*. Nature London 336:727-732

Franz T, Waehneldt TV, Neuhoff V, Wächtler K (1981) Central nervous system myelin proteins and glycoproteins in vertebrates: a phylogenetic study. Brain Res 226:245-258

Gardinier MV, Macklin WB (1988) Myelin proteolipid protein gene expression in jimpy and jimpy^msd mice. J Neurochem 51:360-369

Günther ACLG (1871) Description of *Ceratodus*, a genus of ganoid fishes, recently discovered in rivers in Queensland, Australia. Phil Trans Roy Soc London 161:511-571

Hudson LD, Friedrich VL, Behar T, Dubois-Dalcq M, Lazzarini RA (1989) The initial events in myelin synthesis: orientation of proteolipid protein in the plasma membrane of cultured oligodendrocytes. J Cell Biol 109:717-727

Hyrtl J (1845) *Lepidosiren* *paradoxa*: Monographie. Abhandl Böhm Ges Wiss 3:1-64

Jeserich G, Waehneldt TV (1986) Bony fish myelin: evidence for common major structural glycoproteins in central and peripheral myelin of trout. J Neurochem 46:525-533

Kirschner DA, Inouye H, Ganser AL, Mann V (1989) Myelin membrane structure and composition correlated: a phylogenetic study. J Neurochem in press

Krefft J (1870) Description a gigantic amphibian allied to the genus *Lepidosiren*, from the Wide Bay District, Queensland. Proc Zool Soc London 38:221-224

Kronquist KE, Crandall BF, Macklin WB, Campagnoni AT (1987) Expression of myelin proteins in the developing human spinal cord: cloning and sequencing of human proteolipid protein cDNA. J Neurosci Res 18:395-401

Laursen RA, Samiullah M, Lees MB (1983) Gene duplication in bovine brain myelin proteolipid and homology with related proteins. FEBS Lett 161:71-74

Laursen RA, Samiullah M, Lees MB (1984) The structure of bovine brain myelin proteolipid and its organization in myelin. Proc Natl Acad Sci USA 81:2912-2916

Lees MB, Brostoff SW (1984) Proteins of myelin. In: Myelin, Morell P (ed) Plenum New York p 197-224

Lees MB, Macklin WB (1988) Myelin proteolipid protein. In: Neuronal and glial proteins: structure, function, and clinical application, Marangos P, Campbell I (eds) Academic Press p 267-294

Lemke G, Axel R (1985) Isolation and sequence of a cDNA encoding the major structural protein of peripheral myelin. Cell 40:501-508

Lemke G, Lamar E, Patterson J (1988) Isolation and analysis of the gene encoding myelin protein zero. Neuron 1:73-83

Lin L-FH, Bartlett C (1988) Immunochemical differentiation of highly conserved forms of myelin proteolipid. Comp Biochem Physiol 3:505-509

Linington C, Webb M, Woodhams PL (1984) A novel myelin-associated glycoprotein defined by a mouse monoclonal antibody. J Neuroimmunol 6:387-396

Løvtrup S (1977) The phylogeny of vertebrata. John Wiley & Sons London

Macklin WB, Campagnoni CW, Deininger PL, Gardinier MV (1987) Structure and expression of the mouse myelin proteolipid protein gene. J Neurosci Res 18:383-394

Müller AH (1985) Lehrbuch der Paläozoologie, Vertebraten. Gustav Fischer Jena

Natterer J (1837) *Lepidosiren* *paradoxa*, eine neue Gattung aus der Familie der fischähnlichen Reptilien. Ann Naturhist Mus Wien 2:165-170

Nave K-A, Lai C, Bloom FE, Milner RJ (1987a) Splice site selection in the proteolipid protein (PLP) gene transcript and primary structure of the DM-20 protein of central nervous system myelin. Proc Natl Acad Sci USA 84:5665-5669

Nave K-A, Bloom FE, Milner RJ (1987b) A single nucleotide difference in the gene for myelin proteolipid protein defines the jimpy mutation in mouse. J Neurochem 49:1873-1877

Nussbaum JL, Mandel P (1973) Brain proteolipids in neurological mutant mice. Brain Res 61:295-310

Owen R (1839) On a new species of the genus *Lepidosiren* of Fitzinger and Natterer. Proc Linnean Soc London 1:27-32

Owen R (1841) Description of the *Lepidosiren* *annectens*. Trans Linnean Soc London 18:327-361

Potter NT, Lees MB (1988) Immunochemical characterization of antibodies to the myelin proteolipid protein (PLP). J Neuroimmunol 18:49-60

Puckett C, Hudson L, Ono K, Friedrich V, Benecke J, Dubois-Dalcq M, Lazzarini RA (1987) Myelin-specific proteolipid protein is expressed in myelinating Schwann cells but is not incorporated into myelin sheaths. J Neurosci Res 18:511-518

Quarles RH (1988) Myelin-associated glycoprotein: functional and clinical aspects. In: Neuronal and glial proteins: structure, function, and clinical application, Marangos P, Campbell I (eds) Academic Press p 295-320

Rosen DE, Forey PL, Gardiner BG, Patterson C (1981) Lungfishes, tetrapods, paleontology and plesiomorphy. Bull Am Mus Natur Hist 167:159-276

Schultze H-P (1986) Dipnoans as sarcopterygians. J Morph Supp 1:39-74

Simons R, Alon N, Riordan JR (1987) Human myelin DM-20 proteolipid protein delection defined by cDNA sequence. Biochem Biophys Res Comm 146:666-671

Smith JLB (1939) A living fish of Mesozoic type. Nature London 143:455-456

Sprinkle TJ (1989) $2',3'$-Cyclic nucleotide $3'$-phosphodiesterase, an oligodendrocyte-Schwann cell and myelin associated enzyme of nervous system. CRC Crit Rev Neurobiol 4:235-301

Stoffel W, Hillen H, Giersiefen H (1984) Structure and molecular arrangement of proteolipid protein of central nervous system myelin. Proc Natl Acad Sci USA 81:5012-5016

Vogel US, Thompson RJ (1988) Molecular structure, localization, and possible functions of the myelin-associated enzyme $2',3'$-cyclic nucleotide $3'$-phosphodiesterase. J Neurochem 50:1667-1677

Waehneldt TV, Kiene M-L, Malotka J, Kiecke C, Neuhoff V (1984) Nervous system myelin in the electric ray, Torpedo marmorata: morphological characterization of the membrane and biochemical analysis of its protein components. Neurochem Int 6:223-235

Waehneldt TV, Malotka J, Karin NJ, Matthieu J-M (1985) Phylogenetic examination of vertebrate central nervous system myelin proteins by electro-immunoblotting. Neurosci Lett 57:97-102

Waehneldt TV, Matthieu J-M, Jeserich G (1986a) Appearance of myelin proteins during vertebrate evolution. Neurochem Int 9:463-474

Waehneldt TV, Stoklas S, Jeserich G, Matthieu J-M (1986b) Central nervous system myelin of teleosts :comparative electrophoretic analysis of its proteins by staining and immunoblotting. Comp Biochem Physiol 84:273-278

Waehneldt TV, Matthieu J-M, Jeserich G (1986c) Major central nervous system myelin glycoprotein of the African lungfish (Protopterus dolloi) cross-reacts with myelin proteolipid protein antibodies indicating a close phylogenetic relationship with amphibians. J Neurochem 46:1387-1391

Waehneldt TV, Matthieu J-M, Malotka J, Joss J (1987) A glycosylated proteolipid protein is common to CNS myelin of recent lungfish (Ceratodidae, Lepidosirenidae). Comp Biochem Physiol 88:1209-1212

Waehneldt TV, Malotka J (1989) Presence of proteolipid protein in coelacanth brain myelin demonstrates tetrapod affinities and questions a chondrichthyan association. J Neurochem 52:1941-1943

Waehneldt TV, Malotka J, Jeserich G, Matthieu J-M (1989) Central nervous system myelin proteins of the coelacanth: phylogenetic implications. Environ Biol Fishes in press

PHYLOGENETIC ASPECTS OF MYELIN STRUCTURE

Hideyo Inouye and Daniel A. Kirschner
Neurology Research, Children's Hospital
Department of Neurology, Harvard Medical School
300 Longwood Avenue
Boston, Massachusetts 02115
U.S.A.

Introduction

Proteins are the major determinants of the spacing between membranes in compact, internodal myelin. This conclusion derives from X-ray diffraction and biochemical analyses (Kirschner et al., 1984; Inouye & Kirschner, 1988a,b; Lemke, 1988). For example, the idea that proteins protrude into the spaces between the membrane bilayers and act as struts to maintain their separation is based on the finding that the distance between them is at least twice as wide as that measured for hydrated myelin lipid multilayers. That P0-glycoprotein has a more extended conformation than PLP derives from the observation that inter-membrane spaces in PNS myelin are greater than those in CNS myelin; and this difference in extension may be consistent with the relative sizes of their putative cytoplasmic and extracellular domains as deduced from primary sequence data (Laursen et al., 1984; Stoffel et al., 1984; Lemke & Axel, 1985). Non-specific forces (such as electrostatic and hydration repulsion, and van der Waals attraction) account for many of the changes in membrane packing at the extracellular apposition when pH or ionic strength are altered; however, specific and short-range molecular interactions presumably involving proteins appear to define the *native* separation and to underlie the invariance of the cytoplasmic apposition.

X-ray and biochemical analyses of myelin structure and composition have been applied to a wide phylogenetic series of vertebrates in order to test the extent to which particular proteins account for the membrane packing (Kirschner et al., 1989). The amino acid sequences for these proteins from rodent, rabbit, bovine, and human have been determined from sequencing peptides or deduced from cDNA or genomic clones (see reviews by Campagnoni, 1988; Lemke, 1988; Sutcliffe, 1988). Recently, the genes for MBP and P0 glycoprotein from a phylogenetically older vertebrate -- the shark -- have also been cloned, and the protein sequences deduced (Saavedra et al., 1989). The results from these two perspectives -- the structural and biochemical -- is the subject of this paper. First, we will review and summarize some of the recent phylogenetic results relating myelin structure with its composition; next, we will analyze the sequences of myelin proteins based not only on predicted secondary structure but also on physical-chemical parameters and consensus sequences for certain structural motifs; and finally, we will relate the results from sequence analysis to the differences in myelin membrane spacing.

Materials and Methods

X-ray diffraction. Experimental details on the preparation of myelinated tissue samples, the recording and analysis of their diffraction patterns, and the interpretation of the calculated membrane profiles for myelin are described elsewhere (Kirschner et al., 1984, 1989). Briefly, myelinated tracts or nerves (e.g., CNS: optic nerve, spinal cord; PNS: sciatic or trigeminal nerve, intradural root, or lateral line nerve [from fish]) were dissected as fresh, unfrozen, unfixed tissue from a phylogenetic series including elasmobranchs, teleosts, lungfish, amphibians, and mammals. The tissue was examined by X-ray diffraction immediately after dissection. The spacings of the diffraction spectra were measured directly off the X-ray films and used to calculate the myelin period (d). The intensities of the spectra were measured densitometrically and used to calculate the integrated intensity $I(h)$ for each Bragg order h, the structure factors $F(h)$, and the electron density profile of the membrane. Dimensions measured from these profiles include: the center-to-center distance of the membranes across their cytoplasmic ($2u$) or extracellular apposition (d-$2u$); the distance from one lipid headgroup peak to the next across the aqueous space at the cytoplasmic (cyt) or extracellular apposition (ext); and the distance across the low density hydrocarbon trough and between the lipid headgroup regions (lpg).

Biochemical analysis. Interpretation of the electron density profiles in terms of chemical constituents requires quantitative protein and lipid data on the intact myelin that is used for the X-ray diffraction analysis; however, this is difficult because the highest resolution structural data requires the use of intact myelin whereas the biochemical data is from analysis of isolated membranes. For making preliminary correlations between structural features and molecular components, it may be sufficient to have qualitative data on what lipids and proteins are present. Such information comes from SDS-PAGE and immunoblotting for the major myelin proteins and HP-TLC for the lipids (for details, see Bürgisser et al., 1986; Waehneldt et al., 1986; Ganser et al., 1988; Kirschner et al., 1989).

Sequence analysis. To explain how proteins determine the spaces in myelin requires knowing their detailed molecular conformation in the membrane; however, except for P2 basic protein of PNS myelin (Jones et al., 1988), the structures of the myelin proteins are not yet known. Currently, their structures are modeled by using predictive methods, by comparison with other proteins whose structures are known, and by using available biochemical and spectroscopic data. Given the localization of the major proteins of myelin from the structural/biochemical correlation, it may be possible to relate the electron

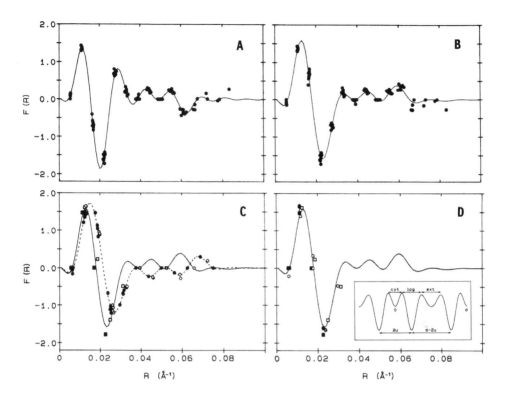

Figure 2. Structure factors plotted as a function of reciprocal coordinate R. (A) Comparison of calculated transform ($F(R)$, *continuous curve*) and observed structure factors ($F(h)$, *data points*) for diffraction data from mammalian PNS myelin from nine different species. The range of periods from 176-185 Å allows the $F(h)$ to sample a single transform corresponding to that for the membrane pair defined relative to the *extracellular* apposition. Using Shannon's sampling theorem and assuming a hypothetical period of 1000 Å, we calculated Fourier transforms for the nine mammals and averaged them to obtain $F(R)$. The clusters of data points correspond to the *1*st, *2*nd, ..., *h*th order structure factors from the different patterns. Phases were assigned according to Kirschner & Ganser (1982). (B) The usual convention in displaying the continuous transform relative to the *cytoplasmic* apposition shows a less satisfactory fit with the observed data. (C) Structure factors for teleost fishes (PNS, *filled circles*; CNS, *open circles*) plotted onto the average continuous transform for mammalian PNS myelin calculated relative to cytoplasmic boundary (*continuous curve*). Note that except for the garfish data (*squares*), that for teleosts does not satisfactorily map onto the continuous transform. *Dashed curve*, calculated transform for goldfish lateral line nerve. (D) Structure factors for elasmobranch (PNS, *filled circles*; CNS, *open circles*) and garfish myelins (*squares*) compared to continuous transform for mammalian PNS myelin. (D, *Inset*) Schematic of electron density distribution calculated from X-ray diffraction data for myelin shows the structural parameters discussed herein.

the averaged transform for mammalian PNS (calculated relative to the cytoplasmic boundary), indicating a narrower spacing at this apposition for the teleosts. Both the garfish and elasmobranch data map onto the calculated mammalian transform (Fig. 2D) suggesting similar cytoplasmic spacings among these three classes and that their differences in myelin

density and widths of these regions in the membrane to the primary sequences of the proteins using their inferred conformation (Inouye & Kirschner, 1989). Sequence analysis consists of searching for a pattern of properties which are defined by chemical and/or physical parameters such as: propensities for α-helix, β-strand, β-turn and coil conformations (Chou & Fasman, 1978; Garnier et al., 1978); hydrophobicity (Eisenberg et al., 1984a; Cornette et al., 1987); and charge density at specified pH. For our analysis, each parameter was averaged over a certain number of residues, its periodic disposition was calculated using a Fourier transform, and correlation coefficients were obtained for pairs of proteins (see Appendix). The computer programs used were written in FORTRAN in our laboratory and run on a VAX 11/780 (VMS operating system).

Results & Discussion

<u>Myelin structure in a phylogenetic series.</u> The periodicity of the myelin, measured from X-ray diffraction patterns, is one indicator of the changes in myelin structure that have occurred during phylogeny (Fig. 1). The periods tend to increase in proceeding towards higher vertebrates; the period for PNS myelin is always greater than for CNS myelin; and the PNS-CNS difference is more marked for mammals and amphibians compared to teleosts and elasmobranchs. The periods for elasmobranchs and the teleost garfish and dipnoan lungfish are exceptional considering their phylogenetic positions. Plots of the structure factors calculated from the diffracted intensities show which of their intermembrane spacings -- i.e., the cytoplasmic or extracellular -- is less variable. When PNS myelins from nine mammalian species are compared (Fig. 2A,B) the data sets map onto a single continuous transform for membrane pairs defined relative to their extracellular apposition, indicating that the spacing at this apposition is less variable in width than at the cytoplasmic apposition. By contrast, the CNS myelins from five mammalian species (not shown) do not show such a trend. For teleost (excluding garfish) PNS and CNS myelins, the structure factors (Fig. 2C) are shifted to larger reciprocal coordinate R compared to

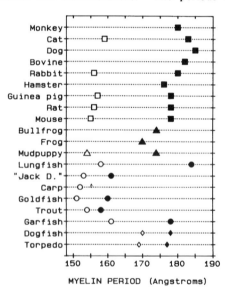

Figure 1. Myelin periods: mammals (*squares*), amphibia (*triangles*), teleosts (*circles*), elasmobranchs (*diamonds*). PNS, *filled symbols*; CNS, *open symbols*. Modified from Kirschner et al. (1989).

Table I

MEASUREMENTS OF THE ELECTRON DENSITY PROFILES

Specimen	PNS						CNS					
	d	$2u$	$d-2u$	cyt	lpg	ext	d	$2u$	$d-2u$	cyt	lpg	ext
Monkey	180	84	96	37	48	47						
Cat	183	83	100	36	49	49	159	79	80	31	48	32
Dog	185	85	100	38	50	47						
Bovine	182	86	96	39	48	47						
Rabbit	180	85	95	38	47	48	156	77	79	31	47	31
Hamster	176	80	96	34	48	46						
Guinea pig	178	81	97	35	48	47	157	77	80	31	47	32
Rat	178	81	97	34	48	48	156	78	78	31	47	31
Mouse	178	81	97	34	49	46	155	76	79	32	44	35
shi/shi	178	81	97	35	48	47						
Bullfrog	174	78	96	34	46	48						
Frog	170	78	92	34	45	46						
Mudpuppy	174	78	96	33	48	45	154	71	83	29	44	37
Lungfish	184	89	95	47	42	53	158	79	79	33	46	33
"J.D."	161	69	92	26	44	47	153	69	84	22	46	39
Goldfish	160	71	89	26	45	44	152	69	83	23	45	39
Trout	158	61	97	24	49	36	154	71	83	§	§	36
Garfish	178	89	89	39	50	39	161	75	86	31	47	36
Dogfish	178	89	89	37	52	37	170	85	85	36	49	36
Torpedo	177	88	89	36	52	37	169	79	90	35	47	40

d, periodicity; $2u$, separation of the membranes at the cytoplasmic apposition; $d-2u$, separation of the membranes at the extracellular apposition; cyt, cytoplasmic space; lpg, distance between lipid polar headgroups; ext, extracellular space; "J.D.", "Jack Dempsey" (*Cichlasoma octafasciatum*).

§ The profile did not show a clear minimum at the cytoplasmic apposition; therefore, cyt and lpg were undetermined.

Myelin protein composition has also been examined in a phylogenetic series (reviewed by Waehneldt et al., 1986; Kirschner et al., 1989). It was found that MBP is present throughout phylogeny in both the PNS and CNS, where its M_r ranges from ~8-20 kDa. P2, a basic protein of the PNS, is not detected in isolated PNS or CNS myelin from elasmobranch, teleost or amphibian, but is present in whole tissue homogenates from elasmobranch and garfish CNS and PNS, and teleost CNS. In mammalian PNS myelin, P2 is a minor protein in rodents, but is a major protein in higher mammals such as bovine and human. P0 or P0-like glycoproteins are detected throughout phylogeny in the PNS, and surprisingly, are also present in the CNS of aquatic vertebrates (elasmobranch and teleost fish). They are sometimes present as a doublet and their M_r ranges from ~22-32 kDa. PLP is always present in the CNS myelin[1] of terrestrial vertebrates (amphibia, reptiles, birds, mammals), and also in certain fish that are regarded as "living fossils" -- e.g., lungfish carry a

---

[1] PLP has also been detected in the PNS of rat, rabbit and human, where it is present in the Schwann cell body but not in the myelin (Puckett et al., 1987).

periods are due to distinct spacings at the extracellular apposition. Measurements from the electron density profiles (Fig. 3) confirm these conclusions drawn from comparisons of the structure factors (Table I).

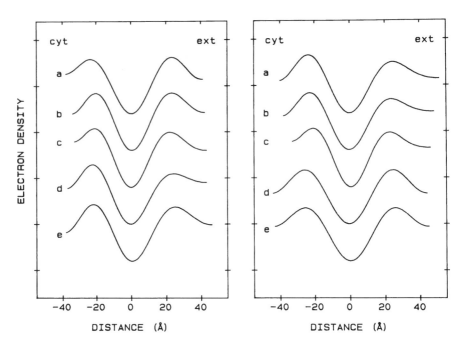

Figure 3. Electron density profiles on a relative scale for myelin membranes of the CNS (*left*) and PNS (*right*). (a) Mammalian (cat optic and sciatic nerve). (b) Amphibian (mudpuppy spinal cord and bullfrog sciatic nerve). (c) Teleost ("Jack Dempsey" spinal cord and goldfish lateral line nerve). (d) Garfish (optic and lateral line nerve). (e) Elasmobranch (*Torpedo* optic and branchial nerve). The profiles were calculated from diffraction data to 0.03 Å^{-1}. Garfish is plotted separately because its diffraction data was not like the other fishes. The profiles are displaced vertically for clarity and aligned at the centers of their hydrocarbon troughs to show that the lipid bilayers are very similar, but that the spaces at the cytoplasmic and extracellular appositions vary. See Table I for a summary of the profiles' dimensions.

Correlation of structural with biochemical results. Myelin lipid composition has been examined in elasmobranch, teleost (including garfish), amphibian and mammal (e.g., Hofteig et al., 1981; Saito & Tamai, 1983; Bürgisser et al., 1986; Kirschner et al., 1989). Whereas neutral lipids show little or no variation among the vertebrate classes (with cholesterol always dominant), the major phospholipids do differ somewhat, and the glycolipids vary substantially. Despite these differences, there are no consistent phylogenetic trends. Moreover, the lipid bilayer thickness is relatively constant (Table I) and there is no apparent correlation with either myelin period or the widths of the spaces between the membranes (Kirschner et al., 1989).

PLP-immunoreactive protein that is glycosylated (Waehneldt et al., 1987), and coelacanth carry a PLP-immunoreactive protein that is unglycosylated (Waehneldt & Malotka, 1989). Thus, there is a significant change in CNS myelin proteins in the fish $\rightarrow$ tetrapod transition. These changes in protein composition appear to relate to changes in the intermembrane spacings (Kirschner et al., 1989). At the cytoplasmic apposition, the narrowest space (~23 Å in teleost) correlates with the presence of a very low M_r basic protein. As this space widens (to ~39 Å in bovine) the 11-13 kDa basic proteins are gradually replaced by higher M_r basic proteins. In amphibians and mammals, myelin containing PLP tends to have a narrower cytoplasmic spacing than myelin containing P0; and this may be consistent with PLP's inferred compact structure (Laursen et al., 1984; Stoffel et al., 1984). Among mammals, myelin containing small amounts of P2 (e.g., rodent) tend to have narrower cytoplasmic spaces than myelin which has significant levels of this protein (e.g., rabbit; Sedzik et al., 1985). Alternatively, the 14 kDa MBP isoform which predominates over the 18.5 kDa form in rodent myelin, may result in a narrower cytoplasmic space. At the extracellular apposition, which varies in width from about 30-50 Å, the tendency is for PLP-containing myelin to have the narrowest spacing, and P0-containing myelin to have wider spacings.

Table II

PROPERTIES OF PHYLOGENETICALLY-RELATED MYELIN PROTEINS

Protein, animal	N	kDa	$\alpha(\%)$	$\beta(\%)$	pI	<H>	Reference
PLP							
human, rodent	276	29.9	30.4	34.4	9.4	0.29	a
bovine	276	29.9	26.1	40.2	9.5	0.30	b
P0							
bovine	219	24.7	25.6	34.7	9.7	-0.03	c
rat	219	24.8	21.5	37.4	9.7	-0.02	d
shark brain	246	27.3	13.8	42.3	10.0	0.00	e
MBP							
human	170	18.5	10.6	17.1	11.5	-0.22	f
rat	127	14.1	6.3	19.7	11.9	-0.30	f
mouse	194	21.4	9.8	19.1	11.5	-0.23	g
mouse	168	18.4	11.3	19.0	11.4	-0.23	g
mouse	127	14.1	7.1	20.5	11.9	-0.30	g
shark brain	155	16.6	19.4	16.8	11.1	-0.22	e
P2							
bovine	131	14.8	46.6	22.9	10.3	-0.12	h
rabbit	131	14.8	47.3	22.9	10.2	-0.10	i

N, number of residues. The α and β contents were estimated by the prediction method of Garnier et al. (1978) (assuming correction factors of zero for helix and β conformations). The pI was obtained from minimum charge calculated as a function of pH, given the intrinsic pK values and the number of ionizable residues. The average hydrophobicity <H> was determined according to the scale proposed by Eisenberg et al. (1984a).

References: **a**, Milner et al. (1985), Diehl et al. (1986), Macklin et al. (1987), Ikenaka et al. (1988), Hudson et al. (1987); **b**, Laursen et al. (1984); **c**, Sakamoto et al. (1987); **d**, Lemke & Axel (1985), Lemke et al. (1988); **e**, Saavedra et al (1989); **f**, Carnegie & Moore (1980); **g**, Newman et al. (1987); **h**, Kitamura et al. (1980); **i**, Narayanan et al. (1988).

Sequence analysis. Table II summarizes certain chemical-physical properties for phylogenetically-related myelin proteins whose sequences are known. Most of the myelin proteins show basic pI values, indicating that they are positively charged at physiological pH. (By contrast, MAG [Arquint et al., 1987] shows a pI ~ 4.9). The average hydrophobicity <H> decreases in the order PLP > MAG > P0 > CNPase > P2 > MBP. (See Vogel & Thompson, 1988, for CNPase sequence.) The total α and β content tends to increase with <H>, with the exception of P2. In the membrane environment, more residues of PLP may be immersed in the lipid bilayer conferring a more compact conformation compared to P0. Among mammalian myelin proteins the tabulated values for the parameters are quite similar, whereas between shark and mammals the values are more different. This is consistent with the similar membrane spaces among mammals compared to elasmobranch.

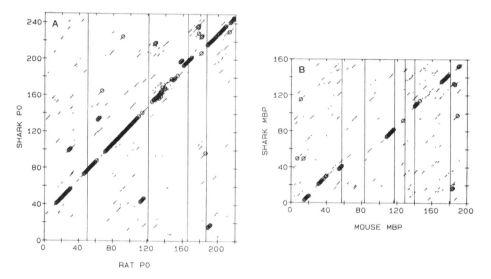

Figure 4. Correlation coefficients for shark _vs_ rodent myelin proteins. The sequences were compared on an all-or-none basis with a window length = 7. Points are for correlation coefficients $\geq 3/7$, and circles for coefficients $\geq 4/7$. The vertical lines indicate the position of the exon-intron boundaries for the rodent protein. (A) P0-glycoprotein. (B) MBP. In mouse, the four MBP isoforms are produced by alternative splicing of mRNA: 21.5 kDa form, all exons are used; 18.5 kDa, exon 2 is deleted; 17 kDa, exon 6 is deleted; and 14 kDa, exons 2 and 6 are not used.

What are the differences in sequence between shark and rodent? Correlation coefficients calculated for the sequences of rat and shark P0-glycoproteins (Fig. 4A) show that the major difference between them is an ~26 residue insertion at the N-terminus in the shark protein. Thus, the putative extracellular domain is larger in shark than rodent. The transmembrane sequence is also longer by 5 residues in shark (148-178) than in rodent (125-150). Physical-chemical parameters plotted as a function of residue number (Fig. 5A,B) show the homologies in greater detail. For P0 the putative extracellular portions for both

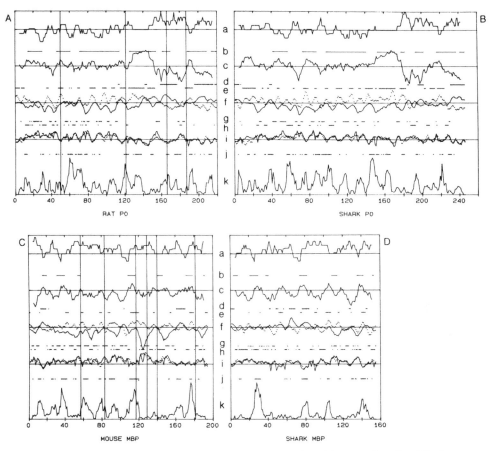

Figure 5. Physical-chemical parameters as a function of residue number for P0 glycoprotein of rat (A) and shark (B) and MBP of mouse (C) and shark (D). The parameters, from the top, are: (a) charge at pH 7 (window length = 9); (b) conserved domains between shark and rodent proteins; (c) hydrophobicity (window length = 9) (Eisenberg et al., 1984a); (d) α-helix; (e) β-strand; (f) α-helix (solid curve), β-strand (dotted curve); (g) turn; (h) coil; (i) turn (solid), coil (dotted) (Garnier et al., 1978); (j) position of the highest intensity for the α-helical amphiphilicity (window length = 7) (PRIFT score, in Cornette et al., 1987); (k) α-helical amphiphilicity curve. The vertical lines show the positions of exon-intron boundaries for the rodent.

rat and shark show consecutive peaks having β-strand propensity, which is similar to the immunoglobulin fold (Fig. 6; Lai et al., 1987; Williams, 1987; Saavedra et al., 1989). The cysteines at positions 21 and 98 for rat, and 48 and 125 for shark are conserved in immunoglobulin. There is also conservation of the glycosylation site in the extracellular domain of P0 (*Asn-Gly-Thr*: 93-95 for rat, 120-122 for shark). The cytoplasmic portions of the P0 proteins show, by contrast, an α-β-α folding pattern.

Based on our earlier analysis of membrane spacings in peripheral myelin as a function of pH, we suggested that deprotonation of *His* residues may be important for the correct

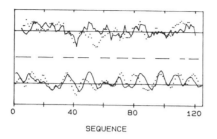

Figure 6. Comparison of the hydrophobicity (top) and β-propensity (bottom) between the extracellular domain of rat P0-glycoprotein (solid line) and the V_H domain of the phosphorylcholine-binding mouse immunoglobulin M603 (interrupted line) (Padlan, 1977). The horizontal bars indicate the β-conformation according to X-ray crystallography.

folding of P0 glycoprotein and the resultant compaction of the extracellular packing under physiological conditions (Inouye & Kirschner, 1988b). We have now found that *His*-52 and -86 for rat and *His*-79 and -106 for shark are in corresponding regions having sequence homology. Because both *His* residues also show a high β propensity, we suggest that deprotonation of these particular histidines may be necessary for the protein to fold into such a conformation. This idea is consistent with the finding of uncharged residues at the histidine positions in the 4th and 7th regions of the β-strands (Fig. 6) in immunoglobulin structures (Padlan, 1977). Immunoglobulin-like folding has also been found in the extracellular domains of neural adhesion molecules, like N-CAM (Cunningham et al., 1987), MAG (Arquint et al., 1987; Lai et al., 1987) and L1 (Moos et al., 1988). Cell-cell interaction may be mediated by homophilic binding between such glycoproteins (Edelman, 1983; Rutishauser, 1984). Although the molecular details of this interaction are not known, such binding may be analogous to the pair-wise association of immunoglobulin constant domains (Schiffer et al., 1985; Santoni et al., 1988). For the cytoplasmic domains of the adhesion molecules, homologies have not yet been reported. We note, however, that in rat and shark P0, and in MAG, there are possible phosphorylation sites with cyclic AMP-dependent kinase (*Arg-Arg-X-Ser*) (Krebs & Beavo, 1979). This suggests that P0 may have a signal transduction role across the myelin membrane.

Correlation coefficients for the sequences of mouse and shark MBPs (Fig. 4B) show that the shark protein can be generated by an ~12 residue deletion at the N-terminus and an ~25 residue deletion (exon 2) from the mouse 21.5 kDa MBP. A comparison of the physical-chemical parameters (Fig. 5C,D) indicates that the conserved regions between mouse and shark are moderately hydrophobic and have higher propensities for the β-strand and α-helix, whereas the variable regions are mostly hydrophilic and have a higher propensity for β-turn and coil (Saavedra et al., 1989). In rodent myelin the 14 kDa species of MBP is the major isoform. The conserved regions of rodent 14 kDa and shark MBP are in the *1*st and *3*rd exons. Therefore, these regions may be where the functional core of MBP is located. Based on consensus sequences of the phosphorylation site with cAMP-dependent (Krebs & Beavo, 1979) or calmodulin-dependent kinases (Pearson et al., 1985), i.e., *Arg-X-X-Ser(Thr)*, three regions in mouse MBP (residues 41-44, 130-133, and 186-189 in 21.5 kDa) and two regions in shark (residues 100-103 and 132-135) may be possible phosphorylation sites. The residues 130-133 in mouse and 100-103 in shark are next to the encephalitogenic determinant *Trp-X-X-X-X-Gln-Arg(Lys)* (Carnegie, 1971) which is conserved in both animals. We also note that the *Pro-Pro-Pro* sequence of mouse (122-124) is not present in shark.

The sequence of bovine PLP differs slightly from that of human and rodent PLPs which have identical sequences. Although proteins having an extensive homology with PLP have not been reported, a sequence motif for nucleotide binding and the phosphotransfer reaction (Hanks et al., 1988) may be present. This site, which is not present in DM-20, is the *Gly-X-Gly-X-X-Gly-....-Lys* at residues 122-127....143(150). A molecular structure for this site has been proposed (Sternberg & Taylor, 1984). It is interesting that the *Lys*-150 is next to the negatively-charged *Asp*-149 to which dicyclohexylcarbodiimide, an inhibitor of proton translocation, binds (Lin & Lees, 1984).

<u>Relating sequence analysis to differences in membrane spacing.</u> Electron density profiles calculated for elasmobranch and rodent myelins show differences in spacing, particularly at the extracellular apposition (Fig. 7). Although there is an ~26 residue long insertion at the N-terminus in shark P0, the extracellular separation in elasmobranch CNS myelin is actually *smaller* by 6-10 Å than that in mouse sciatic nerve myelin. Because the inserted amino-acids are mostly hydrophobic and have a high propensity for the β-conformation, one interpretation of the narrower space is that the extracellular portion of shark P0 is more compact due to the additional β-sheet formation. Another explanation is that there is less electrostatic repulsion in the shark in which the extracellular domain of P0 has a pI value of 6.6, whereas that of rat is 4.8. The cytoplasmic separation in elasmobranch CNS myelin is only 1-2 Å greater than that of mouse sciatic nerve myelin. This is consistent with the similar number of residues forming the α-β-α folding at the C-terminus of P0 in mouse and shark. The slight expansion in the spacing may also be due in part to the higher molecular weight of shark MBP.

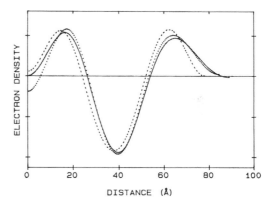

Figure 7. Electron density profiles, on a relative scale, calculated from diffraction data to ~35 Å spacing, for *Torpedo* optic nerve myelin (169 Å; *dotted curve*), rat optic nerve myelin (156 Å; *dashed curve*) and mouse sciatic nerve myelin (178 Å; *solid curve*).

Conclusions

The mechanism underlying membrane-membrane interactions in myelin is of considerable interest because tight apposition between the membranes may be related to the electrophysiological, insulating property of myelin. Phylogenetic studies of myelin by X-ray diffraction and biochemistry have shown that the extracellular apposition may be mediated either by P0 or PLP, and that a larger extracellular spacing is present in P0-containing myelin. The sequence of P0 has been found to be homologous to that of adhesion molecules, and they all share immunoglobulin-like folding. Our comparison between shark and rodent sequence data for myelin proteins reveals that P0 has a similar β propensity curve as the immunoglobulins. A proposed homophilic interaction between P0 molecules may mediate the extracellular apposition and is likely to be hydrophobic. Unlike P0, homologous sequences in other molecules with known structure and function have not yet been found for PLP and MBP. We have proposed that the conserved regions of MBP between shark and mouse may form a functional core; however, its folding is not known. Correlation of further structural and biochemical studies with sequence and homology analyses will clarify the molecular basis of myelin membrane adhesion.

Appendix: Sequence analysis

The distribution of chemical or physical parameter p for a protein having N residues was measured from the individual values of p for each amino acid averaged over a certain number of residues, or window length = $(2w + 1)$. The average value $<P_i>$ at the ith position is written as

$$<P_i> = \sum_j p(i+j), \text{ for } -w \le j \le w \text{ and } (1 + w) \le i \le (N - w).$$

The window size depends on the type of parameter, e.g., $w = 4$ for the hydrophobicity and charge parameters. The values of $<P_i>$ are plotted at the center of the window (Fig. 5a,c).

The periodic disposition of each parameter was determined from the Fourier transform (McLachlan & Stewart, 1976; Bear et al., 1978). The intensity function for the array of assigned parameters $(p_j$, where $k-w \le j \le k+w)$ at frequency ν is given by

$$I_k(\nu) = | \sum_j (p_j - <p_k>) \exp(2\pi i j\nu) |^2,$$

where p_j is parameter p at position j, $<p_k> = \sum_j p_j/(2w+1)$, and $i = \sqrt{-1}$. This intensity function was defined two dimensionally as a function of ν and residue position k. Finer-Moore & Stroud (1984) defined a similar expression when $w = 12$. Setting $2\pi\nu = \delta$ and replacing the exponential term by cosine and sine functions makes this intensity the same as the ones defined by Eisenberg et al. (1984b) and by DeLisi & Berzofsky (1985). The amphiphilicity power plot is calculated from the hydrophobicity scales as a function of residue position and frequency ranging from 0.1-0.5 at steps of 0.01. The intensity of the α-helix amphiphilicity was calculated as a function of residue number with $0.22 \le \nu \le 0.33$ (defined values for α-helix) and averaged to obtain the final curve (Fig. 5k). The residue position for the α-helical amphiphilicity was plotted when the highest intensity $I_k(\nu)$ was obtained in the frequency range 0.22-0.33. These positions may likely be T-cell antigenic sites (DeLisi & Berzofsky, 1985). Although the above studies used the hydrophobicity scales as the parameter, anything else can, in fact, be used (Kubota et al., 1981, 1982; Argos, 1987).

Establishing homologies in protein sequences has usually been based on the mutation data matrix (Dayhoff et al., 1983). In general, the homology between two different arrays of parameters can be evaluated by plotting the correlation coefficients two-dimensionally as a function of residue number. The correlation coefficient (Kubota et al., 1982) is

$$c(i,j) = \{ \sum_k (x(i+k) - <c>) \, (y(j+k) - <c>) \}/ [\{ \sum_k (x(i+k) - <c>)^2 \}\{ \sum_k (y(j+k) - <c>)^2 \}]^{\frac{1}{2}}$$

where $x(i)$ and $y(j)$ are the parameter values at positions i and j of arrays x and y (corresponding to the sequences for two different proteins), $<c>$ is an average parameter values for 20 amino acids, and $-w \leq k \leq w$. (A slightly different form of this equation was presented by Argos, 1987). In our text, however, the sequences were compared on an all-or-none basis (see Fig. 4).

Acknowledgements

We thank Dr. R. Saavedra for providing his manuscript before publication. This work was supported by NIH Grants NS 20824 from the National Institute of Neurological Disorders and Stroke, and was carried out in the Mental Retardation Research Center, Children's Hospital, which is supported by NIH Core Grant HD 06276. The computer analysis described in this paper was carried out in the Image Graphics Laboratory, Children's Hospital.

References

Argos P (1987) A sensitive procedure to compare amino acid sequences. J Mol Biol 193:385-396

Arquint M, Roder J, Chia L-S, Down J, Wilkinson D, Bayley H, Braun P, Dunn R (1987) Molecular cloning and primary structure of myelin-associated glycoprotein. Proc Natl Acad Sci 84:600-604

Bear RS, Adams JB, Poulton JW (1978) Disclosure by Fourier methods of a long-range pattern of non-polar residues in the $\alpha 1(I)$ sequence of collagen. J Mol Biol 118:123-126

Bürgisser P, Matthieu J-M, Jeserich G, Waehneldt TV (1986) Myelin lipids: a phylogenetic study. Neurochem Res 11:1261-1272

Campagnoni AT (1988) Molecular biology of myelin proteins from the central nervous system. J Neurochem 51:1-14

Carnegie PR (1971) Properties, structure and possible neuroreceptor role of the encephalitogenic protein of human brain. Nature 229:25-28

Carnegie PR, Moore WJ (1980) Myelin basic protein. In "Proteins of the Nervous System" (Bradshaw RA & Schneider DM, eds), pp. 119-143, Raven New York

Chou PY, Fasman GD (1978) Empirical predictions of protein conformation. Ann Rev Biochem 47:251-276

Cornette JL, Cease KB, Margalit H, Spouge JL, Berzofsky JA, DeLisi, C (1987) Hydrophobicity scales and computational techniques for detecting amphipathic structures in proteins. J Mol Biol 195:659-685

Cunningham BA, Hemperly JJ, Murray BA, Prediger EA, Brackenbury R, Edelman GM (1987) Neural cell adhesion molecule: structure, immunoglobulin-like domains, cell surface modulation, and alternative RNA splicing. Science 236:799-806

Dayhoff MO, Barker WC, Hunt LT (1983) Establishing homologies in protein sequences. Meth Enzymol 91:524-545

DeLisi C, Berzofsky JA (1985) T-cell antigenic sites tend to be amphipathic structures. Proc Natl Acad Sci USA 82:7048-7052

Diehl H-J, Schaich M, Budzinski R-M, Stoffel W (1986) Individual exons encode the integral membrane domains of human myelin proteolipid protein. Proc Natl Acad Sci USA 83:9807-9811

Edelman GM (1983) Cell adhesion molecules. Science 219:450-457

Eisenberg D, Schwarz E, Komaromy M, Wall R (1984a) Analysis of membrane and surface protein sequences with the hydrophobic moment plot. J Mol Biol 179: 125-142

Eisenberg D, Weiss RM, Terwilliger TC (1984b) The hydrophobic moment detects periodicity in protein hydrophobicity. Proc Natl Acad Sci USA 81:140-144

Finer-Moore J, Stroud RM (1984) Amphipathic analysis and possible formation of the ion channel in an acetylcholine receptor. Proc Natl Acad Sci USA 81:155-159

Ganser AL, Kerner A-L, Brown BJ, Davisson MT, Kirschner DA (1988) A survey of neurological mutant mice. I. Lipid composition of myelinated tissue in known myelin mutants. Dev Neurosci 10:99-122

Garnier J, Osguthorpe DJ, Robson B (1978) Analysis of the accuracy and implications of simple methods for predicting the secondary structure of globular proteins. J Mol Biol 120:97-120

Hanks SK, Quinn AM, Hunter T (1988) The protein kinase familiy: conserved features and deduced phylogeny of the catalytic domains. Science 241:42-52

Hofteig JH, Mendell JR, Yates AJ (1981) Chemical and morphological studies on garfish peripheral nerve. J Comp Neurol 198:265-274

Hudson LD, Berndt JA, Puckett C, Kozak CA, Lazzarini RA (1987) Aberrant splicing of proteolipid protein mRNA in the dysmyelinating *jimpy* mutant mouse. Proc Natl Acad Sci USA 84:1454-1458

Ikenaka K, Furuichi T, Iwasaki Y, Moriguchi A, Okano H, Mikoshiba K (1988) Myelin proteolipid protein gene structure and its regulation of expression in normal and Jimpy mutant mice. J Mol Biol 199:587-596

Inouye H, Kirschner DA (1988a) Membrane interactions in nerve myelin. I. Determination of surface charge from effects of pH and ionic strength on period. Biophys J 53:235-246

Inouye H, Kirschner DA (1988b) Membrane interactions in nerve myelin II. Determination of surface charge from biochemical data. Biophys J 53:247-260

Inouye H & Kirschner DA (1989) Orientation of proteolipid protein in myelin: comparison of models with X-ray diffraction measurements. Dev Neurosci 11:81-89

Jones TA, Bergfors T, Sedzik J, Unge T (1988) The three-dimensional structure of P2 myelin protein. EMBO J 7:1597-1604

Kirschner DA, Ganser AL (1982) Myelin labeled with mercuric chloride: asymmetric localization of phosphatidylethanolamine plasmalogen. J Mol Biol 157:635-658

Kirschner DA, Ganser AL, Caspar DLD (1984) Diffraction studies of molecular organization and membrane interactions in myelin. In "Myelin" (Morell P, ed), pp 51-95, Plenum NY

Kirschner DA, Inouye H, Ganser AL, Mann V (1989) Myelin membrane structure and composition correlated: a phylogenetic study. J Neurochem, in press

Kitamura K, Suzuki M, Suzuki A, Uyemura K (1980) The complete amino acid sequence of the P2 protein in bovine peripheral nerve myelin. FEBS Letters 115:27-30

Krebs EG, Beavo JA (1979) Phosphorylation-dephosphorylation of enzymes. Ann Rev Biochem 48:923-959

Kubota Y, Takahashi S, Nishikawa K, Ooi T (1981) Homology in protein sequences expressed by correlation coefficients. J Theoret Biol 91:347-361

Kubota Y, Nishikawa K, Takahashi S, Ooi T (1982) Correspondence of homologies in amino acid sequence and tertiary structure of protein molecules. Biochim Biophys Acta 701:242-252

Lai C, Brow MA, Nave K-A, Noronha AB, Quarles RH, Bloom FE, Milner RJ, Sutcliffe JG (1987) Two forms of 1B236/myelin-associated glycoprotein, a cell adhesion molecule for postnatal neural development, are produced by alternative splicing. Proc Natl Acad Sci 84:4337-4341

Laursen RA, Samiullah M, Lees MB (1984) The structure of bovine brain myelin proteolipid and its organization in myelin. Proc Natl Acad Sci USA 81: 2912-2916

Lemke G (1988) Unwrapping the genes of myelin. Neuron 1:535-543

Lemke G, Axel R (1985) Isolation and sequence of a cDNA encoding the major structural protein of peripheral myelin. Cell 40:501-508

Lemke G, Lamar E, Patterson J (1988) Isolation and analysis of the gene encoding peripheral myelin protein zero. Neuron 1:73-88

Lin L-F H, Lees MB (1984) Dicyclohexylcarbodiimide-binding sites in the myelin proteolipid. Neurochemical Res 9:1515-1522

Macklin WB, Campagnoni CW, Deininger PL, Gardinier MV (1987) Structure and expression of the mouse myelin proteolipid protein gene. J Neurosci Res 18:383-394

McLachlan AD, Stewart M (1976) The 14-fold periodicity in α-tropomyosin and the interaction with actin. J Mol Biol 103:271-298

Milner RJ, Lai C, Nave K-A, Lenoir D, Ogata J, Sutcliffe JG (1985) Nucleotide sequences of two mRNAs for rat brain myelin proteolipid protein. Cell 42:931-939

Moos M, Tacke R, Scherer H, Teplow D, Früh K, Schachner M (1988) Neural adhesion molecule L1 as a member of the immunoglobulin superfamily with binding domains similar to fibronectin. Nature 334:701-703

Narayanan V, Barbosa E, Reed R, Tennekoon G (1988) Characterization of a cloned cDNA encoding rabbit myelin P2 protein. J Biol Chem 263:8332-8337

Newman S, Kitamura K, Campagnoni AT (1987) Identification of a cDNA coding for a fifth form of myelin basic protein in mouse. Proc Natl Acad Sci USA 84:886-890

Padlan EA (1977) Structural basis for the specificity of antibody-antigen reactions and structural mechanisms for the diversification of antigen-binding specificities. Quart Rev Biophys 10:35-65

Pearson RB, Woodgett JR, Cohen P, Kemp BE (1985) Substrate specificity of a multifunctional calmodulin-dependent protein kinase. J Biol Chem 260:14471-14476

Puckett C, Hudson L, Ono K, Friedrich V, Benecke J, Dubois-Dalcq M, Lazzarini RA (1987) Myelin-specific proteolipid protein is expressed in myelinating Schwann cells but is not incorporated into myelin sheaths. J Neurosci Res 18:511-518

Rutishauser U (1984) Developmental biology of neural cell adhesion molecule. Nature 310:549-554

Saavedra RA, Fors L, Aebersold RH, Arden B, Horvath S, Sanders J, Hood L (1989) The myelin proteins of the shark brain are similar to the myelin proteins of the mammalian peripheral nervous system. J Mol Evol, in press

Saito S, Tamai Y (1983) Characteristic constituents of glycolipids from frog brain and sciatic nerve. J Neurochem 41:737-744

Sakamoto Y, Kitamura K, Yoshimura K, Nishijima T, Uyemura K (1987) Complete amino acid sequence of P0 protein in bovine peripheral nerve myelin. J Biol Chem 262:4208-4214

Santoni M-J, Goridis C, Fontecilla-Camps JC (1988) Molecular modeling of the immuno-globulin-like domains of the neural cell adhesion molecule (NCAM): implications for the positioning of functionally important sugar side chains. J Neurosci Res 20:304-310

Schiffer M, Chang C-H, Stevens FJ (1985) Formation of an infinite β-sheet arrangement dominates the crystallization behavior of λ-type antibody light chains. J Mol Biol 186:475-478

Sedzik J, Blaurock AE, Hoechli M (1985) Reconstituted P2/myelin-lipid multilayers. J Neurochem 45:844-852

Sternberg MJE, Taylor WR (1984) Modeling the ATP-binding site of oncogene products, the epidermal growth factor receptor and related proteins. FEBS Letters 175:387-392

Stoffel W, Hillen H, Giersiefen H (1984) Structure and molecular arrangement of proteolipid protein of central nervous system myelin. Proc Natl Acad Sci USA 81:5012-5016

Sutcliffe JG (1988) The genes for myelin revisited. Trends Gen 4:211-213

Vogel US, Thompson RJ (1988) Molecular structure, localization, and possible functions of the myelin-associated enzyme 2',3'-cyclic nucleotide 3'-phosphodiesterase. J Neurochem 50:1667-1677

Waehneldt TV, Malotka J (1989) Presence of proteolipid protein in coelancanth brain myelin demonstrates tetrapod affinities and questions a chondrichthyan association. J Neurochem 52:1941-1943

Waehneldt TV, Matthieu J-M, Jeserich G (1986) Appearance of myelin proteins during vertebrate evolution. Neurochem Int 9:463-474

Waehneldt TV, Matthieu J-M, Malotka J, Joss J (1987) A glycosylated proteolipid protein is common to CNS myelin of recent lungfish (*Ceratodidae, Lepidosirenidae*). Comp Biochem Physiol 88B:1209-1212

Williams AF (1987) A year in the life of the immunoglobulin superfamily. Immunol Today 8:298-303

COMPONENTS AND STRUCTURES
OF MYELIN

COMPOSITION AND METABOLISM OF MYELIN CEREBROSIDES AND GANGLIOSIDES

Robert K. Yu, Carmen Sato and Megumi Saito
Department of Biochemistry and Molecular Biophysics
Medical College of Virginia
Virginia Commonwealth University
Richmond, VA 23298-0614
U.S.A.

Introduction

Glycolipids are an important class of lipids found in the plasma membrane surface of virtually all vertebrate cells and are particularly abundant in the nervous system (Yu and Saito, 1989). The glycolipid composition is cell specific and is also known to undergo developmental changes as a result of cellular maturation. Recent studies from many laboratories have provided stong evidence that glycolipids not only play an important role in determining the properties of the cell, but also may participate in a variety of cellular functions such as recognition and adhesion. The present paper describes two aspects of myelin glycolipid metabolism, namely, the synthesis of galactocerebrosides and the catabolism of gangliosides with a special emphasis on the role of a newly discovered myelin-associated neuraminidase.

Galactocerebrosides

The myelin membrane is highly enriched in galactolipids, especially galactocerebrosides and their sulfated derivatives, sulfatides (Norton and Poduslo, 1973a). The last step in the galactocerebroside biosynthesis is catalyzed by the enzyme UDP-galactose:ceramide galactosyltransferase (CgalT) (Morrell and Radin, 1969). This enzyme activity rapidly increases in proportion with the rate of myelin formation and then declines after the peak of myelination (Brenkert and Radin, 1972; Costantino-Ceccarini and Morell, 1972). Although the myelin membrane contains 2-hydroxylated as well as non-hydroxylated fatty acid-containing cerebrosides (Nonaka and Kishimoto, 1979), it has not yet been determined whether one or two CgalT enzymes are involved in their synthesis and whether the two cerebroside species are synthesized at the same subcellular site.

The subcellular distribution of CgalT activities was analyzed in oligodendroglial cells (Sato et al., 1988). Oligodendroglia isolated from 18- to 19-day-old Sprague-Dawley rats was homogenized in 0.32 M sucrose and the total cell homogenate was layered on a gradient containing 0.8 M, 1.05 M and 1.3 M sucrose, and centrifuged at 100,000 g for 50 min. Three membrane fractions were collected at the 0.32-0.8 M (F1), 0.8-1.05 M (F2) and 1.05-1.3 M (F3) interfaces. After isolation, the membranes were fixed with 1% OsO_4 in 0.1 M sodium cacodylate, stained with uranyl acetate, and counterstained with lead citrate for electron microscipic observation. CgalT activities were measured by a modification of the procedure described by Cestelli et al. (1979). HFA-CgalT assay was carried out in an incubation medium containing 50 mM Tris-HCl (pH 8), 10 mM $MgCl_2$, 110 μM [^{3}H]UDP-gal (0.1 Ci/mmol) and 40 μl of a liposomal suspension (phosphatidylethanolamine:phosphatidylcholine:HFA-ceramides in a molar ratio of 1:6:0.6) in a final volume of 100 μl. The activity of NFA-CgalT was determined using a liposomal suspension containing phosphatidylcholine and NFA-ceramides in a molar ratio of 3.5:1. After incubation, the radioactivity was determined in HFA-and NFA-cerebrosides separated by HPTLC.

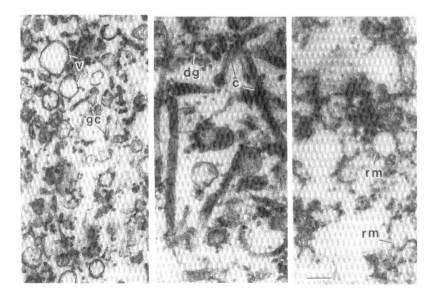

FIG 1. Electron micrographs of oligodendroglial membrane subfractions: **A:** F1 fraction. v, vesicular membrane structures; gc, Golgi cisternae. **B:** F2 fraction. c, rod-shaped structures; dg, dense granules. **C:** F3 fraction. rm, rough microsomes. Bar = 0.1 μm.

Electron micrographs of the oligodendroglial membrane fractions are shown in Fig. 1. Fraction F1 was composed primarily of vesicular membranous structures, and cisternae profiles with distended ends which are considered to be elements derived from the Golgi complex. Fraction F2 contained mainly vesicular and rod-shape structures and clusters of dense granules which according to their sizes were probably free ribosomes. Fraction F3 consisted almost entirely of membranes derived from rough endoplasmic reticulum. These fractions were further characterized by measuring the distribution of thiamine pyrophosphatase (TPP'ase) (Novikoff and Goldfischer et al. 1961; Farquhar et al., 1974), cerebroside sulfotransferase (CST) (Siegrist et al, 1979; Benjamins et al., 1982) and UDP-galactose:N-acetylglucosamine transferase (AGT) (Fleischer et al., 1969) as Golgi markers, and glucose-6-phosphatase (Glu-6-Pase) (Farquhar et al., 1974)) and 5'-nucleotidase (AMPase) (Touster et al., 1970) as markers for endoplasmic reticulum and plasma membranes respectively. As shown in Table 1, F1 was enriched in Golgi-related membranes. This fraction contained 40-50% of the total cell activities of TPP'ase, CST and AGT compared with ¯15% recovery of Glu-6-Pase and AMPase. Approximately 40% of the total cell activity of Glu-6-Pase was recovered in F2 in comparison with only 10 to 15% in the other two fractions.

TABLE 1. Distribution of marker enzymes in membrane subfractions for oligodendroglia.

	RSA				
Fraction	TPP'ase	AGT	CST	Glu-6-Pase	AMPase
F1	10.00	8.12	5.69	2.00	2.27
F2	2.00	2.45	1.65	5.00	0.98
F3	1.42	0.26	ND	2.16	0.99

The results are expressed as relative specific activity (RSA; specific activity in the fraction/specific activity in the whole cell homogenate). Data are mean $\pm$ SEM values from three different membrane preparations. ND, not detectable. Specific activities in the cell homogenate were as follows: TPP'ase, 11.6 nmol of Pi/mg of protein/min; AGT, 2.11 nmol of sugar transferred/mg of protein/min; CST, 2.3 pmol of sulfate transferred/mg of protein/min; Glu-6-Pase, 20 nmol of Pi/mg of protein/min; and AMPase, 2 nmol of Pi/mg of protein min.

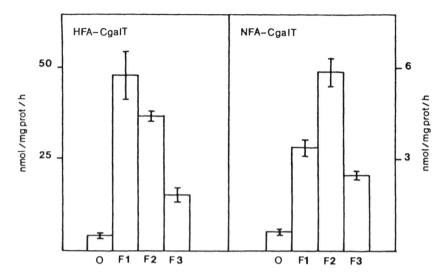

Fig. 2. Distribution of HFA-CgalT and NFA-CgalT activities in membrane subfractions from oligodendroglia. Data are the means ± SEM from 5-6 experiments. The protein content in each fraction was 29.45, 31.67 and 15 µg per mg protein of total cell homogenate for F1, F2 and F3, respectively. O: oligodendroglia.

Thus, F2 appeared to be a fraction enriched in membranes derived from endoplasmic reticulum and might also contain 15% of Golgi elements and less than 10% of plasma membranes.

Figure 2 shows the distribution of HFA- and NFA-CgalT in the membrane subfractions from oligodendroglia. Relative to the total oligodendroglial homogenate HFA-CgalT was enriched 12-fold with a recovery of 34% in F1 and 9-fold with a recovery of 28% in F2. Assay of NFA-CgalT indicated that the highest specific activity (10.5-fold higher than in the whole cells) and recovery (35%) was in fraction F2. The enrichment of NFA-CgalT in F1 was 6-fold with a recovery of 18%. Both enzymatic activities showed a smaller enrichment in fraction F3 which contained 5 and 7% of the total oligodendroglial activity of HFA- and NFA-CgalT, respectively. Thus, fractions F1 and F2 were similar in their enrichment of HFA-CgalT, but F2 was higher in its content of NFA-CgalT.

HFA- and NFA-CgalT activities were also investigated in the myelin membrane isolated by the method of Norton and Poduslo (1973b) as modified by Haley et al. (1981) and Yohe et al. (1983). The results obtained by measuring the galactosyltransferase activity with both hydroxylated and non-hydroxylated ceramides as the acceptor substrates, indicated that purified myelin and oligodendroglial cells had a similar HFA-CgalT activity (Fig. 3).

Interestingly, NFA-CgalT activity in myelin was about 8 times lower than in oligodendroglia.

We have thus investigated the distribution of the UDP-gal:ceramide galactosyltransferase activities in membrane subfractions from oligodendroglia and purified myelin membrane. The presence of CgalT activity has been reported not only in microsomal membranes but also in purified myelin (Costantino-Ceccarini and Suzuki, 1975). However, the information is only available for the HFA-CgalT activity. This is due to the fact that the CgalT was assayed using only HFA-ceramides as the acceptor substrates (Morell and Radin, 1969; Basu et al., 1969).

Subfractionation of the oligodendroglial cells in three membrane fractions resulted in an enrichment of the HFA-CgalT activity in fractions F1 and F2. These fractions also showed an enrichment of markers of Golgi and endoplasmic reticulum, respectively. In agreement with our results, a co-purification of HFA-CgalT activity and Golgi markers was reported after sufractionation of mouse brain (Siegrist et al., 1979) and rat forebrain, brainstem, and subcortial white matter (Benjamins et al., 1982; Costantino-Ceccarini et al., 1982), although the activity was not limited to the Golgi-enriched fraction (Benjamins et al., 1982). In contrast, Singh et al. (1986) found, after

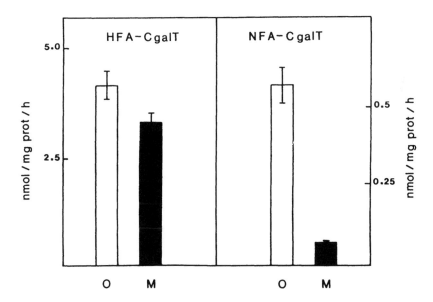

Fig. 3. HFA-CgalT and NFA-CgalT activities in oligodendroglia and myelin. Data are the means ± SEM from 4 to 5 experiments. O: oligodendroglia, M: myelin.

sufractionation of rat white matter, that a fraction enriched in smooth endoplasmic reticulum and Golgi membranes contained only 38% of the activity of HFA-CgalT in rough endoplasmic reticulum. The contradictory results may be due to the fact that the source of tissue was diffierent in each case, resulting in fractions that may contain membranes derived from different types of cells and even myelin contamination. Our results indicate that when non-hydrolated ceramides were used as the acceptor substrates, the oligodendroglia NFA-CgalT activity was highest in the F2 membrane fraction. In addition, oligodendroglia and myelin have a different proportion of both CgalT activities. In fact, the ratio of HFA-CgalT/NFA-CgalT activities in oligodendroglia was 7 compared to a value of 49 for the myelin fraction.

Thus, by using hydroxylated and non-hydroxylated ceramides as acceptor substrates it was possible to distinguish a different intracellular and myelin distribution for the HFA- and NFA-CgalT activities. These differences could not be attributed to different concentrations of endogenous substrates among the different membrane subfractions since for both enzymes, the assay was carried out in the presence of saturating levels of UDP-galactose and ceramides.

These observations raise two intriguing possibilities. First, in relation to the molecular identity of the galactosyltransferase activities, the data may suggest that the two activities may actually exist as two separate enzyme entities. Second, the different intracellular and myelin localization for the synthesis of hydroxylated and non-hydroxylated cerebrosides may result in a different mechanism for their intracellular translocation and probably a specific role for each cerebroside species in the process of myelination.

Gangliosides

The ganglioside composition of CNS myelin has been extensively examined in different species (Cochran et al., 1982) and during development (Suzuki et al., 1967; Ledeen et al., 1973; Yu and Yen, 1975; Cochran et al., 1983), and found to be relatively simple compared to those of neurons, astrocytes or other brain membranes (Yu and Saito, 1989). It also differs significantly from that of the plasma membrane of parent oligodendroglial perikarya. GM1, GD3 and GD1a are the major gangliosides of human oligodendroglia while human myelin contains GM4, GM1 and GD1b, the sum of them amounting to about 80% of the total (Yu and Iqbal, 1979). In rat and bovine brains, GD3 and GM3 are enriched in oligodendroglia but not in myelin (Yu et al., 1989). The ganglioside concentration of myelin continues to increase after active myelination in rat and mouse; and the composition becomes simpler with increasing age with monosialogangliosides predominant (Suzuki et al., 1967; Yu and Yen, 1975). The carbohydrate portion of GM1, including the sialic acid moiety, is replaced continuously by newly synthesized molecules even after the myelination period (Suzuki, 1970; Ando et al., 1984). These findings suggest that gangliosides in myelin and oligodendroglia may be metabolized actively during and after myelination.

Neuraminidase is an enzyme which cleaves sialic acid residues from sialoglycoconjugates and is assumed to be involved in their metabolism. Earlier studies have shown the presence of at least four different neuraminidases in brain tissues: two are soluble (Venerando et al., 1975, 1982) and two are membrane-bound, located in synaptosomal (Schengrund and Rosenberg, 1970) and lysosomal (Gielen and Harprecht, 1969) membranes. The existence of a neuraminidase activity intrinsic to myelin was first discovered by us when we examined interactions between gangliosides and myelin basic protein (MBP) (Yohe et al., 1983). In this study it was found that only GM4 specifically interacted with MBP, becoming more resistant to the action of <u>Clostridium perfringens</u> neuraminidase. On the other hand, MBP neither interacted with polysialogangliosides, nor protected them from the neuraminidase action. Based on these findings, we proposed that myelin might have a neuraminidase activity and it might be responsible for its unique ganglioside pattern having GM4 and GM1 as the major ganglioside components. Subsequently we presented evidence for the presence of a neuraminidase activity in purified rat myelin preparations. Thereafter, the existence of the neuraminidase activity in myelin was confirmed and further characterized using intact or delipidated myelin preparations (Yohe et al., 1986; Saito and Yu, 1986). The enzyme could effectively hydrolyze a non-ganglioside substrate, N-acetylneuramin(2-3)lactitol (NL) (NL-neuraminidase) and fetuin as well as GM3 (GM3-neuraminidase). Interestingly, the enzyme could, though at much slower rates, hydrolyze GM1 and GM2, which are usually resistant to bacterial and viral neuraminidases. On the other hand, GM1 acted as a competitive inhibitor for the hydrolysis of substrates such as GM3 or NL by the enzyme. Based upon time-activity curves on exposure of the enzyme preparation to high temperature or low pH, and Ki values obtained with 2,3-dehydro-2-deoxy-N-acetylneuraminic acid, a specific inhibitor for neuraminidase, it was strongly suggested that the enzyme activities toward the two substrates,

GM3 and NL, were catalyzed by a single enzyme entity.

We recently found that the myelin-associated neuraminidase activity could be effectively extracted from rat myelin with a 1% Triton X-100 solution containing 1 M ammonium acetate at neutral pH (Saito and Yu, 1986). Subsequently, the solubilization and partial purification of myelin-associated neuraminidase was attempted with a purified myelin preparation from rat brains (Saito and Yu, 1989). After delipidation of the myelin preparation with cold acetone and ethanol, successively, the enzyme activity was extracted with 1% Triton X-100-1 M ammonium acetate-10 mM Tris-HCl, pH 7.0, without considerable loss of the activity. The enzyme, however, was not soluble under low ionic strengths; dialysis of the enzyme extract produced a precipitate in which the enzyme activity was recovered. The extract was then applied to a phenyl-Sepharose column equilibrated with 0.2% Triton X-100-2.5 M ammonium acetate-10 mM Tris HCl, pH 7.0. The GM3-neuraminidase activity was eluted with the major peak of protein with continuous gradients of Triton X-100 (0.2 to 2.5%) and ammonium acetate (2.5 to 0 M) (Figure 4). At this point, the enzyme turned soluble under low ionic strengths. The enzyme fraction was dialyzed and applied to a CM-Sepharose column equilibrated with 1% Triton X-100-10 mM Tris-HCl, pH 7.0. When eluting with a continuous gradient of NaCl (0 to 0.7 M) in the buffer, two major peaks of the enzyme activity were obtained (fractions A and B) (Figure 5).

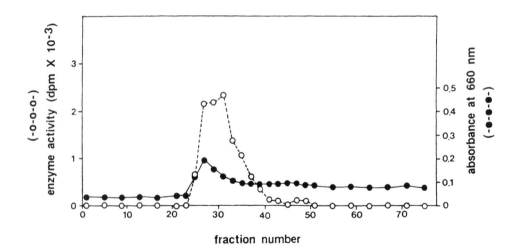

FIG. 4. Phenyl-Sepharose column chromatography of GM3-neuraminidase activity in rat CNS myelin. The protein concentration was measured by the Lowry's method (1951).

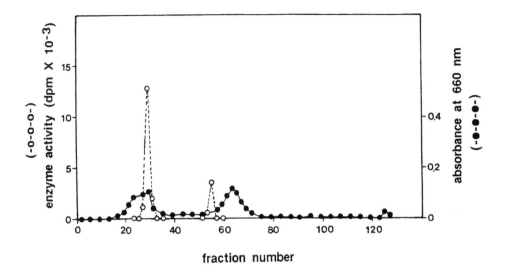

FIG 5. CM-Sepharose column chromatography of GM3-neuraminidase activity.

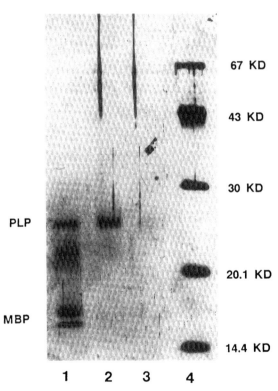

FIG 6. SDS-PAGE of protein fractions. The chromatogram was visualized by the silver straining method. 1, delipidated myelin; 2 and 3, fractions A and B from CM-Sepharose column chromatography; 4, standard protein mixture.

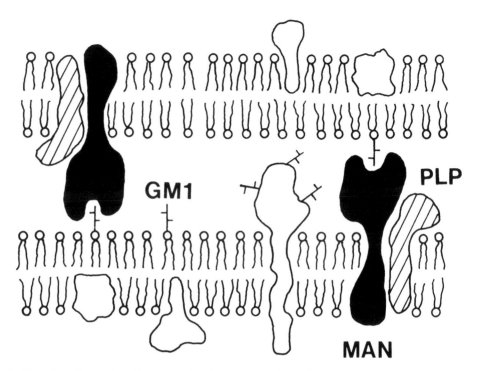

FIG. 7. A schematic diagram showing a possible interaction between myelin associated neuraminidase and GM1. MAN, myelin-associated neuraminidase; PLP, proteolipid protein.

The recovery of the neuraminidase activity in fractions A and B was 35 and 12% of the original activity, respectively. On SDS-PAGE, fraction A showed only one band having the same mobility of proteolipid protein (PLP). A faint protein band with a molecular weight of about 22,000 was observed in fraction B (Figure 6). Since there is evidence that myelin contains a single species of neuraminidase (Saito and Yu, 1986), this finding suggests that the two peaks of the enzyme activity may represent the associated and non-associated forms of the enzyme with a myelin component, most likely PLP. Based on the recovery of the enzyme activity and SDS-PAGE, the magnitude of purification of the enzyme was estimated to be over 1,600-fold for fraction B.

Although the functional role of myelin-associated neuraminidase is not known, two possibilities can be raised. First, the enzyme may regulate the turnover of sialoglycoconjugates in myelin and oligodendroglia. The enzyme may be involved in the compositional changes of gangliosides during and after the myelination period. The enzyme may also participate in the metabolism of sialoglycoproteins such as myelin-associated glycoprotein (MAG) and neural cell adhesion molecules (N-CAM). Second, myelin-associated neuraminidase may function as an adhesion molecule. As described above, myelin-associated neuraminidase can specifically interact with GM1 which is the major ganglioside of the myelin membrane. This interaction may strengthen the surface-to-surface interaction of the myelin membrane. Thus, myelin-associated neuraminidase degrades polysialogangliosides to GM1 and bind GM1 to effect the formation of a multilamellar structure (Figure 7). A similar type of hypothesis has been proposed for glycosylatransferases, which may be involved in intercellular adhesions of cells (Roseman, 1970). Studies are now in progress to assess these possibilities.

Acknowledgements
This work was supported by USPHS Grants NS-11853, NS-23102 and NS-26994.

REFERENCES

Allen J (1963) The properties of Golgi associated nucleoside diphosphatase. I. Cytochemical analysis. J Histochem Cytochem 11:529-541

Ando S, Tanaka Y, Ono Y, and Kon K (1984) Incorporation rate of GM1 ganglioside into mouse brain myelin: Effect of aging and modification by hormones and other compounds. In: Ledeen RW, Yu RK, Rapport MM, Suzuki K (eds) Ganglioside Structure, Function and Biochemical Potential. Plenum, New York, pp 241-248

Basu S, Schultz A, Basu M (1969) Enzymatic synthesis of galactocerebroside from ceramide. Fed Proc 28:540

Benjamins JA, Hadden T, Skoff RP (1982) Cerebroside sulfotransferase in Golgi-enriched fractions from rat brain. J Neurochem 38:233-241

Brenkert A, Radin NS (1972) Synthesis of galactosylceramide and glucosylceramide by rat brain: assay procedures and change with age. Brain Res 35:183-193

Burkart T, Siegrist HP, Herschkowitz NN, Wiesmann UN (1977) 3'-Phosphoadenylsulfate:galactosylceramide 3'-sulfotransferase. An optimized assay in homogenates of developing brain. Biochim Biophys Acta 483:303-311

Cestelli A, White FV, Costantino-Ceccarini E (1979) The use of liposomes as acceptors for the assay of lipid glycosyltransferases from rat brain. Biochim Biophys Acta 572:283-292

Cochran FB, Ledeen RW, Yu RK (1983) Gangliosides and proteins in developing chicken brain myelin. Dev Brain Res 6:27-32

Cochran FB, Yu RK, Ledeen RW (1982) Myelin gangliosides in vertebrates. J Neurochem 39:773-779

Costantino-Ceccarini E, Morell P (1972) Biosynthesis of brain sphingolipids and myelin accumulation in the mouse. Lipids 7:656-659

Costantino-Ceccarini E, Suzuki K (1975) Evidence for the presence of UDP-galactose ceramide galactosyltransferase in rat myelin. Brain Res 93:359-362

Costantino-Ceccarini E, Waehneldt TV, Ginalski H, Bergisser P, Reigner J Matthieu JM (1982) Distribution of lipid synthesizing enzymes, 2',3'-cyclic nucleotide 3'-phosphodiesterase, and myelin proteins in rat forebrain subfractions during developemnt. Neurochem Res 7:1-12

Farquhar MG, Bergeron JJM, Palade GE (1974) Cytochemistry of Golgi fractions prepared from rat liver. J Cell Biol 60:8-25

Fleischer B, Fleischer S, Ozawa H (1969) Isolation and characterization of Golgi membranes from rat liver. J Cell Biol 43:59-79

Gielen W, Harprecht V (1969) Die Neuraminidase-Aktivitat in einigen Regionen des Rindergehirns. Hoppe Seylers Z Physiol Chem 350:201-206

Haley JE, Samuels FG, Ledeen RW (1981) Studies of myelin purity in relation to axonal contaminations. Cell Mol Neurobiol 1:175-187

Hino Y, Asano A, Sato R, Shimizu S (1978) Biochemical studies of rat liver Golgi apparatus. I. Isolation and preliminary characterization. J Biochem (Tokyo) 83:909-923

Ledeen RW, Yu RK, Eng EF (1973) Ganglioside of human myelin: Sialosylgalactosylceramide (G7) as a major component. J Neurochem 21:829-839

Lowry OH, Rosebrough NJ, Farr AL, Randall RJ (1951) Protein measurement with the Folin phenol reagent. J Biol Chem 193:265-275

Morell P, Radin NS (1969) Synthesis of cerebroside by brain from uridine diphosphate galactose and ceramide containing hydroxy fatty acid. Biochemistry 8:506-512

Morre JD (1973) Isolation of Golgi apparatus. Methods Enzymol 22:130-197

Nonaka G, Kishimoto Y (1979) Levels of cerebrosides, sulfatides, and galactosyl diglycerides in different regions of rat brain. Change during maturation and distribution in subcellular fractions of gray matter of sheep brain. Biochim Biophys Acta 572:432-441

Norton WT, Poduslo SE (1973a) Myelination in rat brain: changes in myelin composition during maturation. J Neurochem 21:759-773

Norton WT, Poduslo SE (1973b) Myelination in rat brain: method of myelin isolation. J. Neurochem 21:749-757

Novikoff AB, Goldfischer S (1961) Nucleoside-diphosphatase activity in the Golgi apparatus and its usefulness for cytochemical studies. Proc Natl Acad Sci USA 47:802-810

Roseman S (1970) The synthesis of complex carbohydrates by multiglycosyltransferase system and their potential function in intercellular adhesion. Chem Phys Lipids 5:270-297

Saito M, Yu RK (1986) Further characterization of a myelin-associated neuraminidase: Properties and substrate specificity. J Neurochem 47:632-641

Saito M, Yu RK (1989) Purification of a myelin-associated neurminidase in rat brain. J Neurochem 52:S197

Sato C, Black JA, Yu RK (1988) Subcellular distribution of UDP-galactose: ceramide galactosyltransferase in rat brain oligodendroglia. J Neurochem 6:1887-1893

Schengrund C-L, Rosenberg A (1970) Intracellular location and properties of bovine brain sialidase. J Biol Chem 245:6196-6200

Siegrist HP, Burkart T, Wiesmann, UN, Herschkowitz NN (1979) Ceramidegalactosyltransferase localization in Golgi-membranes isolated by a continuous sucrose graident of mouse brain microsomes. J Neurochem 33:497-504

Singh I, Nolan CE, Hovious J, Figlewitz D, Jungalwala FB (1986) Synthesis and transport of cerebrosides in ER-Golgi complex. Trans Am Soc Neurochem 17:443

Suzuki K (1970) Formation and turnover of myelin gangliosides. J Nerochem 17:209

Suzuki K, Poduslo SE, Norton WT (1967) Ganglioside in the myelin fraction of developing rats. Biochim Biophys Acta 24:604-611

Touster O, Aronson NN, Dulaney JR Jr, Henrickson H (1970) Isolation of rat liver plasma membranes. Use of nucleotide pyrophosphatase and phosphodiesterase I as marker enzymes. J Cell Biol 47:604-618

Venerando B, Goi GC, Preti A, Fiorilli A, Lombordo A, Tettamanti G (1982) Cytosolic sialidase in developing rat forebrain. Neurochem Int 4:313-320

Venerando B, Tettamanti G, Cestaro B, Zambotti V (1975) Studies on brain cytosol neuraminidase. I. Isolation and partial characterization of two forms of the enzyme from pig brain. Biochim Biophys Acta 403:461-472

Yohe HC, Jacobson RI and Yu RK (1983) Ganglioside-basic protein interaction: protection of gangliosides against neuraminidase action. J Neurosci Res 9:401-412

Yohe HC, Saito M, Ledeen RW, Kunishita T, Sclafani JR, Yu RK (1986) Further evidence for an intrinsic neuraminidase in CNS myelin. J Neurochem 46:623-629

Yu RK, Iqbal K (1979) Sialosylgalactosylceramide as a specific marker for human myelin and oligodendroglia: Gangliosides of human myelin, oligodendroglia and neurons. J Neurochem 32:293-300

Yu RK, Saito M (1989) Structure and localization of gangliosides. In: Margolis RU, Margolis RK (eds) Neurobiology of glycoconjugates. Plenum, New York, pp 1-42

Yu RK, Yen SI (1975) Ganglioside in developing mouse brain myelin. J Neurochem 25:223-232

Yu, RK, Macala LJ, Farooq M, Sbaschnig-Agler M, Norton WT, Ledeen RW (1989) Ganglioside and lipid composition of bulk-isolated rat and bovine oligodendroglia. J Neurosci Res 23:136-141.

RECEPTOR ACTIVITY AND SIGNAL TRANSDUCTION IN MYELIN

J.N. Larocca, F. Golly, M.H. Makman, A. Cervone and R.W. Ledeen
Departments of Neurology, Biochemistry and Molecular Pharmacology
Albert Einstein College of Medicine
1300 Morris Park Avenue
Bronx, New York 10461
U.S.A.

INTRODUCTION

External signals detected by surface receptors are translated into a limited repertoire of intracellular second messengers. Important examples of these are inositol 1,4,5 trisphosphate (IP_3) and diacylglycerol, which are produced by phospholipase C-dependent hydrolysis of phosphatidylinositol 4,5-biphosphate (PIP_2) (Berridge, 1984; Nishizuka, 1984). These second messengers, through release of internal Ca^{2+} and activation of protein kinase C, set in motion a cascade of biochemical processes with major consequences to a wide variety of cell types (Berridge, 1986). The large pool of polyphosphoinositides present in myelin (Eichberg and Hauser, 1973) was not previously considered a source of second messengers. However, reports on the presence of enzymes such as phosphoinositide phosphodiesterase (Deshmukh et al, 1982) and protein kinase C (Murray and Steck, 1986) in myelin prompted us to examine this membrane for the presence of receptors which can be linked to the phosphoinositide cycle.

We have found evidence for the presence in myelin of muscarinic cholinergic receptors (Larocca et al, 1987a) and the interaction of these receptors with phospholipase C (Larocca et al, 1987b) and the adenylate cyclase system (Larocca et al, 1987a). In addition our current work has revealed the presence in this membrane of guanine nucleotide regulatory (G) proteins.

Abbreviations:

IP inositol phosphate; IP_2 inositol bisphosphate; IP_3 inositol trisphosphate; PI phosphatidylinositol; PIP phosphatidylinositol phosphate; PIP_2 phosphatidylinositol bisphosphate; DAG diacylglycerol; NMS N-methylscopolamine; QNB quinuclidinyl benzilate; Gpp(NH)p 5'guanylylimidodiphosphate; GTPγS guanosine 5'[α-thio]triphosphate; MBP, myelin basic protein; CNP, 2',3'-cyclic nucleotide 3'-phosphodiesterase.

MATERIALS AND RESULTS

Muscarinic receptor binding in rat CNS myelin.

The presence of muscarinic receptors in CNS myelin was tested with radiolabeled N-methylscopolamine (NMS), quinuclidinyl benzilate (QNB) and pirenzepine. As shown in table I, muscarinic binding to highly purified myelin was considerable, representing 24-38 % of that found in whole brain stem. On the other hand, receptors for several other radioligands which

TABLE 1. Binding of radioligands to purified myelin and whole brain stem membranes.

Radioligand (Receptor Type)	Specific Binding (fmoles/mg protein)	
	Myelin	Brain Stem
^{3}H-p-Aminoclonidine (1 nM) (Alpha$_2$-Adrenergic)	0 (N.D.)	18.4
^{3}H-WB-4101 (1 nM) (Alpha$_1$-Adrenergic)	0 (N.D.)	44.4
^{3}H-Spiroperidol (0.6 nM) (Dopaminergic[D$_2$]-Serotonergic [5-HT$_2$])	0 (N.D.)	23.5
^{3}H-Dala-D-Leu-Enkephalin (7 nM) (Opioid)	0 (N.D.)	13.7
^{3}H-DPAT (1.2 nM) (Serotonergic (5-HT$_{1a}$)	0 (N.D.)	7.1
^{3}H-NMS (0.8 nM) (Muscarinic [M$_1$ + M$_2$])	40.2 $\pm$ 2.2	166 $\pm$ 14
^{3}H-NMS (2.8 nM) (Muscarinic [M$_1$ + M$_2$])	63.1 $\pm$ 6.4	264 $\pm$ 8
^{3}H-QNB (2.5 nM) (Muscarinic [M$_1$ + M$_2$])	101	286
^{3}H-Pirenzepine (5.0 nM) (Muscarinic [M$_1$])	9.5 $\pm$ 2.3	25.0 $\pm$ 1.4

Myelin was purified from rat brain stem. Values are means $\pm$ S.E.M. where indicated. N.D. - Not Detected. Data reproduced from Larocca et al. (1987a) with permission.

were present in whole brain stem could not be detected in myelin. These included alpha$_1$- and alpha$_2$-adrenergic, serotonergic (5-HT$_2$ and 5-HT$_{1a}$), dopaminergic (D$_2$), and opioid receptor binding sites.

The results of saturation studies for the NMS binding sites are shown in Figure 1. Binding was monophasic in brain stem. However, the Scatchard plot of NMS binding in myelin was curvilinear, suggesting the presence of a small number of very high affinity sites in addition to the major component. Brain stem showed only the latter type. Binding data for NMS as well as pirenzepine, a putative antagonist for the M$_1$ receptor subtype, are summarized in Table 2.

Competition studies were carried out to further characterize the NMS binding sites. The muscarinic antagonists atropine and N-methylscopolamine exhibited higher potency than did the agonists oxotremorine,

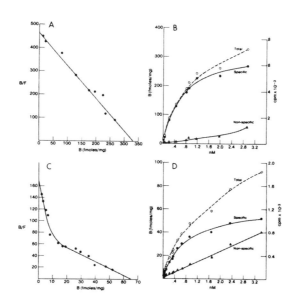

Figure 1. Saturation of ^{3}H-NMS binding sites in brain stem (panels A and B) and in myelin purified from brain stem (panels C and D). A and C are Scatchard plots for saturation data shown respectively in B and D (O---O, total binding; ●---●, specific binding; ▲---▲, non-specific binding). The data represent mean values for the separate experiments summarized in Table 2. Reproduced from Larocca et al. (1987a) with permission.

TABLE 2. Summary of B_{max} and K_D values for radioligand binding to muscarinic receptors in brain stem and purified myelin from brain stem.

Ligand and Preparation	B_{max} (fmol/mg protein)	K_D (nM)
^{3}H-NMS		
Myelin		
Total	73 ± 11 (6)	
(I)	3.5 ± 1.1 (4)	0.030 ± 0.011 (2)
(II)	69 ± 5 (4)	0.75 ± 0.08 (4)
Brain Stem	311 ± 18 (4)	0.54 ± 0.09 (4)
^{3}H-Pirenzepine		
Myelin	18.9 ± 4.2 (3)	4.6 ± 0.9 (3)
Brain stem	75.7 ± 1.3 (3)	6.8 ± 0.9 (3)

Values are means ± SEM obtained from Scatchard analyses of separate saturation experiments (number of experiments in parentheses). Reproduced from Larocca et al (1987a) with permission.

carbachol and acetylcholine. Addition of 100 μM Gpp(NH)p (a non-hydrolysable analog of GTP) to the assay system caused a considerable decrease in the affinity of the agonists. This effect indicated that at least some of these receptors may be coupled to a second messenger system through a G-protein.

Forskolin-stimulated adenylate cyclase activity was also shown to be present in myelin and to be significantly inhibited by carbachol and other muscarinic agonists; in addition, that inhibition was blocked by atropine (Larocca et al, 1987a). These findings indicated the presence in myelin of the M_2 subtype of muscarinic receptor. As mentioned above, the M_1 subtype was suggested by the presence of pirenzepine binding sites. More direct evidence of muscarinic receptor linkage to phospholipase C came from the following studies of phosphoinositide metabolism in myelin.

Carbachol-induced hydrolysis of myelin phosphoinositides: use of brain stem slices.

In these studies, rat brain stem slices labeled _in vivo_ by intracerebral injection of [^{3}H]inositol were incubated with carbachol and changes measured in the inositol phosphates and myelin phosphoinositides. Analysis of the water soluble products (Figure 2A) showed increases in inositol mono- and bisphosphate (IP, IP$_2$) but not inositol trisphosphate (IP$_3$) which was presumably hydrolyzed. Myelin phosphoinositides all showed significant losses in radioactivity (ca 20-38%). These hydrolytic reactions were blocked by 10 uM atropine, (Figure 2B) indicating that the carbachol effect is mediated by a muscarinic receptor.

The increases observed in the inositol phosphates are thought to result from cholinergic stimulation of diverse cells in the brain stem. However, the fact that myelin phosphoinositides showed significant loss of radioactivity is suggestive of effector-mediated phosphoinositide hydrolysis in the myelin membrane itself.

Stimulation of phosphoinositide hydrolysis in isolated CNS myelin.

More direct evidence was obtained using isolated myelin (Golly et al, 1987). In this case myelin labeled _in vivo_ through intracerebral injection of [^{3}H]inositol was stimulated with Gpp(NH)p. This resulted in hydrolysis of polyphosphoinositides, as seen in the enhanced liberation of inositol phosphates (Table 3). When GTPγS was substituted for Gpp(NH)p, similar results were obtained (not shown). These agents by-pass the receptor and stimulate G-proteins directly. The results strongly suggest the presence of a G-protein linked to phospholipase C in myelin.

Studies of myelin G-proteins.

Our current efforts are directed toward the isolation and characterization of G-proteins of myelin. These were quantified in purified bovine CNS myelin by measuring [^{35}S]GTPγS-binding according to Evans et al (1986). The isolated myelin showed as much as 50-60% of the [^{35}S]-GTPγS binding present in the total homogenate of white matter.

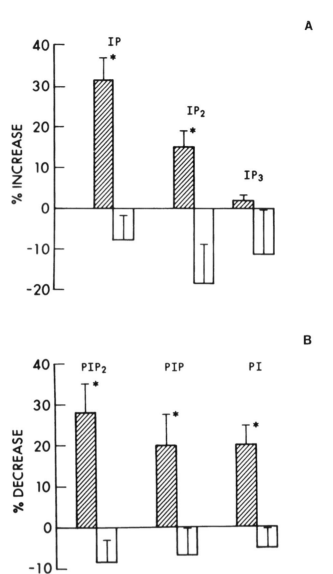

Figure 2. Effects of carbachol and carbachol plus atropine on inositol phosphates (A) and myelin phosphoinositides (B). Brain stem slices previously labeled in vivo with [^{3}H]inositol were incubated 1 h at 37°C with either added buffer (control), carbachol (2 mM) or carbachol (2 mM) plus atropine (10 μM). The atropine was added 5 min before stimulation with carbachol. Results are expressed as % of change with respect to control. Each bar in the histogram represents the mean ± SEM of 4-8 experiments. Open bars represent atropine experiments. *p < 0.05 (Student's two-tailed t-test). Reproduced from Larocca et al. (1987b) with permission.

TABLE 3. Enhancement of phospholipase C in myelin by Gpp(NH)p.

	IP	IP_2	IP_3	$IP_2 + IP_3$
Control	38 ± 5	240 ± 21	100 ± 4	340 ± 30
Gpp(NH)p	50 ± 10	$404 \pm 29*$	$146 \pm 2*$	$550 \pm 44*$

Rats were given intracerebral injections of [^{3}H]myoinositol (60 μCi/rat) and 16-18 h later sacrificed for myelin isolation from pooled brainstems. The purified myelin was incubated 15 min in buffered medium (Tris-HCl, 25 mM, pH 7.5) containing: $CaCl_2$, 1 μM; $MgCl_2$, 5 mM; LiCl 10 mM; leupeptin, 10 μg/ml; antipain 10 μg/ml; DTT, 2 mM; Gpp(NH)p, 1 μM. Controls lacked Gpp(NH)p. The supernatant was analyzed for inositol phosphates as described (Wreggett and Irvine, 1987). Values are DPM per mg protein; analyses were in duplicate. *p < 0.01 (Student's two-tailed t-test). n = 3.

Saturation studies indicated the presence in myelin of at least two populations of GTPγS binding sites with different affinities (Kd 15 and 100 nM; B_{max} 100 and 50 pmol/mg protein, respectively).

We were able to solubilize and partially purify some of these proteins by a procedure involving extraction of myelin with sodium cholate, application to a phenyl sepharose column, and elution with a linear gradient of 0.25-4% (w/v) sodium cholate/500-50 mM NaCl in TED buffer (20 mM Tris-HCl, pH 8.0; 1 mM EDTA; 1 mM DTT) buffer. Assaying each fraction for [^{35}S]-GTPγS binding activity, at least three major peaks were observed (Figure 3). Fractions corresponding to the second major peak were pooled, concentrated by pressure filtration and applied to a DEAE-sephacel column. This was washed with TED buffer containing 0.6% (w/v) CHAPS and eluted with a linear gradient of 0-300 mM NaCl in TED/0.6% (w/v) CHAPS. GTPγS binding activity was resolved into two well-separated peaks (Figure 4). SDS-PAGE showed the presence of two major bands (Mr 40 and 43 kDa) in the fractions corresponding to the second peak (IIb) (Figure 5). These are tentatively identified as Gp (or Gi) and Gs, since they were ADP-ribosylated with pertussis and cholera toxin, respectively.

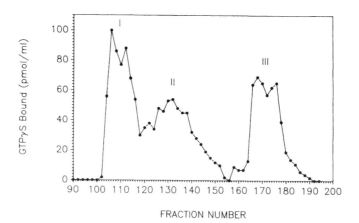

Figure 3. Phenyl sepharose chromatography of myelin cholate extract. The proteins were eluted with a linear gradient of 500-50 mM NaCl/0.25-4% (w/v) Na cholate in TED buffer. Fifty μl (out of 4 ml) from every second tube was assayed for [^{35}S]GTPγS binding activity.

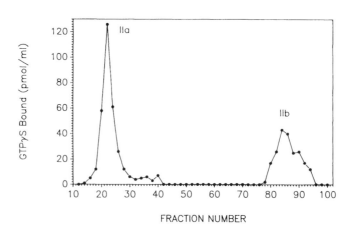

Figure 4. DEAE-Sephacel chromatography of peak I from phenyl sepharose. The proteins were eluted with a linear gradient of 0-300 mM NaCl in TED/0.6% (w/v) CHAPS. Fifty μl (out of 4 ml) from every second tube was assayed for [^{35}S]GTPγS finding activity.

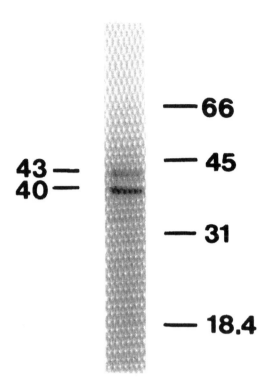

Figure 5. SDS/PAGE of peak IIb from DEAE-sephacel. Proteins were detected with Coomassie blue stain. Numbers at the left side of the photograph indicate the apparent molecular weight in kDa of the proteins described in the text. Numbers at the right side represent the molecular weight of protein standards in kDa.

DISCUSSION

Molecular cloning studies have shown that the muscarinic activity of acetylcholine can be mediated by at least five types of muscarinic receptor. (Liao et al, 1988). Although it is believed by some that the M_1 receptor is a stimulator of phospholipase C and the M_2 receptor an inhibitor of adenylate cyclase, the relative involvement of the individual acetylcholine receptors in each of the above mentioned actions is still controversial. The possibility that at least one of the muscarinic receptors present in myelin is linked to phophoinositide metabolism within that membrane is strongly suggested by the above-mentioned brain stem slice experiments in which carbachol liberated inositol phosphates with increased label while myelin phosphoinositides showed decreased label (Larocca et al, 1987a). This may be further correlated with the presence of pirenzepine binding sites (Larocca et al, 1987a), a putative indicator of the M_1 receptor subtype . Perhaps the strongest evidence for the presence of the phosphoinositide mechanism in myelin is the fact that treatment of isolated myelin with Gpp(NH)p or GTPγS significantly increased the amounts of liberated inositol phosphates.

The particular G-protein which links the muscarinic receptor to phospholipase C has been termed Gp (Cockroft and Taylor, 1987). We have tentatively identified the α subunit of this protein in myelin as the 40 kDa protein detected by SDS-PAGE of fraction IIb from DEAE-Sephacel chromatography (Figure 5). This is based on the observation that pertussis toxin caused ADP-ribosylation of this protein (during treatment of whole myelin) while also blocking phospholipase C activity. G-proteins in general have a heterotrimeric structure with β and γ subunits that are similar or identical in all cases. The β subunit can be resolved into a doublet of about 35 and 36 kDa while the γ subunit migrates at about 8 kDa. It is the α subunits that are distinctive and specify activity in regard to such functions as adenylate cyclase activation (Gsα, Mr 43-45 kDa) or inhibition (Giα, Mr 41 kDa). The well-known ADP-ribosylation reaction occurs on this subunit and this is the protein we have focused on in our isolation procedure. It may be noted that the principal Gα protein of brain has Mr 39-40 kDa and is ADP-ribosylated by pertussis toxin, similar to the 40 kDa Gα protein we have found in myelin.

In addition to our own work, other reports have begun to appear in the literature documenting the presence of G-proteins in myelin. Eichberg and coworkers (1989) describe pertussis toxin-catalyzed ADP ribosylation of a 40 kDa protein in sciatic nerve myelin. Use of 8-azido GTP and labeled GTP (or GTPγS) also revealed GTP-binding proteins in CNS myelin (Chan et al. 1988). Somewhat surprising was the fact that this latter included myelin basic protein (or a band comigrating with it). A more recent study by Boulias and Moscarello (1989) indicated the latter band as well as a 43- and 46 kDa protein are ADP-ribosylated by cholera toxin. This agrees with a brief (abstract) report by (Bernier et al. 1989) indicating that in addition to ADP-ribosylation of MBPs by cholera toxin, the myelin enzyme CNP is ADP-ribosylated by pertussis toxin. Supporting the concept of cholinergic activation of phosphoinositide breakdown was the study of Kahn and Morell (1988) showing that acetylcholine stimulated incorporation of ^{32}P into the inositol phospholipids of myelin.

Questions can be raised concerning the source of the cholinergic signal which would activate the myelin muscarinic receptors. In that regard the study of Vizi et. al. (1978), which indicated that acetylcholine release can occur nonsynaptically from axons through depolarization, may be revelant. Another possibility to consider is transmitter action at a distance, e.g. diffusion of these transmitters over distances to bind to receptors in the parasynaptic mode (Herkenham, 1987). Whatever the source of stimulus, the presence in myelin of muscarinic receptors gives this multilamellar structure the capacity to respond to its environment (especially the axon), in a manner not previously recognized.

ACKNOWLEDGMENTS

This work was supported by NIH grants AG 05554 (M.H.M.), NS-16181 (R.W.L.) and National Multiple Sclerosis Society grant RG 1941-A-1 (J.N.L.).

REFERENCES

Bernier L, Horvath E, and Braun P (1989) Binding proteins in CNS myelin. Am Soc Neurochem 20:254.
Berridge MJ (1984) Inositol triphosphate and diacylglycerol as second messengers. Biochem J 220:345-360.
Berridge MJ (1986) Growth factors, oncogenes and inositol lipids. Cancer Surveys 5:413-430.
Boulias C, and Moscarello MA (1989) Guanine nucleotides stimulate hydrolysis of phosphatidyl inositol bis phosphate in human myelin membranes. Biochem Biophys Res Commun 162:282-287.
Chan KC, and Moscarello MA (1988) Myelin basic protein binds GTP at a single site in the N-terminus. Biochem Biophys Res Commun 152: 1468-1473.

Cockcroft S, and Taylor JA (1987) Flurooaluminates mimic guanosine 5'-[α-thio]triphosphate in activating the polyphosphoinositide phosphodiesterase of hepatocyte membranes. Role for the guanine nucleotide regulatory protein Gp in signal transduction. Biochem J 241:409-414.

Deshmukh DS, Kuizon S, Bear WD, and Brockerhoff H (1982) Polyphosphoinositide mono- and diphosphoesterases of these subfractions of rat brain. Neurochem Res 7:617-626.

Eichberg J, and Hauser G (1973) The subcellular distribution of polyphosphoinositides in myelinated and unmyelinated rat brain. Biochim Biophys Acta 326:210-223.

Eichberg J, Berti-Mattera LN, Day S-F, Lowery J, and Zhu X (1989) Basal and receptor-stimulated metabolism of polyphosphoinositides in peripheral nerve myelin. J Neurochem 52:S24A.

Evans T, Brown ML, Fraser ED, and Northup JK (1986) Purification of the major GTP-binding proteins from human placental membranes. J Biol Chem 261:7052-7059.

Golly F, Larocca JN, and Ledeen RW (1987) Phosphoinositide metabolism linked to muscarinic receptor in purified myelin. Soc Neurosci Abstr 13:727.

Herkenham M (1987) Mismatches between neurotransmitter and receptor localizations in brain: Observations and implications. Neuroscience 23:1-38.

Kahn DW, and Morell P (1988) Phosphatidic acid and phosphoinositide turnover in myelin and its stimulation by acetylcholine. J Neurochem 50:1542-1550.

Larocca JN, Ledeen RW, Dvorkin B, and Makman MH (1987a) Muscarinic receptor binding and muscarinic receptor-mediated inhibition of adenylate cyclase in rat brain myelin. J Neurosci 7:3863-3876.

Larocca JN, Cervone A, and Ledeen RW (1987b) Stimulation of phosphoinositide hydrolysis in myelin by muscarinic agonist and potassium. Brain Res 436:357-362.

Liao C-F, Themmen APN, Joho R, Barberis C, Birnbaumer M, and Birnbaumer L (1988) Molecular cloning and expression of a fifth muscarinic acetylcholine receptor. J Biol Chem 264:7328-7337.

Murray N, and Steck AJ (1986) Activation of myelin protein kinase by diacylglycerol and 4-phorbol 12-myristate 13-acetate. J Neurochem 46:1655-1657.

Nishizuka Y (1984) Turnover of inositol phospholipids and signal transduction. Science 225:1365-1370.

Vizi ES, Gyires K, Somogyi GT, and Ungvary G (1983) Evidence that transmitters can be released from regions of the nerve cell other than presynaptic axon terminal: Axonal release of acetylcholine without modulation. Neuroscience 10:967-972.

Wreggett KA, and Irvine RF (1987) A rapid separation method for inositol phosphates and their isomers. Biochem J 245:655-660.

Myelin Oligodendrocyte Glycoprotein – a Model Target Antigen for Antibody Mediated Demyelination.

S. Piddlesden and C. Linington
Section of Neurology
University of Wales College of Medicine
Heath Park
Cardiff CF4 4XN
United Kingdom

Research on the immunological properties of the myelin membrane have been dominated for the past twenty years by studies on the major protein components of the central nervous system (CNS) membrane, myelin basic protein (MBP) and proteolipid protein (PLP). There is now, however, an increasing interest in the role of the immune response to quantitatively minor protein components of the myelin sheath in the pathogenesis of autoimmune mediated demyelination.

The ability of CNS antigens to provoke a demyelinating autoantibody response following immunisation with CNS tissue in complete Freunds adjuvant (CFA) has been known for many years (Bornstein and Appel, 1961). However, only in the last decade has the dibersity of myelin antigens that may act as targets for a pathogenic demyelinating antibody response been recognised.

Potentially, any structure exposed at the external surface of the myelin sheath may act as a target for antibody mediated demyelination. This includes myelin proteins with extracellular domains, such as the myelin associated glycoprotein (MAG), as well as other cell surface proteins and glycolipids. However, in reality the constraints imposed by Ir gene control, immune suppressor mechanisms and antigen processing limit the repertoire of demyelinating antibody responses that any individual/species may mount. Thus, polyclonal antisera raised against purified PLP and MAG do not exhibit demyelinating activity in vitro (Seil and Agrawal, 1980; Seil et al., 1981), although both these proteins are predicted to have extracellular domains which may be exposed at the external surface of the membrane (Salzer et al., 1987; Hudson et al., 1989). Similarly, although a large proportion of the in vitro demyelinating activity of rabbit anti-CNS or anti-myelin anti-sera can be removed by immunosorption with galactosylceramide (GC: Raine et al., 1976), this is not the case in the guinea pig (Lebar et al., 1976). In this species, immunosorption and biochemical studies tentitively identified the antigen responsible for initiating a complement-dependant demyelinating antibody response as a myelin protein designated M2 (Lebar et al.,1979).

Subsequently, a myelin glycoprotein, MOG – the myelin oligodendrocyte glycoprotein, characterised by a mouse monoclonal antibody (mAb 8-18C5: Linington et al., 1984) was shown to have biochemical characteristics similar to those of the M2 antigen (Lebar et al., 1986). Independantly, the importance of MOG in the pathogenesis of autoimmune mediated demyelination was confirmed by the strong correlation of the anti-MOG antibody titer with the demyelinating activity of sera from guinea pigs immunised with autologous CNS tissue (Linington and Lassmann,1987).

Serendipitiously, the original MOG-specific mouse monoclonal antibody (mAb 8-18C5) also exhibited potent demyelinating activity in vivo following systemic injection into rats with acute T cell mediated EAE (Linington et al.,1988). In this model the immunopathology of antibody mediated demyelination can be studied in the context of inflammatory CNS lesions similar to those see in the human disease multiple sclerosis (MS) (Lassmann et al.,1988). This paper provides a synopsis of our present understanding of the cellular biology of MOG and the role of this antigen in the immunopathology of antibody mediated demyelination in vivo.

Isolation and Characterisation of MOG.
Immunoaffinity purification of MOG from deoxycholate extracts of bovine myelin provides a crude preparation of the antigen with a yield of approximately

3 mg/g myelin protein. Analysis by SDS-PAGE revealed that this preparation consisted of one major Coomassie Blue stained band of apparent molecular weight 55 kDa, together with bands at 100-150 kDa and three components between 20-30 kDa (Figure 1). As previously described, the mAb 8-18C5 recognised both the major component at 55 kDa as well as a minor component migrating just behind PLP with an apparent molecular weight of approximately 26 kDa (Linington et al., 1984). immunoaffinity purification of MOG in the absence of protease inhibitors resulted in a significant loss of the immunoreactive band at 55 kDa and a corresponding increase in the 26 kDa component suggesting the latter is a degradation product. On some SDS-PAGE gels the 55 kDa component was resolved into a closely spaced triplet of mAb 8-18C5 reactive proteins. The significance of this observation is not yet clear.

Western blotting identified the high molecular weight contaminants present in the MOG preparations as mouse IgG eluted from the immunoaffinity column, whilst an anti-PLP peptide anti-serum identified the two remaining lower molecular weight bands as PLP and DM-20 (Figure 1). Attempts to eliminate this contaminating PLP/DM-20 from the MOG preparations have until now proved unsuccessful. Strategies such as washing MOG bound to the mAb 8-18C5 immunoaffinity columns with Triton X-100 in high salt buffers, or SDS have failed to dissociate PLP from MOG and we estimate that PLP/DM-20 accounts for approximately 10-20 % of the protein specifically bound to 8-18C5 – Sepharose 4B and eluted with diethylamine in 0.5 % deoxycholate (pH 11.2). Furthermore, the MOG/PLP aggregates failed to bind to an antiPLP immunoaffinity column. The co-purification of PLP with MOG following solubilisation with either Triton X-100 or sodium deoxycholate may be an artefact of the purification technique, or alternatively may reflect a specific and strong associated of MOG and PLP in vivo.

Figure 1:
Purification of MOG by immunoaffinity chromatography using mAb 8-18C5 coupled to Sepharose 4B.

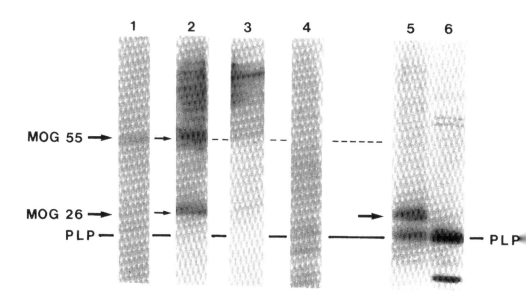

Crude preparations of MOG are prepared by immunoaffinity chromatography of deoxycholate extracts of white matter or purified myelin over mAb 8-18C5 – Sepharose 4B. SDS-PAGE analysis of the material specifically bound to the immunosorbant revealed PLP to be a major contaminant of these MOG preparations. Lane 1 Coomassie Blue stained gel; Lanes 2-4 Western blots of similar gels stained with (2) mAb 8-18C5; (3) goat anti-mouse IgG; (4) rabbit anti-PLP. Lane 5 demonstrates the degradation of MOG 55 to MOG 26 that occurs in the absence of protease inhibitors, lane 6. Myelin. Lanes 5 and 6 are silver stained.

The presence of PLP/DM-20 in these MOG preparations has resulted in some confusion as to the identity of the lower molecular weight MOG component. However, in addition to mAb 8-18C5, a panel of 17 additional monoclonal antibodies have now been raised, against either the rat (Z series) or bovine antigen (Y series), which all recognise epitopes present on both the 55 and 26 kDa forms of MOG, although preliminary ELISA and immunocytochemical studies indicate several different epitopes are involved (Table 1). These new reagents confirm that MOG, as originally defined by the mAb 8-18C5, is indeed a myelin specific antigen with a native subunit molecular weight of approximately 55 kDa.

Immunohistochemistry of CNS tissue sections using these antibodies demonstrate that MOG is associated with CNS white matter and ultrastructual immunocytochemistry has localised the MOG epitope recognised by the 8-19C5 mAb to the external surface of myelin and oligodendrocytes in actively myelinating rats (Linington et al., 1988). However, MOG is not uniformally distributed throughout the compact myelin membrane. Quantitative analysis of the density of immuno-gold particles across transverse sections of myelin sheaths stained with mAb 8-18C5 revealed that MOG is preferentially localised in the outermost lamellae of the sheath (Brunner et al., 1989). This distribution of MOG within the myelin sheath is assumed to reflect its physiological function which is at present unknown. However, in vitro studies have demonstrated that MOG is an antigenic marker for mature oligodendrocytes, being expressed after the major structual myelin proteins (Scolding et al., 1989) and is therefore unlikely to be involved in the initiation of myelinogenesis.

Table 1:
Differential binding of MOG specific mAbs to rat, human and bovine myelin

Antibody	Percent OD Human	405 nm rat	Immunocytochemical staining Pattern
8-18C5	162	259	Myelin
Z12	168	231	Myelin
Z2	147	170	Myelin
Z4	100	176	Myelin
Y2A6	106	140	Myelin/cytoplasmic
Z8	74	246	Myelin
Y5B1	153	36	n. d.
Y3A5	103	43	Myelin
Y2B5	100	42	Diffuse tissue staining
Y5D2	94	47	Myelin/cytoplasmic
Y2A5	92	46	Myelin/cytoplasmic
Y5C1	92	41	n. d.
Y3B3	91	38	Myelin/cytoplasmic
Y3B4	76	22	n. d.
Y5B2	53	35	Myelin/cytoplasmic (PNS ?)
Y5C6	58	7	Myelin
Y4D6	47	17	Myelin/cytoplasmic (PNS ?)
Y6A4	44	18	n. n.

The binding of each mAb to deoxycholate solubilised bovine (B), human (H) and rat (R) myelin was determined by ELISA. Wells were coated with 10 μg/ml myelin protein and the relative binding of the mAbs to the three target antigens assessed with a peroxidase conjugated anti-mouse IgG reagent using ABTS as the substrate. The results are presented in terms of the percentage of the OD 405 nm obtained with bovine myelin. This preliminary screening procedure clearly identifies three distinct groups of antibodies which differ with respect to their binding to MOG from different species, having the relative affinities R > H > B, B > H > R, or B = H > R. Immunocytochemistry of paraffin embedded, paraformaldehyde fixed rat CNS tissue indicates that within these broad groupings further epitopic diversity probably exists, some mAbs staining rat myelin in a manner identical to 8-18C5, whilst others appear to recognise cytoplasmic determinants.

MOG - A model target for antibody mediated demyelination in vivo.

In the rat, the clinical signs of EAE (loss of weight, ataxia and paralysis) occur four to five days after the passive transfer of encephalitogenic MBP specific T cell lines. (BenNun et al., 1981). Histologically, this model of EAE is associated with oedema and inflammation of the CNS, the extent of CNS inflammation and the clinical severity of the disease being proportional to the dose of T cells injected (Lassmann et al., 1988). Unfortunately, demyelination in this T cell mediated model of EAE in the rat is minimal and as such it provides a poor model of immune-mediated CNS demyelination as seen in the human disease, multiple sclerosis (MS). However, systemic injection of mAb 8-18C5 four days after T cell transfer has a profound effect on the histopathology of this disease by initiating extensive demyelination throughout the CNS (Linington et al., 1988; Lassmann et al., 1988). This synergistic animal model of acute antibody-mediated demyelinating EAE (ADEAE) has allowed us to dissect the immune mechanisms underlying the pathogenesis of antibody mediated demyelination in vivo, in particular the relative importance of direct complement mediated myelinolysis and antibody dependant cell-mediated cytotoxicity (ADCC).

A direct role for complement in the pathogenesis of demyelination initiated by mAb 8-18C5 in EAE was first indicated by studies of cultured rat oligodendrocytes in vito. The majority of these cells express MOG on their surface after seven to ten days in vitro and their lysis by mAb 8-18C5 is Fc dependant and complement-mediated (Figure 2). This phenomenon was investigated in vivo by using cobra venom factor (CVF) to decomplement rats with acute EAE before the intravenous injection of mAb 8-18C5 (Linington et al., 1989). CVF acts to rapidly deplete serum of complement components C3/C5, thereby interrupting the complement cascade and inhibiting the formation of membrane attack complex (MAC) and the subsequent lysis of antibody coated targets, In the

Figure 2:
Oligodendrocyte lysis in vitro by mAb 8-18C5 is complement and Fc dependant.

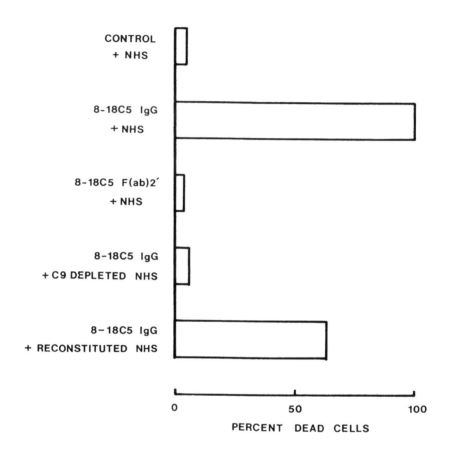

Oligodendrocytes derived from neo-natal rat optic nerve were cultured in vitro for seven days at which time > 85 % of the cultured cells co-express GC, MBP and MOG. Cells were incubated for 30 minutes with media containing (from top to bottom): 20 % normal human serum (NHS) + an irrelevant mouse mAb; NHS + mAb 8-18C5 (300 μg/ml); NHS + mAb 8-18C5 F(ab)2' fragement (300 μg/ml); C9-depleted NHS + mAb 8-18C5 (300 μg/ml) or C9 reconstituted NHS + mAb 8-18C5 (300 μg/ml). Cell death was assessed by propidium iodide staining of the nucleus.

rat a single injection of CVF eliminates the lytic activity of serum complement for approximately 48 to 72 hours (Linington et al., 1989) and repeat injections of CVF can be used to maintain this state for several days.

Initial observations indicated that antibody mediated demyelination in vivo was reduced in those rats with ADEAE and treated with CVF (Linington et al., 1989). However, careful quantitation of the extent of demyelination seen in matched groups of control and CVF treated rats with ADEAE revealed that this reduction was not statistically significant (Piddlesden, in preparation). This observation indicated that complete activation of the complement cascade resulting in the formation of MAC on the myelin surface is not a pre-requiste for antibody mediated demyelination in EAE; demyelination is therefore believed to be mediated by an ADCC response targeted to the myelin sheath by antibody coating its outer surface. This view is supported by ultrastructual studies of demyelinating lesions in this model of associated with monocytes and macrophages, which invade the myelin sheaths, initiate the vesicular dissolution of myelin in their immediate vicinity and phagocytose myelin debris (Lassmann et al., 1988). However the molecular mechanisms responsible for this disruption of the myelin membrane are not yet clear, but are believed to involve the co-operative or synergistic action of a number of potentially destructive agents (proteases, phospholipases, reactive oxygen metabolites and cytotoxins) secreted by the infiltrating monocytes (See Glynn and Linington, 1989).

Although it appears that antibody mediated demyelination can proceed in vivo in the absence of MAC formation, a role for complement in the pathogene-

Figure 3:
Perivascular C9 immunoreactivity in antibody dependant, demyelinating EAE in the rat.

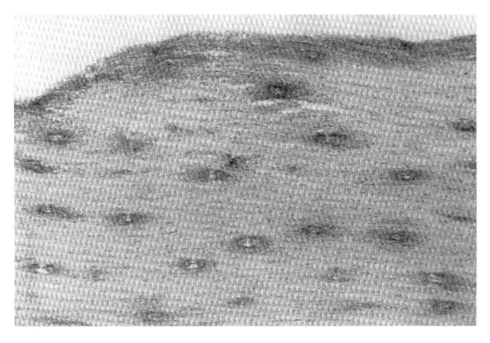

Spinal cord section from a rat with severe clinical EAE, 48 hours after the intravenous injection of mAb 8-18C5. Intense staining for C9 is observed in perivascular and sub-pial areas of the spinal cord corresponding to regions of active antibody mediated demyelination (Linington et al., 1989a), photomicrograph kindly provided by Dr H. Lassmann.

sis of demyelination in ADEAE should not be discounted. Perivascular deposition of C9 in demyelinating lesions is characteristic of antibody mediated demyelination in EAE (Figure 3) and is indicative of the local activation of the complement cascade resulting in the generation of MAC, presumably on the external surface of the myelin sheath and some degree of myelin damage (Linington et al., 1989a). Whilst MAC formation may itself be sublytic, the fixation of C1 and subsequent cleavage of complement components C2, C3, C4 and C5 will generate a number of pro-inflammatory factors that will amplify the local inflammatory response by recruiting additional inflammatory cells into the lesions and activating the infiltrating monocytes. In addition the myelin sheath will be opsonized with C3b for which macrophages express a specific receptor, occupation of which results in macrophage activation (Schorlemmer and Allison, 1976). Thus activation of the early components of the complement cascade will enhance the ability of an ADCC response to mediate demyelination in vivo.

These experiments using MOG and the mAb 8-18C5 to study antibody mediated demyelination in vivo have indicated that in EAE, the presence of anti-myelin antibodies initiated multiple effector mechanisms, both humoral and cellular, which can compromise the integrity of the myelin membrane and lead to demyelination. Furthermore, the animal experiments clearly demonstrate that active antibody mediated demyelination in inflammatory CNS lesions is associated with the localised deposition of C9/MAC. However, the relevance of these observations to human demyelinating disease, and in particular MS remain uncertain (Glynn and Linington, 1989).

Interestingly, consumption of complement components within the CNS (Compston et al., 1987; Sanders et al., 1986), perivascular deposition of MAC (Compston et al., 1989) and MAC bound to myelin debris in the cerebrospinal fluid (Scolding et al., 1989) have all been reported in MS patients. These findings support the contention that complement dependant mechanisms play an important role in de pathogenesis of MS and may indicate the presence of a pathogenis myelin specific antibody response in MS patients.

Alternatively, these observations may reflect the antibody independant activation of complement by myelin (Silverman et al., 1984) or oligodendrocytes (Scolding et al., 1989). However, antibodies to major components of the myelin sheath do occur in the sera of MS patients (Endo et al., 1984; Newcombe et al., 1985) and we are at present investigating the possibility that an antibody response to exposed, and therefore pathogenic, epitopes of MOG may also be represented in MS sera.

Acknowledgements

We thank the Multiple Sclerosis Society, U.K. for financial support and Lisa Dunn for typing the manuscript.

References

Ben-Nun A, Wekerle H, Cohen IR (1981) The rapid isolation of clonable antigen-specific T lymphocyte lines capable of mediating autoimmune encephalomyelitis. Eur J Immunol 11:195

Bornstein MB, Appel SH (1961) The application of tissue culture to the study of experimental allergic encephalomyelitis. I. patterns of demyelination. J Neuropath Exp Neurol 20:141

Brunner C, Lassmann H, Waehneldt TV, Matthieu J, Linington C (1989) Differential ultrastructual localization of myelin basic protein, myelin/oligodendroglial glycoprotein, and 2',3'-cyclic nucleotide 3'-phosphodiesterase in the CNS of adult rats. J Neurochem 52:296

Compston DAS, Morgan BP (1987) Cerebrospinal fluid complement components in multiple sclerosis. In: Lowenthal A, Raus R (eds) Cellular and humoral immunological components of cerebrospinal fluid in multiple sclerosis. Plenum Press, New York, pp 200

Compston DAS, Morgan BP, Campbell AK, Jasanie B (1989) Immunocytochemical localisation of the terminal complement complex in multiple sclerosis. Neuropathol Appl Neurobiol (in press)

Endo T, Scott DD, Stewart SS, Kundu SK, Marcus DM (1984) Antibodies to glycosphingolipids in patients with multiple sclerosis and SLE. J Immunol 132

Glynn P, Linington C (1989) Cellular and molecular mechanisms of autoimmune demyelination in the central nervous system. Critical Reviews in Neurobiol 4:367

Hudson LD, Friedrich VL, Behar T, Dubois-Dalcq M, Lazzarini RA (1989) The initial events in myelin synthesis: Orientation of proteiolipid protein in the plasma membrane of cultured oligodendrocytes. J Cell Biol 109:717

Lassmann H, Brunner C, Bradl M, Linington C (1988) Experimental allergic encephalomyelitis the balance between encephalitogenic T lymphocytes and demyelinating antibodies determines size and structure of demyelinating lesions. Acta Neuropathol 74:566

Lebar R, Boutry JM, Vincent C, Robineaux R, Voisin GA (1976) Studies on autoimmune encephalomyelitis in the guinea pig. J Immunol 116(5):1439

Lebar R, Vincent C, Fischer-Le BourBennec E (1979) Studies on autoimmune encephalomyelitis in the guinea pig III. A comparative study of two autoantigens of central nervous system myelin. J Neurochem 32:1451

Lebar R, Lubetzki C, Vincent C, Lombrail P, Boutry J-M (1986) The M_2 antigen of central nervous system myelin, a glycoprotein present in oligodendrocyte membrane. Clin Exp Immunol 66:423

Linington C, Webb M, Woodhams PL (1984) A novel myelin-associated glycoprotein defined by a mouse monoclonal antibody. J Neuro Immunol 6:387

Linington C, Bradl M, LassmannH, Brunner C, Vass K, (1988) Augmentation of demyelination in rat acute allergic encephalomyelitis by circulating mouse monoclonal antibodies directed against a myelin/oligodendrocyte glycoprotein (MOG). Am J Pathol 130:443

Linington C, Morgan CP, Scolding NJ, Piddlesden S, Wilkins P, Compston DAS, (1989) The role of complement in the pathogenesis of experimental allergic encephalomyelitis. Brain (in press)

Linington C, Lassmann H, Morgan BP, Compston DAS (1989) Immunohistochemical localisation of terminal complement component C9 in experimental allergic encephalomyelitis. Acta Neuropathol 731:1.

Linington C, Lassmann H, (1987) Antibody responses in chronic relapsing experimental allergic encephalomyelitis: Correlation of serum demyelinating activity with antibody titre to the myelin/oligodendrocyte glycoprotein (MOG) J Neuroimmunol 17:61

Newcombe J, Gahan S Cuzner ML, (1985) Serum antibodies against central nervous system proteins in human demyelinating disease. Clin Exp Immunol 59:383

Raine CS, Johnson AB, Marcus DM, Suzuki A, Bornstein MB (1976) Demyelination in vitro: absorption studies indicate that galactocerbroside is a major target. J Neurol Sci 52:117

Salzer JL, Homes WP, Colman DR (1987) The amino acid sequences of the myelin-associated glycoproteins: homology to the immunoglobulin gene superfamily. J Cell Biol 104:957

Sanders MS, Koski CL, Robbins D, Shin ML, Frank MM, Joiner KA (1986) Activated terminal complement in cerebrospinal fluid in Guillian-Barre syndrome and mutiple sclerosis. J Immunol 136:4456

Schorlemmer HU, Allison AC (1976) Effect of activated complement components on enzyme secretion by macrophages. Immunology 31:781

Scolding NJ, Frith S, Linington C, Morgan BP, Campbell AK, Compston DAS (1989) Myelin-oligodendrocyte glycoprotein (MOG) is a surface marker of oligodendrocyte maturation. J Neuroimmunol 22:169

Scolding NJ, Houston A, Linington C, Morgan CP, Campbell AK, Compston DAS (1989a) Oligodenrocytes activate complement but resist lysis by vesicular removal of membrane attack complex. Nature 339:620

Scolding NJ, Morgan BP, Houston A, Campbell AK, Linington C, Compston DAS (1989) Normal rat serum cytotoxicity against synergeneic oligodendrocytes: complement activation and attack in the absence of anti-myelin antibodies. J Neurol Sci 89:289

Seil FJ, Agrawal HC (1980) Myelin-proteolipid protein does not induce demyelinating or myelination-inhibiting antibodies. Brain Res 194:273

Seil FJ, Ouarles RH, Johnson D, Brady RO (1981) Immunization with purified myelin-associated glycoprotein does not evoke myelination-inhibiting or demyelinating antibodies. Brain Res 209:470

Silverman BA, Carney DF, Johnston CA, Vanguri P, Shin ML (1984) Isolation of membrane attack complex of complement from myelin membranes treated with serum complement. J Neurochem, 42:1024

ROLE OF AN ENDOGENOUS MANNOSYL-LECTIN IN MYELINATION AND STABILIZATION OF MYELIN STRUCTURE.

J.-P. ZANETTA, S. KUCHLER, P. MARSCHAL, M. ZAEPFEL, A. MEYER, A. BADACHE,
A. REEBER, S. LEHMANN and G. VINCENDON
Centre de Neurochimie du CNRS
5 rue Blaise Pascal
67084 Strasbourg cedex
France

INTRODUCTION.

The studies of the mechanisms involved in myelination and myelin compaction constitute a broad field of neurobiology since myelin plays an important role of in conductance of nerve impulse both in central and peripheral nervous system. One of the most dramatic demyelinating diseases is multiple sclerosis (MS). The study of demyelination in MS may help in understanding the molecular mechanisms involved in maintenance of myelin structure. Several hypothesis have been presented for explaining myelin compaction. From the studies of **mld** mutant (Doolittle and Schwiekart, 1977), it has been proposed (Jacque et al., 1983; Kimura et al., 1985; Lachapelle et al., 1980; Matthieu 1982; Matthieu et al., 1980a; 1980b; 1981; 1984; Mikoshiba et al., 1987; Quarles, 1984; Waehneldt and Linington, 1980) that myelin basic protein (MBP) is involved in the mechanism of myelin compaction at the cytoplasmic surface of the oligodendrocyte membrane. A similar role has been attributed to proteolipid protein (Duncan et al., 1987). Due to the specific defect of MBP in **mld** mutant there is an absence of the major dense line . It has also been proposed that this adhesion involved interaction of myelin basic protein with negatively charged glycolipids (i.e. sulfatides) present in high amount in these membranes (Zalc et al., 1981). Another possibility was the interaction of MBP with other glycolipids (Ikeda and Yamamoto, 1987). Similarly, the proteolipid protein (PLP) has been suggested to be involved in myelin compaction at the level of extracellular face of the oligodendrocyte (Dautigny et al., 1986). This assumption was based on the observation that myelin of the jimpy mutant, having a specific anomaly of PLP protein (Baumann and Lachapelle, 1982; Hogan and Greenfield, 1984;

Hudson et al., 1982; Nave et al., 1986; Sorg et al., 1986), is not compacted. Furthermore, PLP molecules aggregate very easily. But in **jimpy** mutants, the absence of myelin has also been attributed to the lack of survival of mutant oligodendrocytes.

So far as the interaction between the extracellular surface of myelinating cells are concerned, several hypothesis assumed that myelin glycoproteins were involved in myelin compaction. One of these hypothesis (Webster et al., 1983) was based on the immunocytochemical localization of myelin associated glycoprotein (MAG) in the compact myelin. The proposed model of interaction between MAG molecules suggested that molecules binding to the carbohydrate portion of MAG could make bridges between the glycans of MAG molecules. It was suggested that MAG possesses a site with the property of binding its own glycan. This constituent has been extensively studied (Fujita et al., 1988; Inuzuka et al., 1985; Konat et al., 1987; Matthieu, 1981; Matthieu et al., 1974a and 1981; Quarles , 1984; Quarles et al., 1972; 1973a and 1973b; Sternberger et al., 1979; Trapp et al., 1983; Webster et al., 1983) but this hypothesis could not sustain because the localization of MAG in compact myelin is disputed. In fact, in the PNS, MAG has been localized in periaxonal regions (Trapp et al., 1984a and b).

In the peripheral nervous tissue, a major glycoprotein constituent of myelin is glycoprotein P0 (Kitamura et al., 1976; 1981; Lemke and Axel, 1985; Sakamoto et al., 1987), with a Mr of 29 kDa. This has been proposed as a candidate for ensuring compaction between the external surface of the Schwann cell membrane. This molecule has been clearly demonstrated as a constituent of compact myelin in PNS (Trapp et al,. 1981). P0 glycoprotein has been involved in myelin compaction through two different kinds of interactions. The high hydrophobicity of this molecule was the basis for involving P0 in formation of hydrophobic bonds between external surfaces of Schwann cells (Lemke et al.,1985). The role of the glycan portion of P0 in myelin compaction has been implicated (Lemke et al., 1985) but the mechanism has not been clearly understood.

A number of molecules have been considered in the family of the Cell Adhesion Molecules (CAMs) (Crossin et al., 1986; Edelman, 1985 and 1986; Grumet et al., 1985; Hoffmann et al., 1986 and 1987; Rieger et al., 1986). They have been found to be present in myelin of the central or peripheral nervous system (Martini and Schachner, 1986; Poltorak et al.,1987). Most of the myelin glycoprotein so far identified (MAG and P0) share a common L2/HNK-1 epitope (McGarry et al., 1983; Inuzuka et al., 1984; O'Shannessy et al., 1985), also detected on a glycolipid containing sulfated glucuronic acid residue (Inuzuka et al., 1984). These molecules, termed as CAMs, are potentially involved in myelin compaction although direct experimental evidence is lacking.

Based on different experimental approaches, a mannose-binding protein has been detected in white matter (Bardosi et al., 1988;; Gabius et al., 1988; Kuchler et al., 1987; 1988 and 1989b) corresponding to a molecule initially isolated from the developing rat cerebellum and therefore called "Cerebellar Soluble Lectin" or CSL (Zanetta et al., 1987a). Another carbohydrate-binding protein has been isolated from liver (Colley and Baenziger, 1987a and 1987b) and was called also CSL ("Core Specific Lectin"). Although Cerebellar Soluble Lectin is also present in the liver, the two CSL lectins are unrelated with each other. The Cerebellar Soluble lectin first isolated by Zanetta et al.(1987a), (previously identified as a mannose-binding lectin in 1985 (Zanetta et al.,1985a)) will be refered here as CSL. This paper clearly draws attention to the key role played by CSL in myelination and myelin compaction. We present data and discuss models of the action of this lectin in myelination and myelin compaction.

STRUCTURE AND PROPERTIES OF CSL MOLECULE.

Isolation of CSL.

Lectin CSL was isolated from young rat cerebella by sequential extractions of the tissue. It was concentrated in a fraction specifically solubilized in the presence of 0.5 M mannose with no detergent (Zanetta et al.,1987a). It is separated, on the basis of solubility properties, from another mannose-binding lectin R1 isolated previously (Zanetta et al., 1985a and 1987b). Purified protein bands isolated by preparative gel electrophoresis, were used to produce antibodies in rabbits. Antibodies, in return, were employed to isolate the antigen by immunoaffinity chromatography of the material solubilized in the presence of mannose. The typical profile of a CSL preparation purified by immunoaffinity chromatography is shown in fig. 1a.

Structure of CSL.

With different antibodies, active CSL lectin (possessing agglutinating activity) was obtained as a major doublet band of Mr 33 and 31.5 kDa and having a minor component of Mr 45 kDa (Fig.1a). Antidodies raised in rabbits against individual protein bands reacted with all the three components as revealed by immunoaffinity chromatography. This demonstrated a clear immunochemical relationship between the three components. Amino-acid composition (Zanetta et al., 1987a) showed an homology between 33 and 31.5 kDa components since they differed only by a short polypeptide of about 15-20 aminoacids. It was assumed that 45 kDa protein was a precursor molecule the two other proteins. Similarly the 33 kDa protein (concentrated in intracellular organelles like lysosomes) could be the precursor of the 31.5 kDa protein mainly found in the cytoplasm or extracellularly. The short polypeptide chain responsible for the Mr difference between 33 and 31.5 kDa components of CSL could be a signal sequence for secretion (Zanetta et al., 1987a). However, recent data obtained from cloning of CSL molecules suggested that two mRNA may coexist specific for a 45 kDa and a 33 kDa protein.

Active CSL molecules are found as macromolecular complexes formed of about 40 different identical monomers (Zanetta et al., 1987a) consisting either of the 33 kDa protein or of the 31.5 kDa protein.

Carbohydrate specificity of CSL.

By studying the inhibition of agglutination of red blood cells by various monosaccharides or glycosaminoglycans, glycoproteins or glycopeptides the carbohydrate specificity of lectin CSL was defined (Marschal et al.,1989). Mannose is a relatively poor inhibitor and showed inhibition at 150 mM whereas other monosaccharides did not show any inhibition. Glycosaminoglycans showed poor inhibition , the best being heparin acting at the concentration of 185 µg/ml; the concentration range at which heparin-binding proteins are inhibited). Horseradish peroxidase isoenzyme 8 is a good inhibitor (11 µg/ml) whereas ovalbumin requires much higher concentration (2.5 mg/ml) for inhibition of agglutination. Endogenous glycoproteins B1 and B2 transiently expressed in the cerebellum (Zanetta et al., 1978; Reeber et al., 1981) were good inhibitors (2-3 µg/ml). The glycopeptides isolated from these molecules inhibited at a concentration of 1 µg/ml and corresponded to the best inhibitors yet identified. Studies with glycopeptides of known structures suggest that the first N-acetyl-glucosamine residue of N-glycans is part of the binding site of the lectin. CSL was able to discriminate between oligomannosidic glycans with 5, 6 or 7 mannose residues since glycans containing 6 mannose residues have a much higher affinity.

In the cerebellum and in other parts of the nervous tissue, CSL displayed specificity for a smaller number of glycoprotein subunits than the plant lectin Concanavalin A. This indicates that, in most non transformed cells, the endogenous lectin CSL has a few number of endogenous ligands. This may be the bases for specific cell recognition and adhesion.

CSL AND MYELINATION IN THE CENTRAL NERVOUS SYSTEM.

Presence of CSL in myelin.

CSL is present in oligodendrocytes and in myelin of the central nervous tissue in vivo or in cultures (Kuchler et al., 1987 and 1988). This has been observed by immunocytochemical techniques at the level of electron microscopy. In cultures of oligodendrocytes, CSL lectin is present intracellularly and on the external face of the oligodendrocyte membrane. This seems apparent since compact myelin is formed by the cultured oligodendrocytes. The lectin is also present in the areas of contacts between the different myelin sheaths (Kuchler et al., 1988). The model of oligodendrocyte cultures (Espinosa de los monteros et al., 1986; Knapp et al., 1987; McCarthy and Devellis, 1980; Szuchet et al., 1986) offered the possibility to test the involvement of CSL in myelin compaction. As previously described (Kuchler et al., 1988), very small amounts (4μg/ml) of anti-CSL Fab fragments are able to dissociate compact myelin. Furthermore, the presence of CSL glycoprotein ligands was also demonstrated in these cultures of oligodendrocytes. CSL is also present in myelin found in dysmyelinating mutants (Kuchler et al., in preparation). In mutant mice, CSL displays a non homogeneous distribution and is present only in the areas where myelin shows some normal aspect of compaction and is totally absent from the areas where myelin is fully disorganized.

Oligodendrocyte ligands for CSL.

Two major ligands were found in cultured oligodendrocytes. One has a Mr in the range of 100 kDa and the second one is in the range of 16 kDa (Kuchler et al., 1988). Both glycoproteins are ConA-binding. However, ConA-binding glycoproteins found in these cultures are innumerable. From the experiments performed in oligodendrocyte cultures (Kuchler et al.,1988), it is suggested that all the contacts between external surface of oligodendrocytes in myelin and between various cells of the culture are involving CSL and its ligands.

These results were supported (Fressinaud et al., 1988) by the studies on the adhesion of oligodendrocytes to CSL coated culture dishes (incubations overnight at 4°C with 5µg/ml CSL followed by rinses with PBS). Adhesion on CSL coat induces an immediate and significant increase in the proliferation of oligodendrocytes. This phenomenon was not observed in the astrocytes. These observed variability between proliferative response on the two type of cells could be related to the observation that the population of CSL glycoprotein ligands is much more heterogeneous in astrocytes than in oligodendrocytes. This heterogeneity of CSL ligands results in a multiple meaningful adhesion signal which in turn induces cell responses of contradictory nature. In oligodendrocytes, the adhesion on CSL coated dishes can take place through only two components, which could be endowed with specific signal transduction systems. This could induce cell proliferation.

Studies performed on purified myelin fractions from normal animals indicate that the population of CSL glycoprotein ligands found in myelin preparations was much more complex than in oligodendrocyte cultures (Kuchler et al., in preparation). A doublet glycoprotein band in the region of 100 kDa was seen along with several glycoprotein bands with Mr of 50, 31, 29, 24, 19 and 16 kDa (Fig.1e). Glycoproteins with similar Mr have been previously detected in purified myelin fractions (Linington et al., 1984; Matthieu et al., 1974; Neskovic et al., 1986; Poduslo, 1981; Poduslo et al., 1977 and 1980; Quarles et al., 1979; Roussel et al., 1987; Schluesener et al., 1987; Zanetta et al., 1977a and 1977b). The 16 kDa protein corresponded to that found in oligodendrocyte cultures. In mld and quaking dysmyelinating mutants, the bands at 24 and 19 kDa were absent without apparent modification of the CSL content in myelin (Kuchler et al., in preparation). In contrast, the 16 kDa constituent was unchanged, suggesting that it does not constitute a structural component of compact myelin.

Axonal glycoprotein ligands for CSL.

The 31 kDa component has been recently identified as an axonal glycoprotein present on the whole surface of young axon and disappearing almost completely at the period of synaptogenesis (Kuchler et al., 1989b).

The molecule displayed a sensitivity to the phosphatidylinositol specific phospholipase C (Pierre et al., 1987). It was found to be the only CSL binding ligand found in purified axolemmal membrane fraction isolated from 10 day-old rat forebrain (Fig. 1d and Kuchler et al.,1989b). It corresponds to a glycoprotein transiently present during cerebellar development (Reeber et al., 1981; Zanetta et al., 1985b and 1987c) on the membrane of poorly differentiated neurons (Kuchler et al., 1989b). This molecule seems to be

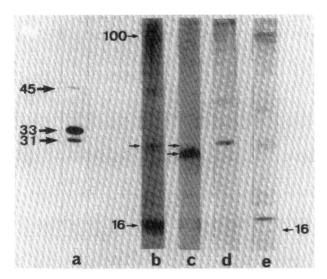

Figure 1. Electrophoretic analysis of CSL and CSL ligands involved in myelination mechanisms (13% polyacrylamide gels in the buffer system of Laemmli (1970).
a) Silver staining of immunopurified purified CSL.
b)-e) CSL ligands revealed by iodinated CSL on blots after blotting (Towbin et al., 1979):
 -b) 1 day-old rat sciatic nerve.
 -c) 60 day-old rat sciatic nerve.
 -d) purified axonal membrane preparation (10 day-old rats).
 -e) myelin preparation from 27 day-old mouse (Norton and Poduslo, 1973).
Several components are present in the region of 30 kDa which are actually unrelated:
 -the two subunits of CSL (33 and 31.5 kDa)
 -the 31 kDa axonal glycoprotein
 -the 29 kDa P0 glycoprotein of PNS myelin.

concentrated in the regions of contacts between oligodendrocytes and axons during the period of myelination. In adult, the only regions where the antigen has been detected correspond to portions of axons where the ensheathment of myelin is no more present (Kuchler et al., 1989b). Tentatively, it could correspond to the lateral loops of the nodes of Ranvier.

Conclusions.

So far as myelin compaction is concerned , bridging of the external surface of the oligodendrocyte membrane by interactions of the polyvalent lectin CSL with glycans of glycoproteins anchored in the oligodendrocyte membrane appears to be a possibility. The basis for this is the localization of CSL and the effect of anti-CSL Fab fragments. The ligands involved in this compaction in vitro could be only MAG and the 16 kDa glycoprotein. If MAG is localized in compact myelin as described by Webster et al. (1983) MAG could be involved in myelin compaction. However, studies of mld or quaking mutant mice, having poorly compacted myelin, show a specific decrease of low molecular weight glycoproteins with no change in MAG and 16 kDa protein. Thus it is likely that mechanism of myelin compaction involving different molecules are the same in vivo and in vitro (The hypothesis is schematized in Fig. 2c).

The localizations of both CSL and 31 kDa axonal glycoprotein in the areas of contact provide support for the hypothesis that they are involved in the initial contact between axon and oligodendrocyte during initial period of myelination (the hypothesis is schematized in Fig. 2d). The postulated glycoprotein of the oligodendrocyte membrane is still not identified and could correspond to MAG or to the 50 kDa glycoprotein which are unchanged in myelin of mld and quaking mutants. The identification of the oligodendrocyte glycoproteins deserves further experiments.

The observation that adhesion of oligodendrocytes to preformed CSL layers induces a rapid proliferation of these cells (Fressinaud et al., 1988) could be of particular importance. In similar experiments performed with neuroblasts from rat hemispheres (Rauvala, personal communication) adhesion on CSL layers is accompanied with rapid neurite outgrowth (similar to that obtained when cells are seeded on the heparin-binding protein P30 (Rauvala et Pihlaskari, 1987). It can be assumed that initial contact between neuroblast and oligodendrocyte surface in vivo could be the signal for the symbiotic development of neurons and oligodendrocytes. This could be the salient property of CSL as a Cell Adhesion Recognition Lectin (CARL).

CSL AND MYELINATION IN THE PERIPHERAL NERVOUS SYSTEM.

Presence of CSL in myelin.

Immunocytochemical detection at the ultrastructural level reveals the presence of CSL in compact myelin of the adult rat sciatic nerve. The lectin is very concentrated intracellularly in Schwann cells and can be detected in low amounts in compact myelin. A mannose-binding protein constituted of two polypeptide chains of Mr 31.5 kDa and 33 kDa has been detected recently in adult pig sciatic nerve (Gabius et al., 1988). It corresponds probably to CSL.

CSL immunoreactivity is detected very early in rat sciatic nerve (postnatal day 1). It is present in oblong cells which become more and more elongated then produce processes surrounding axons at the period of myelination (Kuchler et al., 1989b). Recent studies on Schwann cell cultures demonstrated that CSL is present in Schwann cells in cultures although the levels seem to be less than in oligodendrocytes.

Presence of CSL ligands in developing sciatic nerve.

The studies of CSL glycoprotein ligands during sciatic nerve development (Kuchler et al., 1989b) indicate considerable quantitative variations in CSL glycoprotein ligands. Four major components are present at day 1 with Mr of 100, 50, 31 and 16 kDa (Fig. 1b). The 16 kDa glycoprotein decreases slowly between day 1 and 15, disappearing almost completely in the adult (Fig. 1c). It corresponds to the same molecule which is found in oligodendrocyte cultures and is considerably increased in myelin of mld and quaking mutants. The 31 kDa decreases progressively until the adult where it is present in very small amount. It corresponds to the 31 kDa axonal glycoprotein previously identified in CNS axons (Fig. 1d and Kuchler et al., 1989b). The 50 kDa component decreases progressively with age and does not seem to be specific for the PNS since it corresponds to a similar component in the CNS. The molecule of high Mr has the same Mr as that found in oligodendrocyte cultures and in myelin from the CNS. It decreases considerably with maturation, but is still present in low amounts in the adult sciatic nerve. It is probably MAG. Major modifications were observed for a compound having Mr at 29 kDa. It is absent at day 1 but is clearly detectable at day 5 then increases continuously until adult age. It is a major protein component of adult sciatic nerve and has all the characteristics of P0 glycoprotein (Fig 1d).

All these components binding to CSL lectin are also revealed using the ConA-horseradish peroxidase (HRP) technique. It is, however, not certain that the quantity of CSL binding material is identical to that having ConA-binding properties. In other word, it is not evident that for one given molecular glycoprotein entity all the glycans are CSL binding. In contrast, it is possible that changes in proportion of polypeptide chains binding to CSL occur during development, differently from that changes of polypeptide chain binding to ConA.

Role of CSL in PNS myelin.

Two roles have been advanced for CSL in PNS myelin (Kuchler et al., 1989b). The first one is myelin compaction (Fig 2e). This role of CSL is supported by the presence of CSL in compact myelin and the presence in compact myelin of glycoprotein P0 (Trapp et al., 1981), which is a ligand of this endogenous lectin CSL (Kuchler et al.,1989b). The polyvalent lectin makes bridges between the external surfaces of the Schwann cell membrane by specific binding of glycans of P0 glycoprotein (Fig. 2e). The data available show that both CSL and P0 are actually localized in compact myelin. The cultures of myelinating Schwann cells could probably be useful for demonstration of this hypothesis. Evidently, the number of P0 molecules is by far higher than that of CSL molecules. But it is not necessary that all P0 molecules are involved in a bridging process to ensure a total compaction of myelin. In contrast, few points of attachment could produce a compacted but relatively mobile structure. The involvement of P0 in myelin compaction has been previously advanced (Lemke and Axel, 1985). A role of glycan part has been proposed but also the possibility of compaction through hydrophobic interactions between P0 sequences. The relative importance of these two different postulated mechanisms remains to be estimated.

The second role is similar to that proposed in the CNS myelination i.e. formation of the first contact between axon and myelinating cells. The localization of the 31 kDa glycoprotein in the same areas as in the CNS

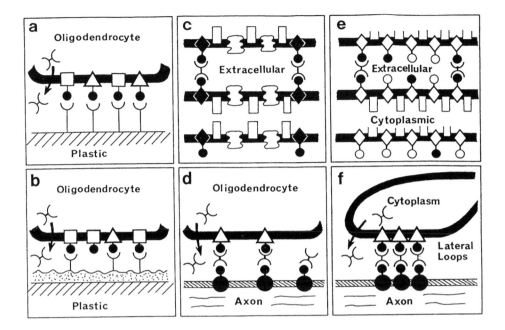

Figure 2. Hypothetical scheme involving CSL in adhesion related to myelination. In all cases, the lectin CSL is externalized by the cells. This property is a commun feature of most soluble lectins (Barondes,1984; Sharon, 1984 and 1987).
-a) Mechanism of adhesion of oligodendrocytes on CSL (⊁⊂)preformed layers. Only two glycoproteins are involved in this binding MAG (△) and the 16 kDa glycoprotein (☐). (●-) represents the glycan of these glycoproteins binding to CSL.
-b) Proposed mechanism for adhesion of oligodendrocyte cultures to the substratum. It is the same as in a). After an initial unspecific adhesion of cells to the substratum, oligodendrocyte binding can be reversed by the action of anti-CSL Fab fragments.
c) Mechanism of myelin compaction in the CNS. CSL make bridges between glycans (●-) of myelin glycoproteins oriented on the extracellular surface. These glycans belong in vivo to minor myelin glycoproteins (◆). MBP (☐) and PLP (⧘) are also represented.
d) Mechanism of the first contact between myelinating cells and axons. It involves CSL bridges between the 31 kDa glycoprotein on the axonal surface (●) and MAG (△). This mechanism could be responsible for the mitogenic properties of axonal membranes for Schwann cells (Wood and Bunge, 1975).
e) Mechanism for myelin compaction in the PNS. CSL make bridges between some P0 molecules (◇). Some of the P0 molecules are not involved in this binding, some of them having glycans without affinity for CSL (O-).
f) Mechanism of adhesion between axon and myelinating cells in the adult myelin of PNS and CNS. 31kDa (●), CSL and probably MAG (△) are involved in the maintenance of this junction in the lateral loops of Ranvier's nodes. This hypothesis is based on the localization of CSL and 31 kDa glycoprotein in this region.

favours a similar role (Fig 2d and 2f). The observation that sensory axons surrounded by non myelinating Schwann cells are still immunoreactive for 31 kDa (direct contact with Schwann cells is not ensured) could be of interest.

CONCLUSIONS.

In the PNS myelin, CSL could play similar roles as in the CNS. Some of the glycoprotein ligands of CSL are identical (MAG, 50 kDa, 16 kDa and 31 kDa axonal) and other are not: PO in PNS instead of low Mr minor constituents in the CNS. It has been reported that molecules of the family of CAMs are involved in maintenance of cell contacts during myelination. Particularly, HNK-1 epitope seems to play important roles in this phenomenum. It is documented by us that several glycoproteins known to react with anti-HNK-1 antibody are ligands for CSL. Thus, it is tempting to hypothesize that CSL is actually a Cell Adhesion Lectin having specific binding for carbohydrate epitopes implicated in cell adhesion. In conclusion, we may say that CSL is possibly one of the "missing link" in the mechanism of cell adhesion.

ACKNOWLEDGMENTS: the authors thank Professor A.N. Malviya for reviewing the manuscript. This work was supported by a grant of the "Association Française contre les Myopathies".

REFERENCES

BARDOSI A, DIMITRI T GABIUS H-J (1988) Endogenous carbohydrate binding proteins in oligodendrogliomas. A histochemical study. Acta Neuropathol 446: 1-7

BARONDES SH (1984) Soluble lectins: a new class of extracellular proteins. Science 223:1259-1264

BAUMANN N, LACHAPELLE F (1982) Neurological mutants. In: Lajtha A (ed) Handbook of Neurochemistry, Second Edition, vol. 2. Plenum press, New York, p 253-279

COLLEY KJ, BAENZIGER JU (1987a) Post-translational modifications of the core-specific lectin. J Biol Chem 262:10296-10303

COLLEY KJ, BAENZIGER JU (1987b) Identification of the post-translational modifications of the core-specific lectin. J Biol Chem 262:10290-10295

CROSSIN KL, HOFFMAN S, GRUMET M, THIERY J-P, EDELMAN GM (1986) Site restricted expression of cytotactin during development of the chicken embryo. J Cell Biol 102:1917-1930

DAUTIGNY A, MATTEI M-G, MORELLO D, ALLIEL PM, PHAM-DINH D, AMAR L, ARNAUD D, SIMON D, MATTEI J-F, GUENET J-L, JOLLES P, AVNER P (1986) The structural gene coding for myelin-associated proteolipid protein is mutated in jimpy mice. Nature 321:867-869

DOOLITTLE DP, SCHWEIKART KM (1977) Myelin deficient, a new neurological mutant in the mouse. J Hered 68:331-332

DUNCAN ID, HAAMMANG JP TRAPP BD (1987) Abnormal compact myelin in the myelin-deficient rat: absence of proteolipid protein correlates with a defect in the intraperiod line. Proc Natl Acad Sci USA 84:6287-6291

EDELMAN GM (1985) Cell adhesion and the molecular processes of morphogenesis. Ann Rev Biochem 54:135-169

EDELMAN GM (1986) Cell adhesion molecules in the regulation of animal form and tissue pattern. Ann Rev Cell Biol 2:81-116

ESPINOSA DE LOS MONTEROS A, ROUSSEL G, NUSSBAUM J-L (1986). A procedure for long-term culture of oligodendrocytes. Dev Brain Res 24:117-125

FRESSINAUD C, KUCHLER S, SARLIEVE LL, VINCENDON G, ZANETTA J-P (1988) Adhesion on a matrix made of the nervous lectin CSL induces increased proliferation of oligodendrocytes. C R Acad Sci Paris 307:863-868

FUJITA N, SATO S, KURIHARA T, INUZUKA T, TAKAHASHI Y, MIYATAKE T (1988) Developmentally regulated alternative splicing of brain myelin-associated glycoprotein mRNA is lacking in the quaking mouse. FEBS Letters 232:323-327

GABIUS H-J, KOHNKE B, HELLMANN T, DIMITRI T, BARDOSI A (1988) Comparative histochemical and biochemical analysis of endogenous receptors for glycoproteins in human and pig peripheral nerve. J Neurochem 51:756-763

GRUMET M, HOFFMAN S, CROSSIN KL, EDELMAN GM (1985) Cytotactin, an extracellular matrix protein of neural and non neural tissues that mediates glia neuron interaction. Proc Natl Acad Sci USA 82:8075-8079

HOFFMAN S, EDELMAN GM (1987) A proteoglycan with HNK-1 antigenic determinants is a neuron-associated ligand for cytotactin. Proc Natl Acad Sci 84:2523-2527

HOFFMAN S, FRIEDLANDER DR, CHUONG C-M, GRUMET M, EDELMAN GM (1986) Differential contributions of Ng-CAM and N-CAM to cell adhesion in different neural regions. J Cell Biol 103:145-158

HOGAN EL, GREENFIELD S (1984) Animal models of genetic disorders of myelin. In: Morell P (ed) Myelin, Second Edition. Plenum Press, New York and London, p 489-534

HUDSON LD, BERNDT JA, PUCKETT C, COZAK CA, LAZZARINI RA (1982) Aberrant splicing of proteolipid protein mRNA in the dysmyelinating jimpy mutant mouse. Proc Natl Acad Sci USA 84:1454-1458

IKEDA K, YAMAMOTO T (1985) Myelin basic protein has lectin like properties. Brain Res 329:105-108

INUZUKA T, QUARLES RH, NORONHA AB, DOBERSEN MJ, BRADY RO (1984) A human lymphocyte antigen is shared with a group of glycoproteins in peripheral nerve. Neurosci Lett 51:105-111

INUZUKA T, QUARLES RH, HEATH J, TRAPP BD (1985) Myelin-associated glycoprotein and other proteins in Trembler mice. J Neurochem 44:793-797

JACQUE C, DELASSALE A, RAOUL M, BAUMANN N (1983) Myelin basic protein deposition in the optic and sciatic nerves of demyelinating mutants quaking, jimpy, trembler, mld and shiverer during development. J Neurochem 41:1335-1340

KIMURA M, INOKO H, KATSUKI M, ANDO A, SATO T, HIROSE T, TAKASHIMA H, INAYAMA S, OKANO H, TAKAMUTSU K, MIKOSHIBA K, TSUKADA Y, WATANABE I (1985) Molecular genetic analysis of myelin-deficient mice: shiverer mutant mice show deletion in gene(s) coding for myelin basic protein. J Neurochem 44:692-696

KITAMURA K, SUZUKI M, UYEMURA K (1976) Purification and partial characterization of the two glycoproteins in bovine peripheral nerve myelin membrane. Biochim Biophys Acta 455:806-816

KITAMURA K, SAKAMOTO Y, SUZUKI M, UYEMURA K (1981) In: Yamatawa T, Osawa T, Handa S (eds) Glycoconjugates. Japan Scientific Societies Press, Tokyo, p 273-274

KNAPP PE, BARTLETT WP, SKOFF RP (1987) Cultured oligodendrocytes mimic in vivo phenotypic characteristics: cell shape, expression of myelin specific antigens and membrane production. Dev Biol 120:356-365

KONAT G, HOGAN EL, LESKAWA KC, GANTT G, SINGH I (1987) Abnormal glycosylation of myelin-associated glycoprotein in quaking mouse brain. Neurochem Int 10:555-558

KUCHLER S, VINCENDON G, ZANETTA J-P (1987) Immunocytochemical localization of an endogenous cerebellar lectin during development of rat cerebellum. C R Acad Sci Paris 305:317-320

KUCHLER S, FRESSINAUD C, SARLIEVE LL, VINCENDON G, ZANETTA J-P (1988) Cerebellar soluble lectin is responsible for cell adhesion and participates in myelin compaction in cultured rat oligodendrocytes. Dev Neurosci 10:199-212

KUCHLER S, ROUGON G, MARSCHAL P, LEHMANN S, REEBER A, VINCENDON G, ZANETTA J-P (1989a) Localization of a transiently expressed glycoprotein in developing cerebellum delineating its possible ontogenetic roles. Neuroscience

KUCHLER S, HERBEIN G, SARLIEVE LL, VINCENDON G, ZANETTA J-P (1989b) An endogenous lectin CSL interacts with major glycoprotein components in peripheral nervous system myelin. Cell Molec Biol in press

LACHAPELLE F, DE BAECQUE C, JACQUE C, BOURRE J-M, DELASSALLE A, DOOLITTLE D, HAUW JJ et BAUMANN N (1980) Comparison of morphological and biochemical defects of two probably allelic mutations of the mouse myelin deficient (mld) and shiverer (shi). In: Baumann (ed) Neurological mutations affecting myelination. North Holland-Elsevier Biomedical press, p 27-32

LAEMMLI UK (1970) Cleavage of structural proteins during assembly of the bacteriophage T4. Nature 227:680-685

LEMKE G, AXEL R (1985) Isolation and sequence of a cDNA encoding the major structural protein of peripheral myelin. Cell 40:501-508

LININGTON C, WAEHNELDT TV (1981) The glycoprotein composition of peripheral nervous system myelin subfractions. J Neurochem 36:1528-1535

LININGTON C, WEBB M et WOODHAMS PL (1984) A novel myelin associated glycoprotein defined by a mouse monoclonal antibody. J Neuroimmunol 6:387-396

MARSCHAL P, REEBER A, NEESER J-R, VINCENDON G, ZANETTA J-P (1989) Carbohydrate and glycoprotein specificity of two endogenous cerebellar lectins. Biochimie 5:53-61

MARTINI R, SCHACHNER M (1986) Immunoelectron microscopic localization of neural cell adhesion molecules (L1, N-CAM, and MAG) and their shared carbohydrate epitope and myelin basic protein in developing sciatic nerve. J Cell Biol 103:2439-2448

MATTHIEU J-M (1981) Glycoproteins associated with myelin in the central nervous system. Neurochem Int 3:355-363

MATTHIEU J-M (1982) Myelin basic protein and the stability of the multilamellar myelin structure. Bull Schweiz Acad med Wiss 101:108

MATTHIEU J-M, DANIEL A, QUARLES R H, BRADY RO (1974a) Interaction of concanavalin A and other lectins with CNS myelin. Brain Res 81:348-353

MATTHIEU J-M, GINALSKI H, FRIEDE RL, COHEN SR, DOOLITTLE DP (1980a) Absence of myelin basic protein and major dense line in CNS myelin of the mld mutant mouse. Brain Res 191:278-283

MATTHIEU J-M, GINALSKI H, FRIEDE RL, COHEN SR (1980b) Low myelin basic protein levels and normal myelin in peripheral nerves of myelin deficient mutant mice (mld). Neuroscience 5:2315-2320

MATTHIEU J-M, GINALSKI-WINKELMANN H, JACQUE C (1981) Similarities and dissimilarities between two myelin deficient mutant mice, shiverer and mld. Brain Res 214:219-222

MATTHIEU J-M, OMLIN F-X, GINALSKI-WINKELMANN H, COOPER BJ (1984) Myelination in the CNS of mld mutant mice: comparison between composition and structure. Dev. Brain Res 13:149-158

MC CARTHY KD, DE VELLIS J (1980) Preparation of separate astroglial and oligodendroglial cell cultures from rat cerebral tissue. J Cell Biol 85: 890-902

MC GARRY RC, HELFAND SL, QUARLES RH, RODER JC (1983) Recognition of myelin associated glycoprotein by the monoclonal antibody HNK-1. Nature 306:376-378

MIKOSHIBA K, OKANO H, INOUE Y, FUJISHIRO M, TAKAMATSU K, LACHAPELLE F, BAUMANN N et TSUKADA Y (1987) Immunohistochemical, biochemical and electron microscopic analysis of myelin formation in the central nervous system of myelin deficient (mld) mutant mice. Dev Brain Res 35:111-121

NAVE K-A, LAI C, BLOOM FE, MILNER RJ (1986) Jimpy mutant mouse : a 74-base deletion in the mRNA for myelin proteolipid protein and evidence for a primary defect in RNA splicing. Proc Natl Acad Sci USA 83:9264-9268

NESKOVIC NM, ROUSSEL G, NUSSBAUM J-L (1986) UDP-Galactose: ceramide galactosyltranferase of rat brain: a new method of purification and production of specific antibodies. J Neurochem 47:1412-1418

NORTON WT, PODUSLO SE (1973) Myelination in rat brain: method of myelin isolation. J Neurochem 21:749-757

O'SHANNESSY DJ, WILLISON HJ, INUZUKA T, QUARLES RH (1985) The species distribution of nervous system antigens that react with anti myelin associated glycoprotein antibodies. J Neuroimmunol 9:255-268

PIERRE M, BARBET J, NAQUET P, PONT S, REGNIER-VIGOUROUX A, BARAD M, DEVAUX C, MARCHETTO S et ROUGON G (1987) Thy-1, Thy-3 and Thymocyte-activating molecule (ThAM): signal transduction T cell markers identified by rat monoclonal antibodies raised against phosphatidyl inositol- specific phospholipase C - solubilized thymocyte surface antigen. In: The T Cell Receptor, Ucla Symposia on Molecular and Cellular Biology, New series, Vol 73. AR Liss Inc, New York, p 293-300

PODUSLO JF (1981) Developmental regulation of the carbohydrate composition of glycoproteins associated with central nervous system myelin. J Neurochem 36:1924-1931

PODUSLO JF, EVERLY JL, QUARLES RH (1977) A low molecular weight glycoprotein associated with isolated myelin: distinction from myelin proteolipid protein. J Neurochem 28:977-986

PODUSLO JF, HARMAN JL, MC FARLIN DE (1980) Lectin receptors in central nervous system myelin. J Neurochem 34:1733-1744

POLTORAK M, SADOUL R, KEILHAUER G, LANDA C, FAHRIG T, SCHACHNER M (1987) Myelin-associated glycoprotein, a member of the L2 / HNK-1 family of neural cell adhesion molecules, is involved in neuron- oligodendrocyte and oligodendrocyte- oligodendrocyte interaction. J Cell Biol 105:1893-1899

QUARLES RH (1984) Myelin-associated glycoprotein in development and disease. Dev Neurosci 6:285-303

QUARLES RH, EVERLY JL, BRADY RO (1972) Demonstration of a glycoprotein which is associated with a purified myelin fraction from rat brain. Biochem Biophys Res Commun 47:491-497

QUARLES RH, EVERLY J, BRADY R (1973a) Evidence for the close association of a glycoprotein with myelin in rat brain. J Neurochem 21:1177-1191

QUARLES RH, EVERLY JL, BRADY RO (1973b) Myelin associated glycoprotein: a developmental change . Brain Res 58:506-509

QUARLES RH, MCINTYRE LJ, PASNAK CF (1979) Lectin binding proteins in central nervous system myelin. Biochem J 183:213-221

RAINE CS (1984) Morphology of myelin and myelination. In: Morell P (ed) Myelin. Plenum press, New York, p 1-50

RAUVALA H et PIHLASKARI R (1987) Isolation and some characteristics of an adhesive factor of brain that enhances neurite outgrowth in central neurons. J Biol Chem 262:16625-16635

REEBER A, VINCENDON G, ZANETTA J-P (1981) Isolation and immunohistochemical localization of a Purkinje cell specific glycoprotein subunit from rat cerebellum. Brain Res 229:53-65

RIEGER F, DANILOFF JK, PINCON-RAYMOND M, CROSSIN KL, GRUMET M, EDELMAN M (1986) Neuronal cell adhesion molecules and cytotactin are colocalized at the node of Ranvier. J Cell Biol 103:379-391

ROUSSEL G, NUSSBAUM J-L, ESPINOSA DE LOS MONTEROS A et NESKOVIC NM (1987) Immunocytochemical localization of UDP-galactose: ceramide galactosyltransferase in myelin and oligodendroglial cells of rat brain. J Neurocytol 16:85-92

SAKAMOTO Y, KITAMURA K, YOSHIMURA K, NISHIJIMA T, UYEMURA K (1987) Complete amino acid sequence of PO protein in bovine peripheral nerve myelin. J Biol Chem 262:4208-4214

SCHLUESENER HJ, SOBEL RA, LININGTON C, WEINER HL (1987) A monoclonal antibody against a myelin oligodendrocyte glycoprotein induces relapses and demyelination in central nervous system autoimmune disease. J Neuroimmunol 12:4016-4021

SHARON N (1984) Surface carbohydrates and surface lectins are recognition determinants in phagocytosis. Immunology Today 5:143-147

SHARON N (1987) Bacterial lectins, cell-cell recognition and infectious disease. FEBS Letters 217:145-157

SORG BJA, AGRAWAL D, AGRAWAL HC, CAMPAGNONI AT (1986) Expression of myelin proteolipid protein and basic protein in normal and dysmyelinating mutant mice. J Neurochem 46:379-387

STERNBERGER NH, QUARLES RH, ITOYAMA Y, WEBSTER HDeF (1979) Myelin-associated glycoprotein demonstrated immunocytochemically in myelin and myelin-forming cells of developing rat. Proc Natl Acad Sci USA 76:1510-1514

SZUCHET S, POLAK PE, YIM SH (1986) Mature oligodendrocytes cultured in the absence of neurons recapitulate the ontogenetic development of myelin membranes. Dev Neurosci 8:208-221

TOWBIN H, STAEHELIN T, GORDON J (1979) Electrophoretic transfer of proteins from polyacrylamide gels to nitrocellulose sheets: procedure and some applications. Proc Natl Acad Sci USA 76:4350-4354

TRAPP BD, ITOYAMA Y, STERNBERGER NH, QUARLES RH, WEBSTER HDeF (1981) Immunocytochemical localization of PO protein in golgi complex membranes and myelin of developing rat Schwann cells. J Cell Biol 90:1-6

TRAPP BD, QUARLES RH, GRIFFIN JW, SUZUKI K (1983) The myelin associated glycoprotein and axon myelinating Schwann cell contact. J Neuropathol Exp Neurol 42:357-368

TRAPP BD, QUARLES RH, GRIFFIN JW (1984a) Myelin-associated glycoprotein and myelinating Schwann cell-axon interaction in chronic B,B'-Iminodipropionitrile neuropathy. J Cell Biol 98:1272-1278

TRAPP BD, QUARLES RH, SUZUKI K (1984b) Immunocytochemical studies of quaking mice support a role for the myelin-associated glycoprotein in forming and maintaining the periaxonal space and periaxonal cytoplasmic collar of myelinating Schwann cells. J Cell Biol 99:594-606

WAEHNELDT TV, LININGTON C (1980) Organization and assembly of the myelin membrane. In: BAUMANN N (ed) Neurological Mutations Affecting Myelination. Elsevier/North-Holland Biomedical Press, Amsterdam, p 389-412

WEBSTER HDeF, PALKOVITS CG, STONER GL, FAVILLA JT, FRAIL DE, BRAUN PE (1983) Myelin associated glycoprotein: electron microscopic immunocytochemical localization in compact developing and adult central nervous system myelin. J Neurochem 41:1469-1479

WOOD PM, BUNGE RP (1975) Evidence that sensory axons are mitogenic for Schwann cells. Nature 256:662-664

ZANETTA J-P, SARLIEVE LL, MANDEL P, VINCENDON G, GOMBOS G (1977a) Fractionation of glycoproteins associated to adult rat brain myelin fractions. J Neurochem 29:827-838

ZANETTA J-P, SARLIEVE LL, REEBER A, VINCENDON G, GOMBOS G (1977b) A protein fraction enriched in all myelin associated glycoproteins from adult rat central nervous system. J Neurochem 29:355-357

ZANETTA J-P, ROUSSEL G, GHANDOUR MS, VINCENDON G, GOMBOS G (1978) Postnatal development of rat cerebellum : massive and transient accumulation of Concanavalin A binding glycoproteins in parallel fiber axolemma. Brain Res 142:301-319

ZANETTA J-P, DONTENWILL M, MEYER A, ROUSSEL G (1985a) Isolation and immunohistochemical localization of a lectin like molecule from the rat cerebellum. Dev Brain Res 17:233-243

ZANETTA J-P, DONTENWILL M, REEBER A, VINCENDON G, LEGRAND CH, CLOS J, LEGRAND J (1985b) ConA-binding glycoproteins in the developing cerebellum of control hypothyroid rats. Develop Brain Res 21:1-6

ZANETTA J-P, MEYER A, KUCHLER S, VINCENDON G (1987a) Isolation and immunochemical study of a soluble cerebellar lectin delineating its structure and function. J Neurochem 49:1250-1257

ZANETTA J-P, DONTENWILL M, REEBER A, VINCENDON G (1987b) Expression of recognition molecules in the cerebellum of young and adult rats. NATO ASI Serie H 2:92-104

MYELIN GENE REGULATION IN THE PERIPHERAL NERVOUS SYSTEM

Joseph F. Poduslo
Molecular Neurobiology Laboratory
Departments of Neurology and
Biochemistry/Molecular Biology
Mayo Clinic and Mayo Foundation
Rochester, MN 55905 USA

INTRODUCTION

The regulatory mechanisms that control myelin gene expression are largely unknown. Further understanding of the transcriptional, translational, and posttranslational regulation of myelin gene expression will provide important information for advancing our unterstanding of human disease where demyelination is prominent as in, for example, the demyelinating peripheral neuropathies. In addition, understanding the complexities associated with the regulation of myelination will further our knowledge of the peripheral nerve during development and after injury, where demyelination and remyelination follow axonal degeneration and regeneration. The paradigms of crush injury and permanent transection of the adult rat sciatic nerve (classical Wallerian degeneration) have proven to be excellent models for evaluating the regulation of myelin gene expression. These experimental animal models of neuropathy are characterized by the presence and absence of axonal regeneration and subsequent myelin assembly (Poduslo, 1984; Poduslo et al., 1985). By evaluating the extent of myelin gene expression after nerve transection, information can be obtained as to the mechanisms by which myelin genes and their products are downregulated. Understanding this down-regulation

P_0 EXPRESSION

This laboratory has been interested in the regulation of myelination for approximately 10 years - an interest that has evolved from studies on the biosynthesis of both protein and lipid components of myelin. The earlier work has focused on the major myelin glycoprotein, P_0, a 29,000 dalton protein which spans the bilayer and has a single complex-type, Asn-linked, oligosaccharide chain (Poduslo, 1984, 1985; Lemke and Axel, 1985; Poduslo and Yao, 1985; Sakamoto et al., 1987). We have demonstrated that the P_0 biosynthesis is affected by perturbation of the environment of the peripheral nerve Schwann cells. For example, if adult rat sciatic nerve is crush-injured, axonal degeneration and demyelination occur ultimately resulting in Schwann cell proliferation, nerve fiber regeneration, and remyelination (Poduslo, 1984). As the crushed nerve remyelinates, increased amounts of P_0 containing the normal complex-type oligosaccharide structure are produced (Poduslo, 1985). In contrast, if the rat sciatic nerve is permanently transected such that axonal regeneration and subsequent myelin assembly are prevented, the level of P_0 expression is reduced both at the level of P_0 mRNA and protein and the extent of oligosaccharide processing is altered (LeBlanc et al., 1987; Gupta et al., 1988; Poduslo, 1984, 1985; Poduslo et al. 1984, 1985a, 1985b). A dramatic reduction occurs in the level of P_0 mRNA with values that are approximately 20% of those found in the adult and 8-10% of those found in a 21-day-old animal (LeBlanc et al., 1987). These results, therefore, indicate that the regulation of the expression of this major myelin glycoprotein occurs predominantly at the transcriptional level in the distal segment of the permanently transected nerve. Similar observations have been made with regard to the expression of the myelin basic proteins (MBPs) and the P_2 protein. The persistence of P_0 and MBP mRNA levels with time after transection injury is surprising and probably relates of the capacity of Schwann cells to respond to reentry of axons with subsequent remyelination. The continued expression of P_O and these other myelin specific genes by Schwann cells after transection injury infers modulation of gene expression by the axon in a constitutive rather than all-or-non manner (LeBlanc and Poduslo, submitted). In addition, post-transcriptional mechanisms have also been postulated with regard to the regulation of the P_O gene (LeBlanc and Poduslo, submitted).

In contrast, we have reported that the steady state levels of P_O and MBP in the distal segment of the crushed sciatic nerve at 35 days after injury, where axonal regeneration and remyelination are permitted, are similar to those found in the normal nerve (Gupta et al., 1988). A major increase in P_O an MBP mRNA occurs between 7 and 14 days after crush injury, although the steady state protein level continues to drop off at this time. Such data suggest potential translational control at 14 days after crush injury which needs to be further evaluated.

In addition, we have demonstrated that the level of newly translated P_O is reduced (Poduslo, 1984, 1985) and that the extent of its oligosaccharide processing is altered after permanent transection (Poduslo, 1985). The majority of the P_O molecules are found to contain the immature $Man_{7-8}GlcNAc_2$ oligosaccharide chain with only a small amount of the glycoprotein maturing to the complex-type. Therefore, in the absence of myelin assembly, the PNS Schwann cells dramatically alter both the biosynthetic rate and posttranslational processing of the myelin component. A question arises from the studies of the transected nerves concerning the ultimate fate of P_O in the absence of myelination. Does the glycoprotein simply build up in the cell where there is no myelination, or is there a mechanism to regulate the cellular level of this component? This question led us to examine the metabolic fate of P_O in the transected nerve assaying the lifetime of the glycoprotein by performing precursor pulse-chase analyses. As opposed to similar studies conducted on crush-injured sciatic nerve, we discovered that the Schwann cells of the transected nerve regulate the intracellular levels of P_O by degrading the glycoprotein through specific delivery to the lysosome 1-2 hours after biosynthesis (Brunden and Poduslo, 1987). This routing occurs sometime after the glycoprotein has progressed to the medial Golgi and seems to require the presence of the oligosaccharide moiety on the protein. We further document that this targeting to the lysosome does not occur by the recognition of mannose-6-phosphate residues, a targeting system responsible for the delivery of acid hydrolases to the lysosome. It is apparent that the level of P_O biosynthesis by Schwann cells of transected nerve is higher than initially believed with the majority of the glycoprotein being targeted to the lysosome for degradation shortly after biosynthesis. Therefore, axonally-deprived, transected nerve Schwann cells not only alter the transcription and subsequent expression of this myelin component, but also alter the intracellular protein processing and targeting. These experimental paradigms can, therefore, be used to address the targeting mechanism of P_O delivery to the lysosome after transection injury, as well as to myelin after crush injury.

GLYCOLIPID EXPRESSION

With regard to Schwann cell expression of myelin specific glycolipids, we have demonstrated that there is a switch from the biosynthesis of galactocerebrosides to form glucocerebrosides and a series of oligohexosylceramides with globoside being a major product after permanent transection injury (Yao and Poduslo, 1988). Such a switch of form glucocerebroside also occurs with explants from neonatal nerve, as well as from purified Schwann cells (Yao et al, submitted). This regulation of glycolipid biosynthesis by Schwann cells has been demonstrated to involve both UDP-galactose:ceramide galactosyltransferase (CGalT) and UDP-glucose:ceramide glucosyltransferase (CGlcT) (Ceccarini and Poduslo, 1989). By measuring enzyme activities of CGalT and CGlcT with time after crush and transection injury of the adult rat sciatic nerve, it was found that a 50% reduction of CGalT specific activity was observed within the first 4 days after both injuries. Such activity remained unchanged at 7 days after injury; however, by 14 days the CGalT activity diverged in the two models. An increase in activity in the crushed nerve was observed by 21 days which reached control values while a further decrease was observed in the transected nerve such that the activity was nearly immeasurable by 35 days.

In contrast, the CGlct activity showed a rapid increase between 1 and 4 days, plateaued to a level that was 3.4 fold greater than the normal nerve, and persisted throughout the observation period in both the crush and transection models. Such data indicate that this shift in myelination-dependent glycolipid biosynthesis is controlled at least partially at these sugar transferases.

These studies suggest multiple independent regulatory mechanisms by which the differential expression of myelin specific genes are controlled. For the glycolipids, the regulation likely occurs in the ER and involves both CGalT and CGlcT. For the P_O and the MBPs, transcriptional control predominates in a constitutive rather than an all-or-nothing manner. A significant level of P_O can be demonstrated to undergo translation with efficient catabolic degradation by lysosomes to account for the decreased level of P_O. Since both transcriptional and post-translational regulation of the expression of P_O is observed, it is hypothesized that similar regulation occurs for other myelin specific proteins.

MAG EXPRESSION

Although the level of the major myelin glycoprotein, P_O, is transcriptionally and posttranslationally regulated within transected nerve Schwann cells, the generality of this mechanism as a means of regulating myelin components is unknown. If a similar degradative mechanism could be demonstrated with other myelin proteins, one would obtain a greater understanding of the methods employed by Schwann cells in controlling myelination. Likewise, if a distinct targeting signal directs the lysosome-bound species of P_O, it would seem that such a signal might route other myelin proteins that have been demonstrated to undergo lysosomal catabolism.

Perhaps the most extensively studied of these myelin glycoproteins is MAG, a 100 kD myelin associated glycoprotein that contains approximately 30% carbohydrate by weight (Quarles, 1983). MAG is a minor constituent of both CNS and PNS myelin with levels of the glycoprotein being less than 1% of total myelin protein (Quarles, 1983). Since MAG is processed in the ER and Golgi prior to normal delivery to myelin, it might follow a degradative pathway similar to that of P_O.

While MAG might be expected to be transcriptionally and posttranslationally regulated within the PNS in a fashion analogous to P_O, sensitive detection methods must be employed to assay the de novo biosynthesis of this myelin glycoprotein. Since MAG is a minor component of both CNS and PNS myelin, the levels of its biosynthesis may be low. Frail et al. (1985) have demonstrated that in vitro translation of either adult rat brain cytoplasmic RNA or sciatic nerve cytoplasmic RNA, followed by immunoprecipitation of the translation products with anti-MAG, resulted in readily detectable levels of the glycoprotein. Less than 25 adult Sprague-Dawley rats were needed to obtain sufficient quantities of sciatic nerve RNA for these experiments. Hence, by increasing the numbers of animals, it should be possible to measure the levels of MAG biosynthesis in vivo in the sciatic nerves of normal, crushed, transected, and 21 day old animals by precursor incorporation into sciatic nerve endoneurial slices (Poduslo, 1984, 1985). [3H]fucose has been used as an efficient label for MAG in brain (Poduslo et al., 1977). In particular, [3H]fucose labeled MAG was the basis for establishing a novel RIA for measurement of MAG (Poduslo and McFarlin, 1978). The precursor incorporation studies will allow evaluation of the role of lysosomal degradation in the regulation of MAG levels within the transected nerve. The elucidation of the levels of steady-state and newly transcribed MAG mRNA within the different models should be feasible with the availability of a MAG cDNA clone (Salzer et al., 1987). All of the above measurements will be particularly manageable if MAG expression is regulated in a manner comparable to P_O.

It is, of course, quite possible that MAG is not transcriptionally or posttranscriptonally regulated like P_O. This would be readily discernible from the experiments comparing the levels of MAG mRNA and in vivo MAG biosynthesis. If MAG is not regulated like other PNS myelin components, the question arises as to the role and location of this glycoprotein. If MAG is located throughout myelin, it would seem likely that it would be regulated like P_O and MBP. In contrast, if MAG is indeed found only in the periaxonal regions of myelin (Trapp et al., 1984), one might envision that the regulation of its expression would not be analogous to myelin components that show a more uniform distribution. This could be particularly true in the crushed nerve and young (21 d) sciatic nerve, where it is conceivable that MAG expression may not be significantly elevated after the deposition of the initial myelin lammellae. This, of course, would be in contrast to the elevated P_O and MBP expression in the crushed and 21 day old nerve. Similarly, the possibility exists that MAG is not targeted to the lysosomes after synthesis in the transected nerve. Again, this possibility may be greater if MAG is localized in periaxonal regions, as the glycoprotein may not be recognized and regulated as a "true" myelin component if this is the case. If the data reveal that MAG is neither transcriptionally, translationally nor posttranslationally regulated like P_O, the generality of myelin regulation may be questionable, and such regulation may depend on the specific localization of the myelin components.

If the MAG mRNA levels and the in vivo translation studies described above reveal that MAG is synthesized in the transected nerve, it becomes important to examine the fate of the glycoprotein. Like P_O, the production of MAG in the transected nerve Schwann cells would seem to serve no purpose, given the lack of myelin assembly in this injury model. Thus, the Schwann cells must "clear" themselves of this unneeded MAG in some way, with a lysosomal delivery system analogous to that observed for P_O being a likely mechanism. To investigate this possibility, 35 day transected endoneurial slices could be incubated withe either [^{3}H]amino acid mixture, [^{3}H]fucose, or [^{3}H]mannose for 1-2 h, followed by a 2-3 h chase in the absence of these isotopes. These precursors will allow MAG to be metabolically labeled in both its polypeptide and oligosaccharide moieties. Under similar conditions, P_O that is labeled during the pulse period is found to be degraded by the end of the chase interval (Brunden and Poduslo, 1987). The fate of MAG in the pulse-chase experiment would be evaluated by immunoprecipitating the glycoprotein with anti-MAG antibody, followed by SDS-PGE analysis (Poduslo and Rodbard, 1980) and visual inspection of the resulting fluorogram for catabolism of the glycoprotein during chase. If the results indicate that MAG is not being degraded within 2-3 hours after synthesis, it would seem that this myelin component is not being posttranslationally regulated like P_O. To insure that the delivery of this glycoprotein is not occuring at a slower pace than P_O, a similar precursor pulse would be performed with an extended chase of up to 6-8 hours. If after this time it is concluded that MAG is not posttranslationally regulated, then this gene would clearly be regulated in a manner different than P_O.

If gel analysis reveals that MAG is catabolised during chase, it would seem probable that it is being routed to the lysosomes in a manner similar to P_O. To assay the involvement of lysosomes in the degradation of MAG, pulse-chase studies could be performed in the presence of the lysosomal inhibitors, NH_4Cl and methionine methyl ester. As described previously (Brunden and Poduslo, 1987), these two inhibitors alter lysosomal activity in differing fashions, so that if both inhibit MAG degradation, it is safe to assume that the degradative organelles are involved in the catabolism of this myelin component. If MAG catabolism is not inhibited by methionine methyl ester, the fate of the glycoprotein could be further analyzed by assaying for excretion of the molecule from the cell into the surrounding medium. This likelihood is slim, given the integral membrane behavior of the glycoprotein.

Likewise, the involvement of the ubiquitin cytosolic degradative system seems unlikely because of the membranous location of MAG throughout its lifetime. If MAG is degraded within the lysosomes of the transected nerve, it becomes imperative to define the site of departure. If MAG routing is similar to that described for P_0, the molecule would exit for degradation after the action of Golgi GlcNAc transferase I, the enzyme that precedes mannosidase II. To evaluate this, pulse-chase studies would be done in the presence and absence of swainsonine (SW), using [^{3}H]fucose as a precursor. SW is an inhibitor of both Golgi mannosidase II and lysosomal mannosidases. While SW inhibits the catabolism of P_0 in the transected nerve, it is not definitively known if this inhibition is a result of SW action on Golgi mannosidase II, lysosomal mannosidases, or both. Regardless of the site of action, evidence of fucosylation in the presence of SW indicates that the molecule has progressed, to the site of GlcNAc transferase I, as it has been demonstrated that this enzyme must act on oligosaccharides prior to fucosyltransferase (Schacter, 1986). Thus, if fucosylated MAG persists following chase in the presence of SW, the glycoprotein is being delivered after the action of GlcNAc transferase I, analogous to that found with P_0.

If all of the above leads us to the conclusion that MAG follows a pathway like that of P_0 in the transected nerve, then it would be of interest to define a targeting signal on MAG. Tunicamycin treatment of transected nerve endoneurial slices results in the formation of a non-glycosylated species of P_0. This species is not targeted to the lysosomes, indicating the need of oligosaccharide for delivery. A similar experiment would be done to evaluate the role of oligosaccharide in the delivery of MAG. If, like P_0, oligosaccharide moieties are necessary for MAG targeting, the subsequent step would be to examine MAG for any unique signal that may have been discovered for P_0. It is unwise to use MAG in initial oligosaccharide characterization studies, since the glycoprotein has 8 potenial Asn-linked sites (Salzer et al. 1987) that would greatly complicate analyses. When a defined oligosaccharide structure is found for the lysosomal species of P_O, a similar lysosomal species of MAG will allow us to investigate the universality of any proposed signal.

ACKNOWLEDGEMENTS

The research described in this review was supported by the National Institute of Health (NS-20551) and the Borchard Fund.

REFERENCES

Brunden KR, Berg CT, Poduslo JF (1987) Isolation of an integral membrane glycoprotein by chloroform-methanol extraction and C_3-HPLC separation. Anal Biochem 164:474-481

Brunden KR, Poduslo JF (1987) Lysosomal delivery of the major myelin glycoprotein in the absence of myelin assembly: Post-translational regulation of the levels of expression by Schwann cells. J Cell Biol 104:661-669

Costantino-Ceccarini E, Poduslo JF (1989) Regulation of UDP-galactose:ceramide glucosyltransferase and UDP-glucose:ceramide glucosyltransferase after crush and transection nerve injury. J Neurochem 53:205-211

Frail DE, Webster H deF, Braun PE (1985) Developmental expression of the myelin associated glycoprotein in the PNS is different from that in the CNS. J Neurochem 45:1308-1310

Gupta SK, Poduslo, JF, Mezei C (1988) Temporal changes in P_O and MBP gene expression after crush-injury of the adult peripheral nerve. Mol Brain Res 4:133-141

LeBlanc AC, Poduslo JF. Axonal modulation of myelin gene expression in the peripheral nerve. (submitted)

LeBlanc AC, Poduslo JF, Mezei C (1987) Gene expression in the presence or absence of myelin assembly. Mol Brain Res 2:57-67

Lemke G, Axel R (1985) Isolation and sequence of a cDNA encoding the major structural protein of peripheral myelin. Cell 40:501-508

Poduslo JF (1984) Regulation of myelination: Biosynthesis of the major myelin glycoprotein by Schwann cells in the presence and absence of myelin assembly. J Neurochem 42:493-503

Poduslo JF (1985) Prost-translational protein modification: Biosynthesis control of mechanisms in the glycosylation of the major myelin glycoprotein by Schwann cell. J Neurochem 44:1194-1206

Poduslo JF, Berg CT, Dyck PJ (1984) Schwann cell expression of a major myelin glycoprotein in the absence of myelin assembly. Proc Natl Acad Sci USA 81:1864-1866

Poduslo JF, Berg CT, Ross SM, Spencer PS (1985) Regulation of myelination: Axons not required for the biosynthesis of basal levels of the major myelin glycoprotein by Schwann cells in denervated distal segments of the cat sciatic nerve. J Neurosci Res 14:177-185

Poduslo JF, Dyck PJ, Berg CT (1985) Regulation of myelination: Schwann cell transition from a myelin maintaining state to a quiescent state after permanent nerve transection. J Neurochem 44:388-400

Poduslo JF, Everly JL, Quarles RH (1977) A low molecular weight glycoprotein associated with isolated myelin: Distinction from myelin proteolipid protein. J Neurochem 28:977-986

Poduslo JF, McFarlin DE (1978) Immunogenicity of membrane surface glycoprotein associated with central nervous system myelin. Brain Res 159:234-238

Poduslo JF, Rodbard D (1980) M_r estimation using sodium dodecyl sulfate-pore gradient electrophoresis. Anal Biochem 101:394-406

Poduslo JF, Yao JK (1985) Association and release of the major instrinsic membrane glycoprotein from peripheral nerve myelin. Biochem J 228:43-54

Quarles RH (1983) Myelin-associated glycoprotein in development and disease. Dev Neurosci 6:285-303

Sakamoto Y, Kitamura K, Yoshimura K, Nishijima T, Uyemura K (1987) Complete amino acid sequence of P_0 protein in bovine peripheral nerve myelin. J Biol Chem 262:4208-4214

Salzer JL, Holmes WP, Colman DR (1987) The amino acid sequences of the myelin-associated glycoproteins: homology to the immunoglobulin gene superfamily. J Cell Biol 104:957-965

Schacter H (1986) Biosynthetic controls that determine the branching and microheterogeneity of protein-bound oligosaccharides. Biochem Cell Biol 64:163-181

Trapp BD, Quarles RH, Suzuki K (1984) Immunocytochemical studies of quaking mice support a role for the MAG in forming and maintaining the periaxonal space and periaxonal cytoplasmic collar of myelinating Schwann cells. J Cell Biol 99:594

Yao JK, Poduslo JF (1988) Biosynthesis of neutral glucocerebroside homologues by Schwann cells in the absence of myelin assembly. J Neurochem 50:630-638

Yao JK, Windebank AJ, Poduslo JF, Yoshino JE. Axonal regulation of Schwann cell glycolipid biosynthesis. (submitted)

GTP-Binding Proteins associated with CNS Myelin

Peter E. Braun and Lise Bernier
Department of Biochemistry
McGill University
Montreal, Canada, H3G 1Y6

Introduction

The unexpected discovery of cholinergic receptor activity (Larocca et al, 1987) in isolated myelin and the abundant, dynamic metabolism of polyphosphoinositides of PNS and CNS myelin (Kahn and Morell, 1988; Eichberg et al 1989) is forcing a reevaluation of the role of this membranous organelle in the nervous system. Implicit in these considerations is the probable participation of GTP-binding proteins (G proteins) in the signal transduction apparatus.

There are two other observations pertinent to possible modes of signalling and targetting that employ GTP-associated mechanisms: Chan et al (1988) have reported the specific and stoichiometric binding of GTP to purified myelin basic protein (MBP); Bernier et al (1989) and Sprinkle and Hancock (1989) have noted possible GTP binding domains in 2',3'-cyclic nucleotide 3'-phosphodiesterase (CNP). We suggest that these latter observations, when viewed in the context of emerging concepts of GTP-mediated regulation of secretion, endocytosis and exocytosis (Bourne, 1988; Mayorga et al, 1989) point to possible modes of intracellular targetting of myelin components in myelinogenesis.

In order to assess the range of GTP-binding proteins associated with CNS myelin and oligodendrocytes we have used several approaches: 1) the ligand-blot overlay (Bhullar and Haslam,1987) to obtain a profile of low molecular weight proteins that bind GTP on nitrocellulose electroblots; 2) G protein-specific antibodies; 3) ADP-ribosylation by cholera and pertussis toxins.

Material and Methods

Mouse and rat brain myelin was isolated by the procedure of Norton and Poduslo (1973). Primary cultures of Oligodendrocytes were provided by Dr. Voon Wee Yong; they were prepared from myelinating rat brain at d 19 and purified by the procedure of Yong et al (1988). A crude membrane fraction of these cells was obtained as follows: After removal of culture medium 18×10^6 cells in dishes were washed once with Hepes buffer (20 mM, pH 7.5) containing KCl (3 mM), $MgCl_2$ (3 mM) and protease inhibitors. Cells were then scraped into the same buffer (10 ml total) and homogenized by hand (15 strokes of the Dounce pestle). Large cell particulates and nuclei were removed by centrifugation at 3000 x g, for 10 min. and total membranes were recovered from the supernatant by centrifugation for 30 min. at 50.000 x g. The membrane-containing pellet was dispersed in sample buffer in preparation for SDS-PAGE.

We used the procedure of Bhullar and Haslam (1987) to detect G_n proteins. Membrane proteins were separated by SDS-PAGE in 15 % minigels, electroblotted onto nitrocellulose and overlaid with labeled GTP (2 μCi α^{32}P-GTP or 4 μCi γ^{35}S-GTP per ml). The overlay solution also contained 10 μM ATP and 50 μM $MgCl_2$ to reduce non-specific binding of GTP to proteins and nitrocellulose.

Proteins were ADP-ribosylated by a modification of the procedure of Chan et al (1988). Membranes (200 μg protein) were incubated at 37 ° with activated cholera toxins (2.5 μg) or pertussis toxin (2 μg) in 200 μl of a buffer containing 100 mM phosphate (pH 7.5), 10 mM thymidine, 1 mM ADP, 0.1 mM GTP, 4 mM ADP-ribose, 2.5 mM $MgCl_2$, 1 mM dithiothreitol, and 0.05 % Triton X-100. The reaction was initated by the addition of ^{32}P-NAD in 5 sequential aliquots of 2 μCi each at 10 min intervals and terminated by the addition of sample buffer in preparation for SDS-PAGE. Proteins were electroblotted onto nitrocellulose and labeled bands visualized by autoradiography.

G proteins were visualized by Western blotting according to the method of Towbin et al, (1979). Antisera prepared against synthetic peptide regions of $G_{i\alpha}$, $G_{o\alpha}$ and $G_{s\alpha}$ were provided by Dr. Alan Spiegel, National Institutes of Health, USA. Anti-*ras*, was obtained from Dupont, Inc.

Results

Immobilized Proteins Detectable by GTP Overlay

In Fig. 1 we show that myelin possesses GTP binding components in the molecular weight range of 18-25 kD. The two major bands (sometimes seen as one broad band) co-migrate with the 21.5 kD MBP and with proteolipid protein (PLP; 25 kD).

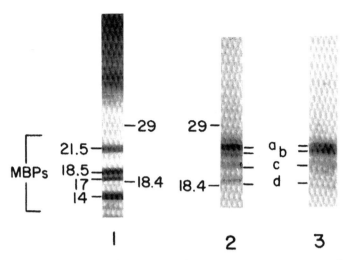

Figure 1. Low molecular weight GTP-binding proteins in CNS myelin. Lane 1, Coomassie blue stained gel of mouse brain myelin proteins; Lane 2, autoradiogram of CNS myelin proteins on a Western blot overlayed with γ^{35}S-GTP; Lane 3, same as lane 2 except that α^{32}P-GTP was the binding ligand. M_r of GTP-binding components: a, ~ 25 kd; b, ~ 24 kd; c, ~ 22 kd; d, ~ 19 kd

In order to ascertain their possible identity with these major myelin proteins we tested the ability of isolated MBPs (prepared by either acid or salt extraction), and of purified PLP to bind GTP by this approach. None of these bound GTP above background levels. In addition (data not shown), the membranous residues from the extractions retained, undiminished, the full complement of these low M_r GTP binding proteins (termed G_n by Bhullar and Haslam, 1987). The specificity of GTP binding was apparent when pre-treatment of the electroblots with ATP or GMP failed to block GTP binding, whereas unlabelled GDP or GTP partially or completely prevented labelled GTP binding (Fig. 2).

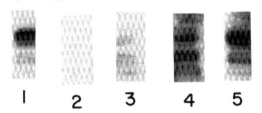

Figure 2. Specificity of nucleotide binding. Electroblots of CNS myelin proteins were preincubated with unlabelled nucleotides for 60 min., followed by an overlay with α^{32}P-GTP in the presence of the same nucleotide. G_n proteins were visualized by autoradiography. Lane 1, control (only labeled GTP); Lane 2, GTP; Lane 3, GDP; Lane 4, GMP; Lane 5, ATP.

Proteins Detectable by ADP-Ribosylation

ADP-ribosylation of polypeptides by cholera or pertussis toxins is often diagnostic of signal transducing G proteins (Gilman, 1987). When we separated the ADP-ribosylated proteins from myelin by SDS-PAGE we noted that the predominant proteins modified by cholera toxin were the MBP isoforms (Fig.3) in agreement with the report of Chan et al (1988).

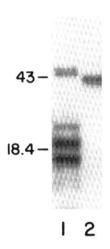

Figure 3. Autoradiogram showing ADP-ribosylation of CNS myelin proteins. Mouse brain myelin proteins were ADP-ribosylated by cholera toxin (Lane 1) or pertussis toxin (Lane 2). After SDS-PAGE, proteins were electroblotted onto nitrocellulose and visualized by autoradiography.

Two polypeptides (M_r 44-46 kD) selectively modified by cholera toxin and the major polypeptide(s) ($M_r \sim 40$ kD) selectively modified by pertussis toxin corresponded to the reported behavior of three well known alpha subunits of signal transducing G proteins (Gilman, 1987). A less prominent component also ADP-ribosylated exclusively by pertussis toxin (not visible in Fig. 3) corresponds in M_r exactly to CNP. In order to verify its identity, we immuno-precipitated the labelled myelin with CNP-specific antibody and demonstrated that the immunecomplex contained ADP-ribosylated CNP (not shown).

Proteins Detectable by Anti-G-Proteins

When electroblots of myelin proteins were exposed to antisera specific for authentic G proteins, we detected the presence of the alpha subunits for G_o, ($\sim$ 40 kD), G_i ($\sim$ 40 kD), and G_s ($\sim$ 45 kD), as well as the proto-oncogene product, ras ($\sim$ 21 kD) (Fig.4).

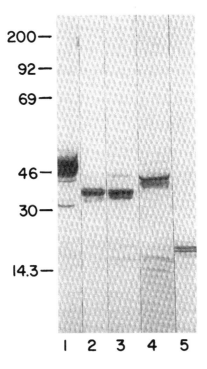

Figure 4. Identification of G proteins by immunoreactivity on a Western blot. Mouse brain myelin proteins were separated by SDS-PAGE in a 11-23 % gel and electroblotted. Primary antisera were used at 1:1000. Lane 1, CNP; Lane 2, G_o; Lane 3, G_i; Lane 4, G_s; Lane 5, *ras*.

Similarly, when we immunostained electroblots of oligodendrocyte membranes with the same antibodies all four categories of GTP-binding proteins were evident (data not shown). By immunostaining the electroblots containing the labeled, ADP-ribosylated proteins, we determined that the $G_{s\alpha}$ corresponded exactly to the cholera toxin modified polypeptides and that $G_{i\alpha}$ and $G_{o\alpha}$ corresponded to the pertussis toxin modified components, as expected from their recognized properties (data not shown; Gilman 1987). It should be noted that the alpha subunits of G_i and G_o are not completely resolved in this gel system and appear as one band, accounting also for the single band observed by pertussis toxin ADP-ribosylation.

Discussion

Our interest in the GTP binding proteins of myelin began, in part, with our discovery (Bernier et al, 1989) independent of that of Sprinkle and Hancock

(1989) that CNP shares at least two regions of primary sequence with known G proteins:

1) the domain gly-x-x-x-ser-gly-lys-ser-thr-leu of CNP is also shared by $G_{s\alpha}$, $G_{i\alpha}$, $G_{t\alpha 1}$, $G_{t\alpha 2}$ and, with a few conservative substitutions by *ras* and the GTP-binding elongation factor in the protein synthesis machinery. This is believed to be the phosphoryl moiety binding site; 2) the domain cys-x-x-x-COOH at the C-terminus, which is the site for pertussis toxin-catalyzed ADP-ribosylation of known G proteins and is absent from almost all other known proteins except those that might be acylated or isoprenylated at the C terminus (Magee and Hanley (1988)). There are in CNP only a few other minor regions of shared homology with G proteins, and overall ancestral homology cannot be claimed. Experimentally, what began as a search for the physiological function of CNP has become a wider investigation of GTP binding proteins associated with myelinating systems. We have observed relatively low levels of direct GTP binding to CNP immobilized on nitrocellulose and of pertussis toxin catalyzed ADP-ribosylation of CNP in myelin. Although these two characteristics might be qualitatively anticipated from considerations of the CNP primary structure their low abundance relative to known GTP-binding proteins may, on the one hand, represent an artifact; on the other hand, most G proteins also do not bind GTP by blot overlays, and their ADP-ribosylation is not always strong because of a variety of extenuating factors. Further investigations of GTP binding to CNP by other approaches, using highly pure CNP prepared by recombinant DNA strategies may shed further light on this matter, and are underway in our lab.

Our failure to verify GTP binding to MBPs by the blot overlay was not altogether unexpected and may reflect loss of native binding domains following SDS-PAGE and electroblotting. Although cholera toxin catalyzes an avid ADP-ribosylation of MBPs, this must be interpreted with caution, since this toxin is a rather promiscuous catalyst, modifying many proteins unrelated to GTP binding. The finding of the adenylate cyclase stimulatory regulator ($G_{s\alpha}$) and the inhibitory regulator ($G_{i\alpha}$) in myelin was anticipated from the work of Larocca et al (1987) who demonstrated cholinergic receptors and adenylate cyclase associated with purified myelin. The presence of $G_{o\alpha}$ is interesting in that not all cell types possess it (Gilman, 1987). Although its exact function remains unknown, there are indications that it may regulate ion-channels (Gilman, 1987), in addition to other possible roles it might have. The same can be said for the proto-oncogene product, *ras;* its presence in oligodendrocytes is expected, given its wide distribution, but its occurrence in myelin is also worthy of note. Although the precise role of *ras* is still unknown, (Barbacid, 1987), recent evidence points to *ras* participation in the inositol phospholipid pathway that is associated with signal transduction (Kamata and Kung, 1988). The persistent association of these four G proteins with myelin membranes despite extensive purification strengthens the case for ligand-responsive receptors or ion-channels in the sheath, presumably localized in the paranodal loops that remain attached to myelin lamellae during the isolation process.

A new class of GTP-binding proteins has recently been described (Bhullar and Haslam, 1987; 1988; Comerford et al, 1989). These are low molecular weight polypeptides ($M_r \sim 18$-27 kD) that bind GTP when immobilized on nitro-cellulose by electroblotting, and are referred to as G_n proteins. There are many G_n proteins in this size range, including *ras,* not all of them binding GTP with the same avidity. We found that both myelin and a crude membrane fraction of oligodendrocytes possessed at least four polypeptide species in this size range: two major GTP binding components (Mr ~ 25 kD and ~ 24 kD) and two minor components (Mr ~ 22 kD and ~ 19 kD). The band at $M_r \sim 19$ kD co-migrated with a minor component of the anti-*ras* immunostaining polypeptides. No function has yet been ascribed to this class of proteins, but it has been inferred that in platelets these proteins may participate in events giving rise to shape change, aggregation and secretion (Nagata et al, 1989); and Bourne (1988) has suggested that small GTP-binding proteins may mediate intracellular membrane traffic leading to secretion and endocytosis. We would like to carry this speculation one step further and suggest that these proteins may be involved in the process of myelin assembly.

References

Ashkenazi A, Ramachandran J and Capon DJ (1989) Acetylcholine analog stimulates DNA synthesis in brain-derived cells via specific muscarinic receptor subtypes Nature 340:146-150
Barbacid M (1987) *ras* genes. Ann Rev Biochem 56:779-828.
Barres BA, Chun LLY and Corey DP (1988) Ion channel expression by white matter glia: type 2 astrocytes and oligodendrocytes. Glia 1:10-30
Bernier L, Horvath E and Braun P (1989) GTP binding proteins in CNS myelin. Trans Amer Soc Neurochem 20:254

Bhullar R and Haslam RJ (1988) Gn-proteins are distinct from *ras* p.21 and other known low molecular mass GTP-binding proteins in the platelet. FEBS Letts 237:168-172

Bhullar RP and Haslam RJ (1987) Detection of 23-27 kDa GTP-binding proteins in platelets and other cells. Biochem J 245:617-620

Bourne HR (1988) Do GTPases direct membrane traffic in secretion? Cell 53:669-671.

Burgoyne RD (1987) Control of exocytosis. Nature 328:112-113

Chan KC, Ranwani J and Moscorello MA (1988). Myelin basic protein binds GTP at a single site in the N-terminus Biochem Biophys Res Comm 152:1468-1473.

Comerford JG, Gibson JR, Dawson AP, Gibson I (1989) RAS p21 and other Gn proteins ore detected in mammalian cell lines by $[\gamma^{35}S]GTP \gamma S$ binding. Biochem Biophys Res Comm 159:1269-1274

Eichberg J, Berti-Mattera LN, Day S-F, Lowery J and Zhu X (1989) Basal and receptor-stimulated metabolism of polyphosphoinositides in Peripheral nerve myelin. J Neurochem 52:524 A

Freissmuth M, Casey PJ and Gilman AG (1989) G proteins control diverse pathways of transmembrane signaling. The FASEB J 3:2125-2131.

Gilman AG (1987) G proteins: transducers of receptor-generated signals. Ann Rev Biochem 56:615-650

Hancock JF, Magee AI, Childs JE and Marshall CJ (1989) All *ras* proteins are polyisoprenylated but only some are palmitoylated. Cell 57:1167-1177

Kahn DW and Morell P (1988) Phosphatidic acid and phosphoinositide turnover in myelin and its stimulation by acetylcholine. J Neurochem 50:1542-1550

Kamata, T. and Kung, H.F. (1988) Effects of *ras*-encoded proteins and platelet-derived growth factor on inositol phospholipid turnover in NRK cells. Proc Natl Acad Sci (USA) 85:5799-5803

Kikuchi A, Yamamoto K, Fujita T and Takai Y (1988) ADP-ribosylation of the bovine brain *rho* protein by botulinum toxin type Cl. J Biol Chem 263:16303-16308.

Larocca JN, Ledeen RW, Dvorkin B and Makman MH (1987) Muscarinic receptor binding and muscarinic receptor-mediated inhibition of adenylate cyclase in rat brain myelin. J Neurosci 7:3869-3876

Magee T and Hanley M (1988) Sticky fingers and CAAX boxes. Nature 335:114-115

Mayorga LS, Diaz R and Stahl PD (1989) Regulatory role for GTP-binding proteins in endocytosis. Science 244:1475-1477

Miller RJ (1988) G proteins flex their muscles (1988) Trends in Neurosci 11:3-6.

Nagata K, Nagao S and Nozawa Y (1989) Low Mr GTP-binding proteins in human platelets: cyclic AMP-dependent protein kinase phosphorylates m22KG(I) in membrane but not c21KG in cytosol. Biochem Biophys Res Comm 160:235-242.

Norton WT, Poduslo SE (1973) Myelination in rat brain: method of myelin isolation. J Neurochem 21:749-757

Ribeiro FAP and Rodbell M (1989) Pertussis toxin induces structural changes in Gα proteins independently of ADP-ribosylation. Proc Nat'l Acad Sci (USA) 86:2577-2581

Santos E and Nebreda AR (1989) Structural and functional properties of *ras* proteins. The FASEB J 3:2151-2163.

Spiegel AM (1987) Signal transduction by guanine nucleotide binding proteins. Molec Cell Endocrin 49:1-16

Sprinkle TJ and Hancock J (1989) Identification of phosphoryl and nucleotide binding sites on CNPase. Trans Amer Soc Neurochem 20:255

Towbin H, Staehlin T and Gordon J (1979) Electrophoretic transfer of proteins from polyacrylamide gels to nitrocellulose sheets: procedure and some applications. Proc Nat Acad Sci USA 76:4350-4354

Ulmer JB and Braun PE (1983) In vivo phosphorylation of myelin basic proteins in developing mouse brain: evidence that phosphorylation is an early event in myelin formation. Dev Neurosci 6:345-355

Yong VW, Sekiguchi S, Kim MW and Kim SU (1988) Phorbol ester enhances morphological differentiation of oligodendrocytes in culture. J Neurosci Res 19:187-194.

MOLECULAR BIOLOGY OF GENES CODING FOR MYELIN PROTEINS

MOLECULAR APPROACHES TO THE STUDY OF THE REGULATION OF MYELIN SYNTHESIS

Arthur Roach
Division of Molecular Immunology and Neurbiology
Mount Sinai Hospital Research Institute
600 University Avenue
Toronto, M5G 1X5
Canada

INTRODUCTION

The process of myelination involves the coordinated expression of specific genes in oligodendrocytes and Schwann cells. Since the formation of myelin is the result of an interaction of one of the above cell types with an appropriate class of axon, it might be predicted that regulatory signals exist which allow for the recognition of the correct axonal type and specify the induction of expression of myelin-specific genes. A number of lines of evidence infer the existence of such signals, but the biochemical identity of the signals, and the mechanisms and pathways through which they act, remain unknown. In this article I shall explore what we can infer about this hypothesized regulation, and present some experimental approaches which may shed light on the questions raised.

The series of events in which Schwann cells proliferate and then divide out, ensheathe and myelinate PNS axons has been well described (Bunge, 1987; Raine, 1984) and will not be repeated here. However this well-studied behaviour provides strong grounds for the inference that a signal from an appropriate axon type produces a program of gene expression which includes the turn on of the myelin-specific genes.

Since Schwann cells taken from a nerve which normally produces no myelin will myelinate fibres when transplanted into a nerve which normally does contain myelinated fibres, it is clear that some component of the environment in which the grafted Schwann cells find themselves can instruct them to produce myelin (for review, see Bray et al., 1981). The instructive signal is likely to be a molecule on the surface of the new axons which invade the graft. Purely a diffusible signal is unlikely since individual fibres of certain classes are reproducibly myelinated while nearby axons of other classes are not. Similarly in the CNS oligodendrocytes are able to reliably identify and myelinate specific classes of axons but do not myelinate others. The possibility that inhibitory signals play a role in this regulation must also be considered. An axon-specific signal which blocks the expression of a myelination program in competent cells which associate with axons could result in similar behaviour. In either case, however, the high degree of specificity with which specific axonal types are recognized and myelinated strongly implies an axon-produced component to the signal.

Interestingly there is convincing evidence that in oligodendrocytes the initial appearance of the ability to express myelin-specific genes can occur in the absence of axons (Zeller et al., 1985; Dubois-Dalcq et al., 1986). However in this experimental system high levels of expression of myelin proteins are not seen, suggesting that some further environmental signals are required for active myelination. Thus newly-born oligodendrocytes may lose responsiveness to astrocyte-produced platelet-derived growth factor (PDGF) and acquire the ability to make several myelin proteins (Raff et al., 1988), but further stimulation, possibly by axons, is required for high level expression.

Since we know some of the genes which are subject to regulation in myelination, an approach starting in the cell nucleus can also be employed to elucidate regulatory pathways. Myelin basic protein (MBP) (Roach et al., 1983), proteolipid protein (PLP, or lipophilin) (Milner et al., 1985; Naismith et al., 1985), P_0 (Lemke, and Axel, 1985), myelin associated glycoprotein (MAG) (Arquint et al., 1987) and 2′,3′-cyclic-nucleotide 3′-phosphodiesterase (CNPase) (Kurihara et al., 1987) are all myelin-specific proteins for whose genes recombinant DNA probes have become available over the past six years. Each of their corresponding genes is a potential starting point from which can be explored the intracellular pathway by which myelin gene expression is regulated. Specifically, recombinant DNA techniques make it possible to identify regions of the genes involved in regulation, and ultimately, the molecules which act on those DNA sequence elements.

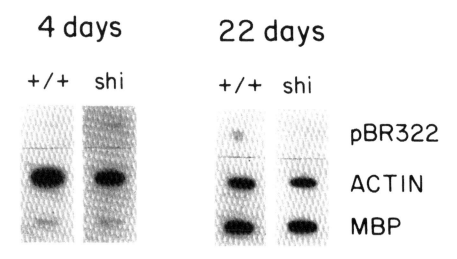

Figure 1. MBP gene transcription run-on assay with +/+ and *shi/shi* mouse brains. 2μg of cloned DNA was attached at each slot to nylon membrane. Membrane strips were hybridized for 72 hours with 2.5 X 10^6 cpm/ml pulse-labeled nuclear run-on probes from day 4 or day 22 +/+ or *shi/shi* brains, and washed in 1X SET, 0.2% SDS at 60C, 1 hr. Actin: a cDNA clone encoding mouse cytoplasmic actin (Minty et al., 1983); MBP: pex-1, containing a 3.3 kb HindIII fragment from the 5' end of the mouse MBP gene.

with radiolabelled RNA precursors under conditions which favour the extension of previously initiated transcripts. The resulting pulse-labelled nuclear RNA is hybridized with membrane-bound cloned DNA to identify the portion of the total which results from transcription of a specific gene. In the experiment shown in figure 1, cytoplasmic actin is included as a control, representing a gene which is abundantly expressed at all ages, and the prokaryotic plasmid pBR322, which is not present in the mouse genome, is a negative control. This experiment shows a striking increase in the rate of MBP transcription relative to actin transcription occurring over the first three weeks postnatally. Other data not shown here show a similar rise when the MBP rate is normalized to methionyl tRNA or cyclophilin gene transcription. Thus it is clear that in the early postnatal period a large increase in MBP transcription rate in the brain is accompanied by a large accumulation of mRNA and rapid synthesis of MBP. Therefore the influences which act on oligodendrocytes to induce myelination appear to lead to an activation of at least one myelin-specific gene, and the corresponding signal pathway must lead to the nucleus.

In addition to numerous available genes for study and comparison, the study of the molecular genetics of myelination is facilitated by several mutations in the mouse and rat which disrupt myelination. While the disruption of many developmental processes of the nervous system in mammals would lead to fatal consequences, myelin seems to be required only postnatally in the mouse, and even some severely hypomyelinated animals can often survive to breed. Some myelination mutations (eg. *shiverer, myelin deficient, jimpy*) have been shown to be in identified genes (Roach et al., 1985; Nave et al., 1986; Popko et al., 1987) while others (e.g. *Trembler, quaking*) appear not to affect directly any of the known myelin genes. Thus the mutations available may serve as tools to help identify regulatory features of already-identified myelin genes on one hand, and to point to previously unknown gene products that are essential for normal myelination on the other.

Amongst the myelin genes, the study of the genes encoding the myelin basic proteins is the most advanced, as a result of the efforts of several groups. Since the initial report of the cloning of a full-length cDNA for rat MBP (Roach et al., 1983), stidies on the binding of nuclear factors to the gene (Miura et al., 1989; Tamura et al., 1988) its subsequent transcription (Roch et al., 1989; Tosic et al., submitted; Wiktorowicz and Roach, manuscript in preparation), the splicing of the primary transcript into mRNAs (Takahashi et al., 1985; deFerra et al., 1985), and the translation of the mRNAs (Campagnoni et al., 1987) have all been reported. In addition two mutations (*shiverer* and *myelin deficient*)have been identified as lesions of the MBP gene (Roach et al., 1985; Akowitz et al., 1987), and the structures of the mutant alleles deduced. The cloned normal gene has been used to "rescue" mutant animals by transgenic methodologies (Readhead et al., 1987; Popko et al., 1987), and the shivering phenotype has been created by a transgenic antisense knockout approach (Katsuki et al., 1988).

RESULTS AND DISCUSSION

1. The transcription rate of the MBP gene is regulated during development.

The first three weeks of postnatal development see a large increase in levels of accumulation of MBP and its mRNA (Carson et al., 1983; Zeller et al., 1984; Sorg et al., 1987) and this may be the result of changes in any of several levels of regulation. To measure the rate of transcription of the MBP gene in development, the transcription run-on technique (Groudine et al., 1981) can be used. Isolated nuclei are pulse-labelled

2. Transcription of the undeleted 5' end of the *shiverer* MBP gene is also developmentally regulated.

The region(s) of the MBP gene required to respond to developmental signals can be deduced by a number of means. However, in vitro studies of the promoter cannot easily yield this information since development is not readily reproduced in vitro. Transgenic animals carrying cloned copies of the MBP gene on a *shiverer* background (Readhead et al., 1987) express MBP with correct developmental timing. Therefore the clone cos138 (Takahashi et al., 1985), which carries the entire 32 kb mouse MBP gene with 4 kb of 5' flanking and 2 kb of 3' flanking material, contains all sequences required for developmental regulation. The *shiverer* mouse also shows correct developmental regulation of transcription of the undeleted 5' portion of the gene as shown in figure 1 (and unpublished work), indicating further that the required sequence elements are in the 5' half of the gene.

When cloned fragments spanning the shiverer deletion breakpoint are used in the transcription run-on assay (figure 2) it is found that transcription of the partially-deleted MBP gene continues right up to the deletion breakpoint.

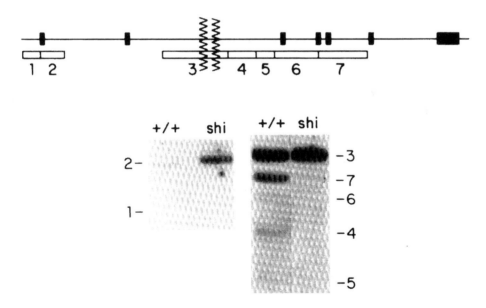

Figure 2. MBP transcription in the *shiverer* mouse continues up to the deletion breakpoint. 2μg of restriction digested DNA from clone pex-1 and pSPGM-1 (Roch et al., 1989) were blotted onto nylon membrane and hybridized with run-on probes from either 22 day +/+ or 22 day *shi/shi* brains. The numbers 1-7 identify the restriction fragments as they occur from 5' to 3' in the gene.

Despite a normal transcription rate, transcripts from the *shiverer* MBP gene do not accumulate to wild-type levels. Figure 3 represents the results of an S1 nuclease protection experiment in which all RNAs which initiate correctly at the 5' end of the MBP gene, regardless of further processing, are detected (Roach et al., 1985). In wild-type brain RNA the majority of these molecules are presumably the highly-abundant MBP mRNAs. However the primary transcript from the partially-deleted *shiverer* MBP gene lacks the normally-used polyadenylation, transcription termination and splice sites, and cannot be processed correctly. As a result, the transcripts which are made appear to be rapidly degraded and accumulate to less than 6% of the wild-type levels.

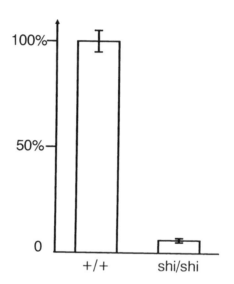

Figure 3. Accumulation of MBP transcripts (5'ends) in 18 day +/+ and *shi/shi* brains. The protected bands from an S1 nuclease protection assay for the 5' ends of correctly initiated MBP transcripts were quantitated by scintillation counting. This figure represents data from Roach et al., 1985.

3. The absence of MBP perturbs regulation of MBP gene transcription.

Transcription rate measurements have suggested an interesting possibility for the regulation of MBP synthesis at later stages of myelination. Several studies have shown a decrease in the accumulation of MBP mRNA after the peak which occurs approximately three weeks after birth in the mouse (Zeller et al., 1984; Sorg et al.,

1987), but the mechanisms leading to this decrease are unknown. Recent studies of transcription rates have shown that a decrease in MBP gene relative transcription rate also occurs after three weeks, falling by greater than three-fold between 3 and 7 weeks. However in the *shiverer* mouse which cannot synthesize MBP, this decrease does not occur, and transcription of the undeleted 5' end of the MBP gene continues at the maximal rate seen at 3 weeks (Wiktorowicz and Roach, in preparation). Thus it may be that the accumulation of mature MBP, or the synthesis of mature, compact myelin, is essential for the normal down-regulation of MBP gene expression at later stages in myelination, suggesting a regulatory feedback loop in myelination. It will be of particular interest to determine whether the genes for other major myelin proteins are similarly regulated, both in the presence and in the absence of mature myelin and MBP.

Figure 4 summarizes the model for expression of the MBP gene in normal and *shiverer* mice that is supported by the experimental results described above and elsewhere (Roach et al., 1985; Molineaux et al., 1986).

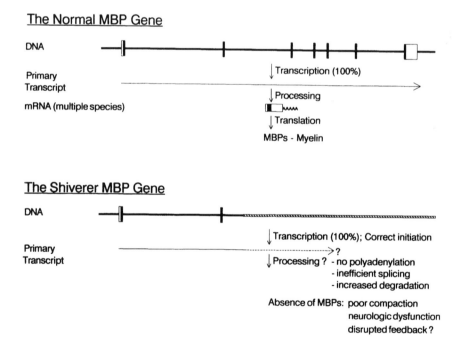

Figure 4. A summary of the expression of the partially-deleted *shiverer* MBP gene. MBP exons 1-7, spanning 32 kb, are indicated by boxes, white = untranslated, black = translated. The striped line represents non-MBP DNA which replaces the 3' end of the gene in *shiverer* mice.

4. A strategy for direct cloning of MBP transcription - inducing factors.

To further analyze the interactions of regulatory signals with the MBP gene, the precise regions of the gene required for transcription can be dissected by assay of deleted MBP gene constructs for transcription-promoting ability in cell free systems or in transfected cells. Both of these approaches have been employed successfully (Miura et al., 1989; Tamura et al., 1988) and the results also suggest one class of transcription factor which may regulate MBP gene expression. A second approach may allow for the cloning of the gene for a factor which promotes MBP transcription in transfected cells, and is outlined in figure 5. The strategy involves creating a special "primed" cell line which carries a selectable marker gene under control of the MBP promoter. It is important that the primed cell line does not express its endogenous MBP gene. The primed cell line is then transfected with human genomic DNA, or an expression library, and selective pressure is applied to allow to grow only those cells which now express the selectable marker gene. Some of the clones obtained may be the result of the integration and expression of a gene which activates transcription from the MBP promoter. Expression cloning strategies incorporating transfection with high molecular weight genomic DNA have been used successfully to clone genes for the nerve growth factor receptor (Chao et al., 1986) and a protein which regulates myogenic differentiation (Davis et al., 1987). A number of potential pitfalls inherent in the selection scheme outlined in figure 5 should not be overlooked, and some of these are described below.

It is possible that multiple additional gene activities are essential to activate the MBP promoter in the primed cell, perhaps with several gene products forming a complex, or with one gene product postranslationally modifying another. The use of several cell lines, including ones with some glial characteristics, would increase the chance of finding a line which requires just one additional gene product. Growth under selective conditions could indicate activation of the MBP promoter but could also be the result of rearrangements which introduce an active promoter - either endogenous or from the transfected DNA - into the vicinity of the selectable gene. Thus specific, predicted MBP-promoted transcripts should be assayed for to ensure that a given clone is the result of trans-activation of the MBP promoter. Finally, it may be difficult to identify the responsible exogenous DNA, if it is not closely linked to a highly repeated *Alu* sequence. This problem may be overcome by the tagging of transfected DNA with prokaryotic sequences instead of relying on closely-linked repeats.

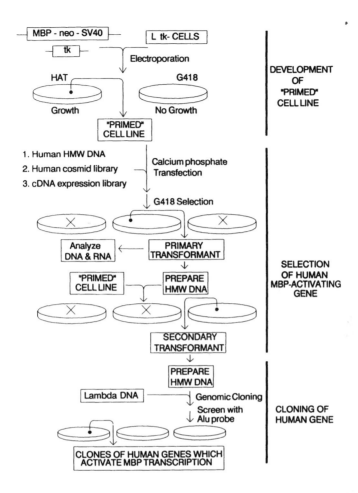

Figure 5. Diagrammatic representation of a strategy for the cloning of intracellular factors which activate MBP transcription. Initially a cell line is selected which carries the neomycin resistance gene under control of the MBP promoter, but does not express the gene. Subsequently a human gene is sought which confers G418 resistance when stably transfected into this "primed" cell line. The human gene responsible is then cloned based on proximity to *Alu* sequences, or marking with cotransfected prokaryotic DNA.

5. Intercellular signals promoting MBP synthesis.

The molecules of neuronal origin which act on, rather than within, oligodendrocytes and Schwann cells to induce the synthesis of myelin components have not been identified. However the availability of cloned probes for several of the major myelin genes provides an assay system for such molecules. Specifically, a fruitful approach may be to culture Schwann cells and oligodendrocytes with neuronal cells and membrane fractions of axonal origin, and test for induction of expression of specific myelin genes. Since early stages of gene expression may be activated without appearance of the protein product in myelin (Puckett et al., 1987), it may be possible to detect activities which regulate myelin synthesis but do not lead to incorporation of proteins into myelin in vitro due to the lack of axons or other environmental requirements.

SUMMARY

The era of recombinant DNA has made available tools for studying the earliest events in differentiation and myelin synthesis. However little has been learned about the signals and mechanisms which lead to the turning on and regulation of myelin-specific genes during the intimate association with selected axons which leads to myelination. Recent evidence summarized above suggests the possibility of a negative feedback mechanism requiring MBP, but further study is required to confirm this model, and determine how many genes in oligodendrocytes it might regulate. The intercellular signals which co-ordinate myelin synthesis with neuronal requirements thus remain elusive. The illumination of these important elements in the control of myelinogenesis may still prove to be amongst the most important contributions to our understanding of the process of myelination.

REFERENCES

Akowitz, A.A., Barbarese, E., Scheld, K., and Carson, J.H. (1987). Structure and expression of myelin basic protein gene sequences in the mld mutant mouse: reiteration and rearrangement of the MBP gene. Genetics 116, 447-464.

Arquint, M., Roder, J., Chia, L.S., Down, J., Wilkinson, D., Bayley, H., Braun, P., and Dunn, R. (1987). Molecular cloning and primary structure of myelin-associated glycoproteins. Proc. Natl. Acad. Sci. USA 84, 600-604.

Bray, G.M., Rasminsky, M., and Aguayo, A.J. (1981). Interactions between axons and their sheath cells. Ann. Rev. Neurosci. 4,127-162.

Bunge, R.P. (1987). Tissue culture observations relevant to the study of axon-Schwann cell interactions during peripheral nerve development and repair. J. exp. Biol. 132, 21-34.

Campagnoni, A.T., Hunkeler, M.J., and Moskaitis, J.E. (1987). Translational regulation of myelin basic protein synthesis. J. Neurosci. Res. 17, 102-110.

Carson, J.H., Neilson, M.L., and Barbarese, E. (1983). Developmental regulation of myelin basic protein expression in mouse brain. Dev. Biol. 96, 485-492.

Chao, M.V., Bothwell, M.A., Ross, A.H., Koprowski, H., Lanahan, A.A., Buck, C.R., and Sehgal, A. (1986). Gene transfer and molecular cloning of the human NGF receptor. Science 232, 518-521.

Davis, R.L., Weintraub, H., and Lassar, A.B. (1987). Expression of a single transfected cDNA coverts fibroblasts to myoblasts. Cell 51, 987-1000.

deFerra, F., Engh, H., Hudson, L., Kamholz, J., Puckett, C., Molineaux, S., and Lazzarini, R.A. (1985). Alternative splicing accounts for the four forms of myelin basic protein. Cell 43, 721-727.

Dubois-Dalcq, M., Behar, T., Hudson, L., and Lazzarini, R.A. (1986). Emergence of three myelin proteins in oligodendrocytes cultured without neurons. J. Cell Biol. 102, 384-392.

Groudine, M., Peretz, M., and Weintraub, H. (1981). Transcriptional regulation of hemoglobin switching in chicken embryos. Mol. Cell. Biol. 1, 281-288.

Katsuki, M., Sato, M., Kimura, M., Yokoyama, M., Kobayashi, K., and Nomura, T. (1988). Converstion of normal behaviour to *shiverer* by myelin basic protein antisense cDNA in transgenic mice. Science 241, 593-595.

Kurihara, T., Fowler, A.V., and Takahashi, Y. (1987). cDNA cloning and amino acid sequence of bovine brain 2',3'-cyclic-nucleotide 3'-phosphodiesterase. J. Biol. Chem. 262, 3256-3261.

Lemke, G., and Axel, R. (1985). Isolation and sequence of a cDNA encoding the major structural protein of peripheral myelin. Cell 40, 501-508.

Milner, R.J., Lai, C., Nave, K.-A., Lenoir, D., Ogata, J., and Sutcliffe, J.G. (1985). Nucleotide sequence of two mRNAs for rat brain myelin proteolipid protein. Cell 42, 931-939.

Minty, A.J., Alonso, S., Guenet, J.-L., and Buckingham, M.E. (1983). Number and organization of actin-related sequences in the mouse genome. J. Mol. Biol. 167, 77-101.

Miura, M., Tamura, T., Aoyama, A., and Mikoshiba, K. (1989). The promoter elements of the mouse myelin basic protein gene function efficiently in NG108-15 neuronal/glial cells. Gene 75, 31-38.

Molineaux, S.M., Engh, H., deFerra, F., Hudson, L., and Lazzarini, R.A. (1986). Recombination within the myelin basic protein gene created the dysmelinating shiverer mouse mutant. Proc. Natl. Acad. Sci. USA 83, 7542-7546.

Naismith, A.L., Hoffman-Chudzik, E., and Riordan, J.R. (1985). Study of the expression of myelin proteolipid protein (lipophilin) using a cloned complementary DNA. Nucleic Acids Res. 13, 7413-7425.

Nave, K.-A., Lai, C., Bloom, F.E., and Milner, R.J. (1986). Jimpy mutant mouse: a 74 base deletion in the mRNA for myelin proteolipid protein and evidence for a primary defect in RNA splicing. Proc. Natl. Acad. Sci. USA 83, 9264-9268.

Popko, B., Puckett, C., Lai, E., Shine, H.D., Readhead, C., Takahashi, N., Hunt, S.W., Sidman, R.L., and Hood, L. (1987). Myelin deficient mice: expression of myelin basic protein and generation of mice with varying levels of myelin. Cell 48, 713-721.

Puckett, C., Hudson, L., Ono, K., Friedrich, V., Beneke, J., Dubois-Dalcq, M., and Lazzarini, R.A. (1987). Myelin-specific proteolipid protein is expressed in myelinating Schwann cells but is not incorporated into myelin sheaths. J. Neurosci. Res. 18, 511-518.

Raff, M.C., Lillien, L.E., Richardson, W.D., Burne, J.F., and Noble,M.D. (1988). Platelet-derived growth factor from astrocytes drives the clock that times oligodendrocyte development in culture. Nature 333, 562-565.

Raine, C.S. (1984). Morphology of myelin and myelination. In Myelin, P. Morell, ed. (New York: Plenum Press), pp. 1-50.

Readhead, C., Popko, B., Takahashi, N., Shine, H.D., Saavedra, R.A., Sidman, R.L., and Hood, L. (1987). Expression of a myelin basic protein gene in transgenic shiverer mice: correction of the dysmyelinating phenotype. Cell 48, 703-712.

Roach, A., Boylan, K., Horvath, S., Prusiner, S.B., and Hood, L.E. (1983). Characterization of cloned cDNA representing rat myelin basic protein: absence of expression in brain of shiverer mutant mice. Cell 34, 799-806.

Roach, A., Takahashi, N., Pravtcheva, D., Ruddle, F., and Hood, L. (1985). Chromosomal mapping of mouse myelin basic protein gene and structure and transcription of the partially deleted gene in shiverer mutant mice. Cell 42, 149-155.

Roch, J.-M., Tosic, M., Roach, A., and Matthieu, J.-M. (1989). The duplicated myelin basic protein gene in mld mutant mice does not impair transcription. Brain Res. 477, 292-299.

Sorg, B.A., Smith, M.M., and Campagnoni, A.T. (1987). Developmental expression of the myelin proteolipid protein and basic protein mRNAs in normal and dysmyelinating mutant mice. J. Neurochem. 49, 1146-1154.

Takahashi, N., Roach, A., Teplow, D.B., Prusiner, S.B., and Hood, L. (1985). Cloning and characterization of the myelin basic protein gene from mouse: one gene can encode both 14 kd and 18.5 kd MPBs by alternate use of exons. Cell 42, 139-148.

Tamura, T., Miura, M., Ikenaka, K., and Mikoshiba, K. (1988). Analysis of transcription control elements of the mouse myelin basic protein gene in HeLa cell extracts: demonstration of a strong NFI-binding motif in the upstream region. Nucleic Acids Res. 16, 11441-11459.

Zeller, N.K., Hunkeler, M.J., Campagnoni, A.T., Sprague, J., and Lazzarini, R.A. (1984). Characterization of mouse myelin basic protein messenger RNAs with a myelin basic protein cDNA clone. Proc. Natl. Acad. Sci. USA 81, 18-22.

Zeller, N.K., Behar, T.N., Dubois-Dalcq, M., and Lazzarini, R.A. (1985). The timely expression of myelin basic protein gene in cultured rat brain oligodendrocytes is independent of continuous neuronal influences. J. Neurosci. 5, 2955-2962.

FROM MYELIN BASIC PROTEIN TO MYELIN DEFICIENT MICE

M. Tosic and J.-M. Matthieu
Laboratoire de Neurochimie
Service de Pédiatrie
Centre Hospitalier Universitaire Vaudois
1011 Lausanne
Switzerland

In order to control diseases affecting myelination, an understanding of normal myelin structure and its formation is essential. Paradoxically, much information about normal development can be obtained through investigations of hypomyelinated mutants.

The study of any mutation starts with the observed phenotype. In general, the first phase of such a study is descriptive analysis of the phenotype at different levels of biological organisation. The aim is characterization of the primary defect and identification of the mutated gene. The second phase starts with the structure of the mutated gene. The objective is the elucidation of its mode of expression through the interactions with other genes and environmental factors which contribute to the abnormal phenotype. The identification of various interactions at different organizational levels provide valuable information about normal development.

FROM MYELIN DEFICIENT PHENOTYPE TO MYELIN BASIC PROTEIN

Myelin deficient (mld) mice appeared spontaneously in the laboratory strain MDB/Dt in 1972 and were described as behavioural mutants (Doolittle and Schweikart, 1977). They are characterized by trembling of hind limbs, occasional seizures in adults, reduced fertility and shortened lifespan. The mutation shows full recessiveness, and it is transmitted in simple mendelian fashion, indicating a change in a single gene. Morphological investigations showed that the CNS is seriously deficient in myelin. Behaviour of the affected animals was similar to a previously described dysmyelination mutant, shiverer (Biddle et al, 1973). Subsequent genetic analysis showed that the two mutants are allelic (Lachapelle et al, 1980) and that the responsible gene is located at the distal end of chromosome 18 (Sidman et al, 1985).

The yield of myelin isolated from young mld brains by conventional methods (Norton and Poduslo, 1973) was only 2%, as compared to that of normal littermates (Matthieu et al, 1984). No significant difference between normal and mld PNS myelin in amount or morphology was observed (Matthieu et al, 1980b).

Light and electron microscopy showed that most of the axons in the CNS of young adult mutants are not myelinated (Matthieu et al, 1980a). The few myelinated fibers are surrounded by thin, uncompacted lamellae. The major dense line is absent (Fig 1). Several authors suggested (Podulso and Braun, 1975; Golds and Braun, 1976) and later directly demonstrated (Omlin et al, 1982) that myelin basic protein (MBP), one of the major myelin protein components, is located in the cytoplasmic surface of the membrane corresponding to the major dense line. The

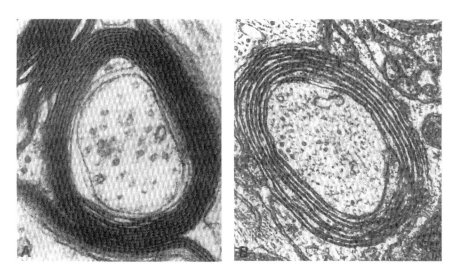

Fig. 1. Myelinated axons in optic nerves from normal (A) and mld (B) mice. Myelin-deficient CNS is charaterized by uncompacted myelin lamellae and the absence of major dense line. A: x 62000; B: x 32000.

absence of this structure in mld CNS myelin suggests that the primary defect lies in the impaired function or absence of MBP.

FROM MYELIN BASIC PROTEIN TO MYELIN BASIC GENE

The analysis of CNS myelin proteins by gel electrophoresis showed only traces of the different MBP isoforms (Matthieu et al., 1980a; Mikoshiba et al., 1987). Immunoblot studies confirmed the drastic reduction in MBP concentrations (Fig 2A). Radioimmunoassay of whole brain homogenates showed that mld MBP concentrations are about 3% that of normal littermates (Matthieu et al., 1980a; Bourre et al., 1980). In vitro protein

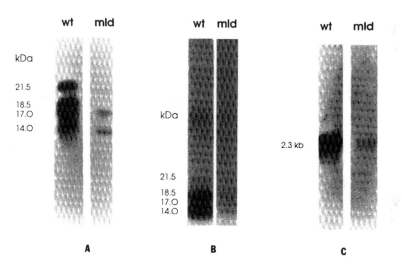

Fig. 2. Myelin basic protein (MBP) and MBP mRNA concentrations in normal (wt) and mld brains
A. Immunoblot analysis of MBP in total brain homogenate.
B. Immunoprecipitated MBP polypeptides transcribed in vitro from brain mRNA.
C. Northern blot analysis of brain mRNA hydridized with cDNA probe coding for rat 14 kD MBP.

synthesis followed by immunoprecipitation confirmed that the primary defect in mld mice is in the repressed synthesis of MBP (Ginalski-Winkelmann et al., 1983; Matthieu et al., 1984). Normal electrophoretic mobility as well as cell-free synthesis of MBP (Campagnoni et al., 1984), indicated that the coding region is probably normal and that the mld mutation affects the regulation of Mbp gene expression.

Gene regulation may occur at the transcriptional and/or translational level. In order to test if the regulation is impaired before or after mRNA translation, isolated mRNA was translated in vitro, and the products were immunoprecipitated (Fig. 2B). Translatable MBP mRNAs in mld extracts presented only about 2% of the amount detected in controls (Roch et al., 1986). Thus, the mutation is already expressed at the mRNA level. The results were confirmed by directly measuring cytoplasmic MBP mRNA concentration in mld and normal brains (Fig. 2C). The concentration of normal sized mRNA is reduced to about 2% when compared to control mice. The intermediate concentration of MBP mRNA in mld heterozygotes suggested that the mutation lies in a cis-acting regulatory element of the Mbp gene itself (Roch et al., 1986).

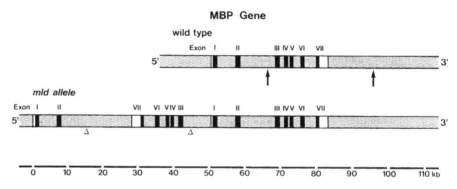

Fig. 3. Organization of wild type and mld allele of myelin basic protein gene in mice. (Adapted from Popko et al. 1988)

MYELIN BASIC PROTEIN GENE

Direct analysis of the Mbpmld allele was made possible only after cloning of the Mbp gene (Roach et al., 1983; de Ferra et al., 1985). Then, several groups reported a partial duplication and inversion of different regions of the Mbp gene within the Mbpmld allele (Akowitz et al., 1987; Okano et al., 1988; Popko et al., 1987; Roch et al., 1989). The complete structure of the

Mbp*mld* allele was published by Popko et al. (1988). The Mbp*mld* is a tandem duplication of the Mbp gene located on chromosome 18 (Okano et al., 1988). In addition, the upstream gene has a large inversion involving exons III to VII and may result in transcription of antisense RNA (Fig 3). An active promoter and some regulatory sequences are conserved in both genes (Okano et al., 1988b; Miura et al., 1989).

FROM MYELIN DEFICIENT ALLELE TO MBP MESSENGER RNA

Most of the mutations which affect regulation are expressed already in the transcription of the gene. Transcription rates for different regions of the Mbp gene, measured directly in isolated nuclei showed no difference between mld and normal brains (Fig 4).

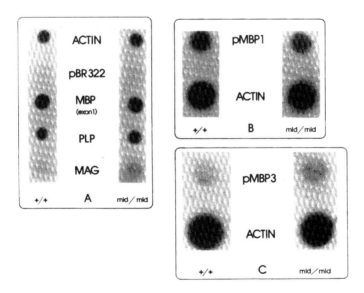

Fig. 4 <u>Transcription analysis of different regions of the Mbp gene in normal (+/+) and mld (mld/mld) mice</u>
A. Transcription rate of Mbp exon I compared to transcription of genes coding for actin, PLP and MAG. pBR 322 was used as a negative control.
B. Transcription rate of all exons coding for the 14 kD MBP compared to transcription rate of the actin gene.
C. Transcription rate of the terminal region of exon VII compared to transcription of the actin gene.

In order to explain the discrepancy between normal transcription rate and drastic reduction in MBP mRNA concentration in the cytoplasm, we measured the concentration of nuclear MBP-specific RNA (Fig 5). MBP-specific RNA concentration in mld nuclei is only slightly reduced compared to control brains, in contrast to the 20-fold difference found for cytoplasmic MBP mRNA. Therefore, mld MBP gene transcripts appeared to be processed abnormally or transported inefficiently to the cytoplasm.

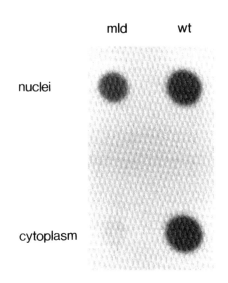

Fig. 5. <u>Dot blot analysis of MBP specific RNA</u>. Total RNA was isolated from nuclei or cytoplasm of normal (wt) and mld brains. Filter was hybridized with a cDNA probe coding for rat 14 kD MBP.

One possible explanation for the arrest of MBP specific RNA in the nuclei can lie in structure and size of the primary transcripts. Uninterrupted transcription of the two adjacent Mbp genes in mld mice would give rise to extremely long RNA molecules with impaired function.

Transcriptional analysis of exon I and the 5' flanking region indicated that the transcription of the downstream gene starts from its own promoter (Fig 6). Nevertheless, the run-through transcription initiated at the upstream promoter reported by Okano et al., (1988 a) cannot be excluded. However, if it does occur in mld mice, it is at an extremely low rate compared to the transcription rate of exon I.

Transcription of the upstream gene which gives rise to antisense RNA, has been reported (Popko et al., 1988); Okano et al., 1988a). We measured the relative transcription rates of

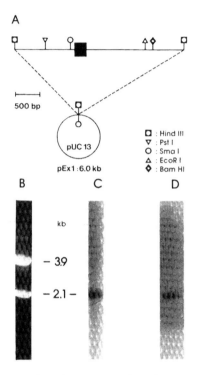

Fig. 6. Northern blot analysis of Mbp 5'end primary transcripts
Plasmid pEx 1 containing 5' flanking region, exon I (filled box) and part of intron I was digested with enzyme Sma I (A), separated by gel electrophoresis (B), transfered to a nitrocellulose filter and hybridized with primay transcripts isolated from wild type (C) and mld (D) brain nuclei.

the linked Mbp genes (Fig 7). Both genes are transcribed at a considerable rate. The transcription of the downstream gene which gives rise to normal protein is reduced compared to the upstream gene. This reduction may be due to transcriptional interference, a phenomenom reported for several genes (Emerman and Temin 1984; Garner et al., 1986). Alternatively, the lower transcriptional rate of the downstream Mbp gene may be due to a structural loss of an enhancer-like element caused by duplication/inversion events. Whatever the mechanism, the reduced transcription rate does not explain the drastically reduced MBP mRNA concentration.

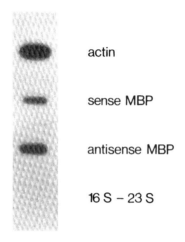

Fig. 7. Slot blot analysis of sense and antisense Mbp-specific primary transcripts. Transcription rates of normal and inverted exon VII of the Mbp gene compared, to that of actin gene. Bacterial 16S and 23S RNAs were used as a negative control.

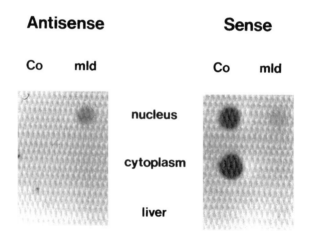

Fig. 8. Dot blot analysis of nuclear and cytoplasmic sense and antisense RNA from normal (wt) and mld brains. Immobilized RNA was hydridized with RNA corresponding to inverted and normal exon VII. Liver RNA was used as a negative control.

Since the upstream gene gives rise to a large antisense RNA fragment complementary to downstream transcripts, the formation of double-stranded RNA may impair normal splicing and transport to the cytoplasm. We tested for the presence of antisense RNA in mld nuclei and cytoplasm. Nuclei contain both sense and antisense RNA and in proportions corresponding to their respective transcription rates (Fig 8). No antisense RNA was found in the cytoplasm. The antisense transcripts are probably not polyadenylated as a consequence of the inverted 3' region, and therefore are most probably not normally spliced or transported to cytoplasm. The disparity of nuclear MBP sense RNA between normal and mld brains is much smaller than that found in cytoplasmic RNA. This observation suggests that much of the sense RNA is not transported to the cytoplasm and translated into protein. The arrest of sense RNA in the nuclear compartment is likely the result of duplex RNA formation, a mechanism previously shown to inhibit gene expression (for review, see van der Krol et al., 1988). Experimental evidence for inhibition of Mbp gene expression was recently reported by Katsuki et al., (1988). Transgenic mice, obtained by introducing the inverted Mbp gene into normal mouse embryos, manifested inhibition of Mbp gene expression, poor myelination and an mld-like phenotype. The demonstration of both sense and antisense RNA in mld brain nuclei, together with inhibitory effect of antisense RNA seen for transgenic mice, strongly suggest that the abnormal posttranscriptional regulation of the Mbp*mld* allele occurs through inhibition by antisense RNA. This inhibition may involve nuclear RNA processing and/or transport of mature RNA to cytoplasm which results in low MBP concentration.

FROM PROTEIN TO PHENOTYPE

The direct effect of Mbp gene expression at higher levels of biological organization is more difficult to analyse due to the additional influences of environment and increasing structural complexity. Nevertheless, some secondary effects of the Mbp*mld* allele have been described.

Mbp*mld* affects other myelin protein genes

Analysis of other myelin proteins showed pleiotropic effects of the Mbp*mld* allele. Among them, the gene encoding for proteolipid protein (PLP) is the least affected. PLP concentration in isolated myelin (Matthieu et al., 1980a), its in vitro synthesis (Matthieu et al., 1984) and its transcription rate (Roch et al., 1989; see Fig 4 A) showed no significant differences, as compared to normal mice. This finding contrasted with studies of shiverer mice. In this allelic mutation, in which a deletion of the Mbp gene causes a complete lack of MBP (Roach et al., 1985), the expression of the Plp gene is drastically reduced (Mikoshiba et al., 1980; Sorg et al., 1986). A comparison of these two mutants indicated that even minimal expression of the Mbp gene fails to disrupt normal Plp gene regulation.

In vitro transcription followed by immunoprecipitation of 2',3'-cyclic nucleotide 3'-phosphodiesterase (CNP) showed significantly increased CNP mRNA concentration in mld brain (Matthieu et al., 1986). CNP activity in mld myelin is four times higher than in normal myelin (Matthieu et al., 1980a). This increase in CNP was confirmed by gel electrophoresis and immunoblotting of myelin proteins (Matthieu et al., 1981; Matthieu et al., 1984). Overexpression of CNP, also detected in shiverer mutant (Mikoshiba et al., 1980) is probably related to the presence of uncompacted myelin and redundant paranodal loops observed in mld and shiverer mice (Matthieu et al., 1984; Mikoshiba et al., 1987). During development, when mld myelin, still in very reduced amounts, is better compacted due to slow accumulation of MBP, CNP specific activities in myelin decrease and reach normal values (Matthieu et al., 1984). This was confirmed by immunostaining during development which showed an inverse relationship between CNP amounts and degree of myelin compaction (Braun et al., 1988). Namely, CNP-specific immunostaining decreases with age while compaction increases. Therefore, CNP overexpression is probably not regulated directly by decreased expression of the Mbp gene, but rather indirectly through reduced amount of Mbp gene products and/or lack of myelin lamellae compaction.

The effect of the Mbp*mld* allele is even more pronounced on the expression of myelin-associated glycoprotein (MAG). MAG concentration in myelin purified from mld brains is reduced (Matthieu et al., 1986). In contrast, in vitro transcription and immunoprecipitation analyses showed very high concentration of translatable MAG mRNA. The accumulation of MAG immunoreactive material in vacuoles, inclusion bodies and rough endoplasmic reticulum of mld oligodendrocytes (Matthieu et al., 1986) indicate that large amounts of MAG are processed, but only small amounts were recovered in myelin. These observations indicated that MAG stability is reduced. This hypothesis was confirmed by high concentration of dMAG, a degradation product of MAG. Furthermore, extremely high MAG turnover, as well as that of sulfatides, indicate that mld myelin is unstable (Matthieu et al., 1986). The effect of the Mbp*mld* allele on MAG expression seems to occur in different ways; directly by increasing MAG gene transcription, as well as indirectly through destabilization of the myelin sheath, which stimulates MAG degradation.

Mbp*mld* effects in peripheral nervous system

Reduced amounts of MBP in the PNS do not seem to affect myelination of peripheral axons. Protein analysis showed normal concentrations (Matthieu et al., 1980b) and synthesis (Matthieu et al., 1984) of P_0, the major PNS myelin protein, as well as normal CNP activity (Matthieu et al., 1980b). The myelin sheath is well compacted with a normal major dense line (Matthieu et al., 1980b). Therefore, MBP is not indispensable in PNS, and its role is probably replaced by P_0 (Kirschner and Ganser, 1980).

MBP IN MYELIN

Based on the composition and structure of myelin in the hypomyelinated mld mutant, several important conclusions about the role of MBP in myelination can be drawn.

1. Composition of the MBP concentrations in whole brain homogenates and isolated myelin from normal (Mbp/Mbp) and heterozygous individuals (Mbp/Mbp*mld*) showed that completely normal CNS myelination can occur even if the MBP concentration is reduced to 50% of the normal level (Roch et al., 1987). Thus "overproduction" of MBP in normal mice could be **a safety factor**, important for such an essential structure.

2. The ability of oligodendrocytes to synthesize membranes which surround axons, even in the absence of MBP, indicates that Mbp gene expression is **not essential for the activation of other myelin protein genes**, and certainly **not an early signal for myelination**.

3. Reduced level of MBP and absence of the major electron dense line in mld mice, confirms the **extrinsic localization of MBP** and its role in compaction of the myelin sheath.

4. High turnover rates of MAG and sulfatides suggest mld myelin instability. Therefore, MBP has an important role not only in compaction, but also in **stabilizing the myelin sheath**.

5. In vitro hydridization, as well as immunocytochemical studies (Roch et al., 1987; Ulrich et al., 1983) show uniform staining in the white matter. All myelinating oligodendrocytes synthesize MBP to the same extend. Therefore, the expression of the Mbp gene is not affected by local environment, but **regulated exclusively at the intracellular level**.

6. Analysis of the major myelin proteins in mld mutants shows that the Mbp gene has **multiple direct and indirect effects** on myelin formation.

7. Finally, the mld defect is another example of **posttranscriptional inhibition** affecting gene expression by formation of **antisense RNA**. This mechanism occurs in several species, and has become an experimental tool useful for the study of gene function.

References

Akowitz AA, Barbarese E, Scheld K, Carson JH (1987) Structure and expression of myelin basic protein gene sequences in the mld mutant mouse; reiteration and rearrangement of the MBP gene. Genetics 116:447-464

Biddle FG, March E, Mitter JR (1973) Mouse News Lett. 48:24

Bourre JM, Jacque C, Delassalle A, Nguyen-Legros J, Dumont O, Lachapelle F, Raoul M, Alvarez C, Baumann N (1980) Density profiles and basic protein measurements in the myelin range of particulate material from normal development mouse brain and from neurological mutants (Jimpy; Quaking; Trembler; Shiverer; and its mld allele) obtained by zonal centrifugation. J Neurochem 35:458-464

Braun P, Sandillon F, Edwards A, Matthieu JM, Privat A (1988) Immunocytochemical localization by electron microscopy of 2',3'-cyclic nucleotide 3'-phosphodiesterase in developing oligodendrocytes of normal and mutant brain. J Neurosci 8: 3057-3066

Campagnoni AT, Campagnoni CW, Bourre JM, Jacque C, Baumann N (1984) Cell-free synthesis of myelin basic proteins in normal and dysmyelinating mutant mice. J Neurochem 42:733-739

de Ferra F, Engh H, Hudson L, Kamholz J, Puckett C, Monineaux S and Lazzarini R (1985) Alternative splicing accounts for the four forms of myelin basic protein. Cell 43:721-727

Doolittle DP, Schweikart KM (1977) Myelin deficient, a new neurological mutant in the mouse. J Hered 68: 331-332

Emerman M, Temin H (1984) Genes with promoters in retrovirus vectors can be independently suppressed by an epigenic mechanism. Cell 39:459-467

Garner I, Minty A, Alonso S, Barton P, Buckingham M (1986) A 5'duplication of the alpha-cardiac actin gene in BALB/c mice is associated with abnormal levels of alpha-cardiac and alpha-skeletal actin mRNA in adult cardiac tissue. EMBO J 5:2559-2567

Ginalski-Winkelmann A, Almazan G, Matthieu JM (1983) In vitro myelin basic protein synthesis in the PNS and CNS of myelin deficient (mld) mutant mice. Brain Res 277:386-388

Golds EE, Braun PE (1976) Organization of membrane proteins in the intact myelin sheath: pyridoxal phosphate and salicylaldehyde as probes of myelin structure. J biol Chem 251:4729-4735

Katsuki M, Soto M, Kimura M, Yokoyama M, Kobayashi K, Nomura T (1988) Conversion of normal behavior to shiverer by myelin basic protein antisense cDNA in transgenic mice. Science 241:593-595

Kirschner DA, Ganser AL (1980) Compact myelin exists in the absence of basic protein in the shiverer mutant mouse. Nature 283:207-210

Lachapelle F, de Baecque C, Jacque C, Bourre JM, Delassalle A, Doolittle D, Hauw J, Baumann N (1980) Comparison of morphological and biochemical defects of two probably allelic mutations of the mouse, myelin deficient (mld) and shiverer (shi). In: Baumann N (ed) Neurological mutations affecting myelination, INSERM Symposium no 14 . Elsevier/North-Holland Biomedical Press, Amsterdam p 27-32

Matthieu JM (1982) Myelin basic protein and stability of the multilamellar myelin structure. Bull Schweiz Akad Med Wiss. 101-108

Matthieu JM, Almazan G, Waehneldt T (1984) Intrinsic myelin proteins are normally synthesized in vitro in the myelindeficient (mld) mutant mouse. Dev Neurosci 6:246-250

Matthieu JM, Ginalski H, Friede RL, Cohen SR, Doolittle DP (1980a) Absence of myelin basic protein and major dense line in CNS myelin of the mld mutant mouse. Brain Res 191:278-283

Matthieu JM, Ginalski H, Friede RL, Cohen SR (1980b) Low myelin basic protein levels and normal myelin in peripheral nerves of myelin deficient mutant mice (mld). Neuroscience 5:2315-2320

Matthieu JM, Ginalski-Winkelmann H, Jacque C (1981) Similarities and dissimilarities between two myelin deficient mutant mice, shiverer and mld. Brain Res 214:219-222

Matthieu JM, Omlin FX, Ginalski-Winkelmann H, Cooper BJ (1984) Myelination in the CNS of mld mutant mice: comparison between composition and structure. Dev Brain Res 13:149-158

Matthieu JM, Roch JM, Omlin FX, Rambaldi I, Almazan G, Braun P (1986) Myelin instability and oligodendrocyte metabolism in myelin-deficient mutant mice. J Cell Biol 103:2673-2682

Mikoshiba K, Aoki E, Tsukada Y (1980) 2', 3'-cyclic nucleotide 3'-phosphohydrolase activity in the central nervous system of a myelin deficient mutant (shiverer). Brain Res 192:195-204

Mikoshiba K, Okano H, Inoue Y, Fujishiro M, Takamatsu K, Lachapelle F, Baumann N, Tsukada Y (1987) Immunohistochemical, biochemical and electron microscopic analysis of myelin formation in the central nervous system of myelin deficient (mld) mutant mice. Dev Brain Res 35:111-121

Miura M, Tamura T, Aoyama A, Mikoshiba K (1989) The promoter elements of the mouse myelin basic protein gene function efficiently in NG 108-15, neuronal/glial cells. Gene 75:31-38

Norton WT, Poduslo SE (1973) Myelination in rat brain: method of myelin isolation. J Neurochem 21: 749-758

Okano H, Ikenaka K, Mikoshiba K (1988a) Recombination within the upstream gene of duplicated myelin basic protein genes of myelin deficient shi*mld* mouse results in the production of antisense RNA. EMBO J 7:3407-3412

Okano H, Miura M, Moriguchi A, Ikenaka K, Tsukada Y, Mikoshiba K (1987) Inefficient transcription of the myelin basic protein gene possibly causes hypomyelination in myelin-deficient mutant mice. J Neurochem 48:470-476

Okano H, Tamura T, Miura M, Aoyama A, Ikenaka K, Oshimura M, Mikoshiba K (1988b) Gene organization and transcription of duplicated MBP gene of myelin-deficient (shi*mld*) mutant mouse. EMBO J 7:77-83

Omlin FX, Webster H de F, Palkovits CG, Cohen SR (1982) Immunocytochemical localization of basic protein in major dense line regions of central and peripheral myelin. J Cell Biol 95:242-248

Poduslo JF, Braun PE (1975) Topographical arrangement of membrane proteins in the intact myelin sheath. J biol Chem 250:1099-1105

Popko B, Puckett C, Hood L (1988) A novel mutation in myelindeficient mice results in unstable myelin basic protein gene transcripts. Neuron 1:221-225

Popko B, Puckett C, Lai E, Shine HD, Readhead C, Takahashi N, Hunt III SW, Sidman RL, Hood L (1987) Myelin-deficient mice: expression of myelin basic protein and generation of mice with varying levels of myelin. Cell 48:713-721

Roach A, Boylan K, Horvath S, Prusiner SB, Hood L (1983) Characterization of cloned cDNA representing rat myelin basic protein: absence of expression in brain of shiverer mutant mice. Cell 34:799-806

Roach A, Takahashi N, Pravtcheva D, Ruddle F, Hood L (1985) Chromosomal mapping of mouse myelin basic protein gene and structure and transcription of the partially deleted gene in shiverer mutant mice. Cell 42:149-155

Roch JM, Braun-Luedi M, Cooper BJ, Matthieu JM (1986) Mice heterozygous for the mld mutation have intermediate levels of myelin basic protein mRNA and its translational products. Mol Brain Res 1:137-144

Roch JM, Cooper BJ, Ramirez M, Matthieu JM (1987) Expression of only one myelin basic protein allele in mouse is compatible with normal myelination. Mol Brain Res 3:61-68

Roch JM, Tosic M, Roach A, Matthieu JM (1989) The duplicated myelin basic protein gene in mld mutant mice does not impair transcription. Brain Res 477:292-299

Sidman RL, Conover CS, Carson JH (1985) Shiverer gene maps near distal end of chromosome 18 in the house mouse. Cytogenet Cell Genet 39:241-245

Sorg BJA, Agrawal D, Agrawal HC, Campagnoni AT (1986) Expression of myelin proteolipid protein and basic protein in normal and dysmyelinating mutant mice. J Neurochem 46:379-387

Ulrich J, Matthieu JM, Herschkowitz N, Kohler R, Heitz PhU (1983) Immunocytochemical investigations of murine leukodystrophies. A study of the mutants "jimpy" (jp) and "myelin deficient" (mld). Brain Res 268:267-274

van der Krol AR, Mol JNM, Stuitje AR (1988) Modulation of eukaryotic gene expression by complementary RNA or DNA sequences. BioTechniques 6:958-975

Proteolipid protein: From the primary structure to the gene

Wilhelm Stoffel
Institut für Biochemie
Medizinische Fakultät
Universität zu Köln
Joseph-Stelzmann-Str. 52
D-5000 Köln 41

Myelinogenesis is a very dramatic and important developmental process of the central nervous system.

It is characterized by the differentiation of oligodendrocytes from the O-2A progenitor cells, the cell growth and proliferation of oligodendrocytes and the expression of oligodendrocyte-specific genes. Their products are responsible for the construction and structure of the highly specialized and ordered structure of the myelin membrane.

The expression of these genes is highly regulated in time and space and occurs on the transcriptional level.

The endeavour of my group at the Cologne Institute of Biochemistry over the past years aims at learning

a) to understand the molecular basis of the myelin membrane, the molecular organization of its constituents, how their properties contribute to compaction of this multilayer system and thereby to the physiological properties of the myelin membrane by understanding its assembly, learn about its antigenic sites,

b) to elucidate by means of recombinant DNA techniques the organization of the main human myelin structural genes, of myelin basic protein (MBP) and proteolipid protein (PLP), and on this basis tackle with cell biological techniques the mechanisms involved in the coordinate regulation of PLP and MBP gene activity and of key enzymes involved in the biosynthesis of the myelin-specific lipids. After all myelin is the membrane richest in lipids (80% of total mass).

c) to analyze mutations of myelin-specific genes in the animal model and in man. The recombinant DNA techniques allow us to dissect myelin proteins and analyze domains essential for function.

I would like to concentrate only on one area of our interest and outline how the reversed molecular biology of myelin

1) started with the structure of PLP protein,

2) continued from the protein to the cDNA and transcriptional level,

NATO ASI Series, Vol. H 43
Cellular and Molecular Biology of Myelination
Edited by G. Jeserich et al.
© Springer-Verlag Berlin Heidelberg 1990

```
Gly Leu Leu Glu Cys Cys Ala Arg Cys Leu Val Gly Ala Pro Phe Ala Ser Leu Val Ala   20
Thr Gly Leu Cys Phe Phe Gly Val Ala Leu Phe Cys Gly Cys Gly His Glu Ala Leu Thr    40
Gly Thr Glu Lys Leu Ile Glu Thr Tyr Phe Ser Lys Asn Tyr Gln Asp Tyr Glu Tyr Leu    60
Ile Asn Val Ile His Ala Phe Gln Tyr Val Ile Tyr Gly Thr Ala Ser Phe Phe Leu        80
Tyr Gly Ala Leu Leu Leu Ala Glu Gly Phe Tyr Thr Thr Gly Ala Val Arg Gln Ile Phe   100
Gly Asp Tyr Lys Thr Thr Ile Cys Gly Lys Gly Leu Ser Ala Thr Val Thr Gly Gly Gln   120
Lys Gly Arg Gly Ser Arg Gly Gln His Gln Ala His Ser Leu Glu Arg Val Cys His Cys   140
Leu Gly Lys Trp Leu Gly His Pro Asp Lys Phe Val Gly Ile Thr Tyr Ala Leu Thr Val   160
Val Trp Leu Leu Val Phe Ala Cys Ser Ala Val Pro Val Tyr Ile Tyr Phe Asn Thr Trp   180
Thr Thr Cys Gln Ser Ile Ala Phe Pro Ser Lys Thr Ser Ala Ser Ile Gly Ser Leu Cys   200
Ala Asp Ala Arg Met Tyr Gly Val Leu Pro Trp Asn Ala Phe Pro Gly Lys Val Cys Gly   220
Ser Asn Leu Leu Ser Ile Cys Lys Thr Ala Glu Phe Gln Met Thr Phe His Leu Phe Ile   240
Ala Ala Phe Val Gly Ala Ala Ala Thr Leu Val Ser Leu Leu Thr Phe Met Ile Ala Ala   260
Thr Tyr Asn Phe Ala Val Leu Lys Leu Met Gly Arg Gly Thr Lys Phe                   276
```

Fig. 1: Amino acid sequence of human proteolipid protein. Cis- and transmembranal domains are underlined.

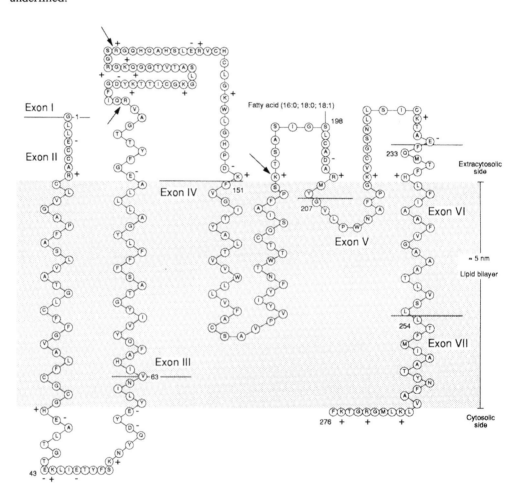

Figure 2: Proposed topology of proteolipid protein. Arrows indicate trypsin cleavage sites.

3) led to the gene structure of PLP, its allocation to the X chromosome and

4) how our knowledge of the gene structure facilitated the elucidation of the genetic defects of two animal models, the jimpy mouse and recently of the md rat in my laboratory.

The extreme hydrophobic nature of proteolipid protein purified from the white matter of bovine and human brain required new methods for the amino acid sequence analysis of the water-insoluble polypeptide.

a) Suitable chemical and endoproteolytic fragmentation (BrCN and tryptophan cleavage);

b) separation by fractional solubilization of hydrophilic and hydrophobic peptides in acidic medium;

c) separation by gel exclusion chromatography in 90% formic acid with gels of different exclusion volumes;

d) HPLC on silica gel with formic acid as solvent;

e) trypsin and chymotrypsin cleavage of lysine- and arginine-modified polypeptides;

f) partial hydrolysis of protein fragments with subsequent derivation of oligopeptides to volatile sialylated polyamino alcohols, separation by capillary GLC and mass spectroscopy yielded the sequence of the 276 a. a. residue long PLP sequence, Fig. 1, with its very canonical structure of an integral membrane protein.

This sequence suggested to us the PLP integration model presented in Fig. 2 in which α–helices of appropriate length span the unusually wide hydrophobic core of the myelin bilayer (5 nm) and two cis-membranal α–helices. Our model bases on biochemical and immunocytochemical analyses.

Disulfide bonds on the extracytosolic side connect the N–terminus with the penultimate hydrophilic sequence at the C-terminus and also the large hydrophilic epitope is linked to cystein. Therefore a clustering of the trans- and cis-membranal helices is most likely, Fig. 3.

Trypsin cleavage (arrows in Fig. 2) yielded only three large fragments when the dissociated myelin membrane layers were exposed to the digestion and the exterior surface attacked by the protease as expected from our model.

Also the plasma membrane of oligodendrocytes exposes only the proposed connecting hydrophilic epitopes for reactions with fluorescence- or gold-labelled antipeptide antibodies.

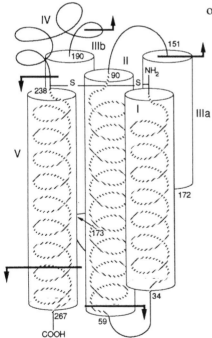

Fig. 3: PLP 3D topology

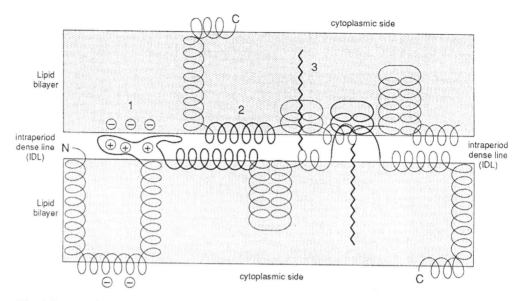

Fig. 4: Proposed stabilizing structures of proteolipid protein contributing to the compaction within the intraperiod dense line: 1 ionic binding forces (anionic polar head groups of complex lipids and cationic side chains of PLP); 2 amphipathic interactions of surface-oriented loops; 3 acyl chains of PLP.

Also in our model the majority of the -SH groups and disulfide bonds within the hydrophilic domains are oriented toward the extracytosolic surface, the others within the first and third intramembranal domain. Five disulfide bonds lead to a compact arrangement. It is also noticeable that the outer surface carries surplus positive charges, that oppositely charged amino acid residues border the transmembrane α-helices and the acylation at Thr^{208} might also add hydrophobic interactions with the apposed next membrane layers, an additional factor which adds to the compacting properties of PLP on both sides of the bilayer (summarized schematically in Fig. 4).

Realizing the limitations of the purely protein-chemical approach we introduced parallel to this work the recombinant DNA techniques to answer the numerous questions related to the differentiation process of myelinogenesis.

We established a rat brain-specific size-fractionated cDNA library of 18-day-old rats and rapidly succeeded in the isolation of PLP- and MBP-specific cDNA clones by screening with oligonucleotides synthesized according to our protein sequence.

The PLP amino acid sequence derived from the nucleotide sequence information was in agreement with the primary peptide structure established in our laboratory, that of Jollès in Paris and Milner in La Jolla. This is also true for the MBP-specific cDNAs isolated in several laboratories including our own.

The PLP-specific cDNA served as a very valuable tool to study the expression of the PLP gene during the myelinization period of the rat and to compare it with that of the myelin basic protein (MBP).

The Northern blot analysis revealed three different sizes of the PLP mRNA: 3.2 kb, 2.4 kb and 1.6 kb, whereas that of the MBP-specific mRNA appears as a smear between 2.0-2.4 kb in the rat. Quantitative densitometric analysis of the autoradiogram of the PLP and MBP mRNA separated by formaldehyde agarose gel electrophoresis clearly indicates the peak of mRNA synthesis of the main myelin proteins around days 20 to 25. It should be mentioned that this developmental program of gene expression is also followed in rat oligodendrocytes in culture between day 2 to 40 in exactly the same pattern, indicating that the oligodendrocyte starts this program independent of axons.

Furthermore the PLP- and MBP-specific cDNAs were the basis for the isolation of the two human genes and the mouse PLP gene from genomic libraries in the EMBL-3 vector and/or from a cosmid library, Fig. 5. The human PLP and MBP gene sequences are outlined in Fig. 6 and 7.

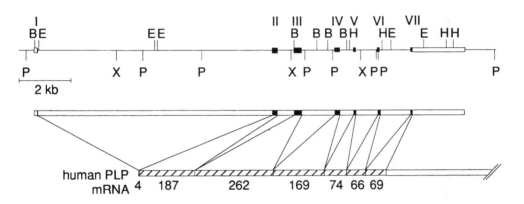

Fig. 5: PLP gene organization and splicing

In the PLP gene seven exons are spread over 17 kb, Fig. 5. The first exon contains no more than four bases of the coding sequences, the ATG of methionine — which is lost in the processing of the primary translation product — and G of Gly[1] of the mature PLP polypeptide. The first intron is about 8 kb long and exon VII contains the more than 2 kb long 3'-untranslated region. Exon II-VI are assigned to structural domains, each of them encodes a hydrophobic trans- or cis-membranal α-helical domain associated with the connecting hydrophilic sequences, Fig. 2.

A retrieval of the available nucleotide sequences of eucaryotic membrane proteins revealed no statistically relevant homology to PLP.

More convincing are the homologies of the coding regions of the rat, mouse, bovine and human PLP gene, Fig. 8.

```
I  AAGAAAATGA AACAATTGGC AGTGAAAGGC AGAAAGAGAA GATGGAGCCC TTAGAGAAGG GAGTATCCCT GAGTAGGTGG GGAAAAGGGG AGGAGAAGGG GAGGAGGAGA GGAGGAGGAA   -205
   AGCAGGCCTG TCCCTTTAAG GGGGTTGGCT GTCAATCAGA AAGCCCTTTT CATTGCAGGA GAAGAGGACA AAGATACTCA GAGAGAAAAA GTAAAAGACC GAAGAAGGAG GCTGGAGAGA   -85
                                                                                    Met G(ly)                                            1
   CCAGGATCCT TCCAGCTGAA CAAAGTCAGC CACAAAGCAG ACTAGCCAGC CGGCTACAAT TGGAGTCAGA GTCCCAAAGA C ATG G gtaagtttcaaaaactttag...               +1

II             (G)ly Leu Leu Glu Cys Cys Ala Arg Cys Leu Val Gly Ala Pro Phe Ala Ser Leu Val Ala Thr Gly Leu Cys Phe Phe Gly Val Ala   29
   ...ttcccttcttcttcccccag  GC TTG TTA GAG TGC TGT GCA AGA TGT CTG GTA GGG GCC CCC TTT GCT TCC CTG GTG GCC ACT GGA TTG TGT TTC TTT GGG GTG GCA   +87

   Leu Phe Cys Gly Cys Gly His Glu Ala Leu Thr Gly Thr Glu Lys Leu Ile Glu Thr Tyr Phe Ser Lys Asn Tyr Gln Asp Tyr Glu Tyr Leu Ile`Asn Va(l)   63
   CTG TTC TGT GGC TGT GGA CAT GAA GCC CTC ACT GGC ACA GAA AAG CTA ATT GAG ACC TAT TTC TCC AAA AAC TAC CAA GAC TAT GAG TAT CTC ATC AAT GT gtaa   +188

   gtacctgccctcccac...
```

III   (Va)l Ile His Ala Phe Gln Tyr Val Ile Tyr [Gly Thr] Ala Ser Phe Phe Phe Leu Tyr Gly Ala Leu Leu Leu Ala Glu Gly Phe Tyr   91 **md** / 75 Pro / C / Ava II
```
   ...ttgtctacctgttaatgcag  G ATC CAT GCC TTC CAG TAT GTC ATC TAT GGA ACT GCC TCT TTC TTC TTC CTT TAT GGG GCC CTC CTG CTG GCT GAG GGC TTC TAC   +273
```

                                                                                                                        DM-20
```
   Thr Thr Gly Ala Val Arg Gln Ile Phe Gly Asp Tyr Lys Thr Thr Ile Cys Gly Lys Gly Leu Ser Ala Thr Va|l Thr Gly Gly Gln Lys Gly Arg Gly Ser Arg|   126
   ACC ACC GGC GCA GTC AGG CAG ATC TTT GGC GAC TAC AAG ACC ACC ATC TGC GGC AAG GGC CTG AGC GCA ACG GTA ACA GGG GGC CAG AAG GGG AGG GGT TCC AGA   +378

   |Gly Gln His Gln Ala His Ser Leu Glu Arg Val Cys His Cys Leu Gly Lys Trp Leu Gly His Pro Asp Lys|                                             150
   |GGC CAA CAT CAA GCT CAT TCT TTG GAG CGG GTG TGT CAT TGT TTG GGA AAA TGG CTA GGA CAT CCC GAC AAG| gtgatcatcctcaggatttt...                       +450

IV             Phe Val Gly Ile Thr Tyr Ala Leu Thr Val Val Trp Leu Leu Val Phe Ala Cys Ser Ala Val Pro Val Tyr Ile Tyr Phe Asn Thr   179
   ...acccatgtcaatcattttag TTT GTG GGC ATC ACC TAT GCC CTG ACC GTT GTG TGG CTC CTG GTG TTT GCC TGC TCT GCT GTG CCT GTG TAC ATT TAC TTC AAC ACC   +537

   Trp Thr Thr Cys Gln Ser Ile Ala Phe Pro Ser Lys Thr Ser Ala Ser Ile Gly Ser Leu Cys Ala Asp Ala Arg Met Tyr G(ly)                            207
   TGG ACC ACC TGC CAG TCT ATT GCC TTC CCC AGC AAG ACC TCT GCC AGT ATA GGC AGT CTC TGT GCT GAT GCC AGA ATG TAT G   gtgagttagggtacgggtgc...         +619

V              (G)ly Val Leu Pro Trp Asn Ala Phe Pro Gly Lys Val Cys Gly Ser Asn Leu Leu Ser Ile Cys Lys Thr Ala Glu   231
   ...gcttttgtgtcttacttag GT GTT CTC CCA TGG AAT GCT TTC CCT GGC AAG GTT TGT GGC TCC AAC CTT CTG TCC ATC TGC AAA ACA GCT GAG gtgagtgggttattt   +693
   gggtt...
```
                 ji /g
```
VI             Phe Gln Met Thr Phe His Leu Phe Ile Ala Ala Phe Val Gly Ala Ala Ala Thr Leu Val Ser Leu                                         253
               Gly Pro Asn Asp Leu Pro Pro Val Tyr Cys Cys Ile Cys Gly Gly Cys Ser Tyr Thr Gly Phe Pro
   ...ctcttttcattttcctgcag TTC CAA ATG ACC TTC CAC CTG TTT ATT GCT GCA TTT GTG GGG GCT GCA GCT ACA CTG GTT TCC CTG gtgagttgactttgaatgat...        +759

VII            Leu Thr Phe Met Ile Ala Ala Thr Tyr Asn Phe Ala Val Leu Lys Leu Met Gly Arg Gly Thr Lys Phe Stop                                 276
               Ser His Leu His Asp Cys Cys His Leu Gln Leu Cys Arg Pro Stop
```

Fig. 6: Exon and inton sequences of the PLP gene. At a. a. 75 in exon III the md defect is shown. The jimpy defect caused by alternative splicing between exons IV and V leads to the loss of exon V and a missense sequence in exons VI and VII (shown in italic). The DM-20 isoprotein arises from the activation of a cryptic splice donor site within the Val[116] codon.

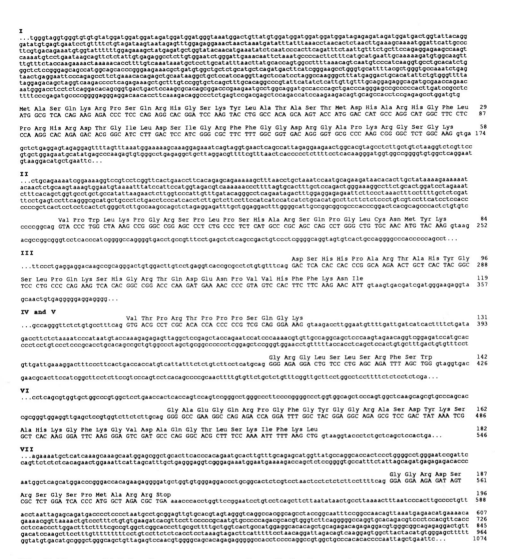

Fig. 7: Exon and joining intron sequences of myelin basic protein

There is a fully conserved amino acid sequence between man and rat and only 22 out of 828 bp (=2.6%) are exchanged, Fig. 8.

In the mouse PLP only two amino acids are exchanged ($Tyr^{91} \to Cys$, $Ser^{125} \to Thr$) but hydrophobicity and hydrophilicity are conserved. In the bovine PLP there are also only 28 different nucleotides (=3.3%) and two different amino acid residues, $Phe^{188} \to Ala$ and $Ser^{198} \to Thr$, which is in agreement with the sequence we obtained by regular protein sequencing. Is is worthy to note that they are located in the hydrophilic domains and $Phe^{188} \to Ala$ at the transition from a hydrophobic to a hydrophilic domain.

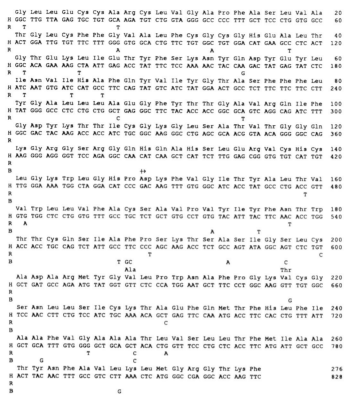

Fig. 8: Comparison of the coding sequences of the human (H), rat (R) and bovine (B) proteolipid protein

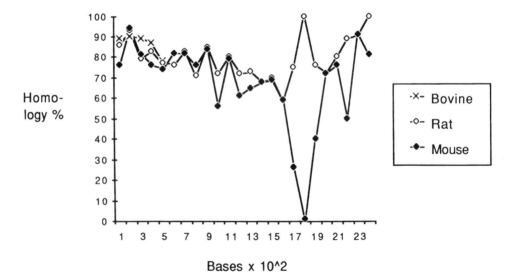

Fig. 9: Chart showing the homology between the bovine, rat und mouse PLP sequence in the 3'–non-coding region

Important features of the 5'- and 3'-noncoding regions of the PLP gene should be outlined.

The putative transcription initiation signal TAAAA at –113 is homologous to the eucaryotic TATA box motif and 61 bp (–174 bp) upstream the CAAT box (5' GTCAATCA 3') homologous to the well-known consensus sequence (5' GGCAAT$\frac{C}{T}$CT 3').

RNA polymerase II starts the transcription usually 20-25 bp downstream from the TATA box with an A followed by a pyrimidine-rich sequence. This position would be at –79 to –85, 28 and 34 bp downstream the TATA box.

The 3'-noncoding region of PLP is 2122 bp long (from stop codon to poly A site), Fig. 9. There is a remarkable homology of the human sequence to the rat and bovine sequence. It is about 90% within the first 200 bp downstream of the stop codon, drops gradually to 55-50% around 1500-1700 bp but increases to more than 90% 200 bp upstream of the poly A site, the last 45 bp are 100% homologous.

Three transcripts are synthesized due to three distinct poly adenylation sites at bp 2049 of the noncoding region leading to the 3.2 kb transcript, at 1319 the 2.4 kb and at 426 yielding the 1.6 kb poly A⁺ RNA.

So far the determinants of the species-specific alternative versus constitutive selection of the endonucleolytic cleavage and subsequent polyadenylation are unknown. The same is true for the alternative RNA splicing which is utilized in the generation of the five isoforms of the mouse and four of the human MBP and two isoforms of main forms of PLP, 30 kDa in size and the 26 kDa DM-20. First indications of a deletion of 35 amino acid residues (116-150) within the highly basic domain of the outer surface of the myelin membrane came from immunological studies with antipeptide antibodies. On the genetic level this deletion is the result of alternative splicing at the cryptic splice donor site within exon III, marked in Fig. 6.

In collaboration with Prof. Grzeschik, Marburg, we allocated the PLP gene to chromosome Xq_{12}–q_{22} by Southern blot hybridization analysis using a suitable *Eco* RI fragment of the human PLP gene and restricted DNA from human-mouse hybrid cell lines (Fig. 10).

Our knowledge about the gene structure and organization of PLP together with its X chromosome localization allowed studies on dysmyelinating X chromosome-linked diseases in animal models such as the jimpy mouse and the myelin-deficient rat and man-known as Pelizaeus-Merzbacher disease, Fig. 6. Nave *et al.* observed a 74 bp defect in the cDNA of jimpy mouse and a missense C-terminal PLP sequence of a 243 residue long polypeptide. A glimpse at the PLP gene structure immediately suggests that the splice acceptor site of intron IV must be mutated and by splicing exon IV and VI with elimination of intron IV to exon V and intron V en bloc a frameshift will occur leading to the jimpy PLP. Indeed the AG dinucleotide at the splice acceptor site of intron IV is mutated to GG, Fig. 6.

The consequence of the synthesis of the jimpy PLP or the lack of functionally active wild type PLP leads to the early death of the oligodendrocyte.

We recently succeeded in the elucidation of the genetic defect in the md (myelin-deficient) rat,

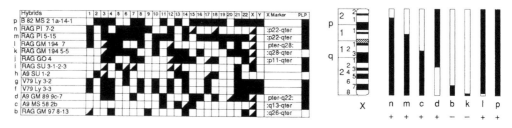

Fig. 10: Assignment of the PLP gene to the X chromosome

a dysmyelinosis phenotypically very similar to the jimpy mouse: severe tremor and seizures lead to an early death of the male sibling between 3 to 6 weeks after birth. Histologically no myelin is found around axons of the CNS and the lack of PLP but also of MBP and myelin-specific lipids is well documented. Several proposals regarding the md defect have been made in analogy to the jimpy defect.

The lack of PLP and MBP mRNA in oligodendrocytes of md rats can elegantly be demonstrated by in situ hybridization of thin sections of md rat brain with [^{35}S]labelled antisense PLP RNA synthesized on the 2.9 kb *Hind* III/*Eco* RI fragment encoding the complete coding sequence of PLP cloned into the pGEM3z system.

The autoradiograms also confirm the previous histology which demonstrated the drastically reduced number of oligodendrocytes.

The genetic defect of the md rat was unravelled by the following strategy:

1) Genomic rat DNA was restricted with several enzymes hybridized with two *Pst* I fragments of our rat PLP cDNA and compared with the restriction pattern of the wild type DNA. The identical pattern excluded gene duplication and gene rearrangements.

2) A md rat genomic library has been constructed in the EMBL3 vector. One clone was isolated which encoded exons IV to VIII. The coding sequences of these exons including their adjacent intron sequences were absolutely homologous to the normal rat. No mutation — neither at the splice donor nor at the acceptor site — was found. Thus a splicing defect analogous to the jimpy defect could be excluded.

Northern blot hybridization of total and poly (A)$^+$ RNA of normal and md rat brain demonstrated that the two 3.2 and 1.6 kb transcripts are identical in length but the md transcript only with a level of 2.7% and 6% respectively of the normal rat transcript.

3) PCR amplifications of genomic DNA of the 5'-untranslated region of the mouse, the md and normal rat and in pairs of exon II-intron II, exon III of md rat, exons III-intron III-exon IV, exons IV-intron IV-exon V, exon V-intron V-exon VI and exon VI-intron VI and exon VII were carried out using the appropriate sense and antisense oligonucleotide primers. The PLP fragments with md rat genomic DNA as template showed no differences in size as compared with the wild type DNA.

For PCR amplification of cDNA synthesized from poly (A)⁺ from 18-day-old male rat brains with an oligonucleotide primer sense homologous to the 5' end of exon I and an antisense primer homologous to the 3' end of exon IV were used and the resulting 980 bp PCR fragment comprising exons I to IV *Sau* 3A and *Rsa* I restricted for subcloning into the *Bam* HI and *Sma* I sites of pUC13 and supercoil sequencing.

The sequence analyses of the subclones revealed an A → C transversion at position 344 within the codon for Thr^{75} mutating threonine to proline. This mutation creates an *Ava* II restriction site. This *Alu* II polymorphism and point mutation was confirmed on the genomic level: by PCR amplification with the appropriate primers a 1180 bp fragment embracing exons II, intron II and exon III was obtained. Its restriction with *Ava* II yielded 960 and 220 bp subfragments only with the md rat genomic DNA as template and the shot gun sequencing analysis of *Alu* I fragments revealed the C transversion as shown on the cDNA level.

4) A 1.1 kb *Eco* RI-*Bam* HI fragment encoding the regulatory region of the PLP gene was isolated from the md genomic library with a 1.45 kb *Bam* HI fragment of the 5'-noncoding region of the mouse PLP gene as probe. It was cloned for sequencing into the pGEM3z vector. There is a strong homology between md and normal rat and mouse between the translation start upstream to bp –950. The following 589 bp of the upstream sequence in the mouse gene is deleted in the rat but followed again by a strongly homologous sequence.

PCR amplification of this upstream region not only clearly documents this deletion but also that no differences are recognizable between the normal and md rat.

The md defect is therefore an A → C transversion a point mutation, leading to the mutation of threonine^{75} to proline within the center of the second transmembranal α-helix. Proline is well known as an α–helix breaker, it leads to the partial reversal of the chain direction facilitated by glycin as neighbour in the chain. The dihedral angels at the proline ring give rise to a kink in the main chain.

We propose that this altered conformation has a crucial impact on the membrane integration of PLP in the assembly process irrespective of the mechanism of the integration of the polytopic PLP. We regard other alterations of myelin components secondary to the expression of mutant PLP. We are presently studying the molecular events leading from this point mutation in the md rat and jimpy mouse to the cell death of oligodendrocytes.

Instead of summarizing let me raise some questions around which our interest is focused presently which are ultimately the molecular approach to the developmental process during myelinogenesis.

A necessary prelude to our understanding of the basis for cell-specific expression is the identification of cell-specific factors required for the gene expression of PLP and MBP but also of the oligodendrocyte-specific lipid synthesizing enzymes, most likely not one single cell-specific but multiple factors. We begin to understand the DNA-binding sites of the factors crucial in PLP and MBP expression in the proximal regulatory region of both genes.

- Are there consensus sequences of the factor-binding sites in enhancer regions of different genes?
- Do oligodendrocyte-specific factors, if they exist, participate in the activation of more than one gene in the oligodendrocyte?
- Why does the oligodendrocyte die if no PLP or a mutant PLP is synthesized? The answer to this question could be far-reaching in our understanding of pathogenic processes.

Acknowledgements

The author gratefully acknowledges the successful and stimulating cooperation with the collaborators appearing as coauthors in the publications listed below. This report gives a summary of the work of our laboratory which is described in detail in the following references.

References

Boisson D, Stoffel W (1989) Myelin-deficient rat (md): A point mutation in exon III (A → C, threonine 75 → proline) of the myelin proteolipid protein causes dysmyelination and oligodendrocyte death. *Embo J.*, in press

Diehl HJ, Schaich M, Budzinski RM, Stoffel W (1986) Individual exons encode the integral membrane domains of human myelin proteolipid protein. *Proc. Natl. Acad. Sci. USA* **83**:9807-9811

Schaich M, Budzinski RM, Stoffel W (1986) Cloned proteolipid protein and myelin basic protein cDNA. *Biol. Chem. Hoppe-Seyler* **367**:825-834

Stoffel W, Hillen H, Schröder W, Deutzmann R (1983) The primary structure of bovine brain myelin lipophilin (proteolipid apoprotein). *Hoppe-Seyler's Zeitschr. Physiol. Chem.* **364**:1455-1466

Stoffel W, Hillen H, Giersiefen H (1984) Structure and molecular arrangement of proteolipid protein of central nervous system myelin. *Proc. Natl. Acad. Sci. USA* **81**:5012-5016

Stoffel W, Hillen H, Schröder W, Deutzmann R (1982) Lipophilin (proteolipid apoprotein) of brain white matter. Purification and amino acid sequence studies of the four tryptophan fragments. *Hoppe-Seyler's Zeitschr. Physiol. Chem.* **363**:1397-1401

Stoffel W, Hillen H, Schröder W, Deutzmann R (1982) Primary structure of the C-terminal cyanogen bromide fragments II, III and IV from bovine brain proteolipid apoprotein. *Hoppe-Seyler's Zeitschr. Physiol. Chem.* **363**:855-864

Stoffel W, Schröder W, Hillen H, Deutzmann R (1982) Analysis of the primary structure of the strongly hydrophobic brain myelin proteolipid apoprotein (lipophilin). Isolation and amino acid sequence determination of proteolytic fragments. *Hoppe-Seyler's Zeitschr. Physiol. Chem.* **363**:1117-1131

Stoffel W, Subkowski T, Jander S (1989) Topology of proteolipid protein in the myelin membrane of central nervous system. *Biol. Chem. Hoppe-Seyler* **370**:165-176

Streicher R, Stoffel W (1989) The organisation of the human myelin basic protein. *Biol. Chem. Hoppe-Seyler* **270**:503-510

DEVELOPMENTAL EXPRESSION OF THE MYELIN PROTEOLIPID PROTEIN GENE

Minnetta V. Gardinier[*] and Wendy B. Macklin
Mental Retardation Research Center
UCLA Medical Center
Los Angeles, CA   90024
USA

Myelinogenesis is a highly regulated developmental process, which is necessary for normal CNS development. Oligodendrocyte differentiation and its subsequent elaboration of a myelin membrane that ensheaths certain axons occur within a very precise time frame - for rodents, from 15 to 25 days of age. Until recently, studies had been restricted to ultrastuctural and biochemical analyses of oligodendrocytes and myelin. Early studies identified the proteolipid proteins (PLP, DM20 protein) and myelin basic proteins (MBP) as two classes of structural proteins which comprise 80% of all proteins found in myelin membranes (Eng et al., 1968; Agrawal et al., 1972; Norton and Poduslo, 1973). Thus during the period of active myelination, the protein synthetic machinery of oligodendrocytes is dedicated largely to the production of these two classes of myelin proteins.

Within the past six years, various investigators have isolated myelin-specific cDNAs, which has led to the study of the molecular genetics of myelinogenesis. MBP cDNA clones were isolated first (Roach et al., 1983; Zeller et al., 1984), rapidly followed by a number of PLP cDNA clones (Dautigny et al., 1985; Gardinier et al., 1986; Milner et al., 1985; Naismith et al.,

---

[*] Present address: Laboratoire de Neurochimie, Service de Pédiatrie, Centre Hospitalier Universitaire Vaudois, 1011 Lausanne, Switzerland

1985). In addition, clones for two other less abundant myelin proteins, myelin-associated glycoprotein (Salzer et al., 1987; Arquint et al., 1987) and 2',3'-cyclic nucleotide 3'-phosphodiesterase (Bernier et al., 1987), were isolated. The availability of these cDNA clones has opened up a new area of investigation by providing important new information regarding the activities of the myelin-specific protein genes (for review, Campagnoni and Macklin, 1988).

THE MOUSE MYELIN PROTEOLIPID PROTEIN GENE

Our laboratory has been involved in the characterization of the mouse PLP gene and its developmental expression. Our initial isolation of a rat PLP-specific cDNA allowed us to demonstrate three PLP mRNA species (3.2, 2.4, 1.5 kb) in rodents, which exhibited peak expression between 14 and 25 days of age (Gardinier et al., 1986). Interestingly, these studies also showed that the three size classes of PLP mRNA were expressed differently in rats and mice. In rats, the 3.2 and 1.5 kb PLP-specific mRNAs predominated; whereas, the 3.2 and 2.4 kb PLP-specific transcripts were most prevalent in mice. The smallest mouse PLP mRNA species was detected as a diffuse band(s) at 1.5-1.6 kb. Milner and coworkers had isolated two full-length rat PLP cDNAs, and demonstrated that the mRNAs differed only in the 3' untranslated region, utilizing different polyadenylation signals (Milner et al., 1985). Thus, the different PLP-specific mRNAs are apparently all capable of encoding PLP.

The structure of the mouse PLP gene was determined, and its exons were sequenced from isolated genomic clones (Figure 1)(Macklin et al., 1987a). The PLP gene spans approximately 17 kb and consists of seven exons. The first intron, separating exons 1 and 2, is 7 kb in length, and exon 1 contains only the 5' untranslated region, a methionine codon, and a guanosine nucleotide for the first codon triplet. Thus, this exon could represent an "all-purpose" exon 1 for any number of transcripts. Multiple mRNA transcription start sites within exon 1 were identified by S1 nuclease protection assays, allowing us to assign putative cap sites and a TATA box. A comparison of this region with the human PLP gene (Diehl et al., 1986) revealed approximately 90% homology. A comparable level of homology was noted in the 2.2 kb 3' noncoding region of rat and mouse PLP exon 7. Within the protein coding segment, all nucleotide differences encoding mouse, rat, and human PLP were conservative, i.e., not a single amino acid change in PLP has occurred among these species. Thus, PLP and its gene appear to be under extremely strong selective pressure.

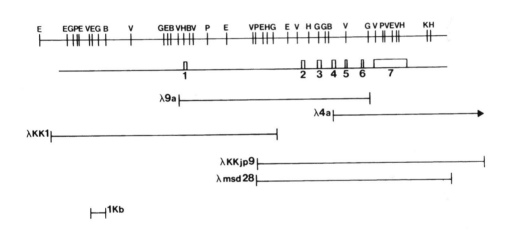

Figure 1. Mouse PLP gene structure and restriction map. Intron regions are indicated by thin lines. Exons 1-7 are indicated by bars. Lambda 9a, 4a, and KK1 represent normal genomic clones; lambda KKjp9 and msd28 represent jp and jp^{msd} genomic clones, respectively. Restriction enzymes: BamHI, B; EcoRI, E; BglII, G; HindIII, H; KpnI, K; PstI, P; PvuII, V

The precise relationship of PLP and the related DM20 protein was identified by analysis of fragments protected in S1 nuclease studies, followed by comparison of these results with the nucleotide sequence of the mouse PLP exons. PLP-related mRNAs containing an internal deletion were mapped using S1 nuclease, and it was shown that these putative DM20 transcripts lacked the 3' end of exon 3 (Figure 2)(Macklin et al., 1987a). Close inspection of the PLP exon 3 sequence revealed that this internal deletion occurred when an alternative donor splice signal within exon 3 was used, removing the 3' half of exon 3 (exon 3b) as intron material. Thus, DM20 mRNAs represent a subpopulation of PLP gene transcripts, containing only exon 3a and all other exons; whereas PLP transcripts utilize exons 3a and 3b (Figure 2). The sequence of DM20 cDNAs isolated by other laboratories confirmed these findings (Nave et al., 1987b; Simons et al., 1987). Immunoblot analyses of PLP and DM20 protein expression during brain development in mice have shown that DM20 protein expression precedes PLP expression at the very earliest stages of myelin formation (Gardinier and Macklin, 1988). As myelin formation progresses, much higher concentrations of PLP, as compared to DM20 protein, accumulate. Thus, these alternatively spliced PLP gene products, PLP and DM20 protein, appear to exhibit a tightly regulated pattern of expression during development.

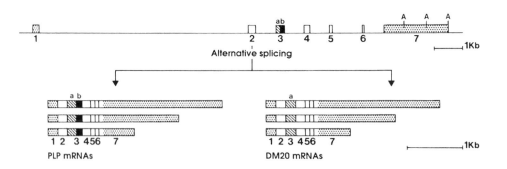

Figure 2. Alternative splicing of the PLP gene. An alternative splice site occurs within exon 3 between exon 3a (hatched area) and 3b (black bar). PLP mRNAs contain both 3a and 3b; DM20 mRNAs utilize 3a. Stippled areas represent noncoding sequence. Alternative polyadenylation sites (A) are also indicated. PLP gene structure is shown as in Figure 1.

PLP GENE EXPRESSION IN DYSMYELINATING MUTANT MICE

Additional information regarding PLP gene expression has been obtained through the study of two X-linked dysmyelinating mutants, jimpy (jp) and myelin synthesis deficient (jp^{msd})(Phillips, 1954; Meier and MacPike, 1970). We established colonies of jp and jp^{msd} mice from breeder pairs kindly provided by Drs. Billings-Gagliardi and Wolf (University of Massachusetts, Worcester, MA, USA). These mice are among the most severely compromised dysmyelinating mutants identified to date, having a lifespan of one month or less. PLP gene defects were suspected in these mutants, because they exhibited a virtual absence of CNS myelin, and PLP is the only known CNS-specific myelin protein. Strong evidence for PLP gene involvement was obtained when the PLP gene was mapped on the human X chromosome to a position syntenic to the jp mutation (Willard and Riordan, 1985; Mattei et al., 1986).

Our earliest PLP-specific mRNA studies showed that jp PLP mRNAs were approximately 100 nucleotides shorter than normal PLP mRNAs (Gardinier, et al., 1986). Using S1 nuclease assays, another laboratory reported a 70 nucleotide deletion in both PLP and DM20 mRNAs (Morello et al., 1986). Sequence analysis of a jp PLP cDNA identified a 74 nucleotide deletion, resulting in a shifted protein reading frame and premature termination of translation (Nave et al., 1986). Isolation and sequence analysis of a jp PLP genomic clone identified a point mutation in the splice acceptor signal within intron 4, immediately preceding exon 5 (Figure 3)(Macklin et al., 1987b; Nave et al., 1987a). Direct comparison of this sequence with the normal mouse PLP gene revealed an AG-->GG transition within the region of the splice signal that is invariant for all splice acceptors identified to date. Thus, aberrant splicing in jp mice eliminates exon 5 from both PLP and DM20 mRNA transcripts. Translation of the jp PLP transcripts would result in PLP containing the normal N-terminal 206 amino acids followed by 34 "nonsense" amino acids at the C-terminus. Immunocytochemical analyses of jp brain revealed accumulations of PLP-reactive material in the Golgi apparatus, suggesting a blockage of intracellular PLP transport (Roussel et al., 1987). We have shown that both PLP and MBP mRNA concentrations are drastically reduced in jp mice as compared to normal littermate controls - <10% and ≤20%, respectively (Gardinier and Macklin, 1988). Thus, a single base change within the PLP gene results in primary effects on PLP gene expression, as well as secondary effects on other myelin protein genes.

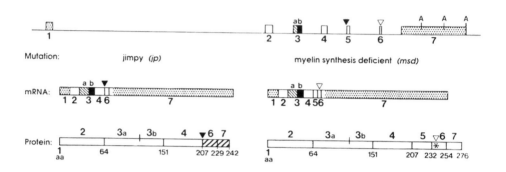

Figure 3. PLP gene and its defects in jp and jp^{msd} mice. A closed arrowhead indicates the jp point mutation in the PLP gene and the splicing defect in the PLP mRNA; the resulting truncated "nonsense" C-terminus is represented by the hatched area encoded by exons 6 and 7. An open arrowhead indicates the jp^{msd} point mutation in the PLP gene and mRNA; an asterisk represents the amino acid substitution within PLP. The amino acid (aa) sequence for PLP is indicated in relation to the exon structure. PLP gene structure is shown as described in Figure 1. Noncoding sequences, alternative splicing and polyadenylation sites are indicated as in Figure 2.

The allelic mutation, jp^{msd}, has provided additional insight into PLP gene regulation. Phenotypically, these mutant mice are indistinguishable from jp mice. Early studies showed that jp^{msd} mice possess about twice as much CNS myelin, as compared to jp mice (Wolf et al., 1983). Correspondingly, PLP/DM20 mRNA levels appeared to be approximately twice as high as their jp counterparts, yet still 15-25% as compared to normal mice (Gardinier and Macklin, 1988). These PLP/DM20 mRNAs appeared to be of normal size. Immunoblot analyses from our laboratory showed that both PLP and DM20 protein were synthesized, albeit at drastically reduced concentrations (<0.5%). Surprisingly, the PLP:DM20 ratio differed dramatically from normal mice. The jp^{msd} mice expressed PLP and DM20 protein in a pattern reminiscent of very young normal mice (7-8 days of age). Thus, the defect in jp^{msd} mice appeared to interrupt the normal developmental program of PLP gene expression.

PLP GENE TRANSCRIPTION IN NORMAL AND MUTANT MICE

Our most recent studies were undertaken to study PLP gene transcription in both normal and dysmyelinating mice. It was necessary to ascertain normal PLP gene activity in nuclear run-on assays. These investigations were then extended to include jp and jp^{msd} mutant mice. Since secondary effects on MBP mRNAs had been observed, MBP gene transcription was also assayed. Whereas our previous Northern and S1 studies reflected steady state RNA levels, these analyses measured de novo RNA synthesis from the PLP and MBP genes during development. Specific gene transcripts were identified by hybridizing a heteogeneous mixture of radioactively labelled transcripts with specific cDNAs immobilized on nitrocellulose filters. CHOB (a "housekeeping" gene) transcription was monitored as an internal standard for sample-to-sample variation. Its activity remains invariant during development (Harpold et al., 1979; Barinaga et al., 1983). Nonspecific hybridization was monitored by inclusion of nitrocellulose dots containing only vector DNA (pUC18, pBR322).

Both PLP- and MBP- specific gene transcription increased during development in normal mice between 7 and 14 days of age (Figure 4). A significant increase in both PLP and MBP gene transcription occured between days 10 and 14 in normal mice. In other studies, both PLP and MBP gene transcription continued to increase until 20-25 days of age (Macklin et al., submitted).

In those studies a small decrease in MBP transcription was noted by 25 days of age, whereas no decrease was seen for PLP transcription through 25 days of age. However, the 2- to 5-fold increase in transcriptional activity is insufficient to explain peak steady state levels for these mRNAs. Whereas the elevated transcription rates remained fairly constant through 20 or 25 days of age, the steady state levels of PLP and MBP mRNAs continue to rise during this period. Thus, the stability of these mRNAs also may play a role in their increased concentrations during active myelination.

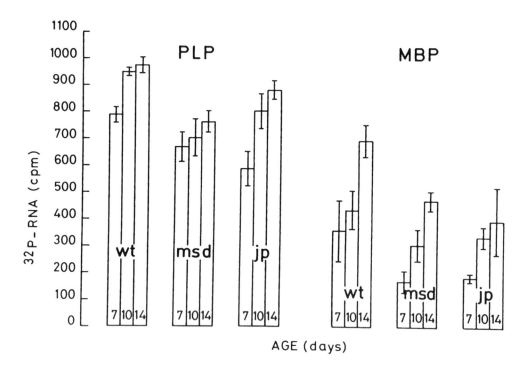

Figure 4. Nuclear run-on analysis of PLP and MBP gene transcription. PLP (left side) and MBP (right side) gene transcription was assayed at 7, 10, and 14 days of age in wild type (wt), jp^{msd} (msd), and jp mice. The data presented are the results from liquid scintillation spectrometry of triplicate individual samples for each time point shown. See Appendix at end of text for protocol details.

In contrast to normal brain development, no significant change in PLP gene transcription was detected through 14 days of age in jp^{msd} mice. Although PLP gene transcription was nearly normal at 7 days of age, little or no significant increase was seen at the later ages. These findings were confirmed in additional studies using nuclei from different animals (7-25 days of age)(Macklin et al., submitted). The combined results of these analyses indicated that jp^{msd} PLP and MBP gene transcription was significantly reduced for days 11-25, as compared to normal gene transcription. Whereas PLP gene transcription showed no developmental increase, MBP gene transcription did show a slight increase during development. Transcription of both PLP and MBP genes was substantially lower in jp mice, as compared to normal littermates. Nonetheless, both PLP and MBP gene activity did increase slightly during development.

In summary, normal mice showed increased levels of PLP- and MBP-specific gene transcription between 7 and 25 days of age, correlating with the period of active myelination. These levels were significantly reduced in jp mice, yet a developmental increase was seen. In contrast, jp^{msd} PLP gene transcription showed little or no developmental increase. Interestingly, the PLP gene defects in jp^{msd} and jp mice appeared to elicit secondary effects on MBP gene transcription, i.e., MBP gene transcription was reduced in both jp and jp^{msd} mice. A developmental increase for jp^{msd} MBP gene transcription was still apparent, despite the lack of a change in PLP gene transcription.

Interpretation of PLP and MBP gene transcription in the dysmyelinating mutants is complicated. The transcriptional activity of PLP and MBP genes increased somewhat during development in jp mice. A definite increase was noted by day 14; however, the activity for both genes decreased by 20 and 25 days of age (Macklin et al., submitted). Since affected jp (and jp^{msd}) males do not survive beyond the first month, additional time points are unavailable. The observations at the later time points (17-25 days of age) can be interpreted two ways: 1) gene activity is reduced, or 2) selective oligodendrocyte loss has occurred in the mutant mice. If the older jp mice possess fewer oligodendrocytes, then the isolated nuclei will contain a higher proportion of non-oligodendrocyte nuclei. The transcription analyses of these nuclei preparations would yield a false indication of reduced PLP and MBP gene activity. Thus, decreased transcription noted at the later time points for jp and jp^{msd} mice could be false indications of PLP or MBP gene activity.

Presently, the data are conflicting regarding premature oligodendrocyte death in jp mice. Early reports indicated that jp oligodendrocyte numbers in corpus callosum (Kraus-Ruppert et al., 1973) and optic nerve (Meier and Bischoff, 1975; Skoff, 1976) were reduced significantly as compared to normal mice. In these studies, cell types were classified on the basis of cellular morphology. Unfortunately, the morphology of jp oligodendrocytes is extremely abnormal, and these investigators noted significant difficulty in positive oligodendrocyte identification. More recent studies by Skoff and coworkers showed that glial cell numbers in optic nerve and spinal cord are comparable between jp and normal mice (Skoff, 1982). In addition, they demonstrated that the total number of glial cells in spinal cord did not differ significantly during development in jp and normal mice (Knapp and Skoff, 1987). No comparable studies are available for jp^{msd} mice. Because of these recent studies, we believe that nuclear run-on assays provide an accurate assessment of PLP and MBP gene activity in jp and jp^{msd} mice.

Further support for the conclusion that these assays provide accurate data comes from differences between PLP transcription data in the two mutant strains and between PLP and MBP transcription data in msd mice. In contrast to the developmental increases in PLP activity seen for normal and jp mice by 14 days of age, no change in PLP gene activity was detected in jp^{msd} mice. After correction of random variations in PLP signal intensity using the corresponding CHOB internal standard, little or no change in PLP gene activity was measured during jp^{msd} development. In contrast, MBP transcription in jp^{msd} mice did show a significant developmental increase between days 10 and 14, diminishing by 20-25 days (Macklin et al., submitted). Despite the developmental increase in MBP gene transcription, transcription was always reduced throughout development, as compared to normal mice. Thus, the differences in PLP and MBP gene activities in older jp^{msd} mice reaffirms our conclusion that oligodendrocyte numbers were comparable between normal and mutant mice. If jp^{msd} mice had fewer oligodendrocytes, then the level of "decreased" transcription should have been comparable in later time points for both PLP and MBP genes. However, PLP gene transcription remained at a constant low level throughout development, while MBP gene transcription exhibited a developmental increase.

Since these studies were completed, our laboratory has confirmed a recent report (Gencic and Hudson, in press) that the jp^{msd} defect is a point mutation residing in exon 6, which causes a conservative amino acid change (Figure 3)(Macklin et al., submitted). In jp^{msd} mice, a single nucleotide change has resulted in the replacement of Ala-241 with a valine residue. It appears that this highly conservative single amino acid substitution defines the genetic defect in jp^{msd} mice, although the possibility cannot be ruled out that a second genetic lesion has occurred independently in jp^{msd} mice. Upon comparison with two computer-generated PLP models (Laursen et al., 1984; Stoffel et al., 1984), we noted that this residue may be in an alpha helix within the lipid bilayer. Perhaps this alpha helix is especially sensitive to amino acid changes. More recently, Hudson and colleagues proposed a significantly different PLP model based on immunofluorescent labelling of cultured oligodendrocytes, protease cleavage mapping, and complement-dependent cytotoxicity assays (Hudson et al., 1989). This model predicts that the C-terminal two-thirds of PLP (and DM20 protein) is located at the extracellular face of the lipid bilayer. Thus, the alpha helix containing the altered amino acid in jp^{msd} PLP would be located on the extracellular surface. If this portion of the PLP molecule is involved in homophilic and/or heterophilic interactions between the bilayers of compacted myelin, these interactions may be disrupted by the amino acid substitution in jp^{msd} PLP and DM20 protein.

The unexpected discovery of this coding region defect in jp^{msd} mice presents an interesting dilemma with regard to the PLP and MBP gene transcription data discussed here. This single nucleotide/amino acid change results in a significant reduction in PLP gene transcription during development. In addition, jp^{msd} PLP- and DM20-specific mRNA concentrations are reduced drastically, as compared to normal mice (Macklin et al., submitted). The expression of these two mRNA populations and their respective translation products throughout development in jp^{msd} mice is very similar to the expression pattern seen in normal mice at 6-8 days of age. Thus, PLP gene expression remains at a low basal level and fails to increase during the normal period of active myelination. If this conservative amino acid substitution is the only genetic defect in these mice, the cascade of events resulting from this amino acid substitution remains to be explained. These events involve not only the PLP gene itself, but also other myelin protein genes.

ACKNOWLEDGEMENTS

The authors wish to thank J.-M. Matthieu for helpful discussions and artwork. This work was supported by a grant from the National Multiple Sclerosis Society (RG1910), a Research Career Development Award (NS1089), and grant NS25304 from the National Institutes of Health.

REFERENCES

Agrawal HC, Burton RM, Fishman MA, Mitchell RD, Prensky AL, (1972) Partial characterization of a new myelin component. J Neurochem 19:2083-2089

Arquint M, Roder J, Chia L-S, Down J, Wilkinson D, Bayley H, Braun P, Dunn R (1987) Molecular cloning and primary structure of myelin-associated glycoprotein. Proc Natl Acad Sci USA 84:600-604

Barinaga M, Yamonoto G, Rivier C, Vale W, Evans R, Rosenfeld G (1983) Transcriptional regulation of growth hormone gene expression by growth hormone-releasing factor. Nature 306:84-85

Bernier L, Alvarez F, Norgard EM, Raible DW, Mentaberry A, Schembri JG, Sabatini DD, Colman DR (1987) Molecular cloning of a 2',3'-cyclic nucleotide 3'-phosphodiesterase: mRNAs with different 5' ends encode the same set of proteins in nervous and lymphoid tissues. J Neurosci 7:2703-2710

Campagnoni AT, Macklin WB (1988) Cellular and molecular aspects of myelin protein gene expression. Mol Neurobiol 2:41-89

Chan DS, Lees MB (1974) Gel electrophoresis studies of bovine brain white matter proteolipid and myelin proteins. Biochem 13:2704-2712

Dautigny A, Alliel PM, d'Auriol L, Pham Dinh D, Nussbaum J-L, Galibert F, Jolles P (1985) Molecular cloning and nucleotide sequence of a cDNA clone coding for rat brain myelin proteolipid. FEBS Lett 188:33-36

Diehl H-J, Schaich M, Budzinski R-M, Stoffel W. (1986) Individual exons encode the integral membrane domains of human myelin proteolipid protein. Proc Natl Acad Sci USA 83:9807-9811

Eng LF, Chao FC, Gerstl B, Pratt D, Tavaststjerna MJ (1968) The maturation human white matter: fractionation of the myelin membrane proteins. Biochemistry 7:4455-4465

Gardinier MV, Macklin WB (1988) Myelin proteolipid protein gene expression in jimpy and jimpy[msd] mice. J Neurochem 51:360-369

Gardinier MV, Macklin WB, Diniak AJ, Deininger PL (1986) Characterization of myelin proteolipid mRNAs in normal and jimpy mice. Mol Cell Biol 6:3755-3762

Gencic S and Hudson LD (to be published) Conservative amino acid substitution in the myelin proteolipid protein of jimpy[msd] mice. J Neurosci

Harpold MM, Evans RM, Salditt-Georgieff M, Darnell JE (1979) Production of mRNA in Chinese hamster cells: Relationship of the rate of synthesis to the cytoplasmic concentration of nine specific mRNA sequences. Cell 17:1025-1035

Hudson LD, Friedrich VL, Behar T, Dubois-Dalcq M, Lazzarini RA (1989) The initial events in myelin synthesis: orientation of proteolipid protein in the plasma membrane of cultured oligodendrocytes. J Cell Biol 109:717-727

Knapp PE, Skoff RP (1987) A defect in the cell cycle of neuroglia in the myelin deficient jimpy mouse. Dev Brain Res 35:301-306

Kraus-Ruppert R, Herschkowitz N, Furst S (1973) Morphological studies on neuroglial cells in the corpus callosum of the jimpy mutant mouse. J Neuropath Exp Neurol 32:197-203

Laursen RA, Samiullah M, Lees MB (1984) The structure of bovine brain myelin proteolipid and its organization in myelin. Proc Natl Acad Sci USA 81:2912-2916

Macklin WB, Campagnoni CW, Deininger PL, Gardinier MV (1987a) Structure and expression of the mouse myelin proteolipid protein gene. J Neurosci Res 18:383-394

Macklin WB, Gardinier MV, King KD, Kampf K (1987b) An AG-->GG transition at a splice site in the myelin proteolipid protein gene in jimpy mice results in removal of an exon. FEBS Lett 223:417-421

Macklin WB, Gardinier MV, Obeso ZO, King KD Altered developmental regulation of the myelin proteolipid protein gene in jimpy^{msd} mice. J Neurochem (submitted)

Marzluff WF, Huang RCC (1984) Transcription of RNA in isolated nuclei. In: Hames BD, Higgins SJ (eds) Transcription and translation - a practical approach. IRL Press, Washington DC, pp 89-129

Mattei MG, Alliel PM, Dautigny A, Passage E, Pham-Dinh D, Mattei JF, Jolles P (1986) The gene encoding for the major brain proteolipid (PLP) maps on the q-22 band of the human X chromosome. Hum Genet 72:352-353

McKnight GS, Palmiter RD (1979) Transcriptional regulation of the ovalbumin and conalbumin genes by steroid hormones in chick oviduct. J Biol Chem 18:9050-9058

Meier C, Bischoff A (1975) Oligodendroglial cell development in Jimpy mice and controls, an electron microscopic study in the optic nerve. J Neurol Sci 26:517-528

Meier C, MacPike AD (1970) A neurological mutation (msd) of the mouse causing a deficiency of myelin synthesis. Exp Brain Res 10:512-525

Milner RJ, Lai C, Nave K-A, Lenoir D, Ogata J, Sutcliffe JG (1985) Nucleotide sequences of two mRNAs for rat brain myelin proteolipid protein. Cell 42:931-939

Morello D, Dautigny A, Pham-Dinh D, Jolles P (1986) Myelin proteolipid protein (PLP and DM-20) transcripts are deleted in jimpy mutant mice. EMBO J 5:3489-3493

Naismith AL, Hoffman-Chudzik E, Tsui L-C, Riordan JR (1985) Study of the expression of myelin proteolipid protein (lipophilin) using a cloned complementary DNA. Nucl Acids Res 13:7413-7425

Nave K-A, Bloom FE, Milner RJ (1987a) A single nucleotide difference in the gene for myelin proteolipid protein defines the jimpy mutation in mouse. J Neurochem 49:1873-1877

Nave K-A, Lai C, Bloom FE, Milner RJ (1986) Jimpy mutant mouse: a 74-base deletion in the mRNA for myelin proteolipid protein and evidence for a primary defect in RNA splicing. Proc Natl Acad Sci USA 83:9264-9268

Nave K-A, Lai C, Bloom FE, Milner RJ (1987b) Splice site selection in the proteolipid protein (PLP) gene transcript and primary structure of the DM-20 protein of CNS myelin. Proc Natl Acad Sci USA 84:5665-5669

Norton WT, Poduslo SE (1973) Myelination in rat brain: changes in myelin composition during brain maturation. J Neurochem 21:759-773

Phillips RJS (1954) Jimpy, a new totally sex-linked gene in the house mouse. Zeitschrift fur induktive Abstammungs- und Vererbunglehre 86:322-326

Roach A, Boylan K, Horvath S, Prusiner SB, Hood L (1983) Characterization of cloned cDNA representing rat myelin basic protein: absence of expression in brain of shiverer mutant mice. Cell 34:799-806

Roussel G, Neskovic NM, Trifilieff E, Artault J-C Nussbaum J-L (1987) Arrest of proteolipid transport through the Golgi apparatus in jimpy brain. J Neurocyt 16:195-204

Salzer JL, Holmes WP, Colman DR (1987) The amino acid sequences of the myelin-associated glycoproteins: Homology to the immunoglobulin gene superfamily. J Cell Biol 104:957-965

Simons R, Alon N, Riordan JR (1987) Human myelin DM-20 proteolipid protein deletion defined by cDNA sequence. Biochem Biophys Res Commun 146:666-671

Skoff RP (1976) Myelin deficit in the Jimpy mouse may be due to cellular abnormalities in astroglia. Nature 264:560-562

Skoff RP (1982) Increased proliferation of oligodendrocytes in the hypomyelinated mouse mutant-jimpy. Brain Res 248:19-31

Stoffel W, Hillen H, Giersiefen H (1984) Structure and molecular arrangement of proteolipid protein of central nervous system myelin. Proc Natl Acad Sci USA 81:5012-5016

Trifilieff E, Skalidis G, Hélynck G, Lepage P, Sorokine O, Van Dorsselaer A, Luu B (1985) Données structurales sur le protéolipide de la myéline de masse moléculaire apparente 20 kDa (DM-20). C R Acad Sci Paris 300:241-246

Vacher-Lepretre M, Nicot C, Alfsen A, Jolles J, Jolles P (1976) Study of the apoprotein of Folch-Pi bovine proteolipid. II. Characterization of the components isolated from SDS solutions. Biochim Biophys Acta 420:323-331

Willard HF, Riordan JP (1985) Assignment of the gene for myelin proteolipid protein to the X chromosome: implications for X-linked myelin disorders. Science 230:940-942

Wolf MK, Kardon GB, Adcock LH, Billings-Gagliardi S (1983) Hypomyelinated mutant mice. V. Relationship between jp and jp^{msd} re-examined on identical genetic backgrounds. Brain Res 271:121-129

Zeller NK, Hunkeler MJ, Campagnoni AT, Sprague J, Lazzarini RA (1984) Characterization of mouse myelin basic protein messenger RNAs with a myelin basic protein cDNA clone. Proc Natl Acad Sci USA 81:18-22

APPENDIX:

The transcription assay was adapted from published protocols (McKnight and Palmiter, 1979; Marzluff and Huang, 1984). Changes are indicated below.

Nuclei isolation: Buffer I was prepared with 1X buffer A/0.32 M sucrose/1 mM DTT/1 mM spermidine (1X buffer A = 5 mM PIPES, pH 7/10 mM NaCl/1 mM EDTA/0.5 mM EGTA/3 mM MgCl$_2$). NP40 was added to 0.5% just prior to the initial centrifugation. Buffer II contained 1X buffer A/1.75 M sucrose/1 mM DTT/1 mM spermidine. Nuclear pellets (approx. 10^8 nuclei in 100 μl) were resuspended in storage buffer containing 25% glycerol and kept in liquid N$_2$ for later use.

Transcription reaction: Transcription reaction mix was prepared using 350 μCi ^{32}P-GTP (per transcription reaction). Samples were incubated at 25° C for 10 min; unlabelled GTP was added to 50 μM, and samples were incubated an additional 15 min at 25° C.

RNA isolation: CaCl$_2$ was added to 5 mM. The transcription reaction was digested with RNase-free DNase (200 U/ml), followed by Proteinase K (200 μg/ml). This material was phenol:chloroform extracted, ethanol precipitated, and trichloroacetic acid precipitated.

Hybridization: Following RNA denaturation at 70° C (15 min), all samples were hybridized at 42° C (48-60 hrs) using equivalent cpm (20-50 X 10^6 cpm). Hybridization buffer contained 50% formamide (Gardinier and Macklin, 1988). DNA nitrocellulose dots contained p27 (PLP; Milner et al., 1985), pMBP (Roach et al., 1983), pCHOB (Harpold et al., 1979), pBR322, or pUC18 (5 ug). Filters were washed once in 2X SSC/0.1% SDS and twice in 0.1X SSC/0.1% SDS, 20 min each at 50° C (PLP, CHOB, pUC18) or 42° C (MBP, pBR322). Filters were analyzed by autoradiography followed by liquid scintillation spectrometry.

THE MYELINATION CASCADE

Greg Lemke, Gerry Weinmaster, and Edwin S. Monuki
Molecular Neurobiology Laboratory
The Salk Institute
P.O. Box 85800
San Diego, CA 92138

Introduction

The elaboration of myelin represents the culmination of a complex developmental interaction between neurons and glia. A variety of experiments with myelinating systems *in vivo* and *in vitro* are consistent with the hypothesis that this process, at least in the peripheral nervous system, is initiated via a contact-dependent differentiation signal conveyed from the surface of neuronal axons to a receptor expressed on the surface of myelin-forming glia. Processing of this differentiation signal must involve a signal transduction apparatus many of whose components are unique to myelinating cells, including the regulatory and structural proteins required for the elaboration of myelin (reviewed in Lemke, 1988).

We have recently begun to temporally dissect the regulated cascade of gene expression that leads to myelination, using cultured Schwann cells dissociated and purified from neonatal rat sciatic nerves. In this *in vitro* system, the myelination cascade appears to be initiated through elevation of the intracellular second messenger cyclic AMP. This cascade requires approximately 24 hours to complete, and culminates in expression of the genes encoding the myelin-specific structural proteins myelin basic protein (MBP) (Roach et al., 1983), protein zero (P_0) (Lemke and Axel, 1985), proteolipid protein (PLP) (Milner et al., 1985), P_2 (Narayanan et al., 1988), and myelin-associated glycoprotein (MAG) (Lai et al., 1987).

Materials and Methods

Reagents. The CM-cellulose fraction of glial growth factor (CM-GGF) was prepared from lyophilized bovine anterior pituitary lobes as described by Lemke and Brockes (1984). Forskolin was purchased from Calbiochem. The following cDNA probes were used: *fos*, the 1089bp AvaI fragment of the *v-fos* clone described by Van Beveren et al. (1983); *jun*, nucleotides 562-1147 of the mouse *c-jun* cDNA described by Lamph et al. (1988); PDGF receptor, the 1161bp HincII fragment of the mouse β type PDGF receptor cDNA described by

Yarden et al. (1986); PLP, the complete rat PLP cDNA described by Milner et al. (1985); P_0, the complete rat P_0 cDNA described in Lemke and Axel (1985).

Schwann cell culture. Rat Schwann cells were prepared by dissociation from neonatal (2-3 day) rat sciatic nerves, and purified by immunoselection using anti-Thy1.1 and complement, as described by Brockes and colleagues (1979). Purified Schwann cell populations were expanded by growth in DMEM supplemented with 10% fetal bovine serum, $2\mu M$ forskolin, and $20\mu g/ml$ CM-GGF, as originally described by Porter and colleagues (1986). For each experiment, Schwann cells were cultured in forskolin and GGF for 3-5 weeks, and then withdrawn from these mitogens by washing cell monolayers twice with DMEM and replacing the medium with DMEM plus 10% FBS alone for 4-5 days prior to the start of the experiment. Forskolin is a fully reversible activator of adenylate cyclase (Seamons et al., 1981). Schwann cells cultured for 3-5 weeks in forskolin and GGF revert to nearly complete quiescence when withdrawn from these mitogens for 4-5 days (Porter et al., 1986). At postnatal day 2-3, many Schwann cells in the rat sciatic nerve already express appreciable levels of major myelin (P_0, MBP, PLP) RNAs (Lemke and Axel, 1985; Lemke and Chao, 1988). Although when placed in culture in the absence of neurons major myelin RNA levels fall dramatically (Lemke and Chao, 1988), low but readily detectable basal levels of these RNAs persist.

RNA isolation and analysis. Total cellular RNA from cultured cells was isolated according to the guanidine/water-saturated phenol procedure described by Chomczynski and Sacchi (1987). RNA was fractionated on 0.8-1% agarose/formaldehyde gels, and Northern blots were performed according to standard procedures, as described previously (Lemke and Chao, 1988). Following transfer to Nytran® membranes (Schleicher and Schuell), RNA was visualized by staining with methylene blue, as described by Herrin (1988). All radiolabeled probes were prepared from double-stranded cDNA fragments, using ^{32}P-$\alpha dCTP$ and a random hexamer priming kit, according to instructions provided by the manufacturer (BRL).

Results and Discussion

Axonal regulation of Schwann cell gene expression

Schwann cells require contact with appropriate axons both for initial induction of myelin-specific genes, and for maintained expression of these genes. We have examined this dependence *in vivo*, by depriving Schwann cells of axonal contact through transection of peripheral (sciatic) nerves; and *in vitro*, by examining myelin gene expression in purified Schwann cells cultured in the absence of neurons. We performed *in vivo* transections in 35 day old rats in which peripheral myelination had proceeded to a relatively advanced stage, and assayed for myelin gene expression in Schwann cells distal to the site of transection (where axons degenerate) at successive days following surgery. *In situ* hybridization signals for mRNAs encoding the major myelin proteins P_0 and MBP fall dramatically in these axon-deprived cells. Significant decreases in the steady-state levels of both the P_0 and MBP mRNAs are evident at 1 day following transection. By 5 days following surgery, *in situ* signals for these mRNAs are near background. Quantitation by slot blot indicates that P_0 and MBP message levels fall by approximately 40-fold (Trapp et al., 1988), demonstrating that Schwann cells exhibit a strong dependence upon axons in order to maintain high steady-state levels of the major myelin mRNAs. As discussed below, this dependence probably reflects alteration in the instantaneous rate of transcription of the major myelin genes, rather than alteration in the turnover rate of the P_0 and MBP mRNAs.

We observed a similarly dramatic dependence upon axonal contact when we compared P_0 and MBP gene expression in 2-3 day sciatic nerve to expression in Schwann cells dissociated from this nerve and cultured in the absence of axons. Cultured Schwann cells deprived of axons express 40 to 50-fold lower levels of P_0 and MBP mRNA than do actively-myelinating Schwann cells in contact with axons (Lemke and Chao, 1988). This low basal level of major myelin gene expression is reached within five days after Schwann cells are placed into culture, and is maintained indefinitely (> 4 months) thereafter.

Cyclic AMP induction of myelin gene expression

In vitro, we have found that the requirement for continuous axonal contact can be largely overridden by any treatment that elevates the level of intracellular cAMP. Thus, each of the major myelin (P_0, MBP, PLP) genes are induced by cAMP analogs, cholera toxin, and the reversible adenyl cyclase activater forskolin (Lemke and Chao, 1988). Induction of these genes can even be achieved by treating cultured Schwann cells with the β-adrenergic agonist isoproterenol (G.L., unpublished observations), since Schwann cell β-adrenergic receptors are coupled to adenyl cyclase. Dose-response experiments indicate that the relative sensitivity of the major myelin genes to induction by cAMP parallels their relative level of expression by Schwann cells in contact with axons. Thus, the P_0 gene is roughly 10-fold more sensitive to induction by forskolin than is the MBP gene, and P_0 is expressed at 5 to 10-fold higher levels than MBP in actively myelinating Schwann cells (Lemke and Chao, 1988). cAMP induction of the major myelin genes is specific, to the extent that TPA, a tumor promoter that activates an independent serine/threonine kinase (protein kinase C), is without effect on Schwann cell expression of P_0 and MBP mRNA. cAMP regulation of major myelin gene expression almost certainly reflects control at the level of transcription of these genes, since the regulatory region of the rat P$_0$ gene is active as a transcriptional activator only when introduced into Schwann cells cultured in the presence of forskolin (Lemke et al., 1988).

Although cAMP induces major myelin gene expression, this induction is not of the form most commonly observed. Unlike cAMP regulation of immediate-early response genes such as those encoding the transcription factors *c-fos* and NGFI-A (Greenburg et al., 1985), induction of the P$_0$, MBP, and PLP mRNAs is neither rapid nor transient. Instead, up-regulated levels of myelin mRNAs are not observed until 18-24 hours following elevation of intracellular cAMP, and once induced, these levels remain high so long as cAMP levels remain elevated (G.W. and G.L., unpublished data). These results suggest that cAMP induction occurs indirectly, via a cascade of intermediate, cAMP-induced regulatory molecules.

cAMP regulation of a glial transcription factor

We have recently identified one such molecule (Monuki et al., 1989): a cAMP-induced protein whose structure is closely related in its putative DNA binding domain to those of the "POU" transcription factors Pit-1/GHF-1 (Ingraham et al., 1988; Bodner et al., 1988), Oct-1 (Sturm et al., 1988), Oct-2 (Clerc et al., 1988), and unc-86 (Finney et al., 1988). This protein, which we have named SCIP (pronounced "skip", for suppressed cAMP inducible POU), is expressed in the developing central and peripheral nervous systems, and at particularly high levels in the rapidly myelinating optic (CNS) and sciatic (PNS) nerves. SCIP is not expressed in liver, kidney, heart, lung, or spleen. Among glial cells, SCIP expression is restricted to myelinating Schwann cells and oligodendrocytes; the SCIP gene is not expressed by cultured astrocytes (Monuki et al., 1989).

Since cAMP in many respects mimics the inductive effect of axons on major myelin gene expression, we examined the kinetics of cAMP induction of the SCIP gene relative to cAMP induction of the major myelin genes. Remarkably, cAMP induces SCIP with neither immediate-early (<15 minutes) kinetics, nor with the delayed (18-24 hour) kinetics characteristic of the major myelin genes. Instead, up-regulation in the level of SCIP mRNA is first apparent at 1 hour following forskolin addition, increases dramatically between 1 and 3 hours, and then more gradually so up to 24 hours. SCIP mRNA levels are stable after this time, so long as intracellular cAMP remains elevated. However, they quickly fall back to baseline when forskolin is withdrawn. Stable cAMP induction of SCIP therefore precedes cAMP induction of the major myelin genes by approximately 12 hours.

This induction follows the Schwann cell-specific, cAMP-mediated repression of the *c-jun* gene however. This ubiquitous transcription factor, which together with *c-fos* forms the AP-1 transcription factor complex (Bohman et al., 1987; Sassone-Corsi et al., 1988), is expressed at unusually high basal levels in quiescent Schwann cells, and is rapidly repressed in response to cAMP elevation. Schwann cell *c-jun* mRNA levels begin tc fall within 15 minutes and reach baseline within 1 hour of cAMP elevation (Monuki et al., 1989). This time course is exactly reciprocal with the cAMP-mediated induction of the SCIP gene. This observation led us to hypothesize that *c-jun* or a similarly-regulated transcription factor might serve to normally suppress transcription of the SCIP gene. To test this hypothesis, we treated Schwann cells with the protein synthesis inhibitor cycloheximide. We observed that this agent is capable of inducing Schwann cell expression of SCIP mRNA, even in the absence of elevated cAMP. Remarkably, pre-treatment with cycloheximide markedly *potentiates* subsequent induction of SCIP mRNA by cAMP. These observations indicate that the SCIP gene is normally suppressed in cultured Schwann cells, and that this suppression is relieved by cAMP. Suppression of SCIP expression could be achieved at one or more biochemical levels. The SCIP gene could be under the control of a transcriptional repressor such as *c-jun*, for example. Alternatively, SCIP mRNA could be actively transcribed yet rapidly degraded, and both cAMP and cycloheximide could function to indirectly stabilize this mRNA (Brawerman, 1989). Although our data do not at the present time allow us to distinguish between these alternatives, the fact that a known transcriptional regulator is repressed immediately prior to the appearance of SCIP mRNA leads us to favor the former model. The fact that cAMP induction of SCIP mRNA occurs even in the presence of cycloheximide demonstrates that all of the proteins required to drive transcription of the SCIP gene must be in place in cultured Schwann cells prior to the elevation of cAMP.

Regulated expression of the major and minor myelin genes

With respect to gene expression, the endpoint of glial differentiation is reached when the genes encoding myelin structural proteins are induced in preparation for the elaboration of myelin. As noted above, cAMP activation of these genes is observed (appropriately) only after 18-24 hours of continuous cAMP elevation, suggesting that prior events, such as the induction of SCIP, are required in order for this activation to occur. Transcriptional activity of both the P_0 (Lemke et al., 1988) and MBP promoters (G.L., unpublished observations) is absolutely dependent upon cAMP elevation for expression in transfected cultured Schwann cells. This dependence almost certainly reflects the cAMP dependence of prior inductive events, such as the induction of SCIP. How this protein and other glial-specific transcription factors regulate myelin gene expression remains to be elucidated. The fact that cAMP induction of SCIP expression precedes major myelin gene expression by ~12 hours in consistent with two related hypotheses: either SCIP acts as inducer of a further set of cell-specific trans-acting proteins that then directly bind to and activate transcription of myelin-specific genes; or alternatively, SCIP both induces these factors and acts in concert with them to activate transcription of myelin-specific genes. We have identified two candidate SCIP binding sites within regions of the rat P_0

promoter that we have found to posses strong transcriptional activity (Lemke et al., 1988: G.L., unpublished observations), and are know asking whether bacterial fusions proteins that contain the SCIP DNA binding (homeobox) domain recognize these sequences.

The picure that emerges from our studies of myelin gene expression in cultured Schwann cells is that of a regulated cascade, initiated by a receptor-driven increase in intracellular cAMP, and culminating in expression of the major and minor myelin genes and the elaboration of myelin. We have tentatively classified the elements of this cascade into four groups, based upon the kinetics of their cAMP regulation. The properties of individual identified members of these groups, some of which are glial-restricted or glial-specific, are summarized in the table below.

cAMP regulated gene	Time of expression in response to cAMP	Regulation in Schwann cells
c-jun	immediate-early	repression (sustained)
SCIP*	early	induction (sustained)
junB	intermediate	induction (transient)
FGF receptor	intermediate	induction (transient)
PDGF receptor	late	induction (sustained)
P_0*	late	induction (sustained)
MBP*	late	induction (sustained)
PLP*	late	induction (sustained)
P_2*	late	induction (sustained)

Little or no cAMP regulation: *c-fos*, *junD*, NGFI-A

*Gene product restricted or unique to myelinating glia.

The dose-response kinetics for three of these cAMP-responsive genes (c-jun, SCIP, and P_2) are summarized in the following graph.

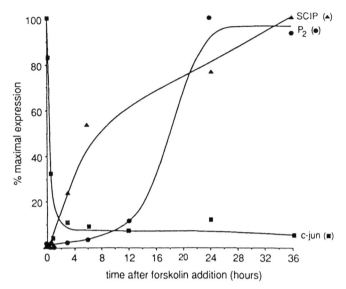

Densitometric analysis of the forskolin response kinetics of *c-jun*, SCIP and P$_2$. These curves were generated from densitometric scans of Northern blot autoradiograms. For each curve, expression values are normalized to the time point exhibiting the most intense Northern hybridization signal.

References

Bodner, M., Castrillo, J.-L., Theill, L.E., Deernick, T., Ellisman, M., and Karin, M. (1988) The pituitary-specific transcription factor GHF-1 is a homeobox-containing protein. Cell 55, 505-518.

Bohmann, D., Bos, T., Admon, A., Nishimura, T., Vogt, P., and Tjian, R. (1987) Human proto-oncogene *c-jun* encodes a DNA binding protein with structural and functional properties of transcription factor AP-1. Science 238, 1386-1392.

Brawerman, G. (1989) mRNA decay: finding the right targets. Cell 57, 9-10.

Brockes, J. P., Fields, K. L., and Raff, M. C. (1979). Studies on cultured rat Schwann cells. I. Establishment of purified populations from cultures of peripheral nerve. Brain Res. 165, 105-118.

Chomczynski, P. and Sacchi, N. (1987) Single-step method of RNA isolation by acid guanidinium thiocyanate-phenol-choloroform extraction. Anal. Biochem. 162, 156-159.

Clerc, R.G., Corcoran, L.M., LeBowitz, J.H., Baltimore, D., and Sharp, P.A. (1988) The B-cell-specific Oct-2 protein contains POU box- and homeo box-type domains. Genes Dev. 2, 1570-1581.

Curran, T., Miller, A.D., Zokas, L., and Verma, I.M. (1984) Viral and cellular fos proteins; a comparative analysis. Cell 36, 259-268.

Dumont, J.E., Jauniaux, J.-C., and Roger, P.P. (1989) The cyclic AMP-mediated stimulation of cell proliferation. Trends in Biochem. Sci. 14, 67-71.

Finney, M., Ruvkun, G., and Horvitz, H.R. (1988) The C. elegans cell lineage and differentiation gene unc-86 encodes a protein with a homeodomain and extended similarity to transcription factors. Cell 55, 757-769.

Greenberg, M.E., Greene, L.A., and Ziff, E.B. (1985) Nerve growth factor and epidermal growth factor induce rapid transient changes in proto-oncogene transcription in PC 12 cells. J. Biol. Chem. *260*, 14101-14110.

Herrin, D.L. (1988) Rapid, reversible staining of Northern blots prior to hybridization. BioFeedback *6*, 196-198.

Ingraham, H.A., Chen, R., Mangalam, H.J., Elsholtz, H.P., Flynn, S.E., Lin, C.R., Simmons, D.M., Swanson, L., and Rosenfeld, M.G. (1988) A tissue-specific transcription factor containing a homeodomain specifies a pituitary phenotype. Cell *55*, 519-529.

Lai, C., Brow, M.A., Nave, K.-A., Noronha, A.B., Quarles, R.H., Bloom, F.E., Milner, R.J., and Sutcliffe, J.G. (1987) Two forms of 1B236/myelin-associated glycoprotein, a cell adhesion molecule for postnatal development, are produced by alternative splicing. Proc. Natl. Acad. Sci. USA *84*, 4227-4341.

Lamph, W.W., Wamsley, P., Sassone-Corsi, P., and Verma, I.M. (1988) Induction of proto-oncogene jun/AP-1 by serum and TPA. Nature *334*, 629-631.

Lemke, G. (1988) Unwrapping the genes of myelin. Neuron *1*, 535-543.

Lemke, G. E., and Brockes, J. P. (1984). Identification and purification of glial growth factor. J. Neurosci. *4*, 75-83.

Lemke, G., and Axel, R. (1985). Isolation and sequence of a cDNA encoding the major structural protein of peripheral myelin. Cell *40*, 501-508.

Lemke, G., and Chao, M. (1988). Axons regulate Schwann cell expression of the major myelin and NGF receptor genes. Development *102*, 499-504..

Lemke, G., Lamar, E., and Patterson, J. (1988). Isolation and analysis of the gene encoding peripheral myelin protein zero. Neuron *1*, 73-83.

Milbrandt, J. (1987) Nerve growth factor rapidly induces a gene which encodes a possible transcriptional regulatory factor. Science *238*, 797-799.

Milner, R. J., Lai, C., Nave, K.-A., Lenoir, D., Ogata, J., and Sutcliffe, J. G. (1985). Nucleotide sequences of two mRNAs for rat brain myelin proteolipid protein. Cell *42*, 931-939.

Monuki, E.S., Weinmaster, G., Kuhn, R., and Lemke, G. (1989) SCIP: a glial POU domain gene regulated by cyclic AMP. Neuron *3*, in press.

Narayanan, V., Barbosa, E., Reed, R., and Tennekoon, G. (1988) Characterization of a cloned cDNA encoding rabbit myelin P_2 protein. J. Biol. Chem. *263*, 8332-8337.

Porter, S., Clark, M. B., Glaser, L., and Bunge, R. P. (1986). Schwann cells stimulated to proliferate in the absence of neurons retain full functional capacity. J. Neurosci. *6*, 3070-3078.

Roach, A., Boylan, K., Horvath, S., Prusiner, S. B., and Hood, L. E. (1983). Characterization of a cloned cDNA representing rat myelin basic protein: absence of expression in *shiverer* mutant mice. Cell *34*, 799-806.

Roger, P.P., Servais, P., and Dumont, J.E. (1987) J. Cell. Physiol. *130*, 58-67.

Ryder, K., Lanahan, A., Perez-Albuerne, E., and Nathans, D. (1989) Jun D: a third member of the Jun gene family. Proc. Natl. Acad. Sci. USA *86*, 1500-1503.

Sassone-Corsi, P., Lamph, W.W., Kamps, M., and Verma, I.M. (1988) fos-associated cellular p39 is related to nuclear transcription factor AP-1. Cell *54*, 553-560.

Seamons, K.B., Padgett, W., and Daly, J.W. (1981) Forskolin: a unique diterpine activator of adenylate cyclase in membranes and intact cells. Proc. Natl. Acad. Sci. USA *78*, 3363-3367.

Sturm, R.A., Das, G., and Herr, W. (1988) The ubiquitous octamer-binding protein Oct-1 contains a POU domain with a homeobox subdomain. Genes Dev. *2*, 1582-1599.

Trapp, B. D., Hauer, P., and Lemke G. (1988). Axonal regulation of myelin protein mRNA levels in actively myelinating Schwann cells. J. Neurosci. *8*, 3515-3521.

Yarden, Y., Escobedo, J.A., Kuang, W.-J., Yang-Feng, T.L., Daniel, T.O., Tremble, P.M., Chen, E.Y., Ando, M.E., Harkens, R.N., Francke, U., Fried, V.A., Ullrich, A., and Williams, L.T. (1986) Structure of the receptor for platelet-derived growth factor helps define a family of closely related growth factor receptors. Nature *323*, 226-232.

Van Beveren, C., van Straaten, F., Curren, T., Müller, R., and Verma, I.M. (1983) Analysis of the FBJ-MuSV provirus and *c-fos* (mouse) gene reveals that viral and cellular *fos* gene products have different carboxy termini. Cell *32*, 1241-1255.

INDEX

A_2B_5 3-17,53,174,283,302,312,317
acetylcholine   408
acetylcholinesterase   162
acid hydrolase   451
action potentials   94
adenylate cyclase   256,282,405,469,535
adhesion signal   438
ADP-ribosylation   405-416,463-471
adrenalectomy   161
adrenoreceptor
  $-\alpha_1$ 108,407
  $-\alpha_2$ 108,407
  $-\beta$ 108
aggregate cultures   283
aggregating cell cultures   155-170
alpha helix   527
alternative splicing   62,65,367,520
amino acid sequence   65,373,505-516
amniotes   368
amphibians   362,377
anti-myelin antibodies   427
antigenic
  -expression   343-359
  -phenotype   9
  -plasticity   356
antipain   106
antisense RNA   497
AP-1 transcription factor complex   536
apposition
  -cytoplasmic   373-387
  -extracellular   373-387
astrocytes   26,36,50,81-97,110,143,155,255-279,287,438,536
  -death of   192
  -perinodal   85
  -reactive   109
  -type 1   4,8,88,150,174,177,257,273,293,317,355
  -type 2   3,11,19,89,273,293,317
astrocytic processes   51,85,132,176

atropine 407

autophosphorylation 203

axolemma 35,59,66,91,260

axonal
  -degeneration 49,452
  -glycoprotein 441
  -growth 143-153,219
  -membrane 82,191
  -mitogens 173
  -signals 344
  -stimuli 130

axons 33,55,66,129-142,143-153,188,191,218,534
  -degenerating 49
  -demyelinated 172
  -premyelinated 84
  -remyelinated 177
  -unmyelinated 8,145

axoplasm 83

axotomy 140,143

basal lamina 88,194,233,244

Bergmann fibers 36

bipotential
  -precurser 286
  -progenitor cell 3-17,317

birds 362

blood vessels 87

blood-brain barrier 40,102,293

bone marrow 107

bony fish 362

bovine growth hormone (bGH) 161

bromodeoxyuridine (BrdU) 7,225-279,297,312

bystander demyelination 102

c-jun 541

carbachol 408

carbohydrate chains 237

carbonic anhydrase (CA) 36,130

catecholamines 108

cell
  -adhesion molecules (CAMs)   434
  -death   42
  -division   299
  -lineage   3-17,23
  -migration   129
  -nucleus   162,219,301
  -proliferation   139,288,438
central nervous system (CNS)   4,21,33,42,61,85,110,143,185,
                                231,255,441,489,505
  -myelin   59-79,161,361-372,373-387,391-404,463-471,522
ceramide galactosyltransferase (CgalT)
  -hydroxylated fatty acid (HFA)   391-404
  -non hydroxylated fatty acid (NFA)   394
ceramides   396
cerebellar cortex   218
cerebellar soluble lectin (CSL)   433-450
cerebellum   36,217,293,437
cerebroside sulfotransferase   36
cerebrosides   35,117,391
cerebrospinal fluid (CSF)   105,427
cerebrum   4,281
c-fos   211,301,536
c-myc   301
charcoal   188,192
chemically defined medium   40,157
chloramphenicol acetyltransferase(CAT)   208
cholera toxin   290,411,466,535
cholesterol   117-142
choline acetyltransferase   162
cholinergic receptors   405-416,469
chondroitinase ABC   240
choroid plexus   36
chromosome   5,195,511
cladogram   369
clathrin   74
cloning strategies   482
ciliary neurotrophic factor (CNTF)   14,317

coated pits   74

cobra venom factor (CVF)   104,423

coelacanth   361-372,379

complement   101-113,423

  -cascade   427

conditioned medium (CM)   4,8,40,104,143-153

connective tissue   147

corpus callosum   258

cross transplantations   195

crush injury   451

cyclic AMP   201-215,256,281-292,302,533-540

  -binding protein (CAP)   206

  -dependent protein kinase   201-215

  -receptor   207

  -response element (CRE)   206

2',3'-cyclic-nucleotide 3'-phosphodiesterase (CNP)   11,
              19-31,59-79,160,281-292,317,380,329-339,361,
              415,363-471,476,498,518

  -mRNA   69,498

cycloheximide   541

cytoplasmic

  -channels   59-79

  -determinants   422

  -loops   36,194

cytoskeleton   231

cytosolic Ca^{2+} 302

degeneration   129-142

  -Wallerian   129-142,143,451

demyelinating diseases   10,30,106,165

demyelination   101-113,115,123,164,176,185-198,417-431,
              433-461

  -area of   188

  -autoimmune mediated   417-431

  -inflammatory   104

  -pathogenesis of   423

  -primary   172

dentate gyrus   220

deoxycholate   421

developmental pattern   40
dexamethasone   160
diacylglycerol   405-416
dibutyryl cyclic AMP (dbcAMP)   249,268,281-292,320,354
differentiation   3,28,36,56,138,162,317
  -signal   533
dividing cells   14
DM-20   63,361-372,383,419,510,517-532
  -cDNA   520
domain
  -C-terminal   65,332
  -cytoplasmic   65
  -extracellular   65,295,380
  -hydrophilic   511
  -immunoglobulin-like   331
  -membrane spanning   332
  -transmembrane   65,295,331
dorsal root ganglia   261
downregulation   8,451,481
edema   107,111,131,423
electrical activity   94
electrogenesis   94
electron density profiles   377
electron microscopy (EM)   69,130,147,186,220,312,490
electrophysiology   49
electroporation   313
ELISA   345,422
elasmobranch   362,377
EMBL3 vector   514
endoglycosidase F   347
endoneurial slices   458
endoplasmic reticulum   393,455
  -smooth   396
  -rough   33,63,67,332,393,498
endosomes   74
endothelial cells   36,107
enucleation   131
epidermal growth factor (EGF)   160,290,255-279,324

epithelial cells   240

ethanol plasmalogen   117

ethidium bromide   171-184

evolution   344,361-372,373-387

exons   70,195,481,489-503,505-516,517-532

experimental allergic encephalomyelitis(EAE)   101-113,223, 423

extracellular matrix (ECM)   166,171,231,244

extracellular signals   324

fatty acids   117

fibroblast growth factor (FGF)   4,7,165,290,255-279,324

  -acidic   7

  -basic   7

fibroblasts   231,257,295

fibronectin   174,236

fishes

  -actinopterygian   361-372

  -cartilaginous   361-372

  -crossopterygian   363

  -sarcopterygian   361-372

  -telostean   343-359,361-372,362

fluorescence-activated cell sorting (FACS)   283

forskolin   207,249,256,290,354,408,533

fos   301,533

free polysomes   62,67,333,393

freeze fracture   84

fura-2   297

G-protein   290,405-416,463-471

galactocerebroside   4,19-31,33-45,51,158,164,174,256,282, 293,351,391,454

galactolipids   21,391-404

ß-galactosidase   209,311-316

galactosylceramide   192,312,418

  -sulfotransferase   162

galactosyltransferase   394,454

gangliosides   3,391-404

garfish   376

gastrointestinal tract   126

GD3 marker   312
gene
  -expression   10,19,25,59,163,201,207,302,451-461,475-
              487,497,522,533-540
  -structure   513
  -transcription   523,525
genetic
  -defect   514
  -lesion   527
genomic
  -clone   521
  -library   514
glia
  -limitans   87
  -promoting factor   257
glial
  -cells   47,81
  -differentiation   537
  -fibrillary acidic protein (GFAP)   4,28,41,53,89,109,
              133,145,163,174,255-279,283,293,317,334,355
  -filaments   85,147
  -growth factor   255-279,533
  -progenitors   28
gliofilaments   53
glioma cells   285
globoside   454
glucocerebrosides   454
glucocorticoids   161,290
glutamic acid decarboxylase (GAD)   41,162
glutamine synthetase (GS)   41,163
glycerol   117
glycerolipids   117
glycerol-3 phosphate dehydrogenase (GPDH)   36,41
glycolipids   9,417,454
glycoproteins   232
  -ConA-binding   437
  -fucosylated   232
  -myelin   355

-sulfated   232
goldfish   143,362
Golgi   63,207,393,456
  -apparatus   132,332,355,522
  -membranes   67
gPLP   361-372
graft   192
grafted cells   311
granule cell layer   218
grey matter   223
growth
  -associate protein (GAP-43)   218
  -cones   55,218
  -factor   4,26,159,226,255-279,294
  -hormone   282
GTP binding proteins   463-471
HeLa cells   329-339
heparin   239
hippocampus   221
histamine   107
HNK-1   27,444
hormones   290
horseradish peroxidase   49,145
human
  -fetuses   261
  -glial cells   255-279
  -myelin   396
  -oligodendroglia   396
  -PLP gene   519
hydrocortisone   161
hydrophobic interactions   518
hypomyelination   64
IGF receptor   281-292
immune complexes   104
immunoaffinity purification   419
immunoblot   225,345,366,500,520
immunocytochemistry   48,61,85,134,255-279,312
immunoglobulin   382

-superfamily 332
immunohistochemistry 421
immunolocalisation 218
in situ hybridisation 6,36,61,295,514,535
in vitro translation 456
inducing factor 286
inflammation 101-113,423
  -perivascular 109
  -submeningeal 106
inositol 1,4,5-triphosphate 405-416
insulin 5,159,281-292
insulin-like growth factors (IGFs) 5,159,281-292
interleukin-1 256
interleukin-2 257,274
intermediate proteins (IP) 343-350,361-372
intermembrane spacings 373-387
intracellular signalling 296
intraperiod line 34,62,194,332
intron 505-516,517-532
iron metabolism 42
isoelectric focusing 236
isolated nuclei 475-487
isoproterenol 535
jimpy 477,517-532
  -mouse 64,513,517-532
  -mutants 434
jun 533-540
kidney 36
lac operon 206
laminin 174
laser irradiation 144
lectin 130,138,433-450
lesions 10,111,171-184,191
  -acute 185
  -irradiated 171-184
  -non-irradiated 171-184
leupeptin 106
lipid bilayer 63,371,527

lipids   22,117-142
liposomes   313
liver   36,122
lower vertebrates   143
lungfish   361-372,378
lymphocytes   103
lymphokines   101,165,290
lysolecithin   104,115,186
lysosomal inhibitors   458
lysosome   436,451-461
macrophages   101-113,129-142,143,191,194,425
 -activated   111
major dense line (MDL)   62,186,311,332,433,490,499
mannose   436
mannose-6-phosphate   454
mannose-binding lectin   435
mannosidases   459
membrane
 -assembly   59
 -potential   53
 -sheets   25,348
mesaxon   64
methionine codon   518
microglia   110,129-142,143
microglial cells   313
microtubules   53,71
migration   314
minigenes   334
minisegments   47-58
mitogens   109,259
mitogenic activity   8
mitosis   8
mld
 -mutant   440
 -myelin   489-503
 -oligodendrocytes   498
monensin   71
monoclonal antibodies   19-31,343-359

monocytes   101,425
monogalactosyldiglyceride   22
monosialogangliosides   397
multiple sclerosis   30,101,110,185,255,361,417–431,433–449
multipolar cells   5
multivesicular bodies (MVB)   73
muscarinic cholinergic receptors   405–416
mutants
  –dysmyelinating   65,433–450,489,517–532
  –hypomyelinated   489
  –quaking   438,477
mutation   491
Müller cells   36
myc   301
myelin   52,59–79,101–113,123,129–142,221,283,343,391–404,
         405–416,417–431,451–461,463–471,517
  –compact   51,59–79,159,332,481,527
  –compaction   70,164,194,373–387,433–450,498,518
  –central   361–372
  –debris   135
  –deficient (mld) mice   489–503
  –deficient rat   41,517
  –gene expression   163,451–465,537–544
  –genes   59,479–491
  –glycoproteins   355,363,365
  –internode   59–79
  –lipids   118,377
  –peripheral   361,381
  –peripheral nerve   332
  –sciatic nerve   414
  –sheath   19,33,61,91,102,145,188,417,499
  –shiverer   191
  –synthesis   126
  –thickness of   188
myelin associated glycoprotein (MAG)   51,59–79,318,400,417,
             434,451–461,476,498,518,533
  –mRNA   457,498
myelin associated neuraminidase   398

myelin basic protein (MBP)   7,19-31,33-45,59-79,103,107,
          159,186,221,223,256,311,317-327,343,382,397,
          417,433,452,465,475-487,489-503,505,533
  -gene   195,475-487,489-503,513,523
  -mRNA   41,67,163,452,480,489-503,514,524,535
  -phosphorylation   223
  -promoter   482,537
  -small   329-339
  -transcription   164,478,524,527
myelin-oligodendrocyte glycoprotein (MOG)   130,164,366,
                                              417-431
myelin proteins
  -36K-protein   343-356,373-387
  -gPLP   361-372
  -intermediate protein (IP)   343-350,361-372
  -myelin associated glycoprotein (MAG)   51,59-79,318,400,
          417,434,451-461,476,498,518,533
  -myelin basic protein (MBP)   7,19-31,33-45,59-79,103,
          107,159,103,107,159,186,221,223,256,311,317-
          327,343,382,397,417,433,452,465,475-487,489-
          503,505,533
  -myelin oligodendrocyte glycoprotein (MOG)   130,164,366,
          417-431
  -P_0-protein   71,329-339,343,361-372,434,451-461,478,499,
          533-541
  -P_2-protein   407
  -proteolipid protein (PLP)   19,41,59-79,317-327,343,361-
          372,400,417,433,465,476,505-515,517,533
myelin-specific genes   475-487
myelinated
  -axons   81,92,145,164
  -fibers   34,82,490
myelination   41,59-79,85,162,191,400,435,475-487,533
  -mutations   477
  -timing of   8
myelinogenesis   26,161,421,463,505,517
myelinogenic cascade   25
N-methylscopolamine   406

Na^+/K^+-ATPase   94

nerve growth factor   226,274

neural cell adhesion molecules (N-CAM)   27,382,400

neuraminidase   391-404

neurite outgrowth   8

neuroblastoma cells   225

neuroepithelial cells   57,294

neurofilaments   55,83

neuron

  -like cells   47-58

  -specific enolase   55

neuronal

  -cell body   83

  -development   8

  -nuclei   219

neurons   8,19,26,36,42,55,92,129,155,220,262,287,441

  -axotomized   143

neutral lipids   118

nitrocellulose   143-153

nodal axoplasm   83

node of Ranvier   3,34,81,97,191

norepinephrine   108

Northern blot   6,8,209,295,491,495,509

nuclear

  -RNA   478,497

  -run-on assays   526

  -signal sequence   207

nuclei   137,494

nucleotide sequences   509

nucleus   55,207

O1 antigen   19-31

O2A

  -cell lineage   294

  -cells   311-316

  -precurser   317

  -progenitors   3-17,19,192,293-307,311-316,505

O4 antigen   3-17,19-31,51

oligodendrocyte

-development 19-31,40,165

-lineage 20,173

-perikarya 135

-phenotype 33-45

oligodendrocytes 3-17,19-31,33-45,47-58,59-79,83,128-142,
150,155-170,171-184,185-198,247-253,255-
279,311,333,361,421,437,463,509,517,536

-cell death of 515

-damaged 191

-dark 33

-grafted 186

-intact 191

-interfascicular 33

-light 33

-mature 24,421

-mld 498

-normal 186

-perineuronal 33

-perivascular 33

-shiverer 186

-subpopulations of 40

oligodendroglia 143,221,392

oligodendroglial process formation 247-283

oligosaccharide

-chain 355

-processing 453

oncogene 317-327

opioid receptor 407

optic nerve 3-17,36,47-58,81-97,129-142,143-153,285,293,
317-327,490,536

-fish 144

-injured 144

oxotremorine 407

P_0 71,329-339,343,361-372,434,451-461,478,499,533-541

-biosynthesis 452

-gene 453,533-541

-mRNA 28,452,535

-promoter 538

-glycoprotein   373-387,434

P_2-protein   378,453,537
paranodal loops   31,66
pathology   106
pECE vector   330
Pelizaeus-Merzbacher disease   513
pepstatin   106
Percoll density gradient   9,312,347
periaxonal
  -cytoplasm   62
  -membrane   59,64
periodicity   52,64
peripheral membrane proteins   332
peripheral nervous system (PNS)   10,34,65,90,132,231,255,
                441,451-461,499,533
  -myelin   62,71,361-372,373-387,489
peripheral neuropathies   451
perivascular
  -infiltration   106
  -staining   333
pertussis toxin   411,466
phagocytic cells   131
phagocytosis   131,140
phenotype   13,38
  -oligodendrocyte   33-45
phenotypic
  -characterization   343-359
  -plasticity   13
phenoxybenzamine   108
pheochromocytoma cells (PC-12)   209
phorbol ester   208,273,247-253
  -TPA   225,247-253
phosphatidylcholine   117
phosphatidylinositol 4,5-bisphosphate (PIP_2)   405-416
phosphatidylinositol kinase   302
phospholipase C   405-416,439
phospholipases   104,425
phospholipids   118

phosphorylation   223
  -sites   65,201-215,382
phylogeny   343-359,361-372,373-387
pig brain   247
pirenzepine   406
plasma membrane   61,394
plasmid
  -pBR 322   312,482
  -recombinant   318
plasmids   330
plasmin   104
plasminogen   104
  -activator   101-113
platelet-derived growth factor (PDGF)   3-17,165,193,225-279,290,293-307,317,331,476
platelets   5,295
point mutation   368,515,521
polysialogangliosides   397
prazosin   108
precursor cells   49,159,195
primed cell line   482
pro-inflammatory factors   427
progenitor cells   3-17,158,293,317
progenitors   19-31,47
proliferation   5,36,129,159,255-279,317,438,452
proligodendrocytes   19-31
promoter   208,479,494
  -regions   211
propanolol   108
proteases   425
protein kinase A   247-253
  -catalytic subunits   201-215
  -regulatory subunits   201-215
protein kinase C   217-230,247-253,275,302,405,535
protein kinase, cAMP dependent   201-215,382
proteinases   101-113
  -acid   106
  -neutral   106

proteoglycans   231
  -chondroitin sulfate   232
  -dermatane sulfate   240
  -sulfated   234
proteolipid protein (PLP)   19,41,59-79,317-327,343,361-372,
        400,417,433,465,476,505-515,517,533
  -amino acid sequence of   506
  -gene   505-515,517-532
  -mRNA   67,163,514,518,524
  -transcription   517-532
protooncogenes   301,467
psychosine   22
pulse-chase experiment   458
Purkinje cell   218
quinuclidinyl benzilate   406
ras   467
reactive gliosis   139
receptor
  -IGF   281-292
  -serotonergic   407
  -T_3 101
  -transferrin   37
recombinant DNA   329-339,476,505
regeneration   8,42,129,137,143-153,219
  -failure of   143
regulatory
  -mechanisms   42
  -signals   482
remyelination   3-17,26,30,117-119,171-184,185-198,252,344,
        451-461
  -failure of   191
  -oligodendrocyte   177
  -Schwann cell   177
  -spontaneous   185
reptiles   362
retina   57,105,145
retinoic acid   226
retrograde transport   73

retroviral
  -infection   311–316
  -vector   317–327
ribonuclear proteins   70
RNA polymerase II   513
Rous sarcoma virus   334
saxitoxin-binding site   84
S1 nuclease   480,519
saltatory conduction   19,35,343
scanning electron microscopy   248
Schmidt-Lanterman incisures   66
Schwann cells   21,34,73,81–97,140,171–184,186,192,344,
    256–279,361,441,451–461,475–477,533–541
  -myelinating   62,73,93,115
sciatic nerve   91,117–142,195,441,451–461,533,536
  -myelin   414
second messenger   290,405,533
secretion   231–246,470
  -products   107
seminolipid   22
sequence homology   204
serine proteinases   106
serotonergic receptors   407
serotonin   108
serum proteins   109
shiverer   3311,475–487,489
  -fibers   188
  -mouse   62,129,185–198,311,333,479,500
  -mutation   333
sialic acid   397
signal transduction   227,231,301,438,469,533
skeletal muscle   94
sodium channels   81–97
  -density of   82
somatomedins   282
Southern blot analysis   319
sphingolipids   117
spinal cord   9,107,171,191

spongioblast 33
squalene 119
  -epoxide 119
ß-strand propensity 381
substrate phosphorylation 201
sugar transferases 455
sulfatides 19-31,35,391,433,598
suppressed cAMP inducible POU (SCIP) 533-541
SV40 T antigen 317-327
SV40 promoter 330
swainsonine 459
synapsin 221
synaptic
  -contacts 218
  -vesicles 221
T cell lines 108,423
T-lymphocytes 101
targeting 454
TATA box 513,519
telencephalon 156
teleosts 343-359,361-372,373-387
tellurium 115-128
  -elemental 115,126
  -intoxication 121
tetrapods 363
tetrodotoxin-sensitive currents 93
thalamus 314
thin layer chromatography 117-142
thyroid hormone 162,290
transcription 206,211,302,454,475-487,494,489-503,523,535
  -factor 211,482,536
  -rate 480,492
  -run-on assay 479
transcriptional
  -interference 497
  -repressor 537
transection injury 451-461
transfection 208,311-316,329-339,482

transferrin   33-45
  -mRNA   36,40
  -receptor   37
transforming growth factor ß   8
transgenic
  -animals   479
  -mice   195,283,497
translation   70,455
transmembranal α-helix   515
transmembrane
  -domain   65
  -protein   295,331
  -sequence   380
transplantation   173
transplants   171-184
trasylol   106
Trembler   195,477
triiodothyronine   162
trout   343-359,363
trypsin cleavage   507
tsul9-5 cell line   317-327
tumor
  -cell lines   295
  -promoters   226
tunicamycin   459
tyrosine kinase   302
ubiquitin   459
UDP-gal: ceramide galactosyltransferase (CGalT)   395,454
UDP-glucose: ceramide glucosyl-transferase (CGlcT)   454
upregulation   8
urokinase   104
vascular permeability   107
vasoactive amines   107
  -intestinal peptide (VIP)   208
vasoconstriction   108
vasodilation   108
vesicular transport   71
vimentin   9,29,53

Wallerian degeneration 129-142,143,451

Western blot 6,345,419

white matter 33,104,107,171,221,223,396,421,507

whole-cell recordings 93

Wolfgram proteins 35

X chromosome 507,514,521

X-irradiation 172

X-ray diffraction 373-387

yohimbine 108

NATO ASI Series H

Vol. 1: Biology and Molecular Biology of Plant-Pathogen Interactions.
Edited by J.A. Bailey. 415 pages. 1986.

Vol. 2: Glial-Neuronal Communication in Development and Regeneration.
Edited by H.H. Althaus and W. Seifert. 865 pages. 1987.

Vol. 3: Nicotinic Acetylcholine Receptor: Structure and Function.
Edited by A. Maelicke. 489 pages. 1986.

Vol. 4: Recognition in Microbe-Plant Symbiotic and Pathogenic Interactions.
Edited by B. Lugtenberg. 449 pages. 1986.

Vol. 5: Mesenchymal-Epithelial Interactions in Neural Development.
Edited by J.R. Wolff, J. Sievers, and M. Berry. 428 pages. 1987.

Vol. 6: Molecular Mechanisms of Desensitization to Signal Molecules.
Edited by T.M. Konijn, P.J.M. Van Haastert, H. Van der Starre, H. Van der Wel, and
M.D. Houslay. 336 pages. 1987.

Vol. 7: Gangliosides and Modulation of Neuronal Functions.
Edited by H. Rahmann. 647 pages. 1987.

Vol. 8: Molecular and Cellular Aspects of Erythropoietin and Erythropoiesis.
Edited by I.N. Rich. 460 pages. 1987.

Vol. 9: Modification of Cell to Cell Signals During Normal and Pathological Aging.
Edited by S. Govoni and F. Battaini. 297 pages. 1987.

Vol. 10: Plant Hormone Receptors. Edited by D. Klämbt. 319 pages. 1987.

Vol. 11: Host-Parasite Cellular and Molecular Interactions in Protozoal Infections.
Edited by K.-P. Chang and D. Snary. 425 pages. 1987.

Vol. 12: The Cell Surface in Signal Transduction.
Edited by E. Wagner, H. Greppin, and B. Millet. 243 pages. 1987.

Vol. 13: Toxicology of Pesticides: Experimental, Clinical and Regulatory Perspectives.
Edited by L.G. Costa, C.L. Galli, and S.D. Murphy. 320 pages. 1987.

Vol. 14: Genetics of Translation. New Approaches.
Edited by M.F. Tuite, M. Picard, and M. Bolotin-Fukuhara. 524 pages. 1988.

Vol. 15: Photosensitisation. Molecular, Cellular and Medical Aspects.
Edited by G. Moreno, R.H. Pottier, and T.G. Truscott. 521 pages. 1988.

Vol. 16: Membrane Biogenesis. Edited by J.A.F. Op den Kamp. 477 pages. 1988.

Vol. 17: Cell to Cell Signals in Plant, Animal and Microbial Symbiosis.
Edited by S. Scannerini, D. Smith, P. Bonfante-Fasolo, and V. Gianinazzi-Pearson.
414 pages. 1988.

Vol. 18: Plant Cell Biotechnology.
Edited by M.S.S. Pais, F. Mavituna, and J.M. Novais. 500 pages. 1988.

Vol. 19: Modulation of Synaptic Transmission and Plasticity in Nervous Systems.
Edited by G. Hertting and H.-C. Spatz. 457 pages. 1988.

Vol. 20: Amino Acid Availability and Brain Function in Health and Disease.
Edited by G. Huether. 487 pages. 1988.

NATO ASI Series H

Vol. 21: Cellular and Molecular Basis of Synaptic Transmission.
Edited by H. Zimmermann. 547 pages. 1988.

Vol. 22: Neural Development and Regeneration. Cellular and Molecular Aspects.
Edited by A. Gorio, J. R. Perez-Polo, J. de Vellis, and B. Haber. 711 pages. 1988.

Vol. 23: The Semiotics of Cellular Communication in the Immune System.
Edited by E. E. Sercarz, F. Celada, N. A. Mitchison, and T. Tada. 326 pages. 1988.

Vol. 24: Bacteria, Complement and the Phagocytic Cell.
Edited by F. C. Cabello und C. Pruzzo. 372 pages. 1988.

Vol. 25: Nicotinic Acetylcholine Receptors in the Nervous System.
Edited by F. Clementi, C. Gotti, and E. Sher. 424 pages. 1988.

Vol. 26: Cell to Cell Signals in Mammalian Development.
Edited by S. W. de Laat, J. G. Bluemink, and C. L. Mummery. 322 pages. 1989.

Vol. 27: Phytotoxins and Plant Pathogenesis.
Edited by A. Graniti, R. D. Durbin, and A. Ballio. 508 pages. 1989.

Vol. 28: Vascular Wilt Diseases of Plants. Basic Studies and Control.
Edited by E. C. Tjamos and C. H. Beckman. 590 pages. 1989.

Vol. 29: Receptors, Membrane Transport and Signal Transduction.
Edited by A. E. Evangelopoulos, J. P. Changeux, L. Packer, T. G. Sotiroudis, and K. W. A. Wirtz. 387 pages. 1989.

Vol. 30: Effects of Mineral Dusts on Cells.
Edited by B. T. Mossman and R. O. Bégin. 470 pages. 1989.

Vol. 31: Neurobiology of the Inner Retina.
Edited by R. Weiler and N. N. Osborne. 529 pages. 1989.

Vol. 32: Molecular Biology of Neuroreceptors and Ion Channels.
Edited by A. Maelicke. 675 pages. 1989.

Vol. 33: Regulatory Mechanisms of Neuron to Vessel Communication in Brain.
Edited by F. Battaini, S. Govoni, M. S. Magnoni, and M. Trabucchi. 416 pages. 1989.

Vol. 34: Vectors as Tools for the Study of Normal and Abnormal Growth and Differentiation.
Edited by H. Lother, R. Dernick, and W. Ostertag. 477 pages. 1989.

Vol. 35: Cell Separation in Plants: Physiology, Biochemistry and Molecular Biology.
Edited by D. J. Osborne and M. B. Jackson. 449 pages. 1989.

Vol. 36: Signal Molecules in Plants and Plant-Microbe Interactions.
Edited by B. J. J. Lugtenberg. 425 pages. 1989.

Vol. 37: Tin-Based Antitumour Drugs. Edited by M. Gielen. 226 pages. 1990.

Vol. 38: The Molecular Biology of Autoimmune Disease.
Edited by A. G. Demaine, J-P. Banga, and A. M. McGregor. 404 pages. 1990.

Vol. 39: Chemosensory Information Processing. Edited by D. Schild. 403 pages. 1990.

Vol. 40: Dynamics and Biogenesis of Membranes.
Edited by J. A. F. Op den Kamp. 367 pages. 1990.

Vol. 41: Recognition and Response in Plant-Virus Interactions.
Edited by R. S. S. Fraser. 467 pages. 1990.

NATO ASI Series H

Vol. 42: Biomechanics of Active Movement and Deformation of Cells.
Edited by N. Akkaş. 524 pages. 1990.

Vol. 43: Cellular and Molecular Biology of Myelination.
Edited by G. Jeserich, H. H. Althaus, and T. V. Waehneldt. 565 pages. 1990.

Printed by Publishers' Graphics LLC